ADVANCES IN CHEMICAL ENGINEERING
Volume 24

ADVANCES IN
CHEMICAL ENGINEERING

Editor-in-Chief
JAMES WEI
School of Engineering and Applied Science
Princeton University
Princeton, New Jersey

Editors

JOHN L. ANDERSON
Department of Chemical Engineering
Carnegie Mellon University
Pittsburgh, Pennsylvania

KENNETH B. BISCHOFF
Department of Chemical Engineering
University of Delaware
Newark, Delaware

MORTON M. DENN
College of Chemistry
University of California at Berkeley
Berkeley, California

JOHN H. SEINFELD
Department of Chemical Engineering
California Institute of Technology
Pasadena, California

GEORGE STEPHANOPOULOS
Department of Chemical Engineering
Massachusetts Institute of Technology
Cambridge, Massachusetts

Volume 24

ACADEMIC PRESS
San Diego   London   Boston
New York   Sydney   Tokyo   Toronto

TP
145
.D7
v.24
1998

This book is printed on acid-free paper.

Copyright © 1998 by ACADEMIC PRESS

All Rights Reserved.
  No part of this publication may be reproduced or transmitted in any form or by any means, electronic or mechanical, including photocopy, recording, or any information storage and retrieval system, without permission in writing from the Publisher.
  The appearance of the code at the bottom of the first page of a chapter in this book indicates the Publisher's consent that copies of the chapter may be made for personal or internal use of specific clients. This consent is given on the condition, however, that the copier pay the stated per copy fee through the Copyright Clearance Center, Inc. (222 Rosewood Drive, Danvers, Massachusetts 01923), for copying beyond that permitted by Sections 107 or 108 of the U.S. Copyright Law. This consent does not extend to other kinds of copying, such as copying for general distribution, for advertising or promotional purposes, for creating new collective works, or for resale. Copy fees for pre-1997 chapters are as shown on the title pages. If no fee code appears on the title page, the copy fee is the same as for current chapters.
0065-2377/98 $25.00

Academic Press
a division of Harcourt Brace & Company
525 B Street, Suite 1900, San Diego, California 92101-4495, USA
http://www.apnet.com

Academic Press Limited
24-28 Oval Road, London NW1 7DX, UK
http://www.hbuk.co.uk/ap/

International Standard Book Number: 0-12-008524-0

PRINTED IN THE UNITED STATES OF AMERICA
98 99 00 01 02 03 QW 9 8 7 6 5 4 3 2 1

*This volume is dedicated to the memories of Gianni Astarita
and Jacques Villermaux*

CONTENTS

Contributors xi
Preface xiii

Kinetics and Thermodynamics in Multicomponent Mixtures
Raffaella Ocone and Gianni Astarita

I. Introduction 2
II. Linear Algebra of Stoichiometry 3
   A. Discrete Description 3
   B. Continuous Description 7
   C. Lumping and Overall Descriptions 13
III. Thermodynamics 14
   A. Phase Equilibria 14
   B. Homogeneous Chemical Equilibria 22
   C. Heterogeneous Chemical Equilibria 28
IV. Kinetics 30
   A. Exact and Approximate Lumping 30
   B. Overall Kinetics 34
   C. Overall Reaction Engineering 49
       Appendix A: Orthogonal Complement 61
       Appendix B 61
       Appendix C: The Feinberg Approach to Network Topology 64
       Appendix D: Mathematical Concepts 66
       References 70

Combustion Synthesis of Advanced Materials: Principles and Applications
Arvind Varma, Alexander S. Rogachev, Alexander S. Mukasyan, and Stephen Hwang

I. Introduction 81
II. Methods for Laboratory and Large-Scale Synthesis 84
   A. Laboratory Techniques 84
   B. Production Technologies 87
III. Classes and Properties of Synthesized Materials 96
   A. Gasless Combustion Synthesis from Elements 96
   B. Combustion Synthesis in Gas–Solid Systems 107
   C. Products of Thermite-Type SHS 115
   D. Commercial Aspects 117

CONTENTS

IV. Theoretical Considerations 120
   A. Combustion Wave Propagation Theory in Gasless Systems 120
   B. Microstructural Models 127
   C. Cellular Models 130
   D. Stability of Gasless Combustion 135
   E. Filtration Combustion Theory 138
   F. Other Aspects 149
V. Phenomenology of Combustion Synthesis 151
   A. Thermodynamic Considerations 152
   B. Dilution 158
   C. Green Mixture Density for Gasless Systems 162
   D. Green Mixture Density and Initial Gas Pressure for
      Gas–Solid Systems 165
   E. Particle Size 169
   F. Other Effects of Combustion Conditions 173
VI. Methods and Mechanisms for Structure Formation 180
   A. Major Physicochemical Processes Occurring during
      Combustion Synthesis 182
   B. Quenching of the Combustion Wave 183
   C. Model Systems for Simulation of Reactant Interaction 190
   D. Time-Resolved X-ray Diffraction (TRXRD) 195
   E. Microstructure of Combustion Wave 197
   F. Concluding Remarks 203
      Nomenclature 206
      References 209

Computational Fluid Dynamics Applied to Chemical Reaction Engineering

J. A. M. Kuipers and W. P. M. van Swaaij

I. Introduction 227
II. Traditional Approaches Followed within Chemical
   Reaction Engineering 228
   A. Role and Types of Modeling in Chemical Engineering 232
   B. Relation with Experimental Work 233
   C. CFD in Chemical Engineering Education 234
III. Computational Fluid Dynamics 234
   A. Definition and Theoretical Framework 236
   B. Numerical Techniques 244
   C. Existing Software Packages 251
IV. Application to Chemical Reaction Engineering 253
   A. Single-Phase Systems 254
   B. Multiphase Systems 265
   C. State of the Art of CFD in Chemical Reaction Engineering ... 280
V. Experimental Validation 282
VI. Selected Applications of CFD Work Conducted at Twente University 287
   A. Two-Fluid Simulation of Gas Fluidized Beds 287
   B. Discrete Particle Simulation of Gas Fluidized Beds 291

C.	Circulating Fluidized Beds	296
D.	Bubble Columns	298
E.	Modeling of a Laminar Entrained Flow Reactor	311
VII.	Conclusion	313
	Nomenclature	316
	References	319

Using Relative Risk Analysis to Set Priorities for Pollution Prevention at a Petroleum Refinery

RONALD E. SCHMITT, HOWARD KLEE, DEBORA M. SPARKS, AND MAHESH K. PODAR

I.	Introduction	330
II.	Summary	331
A.	Developing a Detailed Refinery Release Inventory	331
B.	Identifying Options for Reducing Releases	332
C.	Evaluating and Ranking the Options	333
D.	Discussion of Findings	334
III.	Background	335
IV.	Assembling the Inventory of Refinery Releases	337
A.	Distribution of Releases within the Refinery	338
B.	Distribution of Releases Leaving the Refinery	341
C.	Characterization of Releases	345
D.	Identification of Sources	348
V.	Assessing the Risks Posed by Refinery Releases	351
A.	Air	352
B.	Surface Water	353
C.	Drinking Water	354
D.	Groundwater	354
VI.	Measuring Public Perception of the Refinery	354
VII.	Developing and Evaluating Options for Reducing Releases	355
A.	Option Characteristics	357
B.	Ranking the Options	360
VIII.	Ranking Methods and Results	369
A.	Options for Reducing Total Releases	369
B.	Analytical Hierarchy Process	376
C.	Regulatory Requirements and Project Options	378
D.	Summary of Ranking Results	382
IX.	Implementation of Selected Options: Obstacles and Incentives	384
A.	Background	384
B.	Obstacles	385
C.	General Observations on Five Highly Ranked Options	388
D.	Follow-up to the Original Project	397
	References	398

INDEX ... 401
CONTENTS OF VOLUMES IN THIS SERIAL ... 411

CONTRIBUTORS

Numbers in parentheses indicate the pages on which the authors' contributions begin.

GIANNI ASTARITA*, *Dipartimento di Ingegneria dei Materiali e della Produzione, Universita' di Napoli Federico II, 80125 Naples, Italy* (1)

STEPHEN HWANG, *Department of Chemical Engineering, University of Notre Dame, Notre Dame, Indiana 46556* (79)

HOWARD KLEE, *Environmental Technical Services, Amoco Corporation, Warrenville, Illinois 60555* (329)

J. A. M. KUIPERS, *Department of Chemical Engineering, Twente University of Technology, 7500 AE Enschede, The Netherlands* (227)

ALEXANDER S. MUKASYAN, *Department of Chemical Engineering, University of Notre Dame, Notre Dame, Indiana 46556* (79)

RAFFAELLA OCONE, *Dipartimento di Ingegneria Chimica, Universita' di Napoli Federico II, 80125 Naples, Italy* (1)

MAHESH K. PODAR, *Office of Water, U.S. Environmental Protection Agency, Washington, DC 20460* (329)

ALEXANDER S. ROGACHEV, *Institute of Structural Macrokinetics, Russian Academy of Sciences, Chernogolovka, 142 432 Russia* (79)

RONALD E. SCHMITT, *Environmental Technical Services, Amoco Corporation, Warrenville, Illinois 60555* (329)

DEBORA M. SPARKS, *Environmental Technical Services, Amoco Corporation, Warrenville, Illinois 60555* (329)

W. P. M. VAN SWAAIJ, *Department of Chemical Engineering, Twente University of Technology, 7500 AE Enschede, The Netherlands* (227)

ARVIND VARMA, *Department of Chemical Engineering, University of Notre Dame, Notre Dame, Indiana 46556* (79)

*Deceased.

PREFACE

In chemical manufacturing and oil refining, as well as in environmental considerations of air and water, numerous chemical compounds participate in many chemical reactions. The first article in volume 24 of *Advances in Chemical Engineering,* "Kinetics and Thermodynamics in Multicomponent Mixtures," provides a systematic treatment of this subject. One of the authors is Professor Gianni Astarita of Naples, an outstanding innovator and seminal thinker who recently passed away. The readers must master linear algebra and stoichiometry in the first part before reaching the nuggets in thermodynamics and kinetics.

The thermite process may be the original inspiration of combustion synthesis (CS), a relatively new technique for synthesizing advanced materials from powder into shaped products of ceramics, metallics, and composites. Professor Varma and his associates at Notre Dame contributed the article "Combustion Synthesis of Advanced Materials: Principles and Applications," which features this process that is characterized by high temperature, fast heating rates, and short reaction times.

We have all read about the impact of large-scale computation in fields such as the design of aerospace vehicles, the exploration and production of oil and gas, and the design of new drugs—as fallouts from the national investments in weapons design. Professors Kuipers and van Swaaij of Twente University have applied computation fluid dynamics (CFD) to the analysis and design of chemical reactors, particularly in multiphase flow reactors. The graphic display of their computations will add immeasurably to our understanding of what happens in fluidized beds.

The chemical engineering community needs a broader understanding of actual industrial cases before we can make progress in environmental engineering. This gap is addressed by Roland Schmitt and his associates at Amoco Corporation in the article "Using Relative Risk Analysis to Set Priorities for Pollution Prevention at a Petroleum Refinery," which describes a joint project between Amoco and the Environmental Protection Agency to identify opportunities for preventing pollution at a refinery in Yorktown, Virginia.

These four chapters constitute reviews that will inform the chemical engineering community about important developments in science and technology and will serve as starting points for further advances.

JAMES WEI

KINETICS AND THERMODYNAMICS IN MULTICOMPONENT MIXTURES

Raffaella Ocone[*,1] and Gianni Astarita[†,2]

*Dipartimento di Ingegneria Chimica
†Dipartimento di Ingegneria dei Materiali e della Produzione
Universita' di Napoli Federico II
80125 Naples, Italy

I. Introduction	2
II. Linear Algebra of Stoichiometry	3
A. Discrete Description	3
B. Continuous Description	7
C. Lumping and Overall Descriptions	13
III. Thermodynamics	14
A. Phase Equilibria	14
B. Homogeneous Chemical Equilibria	22
C. Heterogeneous Chemical Equilibria	28
IV. Kinetics	30
A. Exact and Approximate Lumping	30
B. Overall Kinetics	34
C. Overall Reaction Engineering	49
Appendix A: Orthogonal Complement	61
Appendix B	61
Appendix C: The Feinberg Approach to Network Topology	64
Appendix D: Mathematical Concepts	66
References	70

This article reviews the kinetic and thermodynamic behavior of multicomponent mixtures containing a very large number of components. A flurry of activity can be envisaged in recent literature; however, although the needs for lumping in this area have been clearly identified, it is not always easy to catch the links and the relationships among different works. We review various techniques and results showing the logical

---

[1] Present address: Department of Chemical Engineering, University of Nottingham, Nottingham, United Kingdom.
[2] Deceased.

status of the latter and how they can be applied to specific problems. Overall (or global) quantities of interest are identified with reference to industrial problems, and chemical reaction engineering of systems, where one is interested only in the overall kinetics, are presented. Mathematical techniques through which results are obtained can often be cumbersome, and they rely primarily on functional analysis. Such techniques are reviewed in the final section of the article.

I. Introduction

In principle, mixtures containing a very large number of components behave in a way described by the same general laws that regulate the behavior of mixtures containing only a comparatively small number of components. In practice, however, the procedures for the description of the thermodynamic and kinetic behavior of mixtures that are usually adopted for mixtures of a few components rapidly become cumbersome in the extreme as the number of components grows. As a result, alternate procedures have been developed for multicomponent mixtures. Particularly in the field of kinetics, and to a lesser extent in the field of phase equilibria thermodynamics, there has been a flurry of activity in the last several years, which has resulted in a variety of new results. This article attempts to give a reasoned review of the whole area, with particular emphasis on recent developments.

This article is organized as follows. In Section II, we discuss the linear algebra of stoichiometry. Stoichiometry may appear to be a rather trivial subject, but it is in fact far from being so. Thus, it is useful to discuss stoichiometry prior to dealing with the real issues of thermodynamics and kinetics, so that discussion of the latter may be unencumbered by stoichiometric issues. Section III is dedicated to thermodynamics and is divided into three subsections, which deal with phase equilibria, homogeneous chemical equilibria, and heterogeneous chemical equilibria. The first subsection is significantly longer than the other two because of many more recent developments in this area. Section IV is dedicated to kinetics. The first subsection deals with the subject of "lumping": techniques by which a system of order, say, N, can be represented (either exactly or in some well-defined sense of approximation) by a system of order N', with $N' < N$. Again, this subsection, because it deals with a reasonably well-established area, is very concise. The second subsection deals with "overall" kinetics: the kinetic description of some overall quantity of interest (e.g., the total residual sulfur in the hydrodesulfurization of an oil cut) in relation to the underlying true kinetics of individual chemical reactions. Finally, the third subsection is dedicated to the chemical reaction engineering of systems where one is interested only in the overall kinetics.

Section V is dedicated to a few mathematical techniques that are used in the body of the article. We presume the reader has a working knowledge of standard linear algebra, and therefore Section V is restricted to some elementary concepts of functional analysis that are needed.

II. Linear Algebra of Stoichiometry

A. Discrete Description

By *discrete description*, we mean a description of multicomponent mixtures where the number of components considered, though possibly very large, is finite. Consider a mixture containing N ($<\infty$) components A_I, $I = 1, 2, \ldots, N$. Given any physical property f of the components (such as concentration and chemical potential), its value for the Ith component is indicated by f_I; the N ordered values of f_I constitute an N-dimensional vector $\mathbf{f}$. The usual scalar product can be defined for vectors $\mathbf{f}$; we introduce the somewhat unusual notation $\langle \mathbf{f}_1, \mathbf{f}_2 \rangle$ for the scalar product, for reasons that soon become apparent (the $\langle , \rangle$ scalar product is reserved to N-dimensional vectors; we also need vectors of different dimensionality). The notation $\langle , \rangle$ is also used more generally for sums over the range of the I index.

Although chemistry is usually developed in terms of mole units, we use mass units; in most cases, the choice is a matter of taste, but there are some issues that are best dealt with in mass units. So the fundamental component property is their mass m_I, say, the vector $\mathbf{m}$. Let $\mathbf{I}$ be the N-dimensional vector, all components of which are unity. Then the total mass of our system is $\langle \mathbf{m}, \mathbf{I} \rangle$, and this is constant in a closed system, even when, due to chemical reactions, the individual masses m_I are not constant.

We now come to stoichiometry. A chemical reaction may be written as

$$\Sigma \sigma_I A_I = 0, \tag{1}$$

where the σ_I's are the mass-based stoichiometric coefficients, taken as positive for the (arbitrarily chosen) products and negative for the reactants; these form a vector $\boldsymbol{\sigma}$.[3] The mass-based stoichiometric coefficients are the ordinary mole-based ones multiplied by the corresponding molecular weights.

The physical meaning of Eq. (1) is as follows. Let q be an arbitrary scalar, and let M_I be the masses of the components in a closed system at some initial condi-

---

[3] If one multiplies all the σ_I's by the same nonzero scalar Ω, one is describing of course the same reaction. Equation (1) is invariant under multiplication by Ω. The particular choice $\Omega = -1$ shows that the distinction between reactants and products is arbitrary.

tion (these form a vector **M**). Then the values of **m** at any other time can be expressed as

$$\mathbf{m} = \mathbf{M} + \sigma q; \tag{2}$$

that is, the N-dimensional vector $\mathbf{m} - \mathbf{M}$ is uniquely determined by the scalar q. The latter is called the *extent of the reaction* with respect to **M**.

Now consider the case where there are R independent reactions. We discuss the meaning of *independent* shortly. We use the index J to identify reactions, and sums over the range of the J index, including the scalar product of two R-dimensional vectors, are indicated with $\{,\}$ rather than $<,>$. Each of the reactions is identified by its own set of stoichiometric coefficients, so that we will have an $N \times R$ matrix of stoichiometric coefficients σ_{IJ} [compactly indicated as $\boldsymbol{\sigma}$. In the present interpretation, $\boldsymbol{\sigma}$ in Eq. (2) is an $N \times 1$ matrix]. There are now R extents of reaction, which form an R-dimensional vector **q**. The obvious extension of Eq. (2) is

$$\mathbf{m} = \mathbf{M} + \{\boldsymbol{\sigma}, \mathbf{q}\}. \tag{3}$$

The reactions are said to be *independent* if the rank of $\boldsymbol{\sigma}$ is R. This implies that the following equation for the M-dimensional unknown vector $\mathbf{q}^*$ has only the trivial solution $\mathbf{q}^* = \mathbf{0}$:

$$\{\boldsymbol{\sigma}, \mathbf{q}^*\} = 0. \tag{4}$$

Equation (4) implies that **q** in Eq. (3) is unique. In fact, suppose there are two **q**'s, $\mathbf{q}_1$ and $\mathbf{q}_2$, resulting in the same $\mathbf{m} - \mathbf{M}$. Equation (3) would imply that $\{\boldsymbol{\sigma}, (\mathbf{q}_1 - \mathbf{q}_2)\} = 0$, but this implies that $\mathbf{q}_1 = \mathbf{q}_2$.

Although the given definition of independence of chemical reactions is useful for formal manipulations, it is cumbersome to use when one tries to reduce a set of chemical reactions to an independent one. This is, in fact, achieved very simply by the following procedure. First, write down one of the simplest reactions that comes to mind; say this involves only three components, with nonzero stoichiometric coefficients σ_{11}, σ_{21}, and σ_{31}. These constitute a 3×1 matrix of rank 1 (i.e., of rank equal to the number of reactions written down). Next, write a reaction that involves one and only one new component (component 4 in this example). Because σ_{42} is nonzero, the 4×2 matrix has rank 2, and so on. The procedure is guaranteed to work, and it also shows that R is at most $N - 1$ (the first reaction must include at least two nonzero stoichiometric coefficients).

The procedure just illustrated will be referred to in the following paragraphs as the *constructive* procedure. It would appear at first sight that, if one starts with a 3×1 matrix (as most often will be the case), one always ends up with $R = N - 2$. This, however, is not the case, because, for instance, one could have two entirely disjoint matrices $\boldsymbol{\sigma}_1$ ($N_1 \times R_1$, rank $= R_1$) and $\boldsymbol{\sigma}_2$ ($N_2 \times R_2$, rank $=$

R_2) constituting a global $(N_1 + N_2) \times (R_1 + R_2)$ matrix of rank $R_1 + R_2$. Such entirely disjoint sets of reactions are, however, rather special cases: The difference $N - R$, though certainly not always equal to 2, is invariably a small number.

Substituting the condition of mass conservation in Eq. (3) one obtains

$$<\{\sigma, \mathbf{q}\}, \mathbf{I}> = \{\mathbf{q}, <\sigma^T, \mathbf{I}>\} = 0, \quad (5)$$

which must be satisfied for arbitrary $\mathbf{q}$. It follows that $<\sigma^T, \mathbf{I}> = 0$, that is, that the unit vector $\mathbf{I}$ lies in the kernel of σ^T.[4]

Chemistry imposes a stricter constraint than simply conservation of mass: Individual atoms must also be preserved. Let $K = 1, 2, \ldots, P$ be the atomic species present in any one of the components in the system. Let p_{IK} be the grams of atom K per gram of species I; the matrix p_{IK} will be compactly indicated as $\mathbf{p}$. Conservation of atoms can be expressed as

$$<\sigma^T, \mathbf{p}> = \mathbf{0} \quad \text{for all J's and all K's.} \quad (6)$$

We use the standard notation for the inner product in the case of P-dimensional vectors, and more generally for sums over the range of K. Let $\mathbf{i}$ be the P-dimensional vector all components of which are unity. One has $\mathbf{I} = \mathbf{p} \cdot \mathbf{i}$, and hence mass conservation can be written

$$<\sigma^T, \mathbf{p} \cdot \mathbf{i}> = \mathbf{i} \cdot <\sigma^T, \mathbf{p}> = 0. \quad (7)$$

Equation (6) (permanence of atoms) implies Eq. (7) (conservation of mass), but the converse is not true, because $<\sigma^T, \mathbf{p}>$ could simply be orthogonal to $\mathbf{i}$ for Eq. (7) to be satisfied.

Because the vector $\mathbf{m}$ is constrained by the mass conservation requirement $<\mathbf{m}, \mathbf{I}> = \text{const},$[5] the space of possible $\mathbf{m}$ values has $N - 1$ dimensions. If the number of independent chemical reactions, R, is less than $N - 1$, then some vectors $\mathbf{m}$ are not accessible at some assigned $\mathbf{M}$; this, as will be seen, has important consequences in the consideration of heterogeneous chemical equilibria. Now consider the special case where $R = N - 1$, so that indeed all admissible $\mathbf{m}$'s are accessible. Because the kernel of σ contains only the zero vector, there exists an $M \times N$ matrix $\mathbf{A}$ such that

$$\mathbf{q} = <\mathbf{A}, (\mathbf{m} - \mathbf{M})>. \quad (8)$$

---

[4] Should we have chosen to work in the more usual mole units, the vector of molecular weights would have played the role of $\mathbf{I}$. This does not imply any substantial difference, since from a formal viewpoint the crucial issue is that N-dimensional vectors with all components being nonnegative lie in the kernel of σ^T.

[5] Formally, $\mathbf{m}$ lies on the intersection of the hyperplane $<\mathbf{m}, \mathbf{I}> = \text{const}$ with the positive orthant of the N-dimensional components space (Wei, 1962a).

This in turn implies that $\{\mathbf{A}^T, \sigma^T\} = \mathbf{1}$, the $N \times N$ unit matrix. Now let $\mathbf{z}$ be *any* N-dimensional vector lying in the kernel of σ^T. Since $\Omega \mathbf{I}$, with Ω any arbitrary scalar, also lies in that kernel, one can write

$$<\sigma^T, (\mathbf{z} - \Omega \mathbf{I})> = \mathbf{0}. \tag{9}$$

It follows that

$$\begin{aligned}\mathbf{0} &= \{\mathbf{A}^T, <\sigma^T, (\mathbf{z} - \Omega \mathbf{I})>\} = <\{\mathbf{A}^T, \sigma^T\}, (\mathbf{z} - \Omega \mathbf{I})> \\ &= \mathbf{z} - \Omega \mathbf{I};\end{aligned} \tag{10}$$

that is, in this case *only* vectors of the type $\Omega \mathbf{I}$ lie in the kernel of σ^T. This implies that, given P arbitrary scalars Ω_K, $\mathbf{p}$ equals the dyad $\mathbf{I}\Omega$. Say $p_{IK} = \Omega_K$ does not depend on I: It follows that the brute chemical formulas of all species are multiples of each other (the only reactions are isomerization, oligomerizations, or disproportionations). We refer to this special case as a *merization system*. [Note, incidentally, that in a merization system the constructive procedure would need to start with a 2×1 matrix (in mass units, this would of necessity be of the type $1, -1$), which describes a "merization" reaction. Each additional step would also describe a merization reaction. Hence, an alternative, more heuristic proof has been given for the same result.]

Permanence of atoms can also be seen from a different viewpoint. The condition is $<\mathbf{p}^T, \mathbf{m}> = <\mathbf{p}^T, \mathbf{M}>$. Now $<\mathbf{p}^T, \mathbf{M}>$ is a P-dimensional vector $\mathbf{b}$. While the number of components, N, and independent reactions, R, may grow without bounds (and in the next section we illustrate the description where both are allowed to approach infinity), the number of constraints $<\mathbf{p}^T, \mathbf{M}>$ is always finite at P. Finally, it is worth pointing out that in a variety of cases one has additional constraints over and above that of permanence of atoms, such as permanence of benzenic rings or other constraints implied by the chemistry of the reactions involved. Apparently, Wei (1962a) was the first one to identify explicitly the possible existence of constraints other than the permanence of atoms; these can always be expressed as permanence of "something," that is, there are quantities that are invariant as the reactions occur.

Given *any* P-dimensional vector $\mathbf{b}$, the N-dimensional vector $\mathbf{p} \cdot \mathbf{b}$ lies in the kernel of σ^T: $<\sigma^T, \mathbf{p} \cdot \mathbf{b}> = <\sigma^T, \mathbf{p}> \cdot \mathbf{b} = \mathbf{0}$. This, as will be seen, has important consequences in the theory of equilibrium. It is also important to realize that a description of the stoichiometry in terms of the condition $<\mathbf{p}^T, \mathbf{m}> = <\mathbf{p}^T, \mathbf{M}>$ is in many senses equivalent to a description in terms of σ, and in many formal manipulations it may be preferable when the number of species and/or reactions becomes very large, because the dimensions of $<\mathbf{p}^T, \mathbf{M}>$ stay at P no matter how complex the mixture may be.

A more generally applicable method based on linear constraints is as follows (we will make use of this method in most of the following sections). The method is due to Krambeck (1970). Let $\mathbf{m}_1$ and $\mathbf{m}_2$ be two compositions that are both accessible from the initial one. The term $\mathbf{m}_1 - \mathbf{m}_2$ can be regarded as a *reaction vector* in the sense that some linear combination of the reactions considered can bring the system from $\mathbf{m}_2$ to $\mathbf{m}_1$. Now consider all N-dimensional vectors $\mathbf{a}_L$, $L = 1, 2, \ldots, C$, such that $<\mathbf{m}, \mathbf{a}_L> = \text{const}$ and such that $<\mathbf{m}_1 - \mathbf{m}_2, \mathbf{a}_L> = 0$. By "all" we mean that the set is complete, in the sense that there are no other vectors to which some possible $\mathbf{m}_1 - \mathbf{m}_2$ is orthogonal. It is easy to convince oneself that $C = N - R$. The total number of constraints C does not coincide with P, as the simple examples in Appendix A show. The value of C is finite even if both N and R approach infinity. The vectors $\mathbf{a}_L$ (which are N-dimensional ones) form the basis of a vector space that is called the *orthogonal complement* of the space of reaction vectors $\mathbf{m}_1 - \mathbf{m}_2$.

Finally, a slightly different formulation of the linear constraints imposed on the reaction pathways, which is deduced directly from the stoichiometric matrix σ, has been discussed by Slaughter and Doherty (1994). First, one chooses R components (i.e., as many as there are independent reactions), which are called the *reference components* and identified by an R suffix. One obviously has $\mathbf{m}_R = \mathbf{M}_R + \{\sigma_R, \mathbf{q}\}$, where the reduced stoichiometric matrix σ_R is $R \times R$, and hence can be invertible. Because Eq. (4) has only a trivial solution, the reference components may indeed be chosen so that σ_R is invertible. One thus has

$$\mathbf{m} = \mathbf{M} + <\{\sigma, \sigma_R^{-1}\}^T, (\mathbf{m}_R - \mathbf{M}_R)> \tag{11}$$

which can be rewritten as

$$\mathbf{m} - <\{\sigma, \sigma_R^{-1}\}^T, (\mathbf{m}_R)> = \mathbf{M} - <\{\sigma, \sigma_R^{-1}\}^T, (\mathbf{M}_R)> = \mathbf{m}^\sim. \tag{12}$$

It is important to realize that, while $\mathbf{m}^\sim$ is, like the vectors $\mathbf{a}_L$, an N-dimensional vector, $N - R$ of its components are zero.

B. Continuous Description

When the number of components in the mixture becomes very large, a continuous description is in many cases preferable to a discrete one.[6] This is not in any way related to a physically significant difference: In real life, *any* mixture is made up of a finite (if possibly very large) number of components, and hence a continuous description (where the number of components is essentially regarded as infinitely large) is certainly nothing more than a mathematical

---

[6] One may choose simply to consider cases, such as discussed in Appendix A, where in a discrete description N approaches infinity; this leads to a countable infinity of components. It turns out that there are very few instances where the distinction between a countable infinity and an R^1 continuum makes any difference.

artifact.[7] The reason for choosing such an artifact is related simply to the fact that the algebra of integrals is easier to work with than that of finite sums. It is also worthwhile to point out that, once the formalism of a continuous description is used to perform actual numerical calculations, the computer code will necessarily transform back integrals to finite sums, so that the distinction, if it has any conceptual meaning at the level of formal manipulations, becomes a very fuzzy one at the level of numerical calculations.[8]

The earliest formulation of a continuous description is probably due to De Donder (1931): His Chapter 3 is titled "Systèmes Renfermant une Infinité de Constituants." It is worthwhile to quote *verbatim* (except for minor differences in notation):

> nous supposerons que la *masse* de chacun de ces constituants est infiniment petite; prise à l'instant t, la masse du constituant d'indice x sera désignée par $dm(x)$, ou x est une variable qui varie d'une manière continue de x_0 à x_1, deux nombres fixes donnés dans chaque système considéré. . . . Cette masse du constituant d'indice x correspond, en réalité, à l'intervalle compris entre x et $x + dx$.

There are a number of conceptual pearls in De Donder's Chapter 3; here we simply want to point out our italics for the word "masse": De Donder, 30 years before anybody else, recognized that, in spite of what permanence of atoms seems to suggest as the simple way of doing things, a description based on mass (which is preserved when chemical reactions occur) is preferable to a description based on moles (which are not preserved).

De Donder's pioneering contribution was acknowledged[9] in what is nowadays commonly regarded as the fundamental work on the continuous description of complex mixtures (Aris and Gavalas, 1966). In hindsight, the continuous description can be seen as nothing more than the extension to an infinite-dimensional

---

[7] Aris (1968), who certainly cannot be held responsible for a lack of appreciation of mathematical artifacts, when considering the continuous description of kinetics writes: "At first sight it would seem that there is nothing in chemical kinetics that demands more than a thorough understanding of autonomous systems of nonlinear differential equations, for it could be argued that a continuum of reactions such as we have introduced above is an artificial way of describing a discrete, even if fine-grained, situation. However this might not be so, for from a theoretical point of view one would certainly want to investigate the integro-differential equations that arise in the limit of an infinite number of species, and from a practical standpoint this approach has already proved to be of value."

[8] Except for the fact that the numerical code may choose *pseudo-components* (i.e., ways of discretizing the relevant integrals), which have no direct physical meaning. For instance, if x is taken to be (proportional to) the number of carbon atoms in, say, a mixture of hydrocarbons, the computer code, if cleverly programmed, may well choose to discretize x in such a way that the corresponding number of actual carbon atoms is not an integer (see, e.g., Shibata *et al.*, 1987).

[9] Apparently the Aris and Gavalas paper was written with no knowledge of De Donder's work; the latter is quoted in a footnote with the following wording: "We are indebted to Prof. R. Defay for drawing our attention to this reference. The work reported here had been completed when we learned of De Donder's discussion, but little overlap and no conflict of material is found."

Hilbert space of a finite-dimensional normed vector space discrete description. However, there are a number of cases where the extension to an infinite-dimensional Hilbert space cannot be dealt with in too cavalier a fashion, and these will be discussed as the need arises; at this stage, we simply discuss a few essential points about the mathematical formalism connected with a continuous description.

1. Inner Product and Norm

Let f_1 and f_2 be two properties of the species in the system (there may be concentrations, vapor pressures, etc.). If the number of species is finite, one has two vectors $\mathbf{f}_1$ and $\mathbf{f}_2$ in a finite-dimensional vector space (the number of dimensions being the number of species). The inner product $<\mathbf{f}_1,\mathbf{f}_2>$ is generalized to the inner product $<f_1(y)f_2(y)>$ in the Hilbert space for two infinite-dimensional vectors $f_1(x), f_2(x)$ as follows (y is the label x when intended as a dummy variable; we assume that x has been normalized so as to be dimensionless, see later discussion):

$$<f_1(y)f_2(y)> = \int_{x_0}^{x_1} f_1(y)f_2(y)dy. \tag{13}$$

When the vector $f(y)$ does not contain some delta functions (which is possible), its norm is, trivially, $|f_1(x)| = \sqrt{<f_1(y)f_1(y)>}$.[10]

2. Choice of Label and Normalization

In the discrete description of a mixture of N components, different species are identified by an integer subscript, generally chosen as 1, 2, . . ., N. The choice of integer values, and their assignment to individual species, is entirely arbitrary; the only requirement is that the correspondence between indices and species be unequivocal (each index corresponds to only one species, and vice versa). In the continuous description, the *component label x* also has only to satisfy the same requirement. In addition to this, it is generally useful to choose x so that it is dimensionless. For instance, the mass of components with labels between x and $x + dx$ is $m(x)dx$, and unless x is dimensionless $m(x)$ would not have the dimensions of a mass.

---

[10] Krambeck (1994b) observes that the scalar product definition should include the number density function $s(x)$, say,

$$<f_1(y)f_2(y)> = \int_{x_2}^{x_1} f_1(y)f_2(y)dy/s(y).$$

This, in addition to other advantages, would guarantee invariance under rescaling of the label x. Equation (13) from this viewpoint, could be regarded as applying to the case where $s(x) = 1$.

Suppose one has started by choosing some label x^*, which, although satisfying the requirement of uniquely identifying components, is not dimensionless.[11] Suppose also that x^* is restricted to some interval x^*_1, x^*_2 other than $0, \infty$. One first renormalizes the label by defining x^{**} as $(x^* - x^*_1)/(x^*_2 - x^*)$, which ranges between 0 and ∞. Furthermore, for any given problem, some externally imposed mass distribution $M(x^{**})$ is generally available. (This may be, for instance, the initial mass distribution in a batch reactor. Sometimes it is preferable to assign an external concentration distribution rather than a mass distribution.) The quantity M is regarded as already normalized with respect to the total mass of the system, so that the zeroth moment of $M(x^{**})$ is unity. The *average* value of x^{**} in the mixture, x_{AVG}, is $<y^{**}M(y^{**})>$. One now defines a dimensionless normalized label x as $x = x^{**}/x_{AVG}$. This, of course, guarantees that the first moment of the $M(x)$ distribution, $<yM(y)>$, is identically unity.

It is also often useful, given some quantity $u(x)$ and some weighting function $h(x)$, to define an *overall*[12] quantity U as the weighted integral of $u(x)$ over the label range $U = <h(y)u(y)>$ and, should $u(x)$ also depend on some other variable τ, $u = u(x,\tau)$, one, of course, has $U(\tau) = <h(y)u(y,\tau)>$. Often we will restrict our attention to the special case where $h(x)$ is identically equal to unity.[13] Finally, for purposes of analysis it is often useful to choose a specific mathematical form for $M(x)$. A powerful form is that of a *gamma distribution,* which contains only one dimensionless parameter, α:

$$M(x) = \phi(\alpha, x) = \alpha^\alpha x^{\alpha-1} e^{-\alpha x}/\Gamma(\alpha), \tag{14}$$

which, for any value of α, indeed has the property $<M(y)> = <yM(y)> = 1$. Here, $\phi(\alpha, x)$ is unimodal, but apart from this constraint it is a very powerful distribution: It approaches the single-component distribution $\delta(x - 1)$ when α approaches infinity, the exponential distribution $\exp(-x)$ when $\alpha = 1$, and it is admissible (in the sense that all moments $<y^N \phi(\alpha, y)>$ are finite) for any positive α.

We now move to stoichiometry in infinite-dimensional space. The number of independent chemical reactions can, of course, be infinitely large, and one needs to introduce a *reaction label* u that plays for reactions the same role as the com-

---

[11] For instance, if one convinces oneself that the value of some property f (the cases where f is vapor pressure and kinetic constant are discussed later) is all that is needed to characterize components, one could choose f itself as the "initial" label. Of course, f may well be dimensional, and its values may well range in an interval other than $0, \infty$.

[12] The word *lumped* is often used in the literature. However, the word *lumping* has come to mean so many different things that it is perhaps somewhat misleading to use it here. Possibly the first use of the word *overall* in the sense used here is due to Astarita (1989b).

[13] In many real-life problems U is the only quantity accessible to direct measurement and/or the only one of interest. For instance, in a hydrodesulfurization process, $u(x)$ would be the concentration distribution of sulfur-containing species, $h(x)$ the amount of sulfur in species x, and U the total concentration of chemically combined sulfur.

ponent label x plays for components; we use v for the reaction label when it is intended as a dummy variable, and integrals over the range of u will be indicated with curly braces.[14] The mass-based stoichiometric coefficient of component x in reaction u constitutes the stoichiometric distribution $\sigma(x,u)$. The first question to be addressed is this: How does one extend the concept of a set of *independent* reactions? This is answered by requiring that the following equation for the unknown vector $\mathbf{q}^*(u)$ have only the trivial solution $\mathbf{q}^*(u) = 0$:

$$\{\sigma(x,v)\mathbf{q}^*(v)\} = 0; \tag{15}$$

that is, the kernel of $\sigma(x,v)$ contains only the zero vector for all x's.

Now let $m(x)dx$ be the mass of components with label between x and $x + dx$, and let $q(u)$ be the corresponding extent of reaction u with respect to the externally imposed mass distribution $M(x)$. One has, as an obvious extension of Eq. (3),

$$m(x) = M(x) + \{\sigma(x,v)q(v)\} \tag{16}$$

and, if the reactions $\sigma(x,u)$ are independent, two different $q(u)$ cannot result in the same $m(x)$.

Now let $I(x)$ be the function that has values of unity for all x's. The total mass of the system (which does not change as chemical reactions occur) is $<m(y)I(y)>$. It follows that $<\{\sigma(y,v)q(v)\}I(y)> = 0$, and by inverting the order of integration that $\{q(v)<\sigma(y,v)I(y)>\} = 0$. Because the latter must be satisfied for arbitrary $q(u)$, it follows that $<\sigma(y,u)I(y)> = 0$; that is, that the kernel of $\sigma(y,u)$ for all u's contains at least nontrivial vectors of the form $\Omega I(x)$, with Ω any arbitrary scalar. [The kernel of $\sigma(y,u)$ plays a crucial role. Indeed, we will see that at equilibrium points the chemical potential vector is restricted to lie in it.]

Equation (16) shows that, given $q(u)$, one can calculate $m(x) - M(x)$. Two possibilities arise:

1. There are some distributions $m(x) - M(x)$ that are not delivered by *any* $q(u)$.
2. All admissible $m(x) - M(x)$ are delivered by some $q(u)$.

Consider the second possibility first. Since the kernel of $\sigma(x,v)$ contains only the zero vector, two different $q(u)$ cannot deliver the same $m(x)$; it follows that there exists a distribution $A(u,x)$ such that

$$q(u) = <A(u,y)[m(y) - M(y)]>, \tag{17}$$

---

[14] Of course, curly braces will also identify an inner product and a norm in the infinite-dimensional Hilbert space of reactions. While the latter is infinite dimensional just the same as the space of components, one should be careful in generalizing results from finite-dimensional vector spaces, as the discussion in the following shows for some simple problems.

which in turn implies that $\{A(v,x)\sigma(x,v)\} = \delta(x)$. Let $\mathbf{z}(x)$ be any vector lying in the kernel of $\sigma(y,u)$. Since vectors $\mathbf{\Omega}I(x)$ are guaranteed to lie in the same kernel, $<\sigma(y,u)[\mathbf{z}(y) - \mathbf{\Omega}I(y)]> = 0$. It follows that

$$\begin{aligned} 0 &= \{A(v,y)<\sigma(y,v)[\mathbf{z}(y) - \mathbf{\Omega}I(y)]>\} \\ &= <\delta(y)[\mathbf{z}(y) - \mathbf{\Omega}I(y)]> \\ &= \mathbf{z}(y) - \mathbf{\Omega}I(y). \end{aligned} \quad (18)$$

Hence in case 2 *only* vectors of type $\mathbf{\Omega}I(x)$ lie in the kernel of $\sigma(y,u)$. Now consider the requirement of permanence of atoms. Let $\mathbf{p}(x)$ be the P-dimensional vector the components of which are the gramatoms of the atoms 1, 2, ..., P per gram of component x present in species x. Permanence of atoms implies that $<\sigma(y,u)\mathbf{p}(y)> = 0$, that is, that all components of $\mathbf{p}(x)$ lie in the kernel of $\sigma(y,u)$. It follows that, in case 2, the brute chemical formulas of all components are multiples of each other [each component $p_i(x)$ of $\mathbf{p}(x)$ has the form $\Omega_i I(x)$]. Case 2 is an infinite-dimensional merization system. One may notice that the proof of this result[15] represents the extension to the infinite-dimensional case of the proof given in the previous section for the case of a finite-dimensional space.

Let $\mathbf{i}$ be the P-dimensional vector with all components equal to unity, so that $I(x) = \mathbf{i} \cdot \mathbf{p}(x)$. It follows that conservation of mass implies that $\mathbf{i} \cdot <\sigma(y,u)\mathbf{p}(u)> = 0$, a condition which the permanence of atoms $<\sigma(y,u)\mathbf{p}(y)> = 0$ certainly implies. However, the converse is not true, because conservation of mass could be guaranteed by a nonzero $<\sigma(y,u)\mathbf{p}(y)>$ provided it is orthogonal to $\mathbf{i}$. Permanence of atoms implies that $<\mathbf{p}(y)m(y)> = <\mathbf{p}(y)M(y)> = \mathbf{b}$, where $\mathbf{b}$ is a P-dimensional vector. Hence this type of constraint remains a finite-dimensional one even in the continuous description. However, when contrasted with Eq. (16), the constraint $<\mathbf{p}(y)m(y)> = \mathbf{b}$ does not yield $m(y)$ explicitly—there are, in fact, an infinite number of $m(y)$ that satisfy the constraint.

Finally, the orthogonal complement method can be extended to the continuous description. The reaction vectors are $m_1(x) - m_2(x) = \{\sigma(x,v)[q_1(v) - q_2(v)]\}$. The set of basis vectors of the orthogonal complement, $a_L(x)$, $L = 1, 2, \ldots, C$, compactly indicated as $a(x)$, is all those vectors to which all possible reaction vectors are orthogonal:

$$<\mathbf{a}(y)\{\sigma(y,v)[q_1(v) - q_2(v)]\}> = 0, \quad (19)$$

or, exchanging the order of integration,

$$\{[q_1(v) - q_2(v)]<\mathbf{a}(y)\sigma(y,v)>\} = 0. \quad (20)$$

---

[15] The result is due to Astarita and Ocone (1989), and it generalizes an earlier result for discrete mixtures due to Astarita (1976). The result is relevant for the discussion of the Wei and Prater (1962) paper and of the ample subsequent literature that refers back to the Wei and Prater paper.

Equation (20) must be satisfied for all possible reaction vectors; hence, $<\mathbf{a}(y)\sigma(y,v)> = 0$. Now, given any C-dimensional vector $\mathbf{B}$, the distribution $\mathbf{a}(x) \cdot \mathbf{B}$ is orthogonal to $\sigma(y,u)$: $<\sigma(y,u)\mathbf{a}(y) \cdot \mathbf{B}> = <\sigma(y,u)\mathbf{a}(y)> \cdot \mathbf{B} = 0$. Since the set $\mathbf{a}(x)$ is complete, *only* distributions expressible as $\mathbf{p}(x) \cdot \mathbf{B}$ are orthogonal to $\sigma(y,u)$. Now suppose that, from some independent argument, one has ascertained that some distribution $s(x)$, otherwise unknown, is orthogonal to $\sigma(y,u)$; we will see that, under appropriate qualifying conditions, the equilibrium distribution of chemical potentials, $\mu(x)$, falls into this category. Then there must exist a C-dimensional vector $\mathbf{B}$ such that $s(x) = \mathbf{a}(x) \cdot \mathbf{B}$. Thus the search for an unknown function $s(x)$ (an infinite-dimensional vector) is reduced to the search for a C-dimensional vector.

C. Lumping and Overall Descriptions

As discussed in an earlier footnote, the word *lumping* has come to mean a number of different things. In this article, we use this word with a very restricted meaning: *Lumping* means the attempt to describe a finite-dimensional system (i.e., one described in a discrete way) of dimensions, say, N, by a lower order system of dimensions N^*: Several groups of components (and/or of other quantities such as reactions) are "lumped" into single *pseudocomponents* (or *pseudoreactions*). The lower order system is to be either an *exact* equivalent of the original system or at least, in some sense to be made precise, an *approximate* equivalent of it.

Lumping of a discrete system into a lower order one is discussed, for the case of kinetics, in Section IV, A. In that section we also show that it is possible to lump a continuous system into a finite-order one. The important point is that, in dealing with discrete systems with large (or countably infinite) numbers of components, or with continuous systems, it is very useful to reduce the dimensionality of the composition description for practical calculations. For example, one could consider a subset of similar compounds, such as stereoisomers, to be a single compound, or "lump," with the properties of the racemic mixture. More generally, we can project the composition vector onto a lower dimensional subspace that is not simply a partitioning of the compounds into subsets. One could also reduce dimensionality by some nonlinear parameterization rather than by linear projection, but we do not consider that case here.

An entirely different meaning is assigned in the following to the word *overall*. When we use that word we mean that one is not interested in the fine-grained structure of the system, but only in some gross overall property—such as the total concentration of all species of a certain type. Problems of "overall" behavior may, in principle, be discussed on the basis of both a continuous and a discrete de-

TABLE I
SUMMARY OF NOTATION

	Components	Reactions	Atoms	Constraints
Index (discrete)	I	J	K	L
Label (continuous)	x	u		
Dummy label	y	v		
Dummy label (in functionals)	z	w		
Total number	N (discrete)	R (discrete)	P	C
Sum on index (integral on label)	$<,>$	$\{,\}$	$\cdot$	Σ
Unit vector	$\mathbf{I}$	$\mathbf{i}$		
Matrix		σ_{IJ}	p_{IK}	

scription, but it turns out that they are often more easy to deal with in the continuous description. This has, unfortunately, led to a nested couple of misunderstandings in the literature. First, the terms *lumping* and *overall* have been used as if they were interchangeable, which they are not. Second, *lumping* has been associated (somewhat loosely) with discrete, and *overall* with continuous. In fact, both types of problems (lumping and overall) can be described with both (discrete and continuous) formalisms, and the choice of the formalism is often largely a matter of taste. When it is not a matter of choice, the decision is dictated by the specific problem being considered.

The idea of overall kinetics could be regarded as just an extreme case of lumping where the projection is onto a one-dimensional subspace, such as the unit vector. In this case the overall lump is just the total mass of the system or of some subset of the compounds. However, overall kinetics (and/or thermodynamics) are in no way exact or approximate lumping procedures, since the "lump" may well behave in a way that is totally different from that of the original system.

Be that as it may, in either case (lumping or overall) the procedure can be followed regardless of whether the system is discrete or continuous. Lumping a continuous system onto a finite dimensional subspace can give good approximations or even be exact in some cases. Our notation is summarized in Table I.

III. Thermodynamics

A. Phase Equilibria

We assume the reader is familiar with the theory of phase equilibria in mixtures containing a few components. As the number of components increases, one again reaches a point where a continuous description might be preferable. It is relatively straightforward to extend the classical theory to a continuous description, as seen later. However, a number of somewhat unexpected problems arise both in the solution of actual phase equilibrium problems and in a purely conceptual sense. These problems are discussed in some detail.

1. Convexity

We begin with what is logically the first problem to be analyzed, that is, the question of whether a system of assigned average composition will in fact, at equilibrium, exist as a single phase, or whether phase splitting will in fact occur. This problem is a well-known one for a two-component mixture, which may phase split if there is a spinodal region. In the simple two-component case, the analysis is restricted to the curvature of the Gibbs free energy curve. More generally, the problem is related to the convexity of the free energy hypersurface. The analysis presented here is based on work by Astarita and Ocone (1989).

Consider a system where some mole fraction distribution $X^F(x)$ is assigned, with $<X^F(y)> = 1$. Suppose the system exists in a one-phase condition at some pressure p and temperature T. One can define a (Gibbs) free energy of mixing per mole, G^{MIX}, which is the difference between the actual free energy of the one-phase mixture and what one would calculate by linearly adding the free energies of the pure components, each multiplied by the corresponding mole fraction. In an ideal solution, $G^{MIX} = G^{ID}$, where (some subtlety is involved with the following equation that is discussed in the appendix to this section)

$$G^{ID} = RT<X^F(y) \ln[X^F(y)]> < 0. \tag{21}$$

In a nonideal solution, $G^{MIX} = G^{ID} + G^{EX}$, where the excess free energy of mixing G^{EX} at any given temperature and pressure depends on the composition. Say its value is given by a functional $\mathbf{G}[\ ]$ of $X^F(y)$, which also depends parametrically on p and T:

$$G^{EX} = \mathbf{G}[X^F(z); p, T] \tag{22}$$

and, correspondingly,

$$G^{MIX} = G^{ID} + G^{EX} = \mathbf{G}^*[X^F(z); p, t]. \tag{23}$$

Now suppose the system in fact exists as N separate phases, with $K = 1, 2, \ldots, N$ the phase index; let α_K be the number of moles of phase K per mole of the system. The α_K's must satisfy the mass balance condition

$$\Sigma \alpha_K = 1. \tag{24}$$

Let $X^K(x)$ be the mole fraction distribution in phase K. The X^K's must satisfy the mass balance condition

$$\Sigma \alpha_K X^K(x) = X^F(x). \tag{25}$$

In the multiphase condition, the total free energy of mixing per mole of system is

$$\Sigma\alpha_K \mathbf{G}^*[X^K(z); p,T]; \qquad (26)$$

should the system exist as a single phase, it would be:

$$\mathbf{G}^*[X^F(z); p,T]. \qquad (27)$$

It follows that the system will in fact exist as a single phase if, for *any* choice of the scalars α_K and the distributions $X^K(x)$ satisfying Eqs. (24) and (25), one has

$$\mathbf{G}^*[X^F(z); p,T] < \Sigma\alpha_K \mathbf{G}^*[X^K(z); p,T]. \qquad (28)$$

When Eq. (28) is satisfied at some given $X^F(x)$, the functional $\mathbf{G}^*[\ ]$ is said to be *globally convex* at that composition. If $\mathbf{G}^*[\ ]$ is globally convex at *all* possible compositions, it is said to be *unconditionally convex*. Clearly, if $\mathbf{G}^*[\ ]$ is unconditionally convex, the system will exist in a one-phase condition at all possible compositions, and the theory of homogeneous chemical equilibria discussed in the next section applies. In particular, if no phase splitting is possible, the functional giving the chemical potentials as depending on composition is invertible: There cannot be two (or more) different compositions corresponding to the same chemical potential distributions, since these would correspond to two (or more) different phases in equilibrium with each other.

If, at some given composition $X^F(x)$, $\mathbf{G}^*[\ ]$ is not globally convex, then there is at least one combination of more than one phase satisfying the mass balance constraints [Eqs. (24) and (25)] that corresponds to a smaller free energy than the one-phase condition. This means that the one-phase condition is unstable to *large* composition perturbations—there is no requirement that the $X^K(x)$'s of the equilibrium phases should be "close" to $X^F(x)$. Physically, that means that phase splitting may need to occur by *nucleation*. It is therefore of interest to consider infinitesimal stability of the one-phase condition. Again, we consider several phases in amounts satisfying Eq. (24), but we now require all of them to have *almost* the same composition as the average $X^F(x)$. Thus, we write

$$X^K(x) = X^F(x) + \delta X^K(x) \qquad (29)$$

with the $\delta X^K(x)$'s satisfying the following mass balance constraint:

$$\Sigma\alpha_K \delta X^K(x) = 0. \qquad (30)$$

The free energy of mixing per mole of each phase is now that of the feed mixture plus the deviation $<\mu(y)\delta X^K(y)>$, and hence the condition of infinitesimal stability is

$$\Sigma\alpha_K <\mu(y)\delta X^K(y)> > 0. \qquad (31)$$

The functional $\mathbf{G}^*[\ ]$ is said to be *locally convex* at $X^F(x)$ if Eq. (31) is satisfied. If $\mathbf{G}^*[\ ]$ is locally convex everywhere, it is also unconditionally convex. The set of compositions at which **G***[ ] is not locally convex is called the *spinodal set*. (The reader should check that this definition is the obvious extension of the definition of a spinodal region in, say, a two-component mixture.)

It is important to realize that the convexity conditions only refer to the free energy of mixing. The actual free energy of the mixture equals the free energy of mixing plus the linearly additive term $<G^0(y)X^F(y)>$, where $G^0(x)$ is the pure-component free energy of component x. The term does not contribute anything to the properties of convexity, since $<G^0(y)X^F(y)>$ is neither convex nor concave.

Equation (21) shows that G^{ID} is unconditionally convex. It also shows that in the neighborhood of the "corner" compositions $X^F(x) = \delta(x - x^*)$ (i.e., very dilute solutions in a "solvent" x^*), the local convexity of G^{ID} is in fact infinitely large. It follows that in such neighborhoods also **G***[ ] is globally convex—the spinodal region cannot reach the corners (which is another way of saying that solubility can never be exactly zero). Now if a spinodal set exists, where **G***[ ] is locally not convex (and hence also globally not convex), it seems reasonable that a set of compositions should exist that is, in a vague sense, "between" the corners and the spinodal set, which are locally convex but are not globally convex. That set of compositions is called the *metastable set*.

A feed composition in the metastable set is stable to infinitesimal composition disturbances but is unstable to finite ones; hence, for such a composition phase splitting can only occur by nucleation, and not simply by Brownian motion (which, at most, supposedly results in infinitesimal composition disturbances). Hence, a metastable composition may be observed as a one-phase system in the laboratory: Superheated liquids and subcooled vapors are elementary one-component examples. In contrast with this, one-phase spinodal compositions, by virtue of being unstable to infinitesimal perturbations, will never be observed in the laboratory.

2. Raoult's Law

We begin the analysis of phase equilibria (for those cases where phase splitting does in fact occur) with the simplest possible case, that of vapor–liquid equilibrium where both the liquid and the vapor phase are ideal, so that Raoult's law applies (Astarita, 1989; Bowman, 1949; Edminster, 1955). In this case the only parameter that completely characterizes every individual component is its vapor pressure b at the temperature considered; hence, one begins by using b itself as the (dimensional) component label. Let $X^L(b)$ and $X^V(b)$ be the mole fraction distributions in the liquid and vapor phases, respectively, and let p be the total pressure. Obviously, the zeroth moments of both distributions are unity. The continuous form of Raoult's law is:

$$pX^V(b) = bX^L(b). \tag{32}$$

It follows, trivially, that, provided vapor–liquid equilibrium is indeed possible at the assigned values of temperature and pressure, pressure p is the first moment of the liquid-phase mole fraction distribution. It is therefore natural to define the dimensionless label x as b/p, that is, as the ratio of the vapor pressure to the total pressure. This now requires the first moment of $X^L(x)$ to be unity, $<yX^L(y)> = 1$. Raoult's law takes the form

$$X^V(x) = xX^L(x). \tag{33}$$

As simple as these equations are, they result in some unexpected problems. Consider first the case where $X^L(b)$ is assigned. The practical problem becomes the determination of the bubble point at some given temperature. The pressure corresponding to the bubble point is $<bX^L(b)>$—that part is easy. One can now renormalize to $x = b/p$, and use Eq. (33) to calculate the vapor phase composition $X^V(x)$. Notice, however, that $X^V(x)$ will not have the same functional form as $X^L(x)$; while the latter has a first moment equal to unity, $<yX^L(y)> = 1$, the former has (-1)th moment equal to unity, $<X^V(y)/y> = 1$. If, for instance, one uses a gamma distribution, $X^L(x) = \phi(\alpha,x)$, $X^V(x)$ will not be given by a gamma distribution.

Now consider the converse case where $X^V(b)$ is assigned. The practical problem here is the determination of the dew point at some assigned temperature. The dew point pressure is now $p = 1/<X^V(b)/b>$. Renormalizing the label as $x = b/p$, one can use Eq. (33) to calculate $X^L(x)$, but this will not have the same functional form as $X^V(x)$. If one insists that the liquid phase mole fraction distribution should be given by a gamma distribution $\phi(\alpha,x)$, one needs to assign $X^V(x)$ as a modified gamma distribution $\Phi(\alpha,x)$:

$$\Phi(\alpha,x) = x\phi(\alpha,x). \tag{34}$$

Notice that the (-1)th moment of $\Phi(\alpha,x)$ is unity, but its first moment is not. Also, $\Phi(\alpha,x)$ is legitimate for $\alpha > 0$, just as $\phi(\alpha,x)$ and hence $\Phi(\alpha,0) = 0$ whatever the value of α: There cannot be a finite mole fraction of components with vanishingly small vapor pressure in the vapor phase. [In fact, $\phi(\alpha,x)$ as a mole fraction distribution is legitimate only for $\alpha \geq 1$; see Appendix B to this section. The argument above holds *a fortiori*.]

The dew point and bubble point calculations do not present peculiar problems, but the flash calculation does. Let $X^F(x)$ be the mole fraction distribution in the feed to a flash, and let α be the vapor phase fraction in the flashed system. The mass balance is:

$$X^F(x) = \alpha X^V(x) + (1 - \alpha)X^L(x), \tag{35}$$

which is of course a special case of Eq. (25).

If the assigned pressure, p, satisfies $p < 1/<X^F(b)/b>$, the flashed condition will be all vapor; if it satisfies $p > <bX^F(b)>$, it will be all liquid. A nontrivial flash problem arises when $<bX^F(b)> \geq p \geq 1/<X^F(b)/b>$. The borderline cases $<bX^F(b)> = p$ and $1/<X^F(b)/b> = p$ give rise to the bubble point and dew point calculations ($\alpha = 1$ and $\alpha = 0$), respectively. We are not interested in the case of $0 < \alpha < 1$.

One could still, of course, insist that $X^L(x)$ be a gamma distribution $\phi(\alpha,x)$ and, correspondingly, that $X^V(x) = \Phi(\alpha,x)$. However, with any nontrivial value of α, the corresponding $X^F(x)$ could not be either a gamma or a modified gamma distribution, but only a linear combination of those. *It is the mass balance condition in Eq. (24) that causes the problem* (Luks *et al.*, 1990; Sandler and Libby 1991; Shibata *et al.*, 1987).

It is often useful to define a partition coefficient $K(x)$ as follows:

$$K(x) = X^V(x)/X^L(x). \tag{36}$$

In the case where Raoult's law applies, and x has been normalized as b/p, the partition coefficient $K(x)$ is simply x [see Eq. (33): more generally, $K(x)$ depends only on x and is made equal to it by renormalization]. However, as seen later, in general $K(x)$—whatever the dimensionless label x may be—depends on the whole mole fraction distribution in either phase.

3. Nonideal Liquid Solutions

If the liquid phase is not an ideal solution, one cannot use b/p as the dimensionless label, since there might very well be more than one component endowed with the same vapor pressure but otherwise behaving differently from the viewpoint of phase equilibrium. Let x be an appropriate dimensionless label (in the sense that no two components with the same x behave differently in any significant way). One may define the *activity coefficient* of component x in the liquid phase, $\tau(x)$, as follows. Let $\mu(x)$ be the chemical potential of component x (which, at equilibrium, is the same in the liquid and vapor phases). Then

$$\mu(x) = RT \ln[X^L(x)\tau(x)]. \tag{37}$$

Now, if the vapor phase is an ideal gas, the equilibrium equation becomes

$$pX^V(x) = b(x)\tau(x)X^L(x) \tag{38}$$

so that $p = <b(y)\tau(y)X^L(y)>$, that is, the equilibrium pressure is *not* (proportional to) the first moment of $X^L(x)$. It is clear that the problems discussed in the previous section become even more severe.

The flash problem is of course still subject to the mass balance condition in Eq. (35). The bubble point pressure ($\alpha = 0$) is $<b(y)\tau(y)X^F(y)>$, and this is relatively easy to calculate because the activity coefficients $\tau(x)$ are those at the feed composition $X^F(x)$ and they are at worst very weak functions of pressure. The situation is different for the dew point pressure ($\alpha = 1$), which is $1/<X^F(y)/\tau(y)b(y)>$.

The activity coefficients are those at the liquid composition, which needs to be calculated from Eq. (38) with $pX^F(x)$ on the left side.

In principle, in addition to the dew point and bubble point cases, there may be another case where the composition of at least one of the two phases is known: the case where $X^F(x)$ is an *azeotropic* composition, so that Eq. (35) is trivially satisfied with $X^F(x) = X^V(x) = X^L(x)$. Equation (38) shows that, at an azeotropic composition, $b(x)\tau(x) = p$ is a constant (it has the same value at all x's). Because $\tau(x)$ depends on the liquid phase mole fraction distribution $X^L(x)$, there may well be one particular composition at which this rather stringent condition is satisfied.

One could still *define* a partition coefficient $K(x)$ as $X^V(x)/X^L(x)$; however, Eq. (38) shows that $K(x)$ depends on the whole liquid phase mole fraction distribution, as does $\tau(x)$. Even more generally, if one relaxes the condition that the vapor phase should be ideal [so that fugacity coefficients would appear on the left side of Eq. (38)] or if one considers any two-phase system (for instance, a liquid–liquid system where nonunity activity coefficients would need to be considered for both phases), $K(x)$ would depend on both distributions. Say its value is delivered by the following functional:

$$K(x) = \mathbf{K}[X^V(z), X^L(z); x, p, T]. \tag{39}$$

This formulation, while of absolutely general validity, is so complicated that approximate methods of solution of phase equilibrium problems have to be developed. A few essential aspects of these approximate methods are discussed in the next subsection.

4. Approximate Methods

Cotterman and Prausnitz (1991) have reviewed the approximate methods that have been used in the solution of phase equilibrium problems within the framework of a continuous description (or, often, a semicontinuous one, where a few components are dealt with as discrete ones). These methods are based on relatively trivial extensions of classical methods (which make use of Gibbs free energy equations, equations of state, and the like) to a continuous description.

Two approximation methods are, however, peculiar to the continuous description: the so-called *method of moments,* and the *quadrature method.* Some basic conceptual issues about these two methods are discussed next.

In the method of moments (Cotterman *et al.,* 1985; Luks *et al.,* 1990), as applied for example to a flash calculation, a functional form for $X^V(x)$ and $X^L(x)$ is assumed. This will in general contain one or more parameters. One then writes an appropriate model for the chemical potentials in the two phases, and one calculates the values of the parameters by requiring higher and higher moments of the distributions to match the equilibrium condition (equality of chemical potentials in the two phases).

The problem with this method is that, as discussed in Section III,B,2, the functional form of the distributions in the two phases will be different, and both will in general be different from the functional form for the feed (except in the dew point and bubble point limits, where one of the two distributions, but not the other one, will be of the same functional form and, in fact, equal to the distribution in the feed). Hence in this classical formulation the method of moments produces what has been called the *conservation of mass problem* (Luks et al., 1990): the mass balance equation, Eq. (35), can be satisfied only approximately.

The difficulty illustrated above has been discussed by Sandler and Libby (1991), who propose an alternate procedure for the method of moments as applied to a flash calculation. First, one chooses a functional form for one of the two distributions, with which one to choose being suggested by the particular problem one is trying to address. Next, one uses the mass balance condition, Eq. (35), to calculate the functional form for the mole fraction distribution in the other phase. The method of moments (or other solution procedures) can then be used to determine the parameters from the equilibrium condition. Following this procedure, no mass conservation problem arises.

The quadrature method is based on the following approach (Shibata et al., 1987). Given a mixture with N components, where N is a fairly large number, one may want to describe it as mixture with N' ($<N$) pseudocomponents where N' is some manageably small number, and where one hopes that the description in terms of the pseudocomponents reproduces to within some acceptable degree of approximation the behavior of the real mixture. This is easily recognized as a problem of approximate lumping, and the question that arises is the following one: Given that one has decided what N' might be, how does one choose the pseudocomponents so that the level of approximation of the lumping scheme is as good as can be achieved?

The answer to this question is in fact made easier if one allows N to approach infinity, and therefore one works with a continuous description. One sets up the continuous formulation of the phase equilibrium problem, which will certainly contain (weighted) integrals of the relevant distribution functions. These integrals must be represented by the sum of N' terms, which reduces to a quadrature problem: Given an integral over x that has to be represented by a sum of N' terms, how do we choose the best N' values of x so that a discrete quadrature [say $<f(y)>$ is approximated with $\Sigma W_K f(x_K)$, with W_K the appropriate weighting constants, and the sum being intended for $K = 1, 2, \ldots, N'$] approximates as well as possible the value of the integral?

Formulated in this way, the problem reduces to one originally considered by Gauss (see, e.g., Stroud and Secrest, 1966). Shibata et al. (1987) review and extend methods for doing that. The important point is that theorems are available that can be used to determine the best quadrature points and the corresponding weighting constants, while in the case of a discrete description the choice of pseudocomponents is mostly a guessing game. There are, however, two prices to be paid when choosing this approach.

First, pseudocomponents determined by the quadrature method may well be unrealistic ones; for instance, if the label x is (proportional to) the number of carbon atoms, pseudocomponents may well correspond to noninteger x_K values. This may be aesthetically unpleasant, but it does not represent a real problem. More seriously, the appropriate pseudocomponents obviously depend on the composition and, hence, in a repeated calculation such as is required in a distillation tower, pseudocomponents will need to be different at each step. This puts out of tilt the mass balance equations that are coupled to the equilibrium ones, and, even if this problem could be circumvented (as, at least in principle, it can), the procedure would certainly not be applicable to existing software for distillation column calculations.

B. HOMOGENEOUS CHEMICAL EQUILIBRIA

In this section, we discuss the general problem of chemical equilibrium under the assumption that the system considered is guaranteed to exist in a one-phase (homogeneous) condition at all possible compositions. We initially follow the logical approach discussed in Chapter 3 of Astarita (1989a), and we are therefore very concise in the first subsection, referring the reader to that work for details.

1. The Classical Theory

We begin by analyzing the discrete case, where the mass vector at time t, $\mathbf{m}(t)$, is given by Eq. (3), which is rewritten here:

$$\mathbf{m}(t) = \mathbf{M} + \{\boldsymbol{\upsilon}, \mathbf{q}(t)\}. \tag{40}$$

This implies that all quantities (such as pressure, and free energy) which are functions of composition (i.e., of $\mathbf{m}$), can in fact be regarded as functions of $\mathbf{q}$. In particular, pressure p depends in principle on volume, temperature, and composition. We can write

$$p = f(V, T, \mathbf{q}). \tag{41}$$

Since the system is homogeneous, the function $f(\ )$ is invertible for volume. (This is not true for systems that may split into two or more phases. For instance, H_2O at 100°C and 1 atm may have, at equilibrium, any volume between that of the liquid and that of steam.) This in turn implies that the state of the system, which in principle is the set $V, T, \mathbf{q}$, may also be taken to be $p, T, \mathbf{q}$. The latter is more generally useful, and the following discussion is based on that choice. Partial derivatives are indicated as δ/δ, with the understanding that the independent variables are indeed p, T, and q.

In particular, the partial derivative of the Gibbs free energy G with respect to the vector of extents of reaction **q** is called the affinity, θ:

$$\theta = \delta G/\delta \mathbf{q}. \tag{42}$$

As the chemical reactions proceed, **q** changes in time; its rate of change is called the *rate of reaction* vector **r**:

$$d\mathbf{q}/dt = \mathbf{r}(p,T,\mathbf{q}). \tag{43}$$

At some assigned values of p and T, the equilibrium extent of reaction, $\mathbf{q}^*$, is such that the corresponding rate of reaction is zero:

$$\mathbf{q}^*(p,T) : \mathbf{r}(p,T,\mathbf{q}^*(p,T)) = 0. \tag{44}$$

If one is able to calculate the value of $\mathbf{q}^*$ for any given values of **p** and **T**, one obviously can calculate the equilibrium composition from Eq. (40). Now the second law of thermodynamics implies that, at constant pressure and temperature, the Gibbs free energy can never increase, which for the case at hand reduces to the following condition:

$$\{\boldsymbol{\theta},\mathbf{r}\} \leq 0. \tag{45}$$

At equilibrium $\{\boldsymbol{\theta},\mathbf{r}\} = 0$; hence, it has a maximum there. It follows that its differential is zero at equilibrium, and, since at equilibrium $\mathbf{r} = 0$ by definition, the equilibrium affinity $\boldsymbol{\theta}^*$ is such that, for any infinitesimal displacement from equilibrium $d\mathbf{q}$ resulting in some infinitesimal reaction rate $d\mathbf{r}$,

$$\{\boldsymbol{\theta}^*,d\mathbf{r}\} = 0. \tag{46}$$

Given an infinitesimal displacement $d\mathbf{q}$ from equilibrium, one obtains from Eq. (43), with $\mathbf{B} = \delta\mathbf{r}/\delta\mathbf{q}$

$$d\mathbf{r} = \{\mathbf{B},d\mathbf{q}\}. \tag{47}$$

It follows that condition (46) reduces to

$$\{d\mathbf{q},\{\mathbf{B}^T,\boldsymbol{\theta}^*\}\} = 0. \tag{48}$$

Because this has to hold for arbitrary $d\mathbf{q}$, one has

$$\{\mathbf{B}^T,\boldsymbol{\theta}^*\} = 0. \tag{49}$$

Now let **D** be the set of vectors lying in the kernel of $\mathbf{B}^T$; Eq. (49) implies that $\boldsymbol{\theta}^*$ lies in **D**. Truesdell (1984) calls "weak" an equilibrium point where $\boldsymbol{\theta}^*$ is nonzero. We will restrict attention to the (likely) case where **D** contains only the zero vector, that is, to the case where only strong equilibria exist (for the discrete case, **D** contains only the zero vector provided det**B** is nonzero). For that case, the equilibrium condition becomes

$$\boldsymbol{\theta}^* = 0. \tag{50}$$

It is worthwhile to observe for future reference that, if indeed the equilibrium points are strong ones, det **B** is nonzero, and hence **r(q)** is invertible at equilibrium. We will also make the (likely) assumption that the systems considered are, in Coleman and Gurtin's (1967) terminology, strictly *dissipative,* that is, that Eq. (45) is satisfied as an equality only at equilibrium points.

Since **Θ** is a function of **q**, Eq. (50) by itself does not exclude the possibility of the existence of more than one equilibrium value **q***. However, should more than one **q*** exist, the Gibbs free energy hypersurface would necessarily include a spinodal region, and this is impossible for systems that are homogeneous over the whole composition space. Hence, in such systems the equilibrium composition is guaranteed to be unique.

Given the definition of θ in Eq. (42), the relationship between **m** and **q** in Eq. (40), and the definition of the (mass-based) chemical potential vector $\mu = \delta G/\delta \mathbf{m}$, Eq. (50) reduces to the following (classical) form:

$$<\sigma^T, \mu^*> = 0; \quad (51)$$

that is, the vector of equilibrium chemical potentials lies in the kernel of σ^T. As was discussed in Section II,A, given any P-dimensional vector **b,** the N-dimensional vector **p · b** lies in the kernel of σ^T; the same argument applies to whatever set of linear constraints is imposed on the system, and that, as also discussed in Section II,A, is likely to be a low-order set. Thus the search for an N-dimensional vector μ^* reduces to the search for a C-dimensional vector **b**.

The concise analysis just given can be easily generalized to a continuous description. The distribution of affinities, $\theta(u)$, is defined as

$$\theta(u) = \delta G/\delta q(u), \quad (52)$$

while the second law of thermodynamics takes the form

$$\{\theta(v)r(v)\} \leq 0. \quad (53)$$

Both the affinity distribution and the rate of reaction distribution are uniquely determined (at fixed p and T) by the distribution of extents of reactions, and thus one can write

$$\theta(u) = \Theta[q(w);u], \quad (54)$$

$$r(u) = \mathbf{R}[q(w);u], \quad (55)$$

where both $\Theta[\;]$ and $\mathbf{R}[\;]$ are functionals of the extent of reaction distribution $q(u)$ that also depend parametrically on the reaction label u.

Now let $\delta q(u)$ be an arbitrary (and in fact infinitesimal) displacement of $q(u)$ from the equilibrium distribution $q^*(u)$; and let $B(v;u)$ be the functional derivative of $\mathbf{R}[\;]$ at $q^*(u)$. The corresponding (infinitesimal) $\delta r(u)$ is given by

$$\delta r(u) = \{B(v;u)\delta q(v)\}, \quad (56)$$

which is evidently the continuous generalization of Eq. (47). The generalization of Eq. (48) is

$$\{\theta[q^*(w);v]\{B(v';v)\delta q(v')\}\} = 0. \tag{57}$$

This has to hold for arbitrary $\delta q(u)$. Hence,

$$\{\theta[q^*(w);v]B(v,u)\} = 0. \tag{58}$$

The argument (and the assumptions) following Eq. (49) can now be duplicated to yield the result that, if the set **D** of functions $s(v)$ to which $B(v,u)$ is orthogonal contains only the zero function (i.e., if attention is restricted to systems for which only strong equilibria are possible), the equilibrium condition is

$$\theta^*(u) = 0 \tag{59}$$

and that, since the kernel of $B(v,u)$ contains only the zero function, the functional $R[\,;\,]$ can be inverted at $q^*(u)$. This leads to the following generalization of Eq. (51):

$$<\sigma(y,u)\mu^*(y)> = 0. \tag{60}$$

It is important to realize that *any* distribution of chemical potentials satisfying Eq. (51) (in the discrete case) or Eq. (60) corresponds to an equilibrium composition. Of course, given an initial value **M** or $M(y)$, infinitely many of such equilibrium compositions will not satisfy the mass balance condition [Eq. (40) or its equivalent continuous form].

2. The Orthogonal Complement Method

In this and the next subsection, we present a concise summary of the recent analysis given by Krambeck (1994a). As was discussed in Section II,A, the orthogonal complement of the reaction space is C dimensional: The vectors $\mathbf{a}_L$ are a basis for the orthogonal complement, and they are such that, given two values $\mathbf{m}_1$ and $\mathbf{m}_2$, which are both stoichiometrically accessible from some initial composition **M** (so that some linear combination of the reactions may in fact bring the system from $\mathbf{m}_1$ to $\mathbf{m}_2$), $<(\mathbf{m}_1 - \mathbf{m}_2), \mathbf{a}_L> = 0$ for all L's, $L = 1, 2, \ldots, C$. This implies that

$$<\mathbf{a}_L, \{\sigma,(\mathbf{q}_1 - \mathbf{q}_2)\}> = 0, \tag{61}$$

which is equivalent to

$$\{(\mathbf{q}_1 - \mathbf{q}_2), <\sigma^T, \mathbf{a}_L>\} = 0. \tag{62}$$

Because Eq. (62) has to hold for arbitrary $\mathbf{q}_1 - \mathbf{q}_2$, one obtains

$$<\sigma^T, \mathbf{a}_L> = 0; \tag{63}$$

that is, all of the $\mathbf{a}_L$'s lie in the kernel of $\boldsymbol{\sigma}^T$ (and only linear combinations of them lie in it). When this is compared with Eq. (51), one concludes that $\boldsymbol{\mu}^*$ lies in the orthogonal complement, that is, that $\boldsymbol{\mu}^*$ can be expressed as

$$\boldsymbol{\mu}^* = \sum_{1}^{C} \alpha_L \mathbf{a}_L. \tag{64}$$

The coefficients α_L play the role of the usual Lagrange multipliers in the determination of the minimum of Gibbs free energy. *Any* set of α_L's determines an equilibrium point (but infinitely many of these are not accessible from the initial composition). The great advantage of the orthogonal complement method is that the number of linear constraints, C, is invariably small, regardless of how large N and R might be.

The method is easily generalized to a continuous description, where the basis vectors of the orthogonal complement are C distributions $a_L(x)$; the analog of Eq. (64) is simply

$$\mu^*(x) = \sum_{1}^{C} \alpha_L a_L(x). \tag{65}$$

The strength of the orthogonal complement method, in the case of homogeneous systems, lies also in the following fact. If the system is guaranteed to be homogeneous at all possible compositions, the (constant pressure and temperature) $\mu(\mathbf{m})$ function is invertible: Should there be two different values of $\mathbf{m}$ resulting in the same μ, these would correspond to two different phases. Equations (64) and (65) show that the surface of equilibrium points is a C-dimensional linear subspace in chemical potential space. Finding the equilibrium composition for any given $\mathbf{M}$ simply involves mapping this subspace into composition space, and this is easily achieved if $\mu(\mathbf{m})$ is invertible, as it is for homogeneous systems. Krambeck (1994) discusses examples, in both the discrete and continuous formulation: the general case of ideal gas mixtures and the specific one of olefin oligomerization.

3. Behavior in a Neighborhood of Equilibrium

The restriction to strong equilibria implies that $\mathbf{r}(\mathbf{q})$ (or, in the continuous formulation, $\mathbf{R}[\,,]$) is invertible near equilibrium. Now consider the vector $\mathbf{f}$ defined as $d\mathbf{m}/dt$, say,

$$\mathbf{f} = \{\boldsymbol{\sigma}, \mathbf{r}\}. \tag{66}$$

The vector $\mathbf{f}$ depends on $\mathbf{m}$, and, since in a homogeneous system $\mu(\mathbf{m})$ is invertible, one may regard $\mathbf{f}$ as a function of μ:

$$\mathbf{f} = \mathbf{f}(\mu). \tag{67}$$

Let $\mathbf{C}$ be the $N \times N$ matrix $d\mathbf{f}/d\boldsymbol{\mu}$ at equilibrium. Krambeck (1970) has shown that the condition of detailed balancing[16] is equivalent to the requirement that $\mathbf{C}$ be symmetric:[17]

$$\mathbf{C} = \mathbf{C}^\mathrm{T}. \tag{68}$$

Since $\boldsymbol{\mu}(\mathbf{m})$ is invertible, $\mathbf{J} = d\boldsymbol{\mu}/d\mathbf{m}$ possesses an inverse $\mathbf{J}^{-1}$ at equilibrium. Let $\mathbf{K}$ be $d\mathbf{f}/d\mathbf{m}$. One has, at all strong equilibrium points,

$$\mathbf{C} = <\mathbf{K},\mathbf{J}^{-1}>. \tag{69}$$

Notice that, if all the mass rates f_I are multiplied by the same (but otherwise arbitrary) function of $\mathbf{m}$, the result still holds; in other words, linearity near equilibrium may be the quasilinearity of a uniform kinetics of the type that is discussed in Section IV,B.

Wei and Prater (1962) analyzed the case of monomolecular reactions, for which in general

$$\mathbf{f} = <\mathbf{K},\mathbf{m}>. \tag{70}$$

It is now useful to define the following modified inner product $[,]$ (where $\mathbf{x}$ and $\mathbf{y}$ are N-dimensional vectors):

$$[\mathbf{x},\mathbf{y}] = <\mathbf{x},<\mathbf{J},\mathbf{y}>>. \tag{71}$$

One can then show that $\mathbf{K}$ is self-adjoint with respect to $[,]$:

$$[\mathbf{x},<\mathbf{K},\mathbf{y}>] = [<\mathbf{K},\mathbf{x}>,\mathbf{y}]. \tag{72}$$

Thus, $\mathbf{K}$ has a complete set of eigenvectors that is mutually orthogonal with respect to $[,]$, and has real eigenvalues. This is the essential result for the Wei and Prater (1962) analysis of reaction pathways in monomolecular systems. The latter

---

[16] The condition of detailed balancing is that each individual reaction should be at equilibrium. Consider, for instance, a system of three isomers, A, B, and C. It is conceivable that A only reacts to form B, B only to form C, and C only to form A. If the three kinetic constants are equal, the equilibrium condition would be that the concentrations of all three isomers are the same. However, this would not satisfy the detailed balancing condition, since the reaction between A and B would not by itself be at equilibrium (equilibrium is only attained through a cycle).

[17] The symmetry of $\mathbf{C}$ is seen to be a special case of the Onsager reciprocal relations. However, it is somewhat misleading to regard the result from that viewpoint: The symmetry is a consequence of the requirement of detailed balancing (and of mass action kinetics), not a blind application of Onsager's relations. In his 1970 paper, Krambeck included the following footnote concerning this point: "While the requirement that the matrix $\mathbf{C}$ be symmetric at equilibrium is strongly suggested by the Onsager–Casimir reciprocal relations for homogeneous systems as presented on page 123–124 of Truesdell (1969), it cannot be derived directly from them unless they are strengthened so as to cover not only strictly linear systems but also *linear approximations* in the neighborhood of equilibrium." The discussion in Truesdell to which Krambeck refers makes it abundantly clear that in the alleged Onsager–Casimir theorems there is nothing to suggest that it is necessarily $\mathbf{C}$ that should be symmetric.

analysis has been strengthened by Krambeck (1984b). Even stronger constraints on reaction pathways, based on the assumption that every individual term $\theta_J r_J$ in Eq. (45) (and not only their sum) is nonpositive, have been established by Shinnar and Feng (1985). In this regard, see also the discussion in Section IV,B,4,b.

While Eq. (70) holds *in general* for monomolecular systems, it holds only in a neighborhood of equilibrium for the general case considered by Krambeck (1994):

$$\delta f = <\mathbf{K}, (\mathbf{m} - \mathbf{m}^*)>. \tag{73}$$

However, in view of Eq. (69), $\mathbf{K}$ in Eq. (73) is again self-adjoint with respect to [,]. The results are rather easily generalized to the case of continuous mixtures.

C. Heterogeneous Chemical Equilibria

It is of course well known that in nature heterogeneous chemical equilibria are possible: systems in which chemical reactions may take place, and which at equilibrium will exist in more than one phase. The question that arises is the analysis of the conditions under which heterogeneous chemical equilibria are possible in multicomponent mixtures. In this section, we follow closely the analysis that was recently presented by Astarita and Ocone (1989); earlier work on the subject is due to Caram and Scriven (1976) and Astarita (1976).

In systems where heterogeneous chemical equilibria prevail, both chemical and phase equilibrium conditions must be simultaneously satisfied. In practice, this means that the chemical equilibrium condition—Eq. (51) in the discrete description, and Eq. (60) in the continuous one—must be satisfied in one phase, and the phase equilibrium condition [μ or $\mu(x)$ to be the same in all phases] must be satisfied; this clearly guarantees that the chemical equilibrium condition is automatically satisfied in all phases.

It is in the area of heterogeneous chemical equilibria that a formulation in mass units turns out to be definitely preferable to a mole-based one. This can best be seen by considering a system where only components A and B are present, and a simple reaction such as the following one might take place:

$$A = 2B. \tag{74}$$

Let $1 - q$ be the number of moles of A, and consequently $2q$ be the number of moles B, and $1 + q$ the total number of moles. Since q determines the composition of the system, G^{MIX} is a unique and, in principle, known function of q, $G^{MIX} = w(q)$. In Section III,A, we showed that, in the absence of chemical reactions, statements about the Gibbs free energy are only significant to within a term which is linear in the composition. This is not true when chemical reactions can take place, because, for instance in the case at hand, the pure component Gibbs free energy of component B is given by

$$2G_B{}^0 = G_A{}^0 + G^S, \tag{75}$$

where G^S is the standard free energy of reaction which, in principle, is known.

The Gibbs free energy per unit mole of the mixture, g, can now be calculated as

$$g = (G_A^0 + qG^S)/(1 + q) + w(q), \qquad (76)$$

and one can try to find a minimum of g by imposing that

$$dg/dq = 0 = dw/dq + (G^S - G_A^0)/(1 + q)^2. \qquad (77)$$

This, however, does not determine the value of q at which dg/dq is zero, since the value of G_A^0 is arbitrary. The problem, as can be easily checked, does not arise in the case of an isomerization reaction $A = B$, for which the equivalent of Eq. (77) is

$$dg/dq = G^S + w(q), \qquad (78)$$

which determines the value of q, which minimizes the Gibbs free energy unequivocally. This is because in the case of an isomerization reaction the total number of moles is preserved as the reaction takes place. Since in mass units the total mass is preserved anyhow, in a mass units formulation the problem does not arise, and the location of minima of g in the reaction subspace is unequivocal.

We now move to a general formulation for the discrete case, where $g(\mathbf{q})$ has to be intended as the Gibbs free energy per unit mass that would prevail should the system exist in a one-phase condition. Clearly, given a value of $\mathbf{M}$, $g(\mathbf{q})$ is defined only for those compositions that are stoichiometrically accessible from $\mathbf{M}$, that is, only within the reaction subspace of composition space (i.e., it is not defined in the orthogonal complement). Of course, within the reaction subspace there may well be more than one minimum. In our 1989 work, we have proved the following theorems:

1. The absolute minimum of g in the reaction subspace lies in a homogeneous region of it (i.e., at a point that is unconditionally convex).
2. If there is more than one minimum of g in the reaction subspace, there is a spinodal region in it.
3. All minima that are not the absolute one lie in a metastable region of the reaction subspace (except for the freak case where more than one of such minima corresponds to exactly the same value of g, which is at most a singular point in parameter space).

The three theorems are easily proved by considering two local minima of g, $\mathbf{q}_1$ and $\mathbf{q}_2$, and the linear trajectory connecting them, $\mathbf{q} = \alpha\mathbf{q}_1 + (1 - \alpha)\mathbf{q}_2$.

It follows from these theorems that (again except for the freak cases mentioned), as long as one is restricted to the reaction subspace, heterogeneous chemical equilibria are impossible—the true absolute minimum is homogeneous, and it will eventually be reached at least by nucleation of new phases and mass transfer.

However, one is in general not restricted to the reaction subspace: the system considered may very well split into more than one phase in a direction orthogonal to the reaction subspace, that is, in the orthogonal complement of it. This is impossible in the (exceptional) cases where the system considered is a merization one, in the sense discussed in Section II. In all systems where the orthogonal complement has more than one dimension (i.e., in systems subjected to at least one linear constraint in addition to mass balance), phase splitting in one of the directions of the orthogonal complement is possible, and heterogeneous chemical equilibria may prevail. However, since the dimensionality of the orthogonal complement is likely to be small, the number of phases present at a heterogeneous chemical equilibrium point is also likely to be small.

The analysis discussed given here rather trivially extended to a continuous description, as Astarita and Ocone (1989) have shown.

IV. Kinetics

A. Exact and Approximate Lumping

As discussed in Section III,C, the concept of "lumping" is related to the idea that it might be possible, in analyzing the kinetic behavior of a multicomponent mixture, to substitute for the real system (which involves a very large, possibly infinitely large number of chemical reactions) a lower order system with a finite number of pseudocomponents which, in some sense to be made precise, behaves either exactly or approximately in the same way as the original system. The concept was originated by the works of Wei and Kuo and of Kuo and Wei in 1969; its fundamental mathematical structure is based on the older works of Wei (1962a) and of Wei and Prater (1962).

Consider first a discrete *monomolecular* system in which every species present can transform to any other one by a first-order reaction. In such a system, the kinetic equation at all possible compositions is given by Eq. (70), which is repeated here:

$$d\mathbf{m}/dt = -<\mathbf{K},\mathbf{m}>, \qquad (79)$$

where $\mathbf{K}$ is an $N \times N$ matrix of pseudokinetic constants. [In Section III,B, Eq. (70) was written without the minus sign; here we follow as closely as possible the Wei and Kuo terminology.]

The matrix $\mathbf{K}$ must satisfy several constraints, which are best illustrated by considering a simple three-component system. The rate of formation of component 1 from components 2 and 3 is $k_{21}m_2 + k_{31}m_3$, where the k's are true kinetic constants; hence for the off-diagonal elements of **K** one has the constraint

$$K_{IJ} = -k_{IJ} \leq 0 \quad \text{(off-diagonal).} \tag{80}$$

Conversely, the rate of consumption of component 1 is $(k_{12} + k_{13})m_1$, and hence for the diagonal elements:

$$K_{II} = \Sigma k_{IJ}, \tag{81}$$

where the sum is intended over J different from I. Given Eq. (80), one sees that the entries in each column of **K** sum to zero, so that Eq. (81) can be written as

$$\langle \mathbf{K}^T, \mathbf{I} \rangle = 0. \tag{82}$$

Finally, there must exist at least one stationary point **m*** (corresponding to chemical equilibrium) such that

$$\langle \mathbf{K}, \mathbf{m}^* \rangle = 0. \tag{83}$$

We now wish to substitute for this system a lower order one of $N\hat{}(<N)$ pseudocomponents, the masses of which are linear combinations of the masses of the original components. In general, we want to introduce an $N\hat{} \times N$ "lumping matrix" **L** such that

$$\mathbf{m}\hat{} = \langle \mathbf{L}, \mathbf{m} \rangle. \tag{84}$$

In this section, we indicate with [,] the sum over the $I\hat{}$ index, $I\hat{} = 1, 2, \ldots, N\hat{}$. Since the ^ system is of lower order, there must exist couples of different **m**'s, **m**$_1$ and **m**$_2$, which result in the same **m**$\hat{}$, say, $\langle \mathbf{L}, \mathbf{m}_1 \rangle = \langle \mathbf{L}, \mathbf{m}_2 \rangle$. Such couples will be called **L**-equivalent, and indicated with the notation $\mathbf{m}_1 \approx \mathbf{m}_2$.

Now one may want to impose another constraint, namely, that each of the original N species appears in only one of the $N\hat{}$ species; hence, every column of **L** has to be a unit vector of the type $0, 0, \ldots, 1, \ldots, 0$. This is called "proper" lumping. Lumping that is not proper, however, is possible; see Section II,C.

So far, lumping has been defined, but nothing has been said concerning the dynamic behavior of the ^ system. Now we come to the definition of "exact" lumping: A system is said to be exactly "lumpable" by the matrix **L** if there exists an $N\hat{} \times N\hat{}$ matrix **K**$\hat{}$, enjoying the same properties as **K** does (i.e., off-diagonal elements of **K**$\hat{}$ are nonpositive, $[\mathbf{K}\hat{}^T, \mathbf{I}\hat{}] = 0$, and there exists at least an $\mathbf{m}\hat{}* = \langle \mathbf{L}, \mathbf{m}^* \rangle$ for which $[\mathbf{K}\hat{}, \mathbf{m}\hat{}*] = 0$), such that

$$d\mathbf{m}\hat{}/dt = -[\mathbf{K}\hat{}, \mathbf{m}\hat{}]. \tag{85}$$

A corollary of the definition is that any two **L**-equivalent mass distributions have the same rate in the ^ system; that is, if $\mathbf{m}_1(0) \approx \mathbf{m}_2(0)$, then $\mathbf{m}_1(t) \approx \mathbf{m}_2(t)$ at all later times $t > 0$. This property can also be used as a definition of an exactly lumpable system.

Now if Eq. (85) holds, then

$$d\mathbf{m}\hat{}/dt = -[\mathbf{K}\hat{},<\mathbf{L},\mathbf{m}>]. \qquad (86)$$

On the other side, one also has

$$d\mathbf{m}\hat{}/dt = <\mathbf{L},d\mathbf{m}/dt> = -<\mathbf{L},<\mathbf{K},\mathbf{m}>>, \qquad (87)$$

which implies the fundamental result of Wei and Kuo (1969), namely, that in order to have exact lumping,

$$[\mathbf{K}\hat{},\mathbf{L}] = <\mathbf{L},\mathbf{K}>. \qquad (88)$$

Many of the Wei and Kuo results are strengthened by the result that Eq. (88) also applies to analytical functions of $\mathbf{K}$ (in particular, the extension to "uniform systems" discussed in Section III,B follows from this result). Given any matrix-valued analytical function of $\mathbf{K}$, $\mathbf{F}(\mathbf{K})$ (where $\mathbf{F}$ is a square matrix of the same order as its square-matrix argument), one can show that

$$[\mathbf{F}(\mathbf{K}\hat{}),\mathbf{L}] = <\mathbf{L},\mathbf{F}(\mathbf{K})>. \qquad (89)$$

Wei and Prater (1962) showed that $\mathbf{K}$ is always diagonalizable. It follows that if $\mathbf{x}$ is any eigenvector of $\mathbf{K}$, the corresponding eigenvalue α is real [see the discussion following Eq. (72)], where

$$<\mathbf{K},\mathbf{x}> = \alpha\mathbf{x}. \qquad (90)$$

Transforming to the ^ system, one has

$$[\mathbf{K}\hat{},<\mathbf{L},\mathbf{x}>] = <\mathbf{L},<\mathbf{K},\mathbf{x}>> = \alpha<\mathbf{L},\mathbf{x}>, \qquad (91)$$

which can be satisfied only in two cases: either $<\mathbf{L},\mathbf{x}> = 0$, or $<\mathbf{L},\mathbf{x}>$ is an eigenvector of $\mathbf{K}\hat{}$ with α being the corresponding eigenvalue.

As already observed by Wei and Kuo (1969), this result really simply means that, as one lumps the system into a lower order one by way of the L-transformation, several eigenvectors necessarily "vanish," in the sense that the corresponding eigenvectors in the ^ system are zero. In the transformed system, the $\mathbf{K}\hat{}$ matrix can have at most $N\hat{}$ nonzero eigenvectors. The original eigenvectors, which do not vanish, preserve the same eigenvalues in the transformed system.

Wei and Kuo (1969) have shown a direct construction of the matrix $\mathbf{K}\hat{}$ from the knowledge of the eigenvectors and eigenvalues of $\mathbf{K}$. A simpler construction (for proper lumping) was proposed by Coxson and Bischoff (1987b): Let $\mathbf{S}$ be the diagonal $N\hat{} \times N\hat{}$ matrix whose elements are the inverses of the number of species in each "lump," that is, in each ^ pseudospecies. Then $\mathbf{K}\hat{} = [[<\mathbf{L},\mathbf{K}>,\mathbf{L}^T],\mathbf{S}]$.

The distinction of the original eigenvectors into a set of vanishing ones and a set of eigenvalue preserving ones forms the basis of the Ozawa (1973) analysis of exact lumping.

For further details on the Wei and Kuo analysis of exact lumping, and for alternate mathematical descriptions of the same problem, the reader is referred to the original literature. We move here to a brief discussion of an extension due to Bailey (1972), who considered the case where the original system is a continuous one, so that Eq. (79) is written as

$$dm(x)/dt = -<K(x,y),m(y)>. \tag{92}$$

Now suppose that one still wants to lump the system into a finite, $N^\wedge$-dimensional one. This can be achieved by defining an $N^\wedge$-dimensional set of $L_I(x)$ distributions, $\mathbf{L}(x)$, such that

$$\mathbf{m}^\wedge = <\mathbf{L}(y), m(y)>. \tag{93}$$

For exact lumping, one now again requires Eq. (85) to hold. Bailey (1972) has shown that the analog of Eq. (88) is, for the case at hand,

$$<\mathbf{L}(y), K(x,y)> = [\mathbf{K}^\wedge, \mathbf{L}(x)]. \tag{94}$$

Another type of extension of the Wei and Kuo results is to bilinear kinetic forms, an extension that was originally discussed by Li (1984), and later analyzed in much more detail by Li and Rabitz (1989, 1991a,b,c). Practical and experimental aspects of the general problem of exact lumping, some of which had already been discussed by Wei and Kuo (1969) and by Kuo and Wei (1969), are excellently reviewed by Weekman (1979).

An important consideration about exact lumping is the following one, which is due to Coxson and Bischoff (1987a). If one considers reactor types other than a batch (or, equivalently, a plug flow) reactor, exact lumping carries over. In other words, the dynamics of the reduced system behave as if they were representative of true intrinsic kinetics. (This, as discussed in Section IV,C, is not true for overall kinetics, which may be regarded as nonexact lumping.) A somewhat similar result was proved by Wei and Kuo (1969) for the case of reactions with diffusion, such as occurs in porous catalysts.

The constraints required for exact lumping are invariably very strict, and it is only natural that works have been published where the concept of "approximate" lumping is introduced. In its simplest possible form, approximate lumping is achieved when Eq. (88) [or its semicontinuous analog, Eq. (94)] holds only in some appropriately defined approximate sense, rather than exactly.

It is in the area of approximate lumping that the distinction between the meaning of *lumping,* as used in this article, and of *overall* kinetics, possibly becomes a fuzzy one. For instance, the analysis of many parallel irreversible first-order reactions, to be briefly discussed in the next section, has been called *lumping* in the literature, though it is not in any sense an exact (Wei–Kuo) lumping, nor an approximate one: One chooses to consider only one lump, as given later by Eq. (97) but the qualitative kinetic behavior of the "lump" in no way resembles that of the original system (except when the latter consists in fact of only one species). The lump does not even follow first-order kinetics, let alone respect the other constraints required of the matrix **K^** by the Wei and Kuo definition of an exactly lumpable system.

B. Overall Kinetics

In this section, we attack the problem of kinetics in multicomponent mixtures, and we dedicate attention mostly to the case where one is only interested in, or may only be able to determine experimentally, some overall concentration of species of a certain class, such as sulfurated compounds in an oil cut during a hydrodesulfurization process. The presentation is given in terms of a continuous description; special cases of the corresponding discrete description are discussed as the need arises. Instead of working with the masses of individual species, we will work with their mass concentration distribution $c(x)$. In the case of a batch reactor, the distinction is irrelevant, but in the case of a plug flow reactor the concentration-based description is clearly preferable. The discussion is presented in purely kinetic terms for, say, a batch reactor.

It is useful to begin with an extremely simple example. Suppose one has a mixture with only two reactants, A_1 and A_2, both of which react irreversibly with first-order kinetics. A semilog plot of c_I ($I = 1, 2$) versus time (see Fig. 1) would be linear for both reactants. However, suppose that for some reason one is able to measure only the overall concentration $c_1 + c_2$; this is also plotted in Fig. 1. the curve is nonlinear, and it seems to indicate an overall kinetics of order larger than unit. This shows that the overall behavior of the mixture is qualitatively different

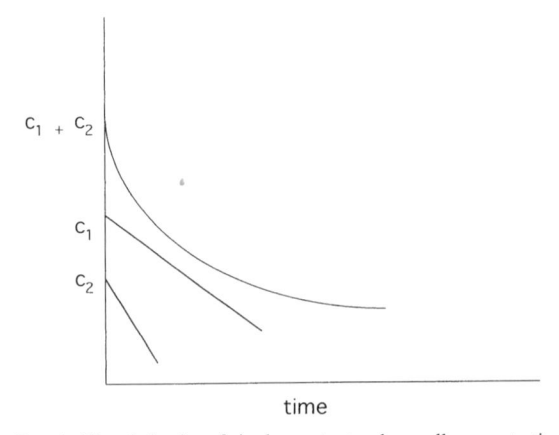

FIG. 1. Time behavior of single reactant and overall concentrations.

from the behavior of individual reactants: Experiments with model compounds (say, with only A_1) do not give information on the overall behavior; conversely, overall experiments (where one measures $c_1 + c_2$) may well mask the actual intrinsic behavior of individual reactants.

We now move to the general case of a continuous description; the pragmatical usefulness of such a type of description in real-life kinetic problems has been discussed by Krambeck (1991a,b). Let $c(x,0)$ be the initial distribution of reactant concentrations in the batch reactor, and let $c(x,t)$ be the concentration distribution at any subsequent (dimensional) time t. We assume that both the label x and the concentration c have already been normalized so that $<c(y,0)> = <yc(y,0)> = 1$. Furthermore, we assume that a (dimensional) frequency factor $k(x)$ can be identified, and that x has been normalized so that $k(x) = k^*x$, where k^* is the average value of $k(x)$ at $t = 0$. One then normalizes the time scale as well by defining the dimensionless time t as k^*t. The "overall" concentration $C(t)$ is defined with a weighting function that is identical to unity, $C(t) = <c(y,t)>$, $C(0) = 1$.

The question that arises is that of the description of the *intrinsic kinetics*, that is, of the constitutive equation for $c_t(x,t)$ in a batch reactor (we indicate partial derivatives with a subscript). Special cases that have been dealt with in the literature are discussed later.

1. Irreversible First-Order Kinetics

This case was analyzed in 1968 by Aris, based on a formalism introduced more generally in an earlier paper (1965a). Essentially the same problem was analyzed again by Bailey in 1972. Because the reactions are irreversible, $c_t(x,t)$ is negative unless $c(x,t)$ is zero. Because the reactions are first order, the only relevant parameter of any species x is the kinetic constant $k(x)$, and this has been reduced by renormalization to k^*x. Hence, the kinetic equation in a batch reactor has the form:

$$-c_t(x,t) = xc(x,t). \tag{95}$$

This integrates (trivially) to

$$c(x,t) = c(x,0) \exp(-xt). \tag{96}$$

The overall concentration $C(t)$ is given by

$$C(t) = <c(y,0) \exp(-yt)>; \tag{97}$$

that is, $C(t)$ is a Laplace transform of $c(y,0)$, with the dimensionless time t playing the rather unusual role of the transform parameter. Notice that $-dC/dt$ (the overall rate of consumption) equals $<yc(y,0) \exp(-yt)>$; that is, it is equal to the first moment of $c(x,t)$.

Now suppose $c(x,0)$ is the gamma distribution $\phi(\alpha,x)$. One can now eliminate $c(x,t)$ between the equations to obtain an *alias* (Aris and Astarita, 1989a), an explicit relationship between $-dC/dt$ and C:

$$-dC/dt = C^{(\alpha+1)/\alpha}. \tag{98}$$

Equation (98) should be discussed in some detail. When α approaches ∞ (which is the case where only reactant $x = 1$ is present in the mixture), it correctly predicts the single-reactant result $dC/dt = -C$. This is worded by saying that the alias satisfies the single-component identity (SCI): When the initial concentration distribution approaches a delta function, one recovers the result for a single component. It is clear that the SCI requirement must be satisfied for *every* kinetic equation, not only the linear one. In fact, one can generalize the requirement to that of the discrete component identity—when the initial concentration distribution approaches the sum of N distinct delta functions, one must recover the corresponding discrete description for N components (Aris, 1991a).

It is also of interest that the apparent overall order of reaction given by Eq. (98) is $(\alpha + 1)/\alpha$; that is, it is always *larger* than the intrinsic order of unity, except when α approaches infinity. When the intrinsic kinetics are linear (first order) the apparent overall order of reaction is larger than the intrinsic one; this, as will be seen, is not necessarily the case when the intrinsic kinetics are nonlinear. Physically, the point is easily understood by considering that, as time goes by, the components with the largest kinetic constants disappear first, and hence that the average kinetic constant of the mixture [which is proportional to the first moment $<yc(y,t)>$] decreases. It follows that, from an overall viewpoint, the overall kinetic constant decreases as the overall concentration decreases; this is typical of reactions of order larger than one.

It is also worthwhile to notice that the overall order of reaction is 2 when $\alpha = 1$, that is, with an exponential distribution. The only value of α for which $\phi(\alpha,0)$ is finite is $\alpha = 1$. This result can be generalized to any initial distribution by simply making use of the limit properties of the Laplace transform: $C(t)$ at large times becomes proportional to $t^{-\Omega}$ if and only if $c(x,0)$ at small x becomes proportional to $x^{\Omega-1}$, $\Omega > 0$. Now if $0 < \Omega < 1$, $c(0,0) = \infty$, and if $\Omega > 1$ $c(0,0) = 0$. It follows that $c(0,0)$ is finite only if $\Omega = 1$, in which case $C(t)$ at large times becomes proportional to $1/t$, which is the behavior typical of sound-order kinetics (Krambeck, 1984b). The question of the large-time asymptotic behavior of the overall kinetics is discussed later in more general terms.

The discrete equivalent of the Aris (1968) result was given by Luss and Hutchinson (1970). The concentration of the Ith component at time t is $c_I(t) = c_I(0) \exp(-k_I t)$, and the overall concentration is $C(t) = <\mathbf{c}(t),\mathbf{I}>$; if the k_I's are arranged in ascending order $C(t)$ is a Dirichlet series of type K_I. Properties of this series relevant to the case at hand are discussed by Aris (1991a). It is easy to convince oneself that the algebra of the discrete description can easily be more cumbersome than that of the continuous one.

2. Irreversible Nonlinear Kinetics

When the intrinsic kinetics are nonlinear, some interesting problems arise that are best discussed first with a discrete description. A possible assumption for the kinetics is that they are *independent,* that is, that the rate at which component I disappears depends only on the concentration of component I itself (this assumption is obviously correct in the first-order case). The difficulties associated with the assumption of independence are best illustrated by considering the case of parallel nth order reactions, which has been analyzed by Luss and Hutchinson (1971), who write the kinetic equation for component I as $-dc_I/dt = k_I c_I^n, I = 1, 2, \ldots, N$, where c_I is the (dimensional) concentration of component I at time t. The total initial concentration $C(0)$ is $<\mathbf{c}(0),\mathbf{I}>$, and this is certainly finite. Now consider the following special, but perfectly legitimate case. The value of $C(0)$ is fixed, and the initial concentrations of all reactants are equal, so that $c_I(0) = C(0)/N$. Furthermore, all the k_I's are equal to each other, $k_I = k^*$. One now obtains, for the initial rate of decrease of the overall concentration:

$$-dC(0)/dt = k^*C(0)^n N^{1-n}, \qquad (99)$$

which implies that, as the number of components N approaches ∞, $-dC(0)/dt$ approaches zero if $n > 1$, and it approaches infinity if $n < 1$. Clearly both conclusions do not make sense; the paradox is harder to prove but appears also for any initial distribution of concentrations and kinetic constants, provided $C(0)$ is finite, as in practice will always be the case. Indeed, Luss and Hutchinson (1971) state explicitly that serious problems arise if one tries a continuous description—one where N approaches ∞. The mathematical difficulties connected with the assumption of independent kinetics in the continuous description have been discussed by Ho and Aris (1987).

Ho *et al.* (1990) have presented an approach to the description of independent kinetics that makes use of the "method of coordinate transformation" (Chou and Ho 1988), and which appears to overcome the paradox discussed in the previous paragraph. An alternate way of disposing of the difficulties associated with independent kinetics is intrinsic in the two-label formalism introduced by Aris (1989, 1991b), which has some more than purely formal basis (Prasad *et al.,* 1986). The method of coordinate transformation can (perhaps in general) be reduced to the double-label formalism (Aris and Astarita, 1989a). Finally, the coordinate transformation approach is related to the concept of a number density function $s(x)$, which is discussed in Section IV,B,5.

Be that as it may, there is a largely heuristic argument in support of the idea that the assumption of independent kinetics is reasonable only for first-order reactions. If the reactions are first order, the probability of a given molecule to undergo the reaction is a constant, so that the number of molecules of any given reactant that reacts per unit time is simply proportional to how many such molecules there are. For *any* nonlinear kinetics, the probability depends on the environment, and in the case of very many reactants each one of which is present in very small concentrations, the environment is essentially made up of molecules of components other than the one under consideration. Hence it would seem unlikely that the rate at which a given reactant disappears depends only on the concentration of that reactant.

Remaining still in the framework of a discrete description, the environment of reactant I at time t is presumably fully described by the vector $\mathbf{c}(t)$. This suggests that the rate of consumption of reactant I at time t can be written as

$$-dc_I/dt = k_I c_I F_I(\mathbf{c}), \qquad (100)$$

where k_I is a frequency factor and $F_I(\mathbf{c})$ is a dimensionless measure of the influence of the environment on the probability of molecules I to undergo the reaction. Kinetics of the type of Eq. (100) will be called *cooperative* in the sense that the rate of reaction of component I depends on the whole instantaneous distribution of concentrations $\mathbf{c}(t)$.

Equation (100) is easily generalized to the dimensionless continuous description:

$$-c_t(x,t) = xc(x,t)\mathbf{F}[c(z,t);x], \qquad (101)$$

where $\mathbf{F}[\,;\,]$ is a functional of the concentration distribution at time t, which also depends parametrically on the component label x. The value of $\mathbf{F}[\,;\,]$ is identically unity in the case of irreversible first-order kinetics.

In the very general form of Eq. (101), cooperative kinetics are very hard to deal with mathematically. However, a very strong (but also powerful) assumption of "uniformity" was introduced by Astarita and Ocone (1988): The functional $\mathbf{F}[\,;\,]$ does not depend on the component label x, so that the continuous description of uniform cooperative kinetics is

$$-c_t(x,t) = xc(x,t)\mathbf{F}[c(z,t)]. \qquad (102)$$

The assumption of uniformity is in fact justified for some realistic kinetic schemes, such as Langmuir isotherm catalyzed reactions, Michaelis–Menten kinetics, and others (Aris, 1989; Cicarelli *et al.*, 1992). The assumption bears a more than superficial analogy with those systems termed *pseudomonomolecular* by Wei and Prater (1962). Mathematically, it is a very powerful assumption: By crossing out the dependence of $\mathbf{F}[\,;\,]$ on x, its value has been reduced from an infinite-dimensional vector (a function of x) to a scalar. This simplification makes Eq. (102) a quasilinear one, and it can be integrated explicitly by introducing a warped time scale $\tau(t)$, $\tau(0) = 0$. The solution, as can be verified by inspection, is

$$c(x,t) = c(x,0) \exp(-\tau(t)), \qquad (103)$$

where $\tau(t)$ is delivered by the following first-order differential equation [which can always be solved for $t(\tau)$ by direct quadrature]:

$$d\tau/dt = \mathbf{F}[c(z,0)\exp(-\tau(t))]. \tag{104}$$

Notice that the overall concentration $C(t) = \langle c(y,t)\rangle$ is again a Laplace transform of the initial distribution $c(x,0)$, with τ playing the role of the transform parameter. The overall rate of reaction $-dC/dt$ is given by

$$-dC/dt = \langle yc(y,t)\rangle\mathbf{F}[c(z,0)\exp(-\tau(t))]. \tag{105}$$

Stronger results can be obtained by assigning specific forms to the functional $\mathbf{F}[\ ]$. Let $Q(t)$ be a weighted overall concentration at time t, $Q(t) = \langle K(y)c(y,t)\rangle$, where $K(x)$ is an appropriate weighting function. In many realistic cases the functional $\mathbf{F}[\ ]$ can be expressed as a function of $Q(t)$, $\mathbf{F}[\ ] = F(Q(t))$. It is also necessary to assign a specific form to the weighting function $K(x)$, and the following one has been proposed by Astarita (1989b) for use in conjunction with $\phi(\alpha,x)$ for $c(x,0)$:

$$K(x) = K^*\Gamma(\alpha)(\alpha x)^{\beta}/\Gamma(\alpha + \beta), \tag{106}$$

which is already normalized so that K^* is indeed the average value of $K(x)$ in the initial mixture. The case where $c(x,0) = \phi(\alpha,x)$ and $K(x)$ is given by Eq. (106) has been termed that of the α, β, Γ distribution by Aris (1989). The interesting point is that, for an α, β, Γ distribution, it is always possible to eliminate $c(x,t)$ between the equations so as to obtain an alias, that is, a function $R(\ )$ satisfying $-dC/dt = R(C)$ (Aris and Astarita, 1989a,b):

$$R(C) = F(K^*C^{(\alpha+\beta)/\alpha})C^{(\alpha+1)/\alpha}. \tag{107}$$

It is also possible to solve explicitly the inverse problem. That is, given a function $R(C)$ (which may well have been measured experimentally by working with the complex mixture), one can determine a function $F(\ )$ that will generate it:

$$F(Q) = (Q/K^*)^{-(\alpha+1)/(\alpha+\beta)}R((Q/K^*)^{\alpha/(\alpha+\beta)}). \tag{108}$$

It is interesting to consider the special case of Langmuir isotherm kinetics (Astarita, 1989b) where $F(Q) = 1/(1 + Q)$. One obtains

$$-dC/dt = C^{(\alpha+1)/\alpha}/(1 + K^*C^{(\alpha+\beta)/\alpha}). \tag{109}$$

Let's discuss Eq. (109) in some detail. First of all, it satisfies the SCI requirement, because when α approaches ∞ it yields $-dC/dt = C/(1 + K^*C)$. The intrinsic kinetics are always of nonnegative order, and approach order zero only when the reactant concentration approaches ∞. In contrast with this, Eq. (109) yields an apparent overall order of reaction that may well be negative at high C if $\beta > 1$.

Physically, the reason for this behavior is as follows. When $\beta > 0$, reactants with higher kinetic constant also have higher adsorption constant; if $\beta > 1$, the adsorption constant grows more rapidly than the kinetic constant. The reactants with the larger kinetic constant disappear first, and, while this makes the average kinetic constant a decreasing function of time, it also results in the disappearance of reactants that compete fiercely for active sites, thus making the catalyst more available. When $\beta > 1$, the latter effect is more important than the former one, and the apparent overall order of reaction at large C becomes negative.

3. An Aside on Reversibility

Consider again the general formulation of cooperative uniform kinetics embodied in Eq. (102). Suppose that the value of $\mathbf{F}[\ ]$ must be nonnegative and must be zero only if the argument function $c(z,t)$ is zero. Then Eq. (102) describes irreversible kinetics, since the rate of consumption $-c_t(x,t)$ becomes zero only when $c(x,t) = 0$. However, Eq. (102) can also describe reversible kinetics. Let $c^*(x)$ be the equilibrium concentration distribution that will eventually be attained from the initial distribution $c(x,0)$, and let $\mathbf{F}[c^*(z)] = 0$, so that the (reversible) equilibrium condition is satisfied at some nonzero $c^*(x)$. Equation (102), however, can only describe a special form of approach to equilibrium. In fact, suppose that $\mathbf{F}[c(z,0)]$ is positive (or negative). Equation (102) now implies that the concentration of *all* reactants decreases (or increases) until $c(x,t)$ becomes equal to $c^*(x)$. This implies two things:

1. That $c^*(x)$ is smaller (or larger) than $c(x,0)$ for all x's.
2. That equilibrium is approached monotonically; the sign of $c_t(x,t)$ cannot change

Of course, both restrictions 1 and 2 are trivially satisfied in the case of irreversible reactions, but they may not be unrealistic for some category of reversible ones. Restriction 1 could perhaps be relaxed by extending the range of the label x to negative values. The description of reversible cooperative uniform kinetics has not been attempted in the literature.

Krambeck (1994b) has observed that this way of describing reversibility is probably not physically meaningful, because, for instance, detailed balancing would not be satisfied. On the other hand, one could add a nonunity multiplying function to any example of reversible continuous linear kinetics, thus obtaining reversible uniform kinetics.

4. Reaction Networks

So far, only degenerate reaction networks have been considered, where reactants of, say, type A, react irreversibly (or, within rather strict specifications, reversibly) to some otherwise unspecified products. In a mixture containing very many reactants, much more structure than that may exist as far as the topology of reaction networks is concerned.

a. Parallel and Sequential Reactions. We begin the discussion by considering two very simple cases: that of sequential reactions where species of type A may react to form species of type B, and the latter may in turn react to form species of type C [say, in a continuous description, $A(x) \to B(x) \to C(x)$], and that of parallel reactions $C(x) \leftarrow A(x) \to B(x)$ where the original reactants of type A may react to form both B-type and C-type products. Furthermore, we begin by considering the simple case where the intrinsic kinetics are linear.

Let $c_A(x,t)$ and $c_B(x,t)$ be the concentration distribution of the initial reactants of type A, and of the desired products of type B. [The concentration distribution of type C components is obtained from an elementary mass balance, $c_C(x,t) = c_A(x,0) - c_A(x,t) - c_B(x,t)$.] Let $C_A(t) = <c_A(y,t)>$ and $C_B(t) = <c_B(y,t)>$ be the corresponding overall concentrations [with, again, $C_C(t) = 1 - C_A(t) - C_B(t)$]. Attention is restricted to the case where $c_A(x,0)$ is the gamma distribution $\phi(\alpha,x)$. For the sequential case (which in the context of a discrete description was discussed in 1974 by Golikeri and Luss), let x be normalized so that it is proportional to the kinetic constant of the first step. For the parallel case (which in the context of a discrete description was discussed in 1975 by Luss and Golikeri), let x be normalized to be proportional to the sum of the kinetic constants for both steps. Then $c_A(x,t)$ and $C_A(t)$ are obviously given by

$$c_A(x,t) = c_A(x,0) \exp(-xt), \tag{110}$$

$$C_A(t) = [1 + t/\alpha]^{-\alpha}, \tag{111}$$

where Eq. (111) is intended in the limiting sense $C_A(t) = -\ln t$ when $\alpha = \infty$ (single component).

For the sequential case (Aris, 1989), let $\sigma(x)$ be the (dimensionless) kinetic constant for the second step. One obtains

$$c_B(x,t) = \phi(\alpha,x)(e^{-xt} - e^{-\sigma(x)t})/(\sigma(x) - 1). \tag{112}$$

This equation can, of course, be integrated over x for any $\sigma(x)$. In the particularly simple case where $\sigma(x) = \sigma$ (a constant), one obtains (Aris, 1989):

$$(\sigma - 1)C_B(t) = (1 + t/\alpha)^{-\alpha} - (1 + \sigma t/\alpha)^{-\alpha} \tag{113}$$

which satisfies the SCI. The term $C_B(t)$ always goes through a maximum at some value of t, and hence a selectivity problem [a value of t for which $C_B(t)$ is maximized] exists both for the single component and for the mixture cases. Of course, the value of t which maximizes the selectivity still depends on α (i.e., the selectivity problem has a different solution in the mixture and in the single reactant cases).

The parallel case has been discussed by Astarita and Ocone (1991). Let $\sigma(x)$ be proportional to the kinetic constant of the $A \to C$ step, with $0 \leq \sigma(x) \leq x$. One obtains

$$c_B(x,t) = \sigma(x)\phi(\alpha,x)(1 - e^{-xt})/x. \tag{114}$$

The case where $\sigma(x)$ is a constant is trivial, but all other cases are not. For instance, suppose that $\sigma = x^2 e^{-x}$, and let $(\alpha + 1)/\alpha = m$. One obtains

$$C_B(t) = m^{\alpha+1} - (m/(1 + t/m))^{\alpha+1}. \tag{115}$$

It follows that the overall selectivity $C_B(t)/C_C(t)$ is not constant in time, contrary to what happens in the single-component case. Astarita and Ocone (1991) also discuss the uniform cooperative case of parallel reactions, and the analogous analysis of the sequential case does not present substantial problems.

b. Bimolecular Reactions. Another problem where the reaction network has some topological structure is that of a system of bimolecular reactions, where every reactant may react irreversibly with any other one. In the dimensionless continuous description, this yields (Scaramella *et al.,* 1991):

$$-c_t(x,t) = c(x,t)<b(x,y)c(y,t)>, \tag{116}$$

where $b(x,y)$ is the dimensionless kinetic constant of the reaction between species x and y; $b(x,y)$ is obviously symmetric, $b(x,y) = b(y,x)$, and it has been normalized so that its average value in the initial mixture is unity:

$$<<b(y,y')c(y,0)>c(y',0)> = 1. \tag{117}$$

The cooperative kinetics described by Eq. (116) are not uniform. A reasonable approximation is to assume that, in the discrete description, the kinetic constant $k_{II'} = (k_{II}k_{I'I'})^{1/2}$, which in the continuous description reduces to $b(x,y) = xy$. This reduces Eq. (116) to cooperative uniform kinetics and, if $c(x,0) = \phi(\alpha,x)$ the solution is obtained by standard techniques as

$$-dC/dt = C^{2(\alpha+1)/\alpha}. \tag{118}$$

This equation is a generalization of the Aris (1968) result given in Eq. (98): The apparent overall order is $(\alpha + 1)/\alpha$ times the intrinsic order of 2. Scaramella *et al.* (1991) have also analyzed a perturbation scheme around this basic solution, which, incidentally, lends itself to a solution via reduction to an integral Volterra equation of the same type as discussed in Section IV,C with regard to a plug flow reactor with axial diffusion. Scaramella *et al.* have also shown that if $b(x,y)$ can be expressed as the sum of M products of the type $b(x)B(y) + B(x)b(y)$, which is a fairly general representation for a symmetric function, the solution can be obtained in terms of M warped times, which are solutions of M coupled ordinary linear first-order differential equations of the type of Eq. (104). As far as we know, this and the one given later in Eq. (123) are the only examples in the literature where a nonuniform cooperative kinetic mechanism has been approached via the introduction of a warped time technique.

An alternate approach to bimolecular systems has been developed in a series of papers by Ho and his coworkers (Chou and Ho, 1989; Ho, 1991a,b; Li and Ho, 1991a,b; White et al., 1994). These authors chose a discrete description, but their analysis can easily be couched in a continuous one, so as to emphasize the comparison with the results given earlier. We base our presentation essentially on the White et al. paper, which is the last one of the series and includes the results of previous papers as special cases.

The emphasis of the Ho approach is on the *long time* asymptotic behavior of the system; because all the reactions are irreversible, the long time behavior is simply that $C(t)$ and $c(x,t)$ approach zero. However, Ho and coworkers define the mole fraction distribution $X(x,t) = c(x,t)/C(t)$, and this is restricted by the obvious requirement that $<X(y,t)> = 1$, so that X necessarily stays finite at least over some range of the label x (as we will see, that range may well reduce to a point, i.e., to a single component).

Beginning with the general kinetic equation, Eq. (116), integration over the label range yields

$$-dC/dt = <c(y,t)<b(y,y')c(y',t)>. \tag{119}$$

Substituting the definition of $X(x,t)$ and rearranging yields

$$-dC/dt = \Omega(t)C^2, \tag{120}$$

where

$$\Omega(t) = <X(y,t)<b(y,y')X(y',t)>>. \tag{121}$$

Should $\Omega(t)$ approach some nonzero limiting value $\Omega(\infty)$, as it can under appropriate conditions, the overall apparent order of reaction would approach 2—contrary to Eq. (118) [which holds if $b(x,y) = xy$]. This apparent contradiction is clarified later. One can now differentiate the definition of $X(x,t)$ and rearrange to obtain

$$-X_t(x,t) = CX(x,t)(<b(x,y)X(y,t)> - \Omega(t)). \tag{122}$$

It is useful to define a warped time $\tau(t)$ as

$$\tau(t) = \int_0^t C(t')dt'. \tag{123}$$

Because $C(t)$ is positive, $\tau(t)$ approaches ∞ when t does so, and is monotonous. With this, one obtains the following autonomous integrodifferential equation:

$$-X_t(x,t) = X(x,t)(<b(x,y)X(y,t)>$$
$$-\Omega(t)) = G[X(z,t)], \qquad (124)$$

where $G[\ ]$ is a known functional.

It is now of interest to ask oneself whether there are one or more mole fraction distributions $X\hat{}(x)$ that are stationary points, in the sense that $G[X\hat{}(z)] = 0$. First consider the case where the stationary point corresponds to the survival of only one component, say, $X\hat{}(x) = \delta(x - x^*)$. Correspondingly, $\Omega = b(x^*,x^*)$, and $<b(x,y)X\hat{}(y)> = b(x^*,x^*)$: Any such point is a stationary one. (Of course, infinitely many of such points will not in fact be reached unless one starts from them.) Now consider the special case where the kinetics are uniform, $b(x,y) = xy$. One obtains

$$G[X(z)] = -X(x)[x<yX(y)> - <yX(y)>^2] \qquad (125)$$

so that, since $<yX(y)>$ is always positive, at any stationary point $<yX\hat{}(y)> = x$ for all x's for which $X(x)$ is nonzero. This is clearly possible only if $X\hat{}(x) = \delta(x - x^*)$. Clearly, the surviving component is the most refractory one, $x = 0$, and that component does not react. The result is related to the strong degeneracy of the xy form for $b(x,y)$: If regarded as a continuous linear operator, xy transforms *any* function $X(x)$ into $<yX(y)>x$; we say xy has rank one. More general forms for $b(x,y)$ furnish a nontrivial geometry of possible reaction pathways $X(x,\tau)$, with nontrivial stationary points that may be stable or unstable; if more than one stable stationary point is possible, which one is reached depends on the initial distribution $X(x,0)$. White et al. (1994) also show that such stationary points are not approached in a spiraling way—in some vague sense, a principle of microscopic reversibility holds. Once a nontrivial stable stationary point has been reached, Ω stays at some constant value $\Omega(\infty)$, and the overall order of reaction becomes 2 again. [The degeneracy of the xy form for $b(x,y)$ can be seen also from a different viewpoint: $\Omega(\infty)$ can be zero if and only if $b(x,x)$ is zero for some x (the proof is easy). When $b(x,y) = xy$, the condition is satisfied: $b(0,0) = 0$. When $\Omega(\infty) = 0$, the overall order never becomes 2, and this is the case for uniform kinetics.]

c. Cracking. Cracking reactions have a moderately complex topology: Components with any given molecular weight within the admissible class can be formed by cracking of (almost) all higher molecular weight components, and they may in turn crack to (almost) all components of lower molecular weight. The qualification "almost" is related to the fact that there may be a lowest molecular weight formed by the cracking reactions within the class considered. For instance, in the enzymatic hydrolysis of cellulosic materials, where the end product is glucose (the "monomer"), only the dimer and higher oligomers are formed by the action of certain enzymes, while other enzymes catalyze the cracking of the dimer (see, e.g., Cicarelli *et al.*, 1992, and the literature quoted therein).

In cracking reactions, it is natural to choose as the starting label the number of monomers in the oligomer considered. Next, one may renormalize the label as the initial one divided by its average value in the initial mixture: This produces an x that, whatever the initial concentration distribution may be, makes the first moment of the initial distribution equal to unity (the zeroth moment already being unity by a trivial normalization of concentrations). A continuous description is preferable if the initial average number of monomers per oligomer is large as compared to unity. This choice of labeling, of course, makes it impossible to normalize labels so as to be proportional to the kinetic constant(s) of the cracking reactions which any given oligomer may undergo. However, it is reasonable to make the following assumptions:

1. The overall cracking kinetic constant (or, in the case of nonlinear kinetics, the overall frequency factor) is the same for all oligomers.
2. The probability of any given bond within an assigned oligomer being cleaved is the same for all cleavable bonds.
3. The kinetics are either linear or of the cooperative uniform kind described by Eq. (102).

With these assumptions, the kinetic equation becomes:

$$c_t(x,t) = [\int_{x+\epsilon}^{\infty} 2c(y,t)dy/y - c(x,t)]\mathbf{F}[c(z,t)], \qquad (126)$$

where ϵ is the minimum value of x for which the kinetics are in the class considered (in the simplest case where only the monomer does not belong to the class considered, ϵ is the inverse of the average initial value of monomers in the oligomers in the mixture). The following analysis has physical significance only if $\epsilon \ll 1$. Components with $x \leq \epsilon$ will be called *end products;* in the simplest case referred to earlier, the only end product is the monomer.

In Eq. (126) the first term within the square brackets represents the rate of formation of species x from higher molecular weight ones; the second term represents its rate of cracking to lower molecular weight products. With $\epsilon = 0$ and $\mathbf{F}[\ ]$ identical to unity (first-order intrinsic kinetics), Eq. (126) reduces to the form given by Aris and Gavalas (1966) in the part dealing with thermal cracking of their pioneering work on the continuous description of mixtures.

Again, it is useful to define a warped time $\tau(t)$, $\tau(0) = 0$, which is delivered by the following differential equation:

$$d\tau/dt = \mathbf{F}[c(z,t)] \qquad (127)$$

so that Eq. (126) reduces to

$$c_\tau(x,\tau) = [\int_{x+\epsilon}^{\infty} 2c(y,\tau)dy/y - c(x,\tau)]. \tag{128}$$

An important point about Eq. (128) is that it shows that, in the warped time scale τ, the behavior is entirely independent of the form of the functional $\mathbf{F}[\ ]$; the latter influences only the relationship between the warped time τ and the actual time t. The case of first-order intrinsic kinetics (Aris and Gavalas, 1966) is recovered by simply setting $\tau = t$.

Cicarelli *et al.* (1992) have developed the solution of Eq. (128) by a perturbation expansion, with ϵ the perturbation parameter. They consider the special case of Langmuir isotherm kinetics, where $\mathbf{F}[\ ] = 1/(1 + <K(y)c(y,t)>)$. At the zero-order level, $C = \exp(\tau)$. This result simply reflects the fact that, in the distorted time scale τ where the kinetics are linear, two species are formed from one at every reaction step, and hence the total concentration grows exponentially. This, however, does not include the fact that end products are being formed, and thus disappear from the spectrum of concentrations (at the zero-order level, $\epsilon = 0$ and no end products are formed). The critical warped time τ_C at which the zero-order approximation breaks down is estimated as $-\ln \epsilon$; that is, it is well in excess of unity. Even for linear kinetics, there is an induction time significantly longer than the inverse of the kinetic constant during which very few end products are formed (this is even more true for nonlinear kinetics of the type considered). The solution can be obtained formally at all levels of perturbation; the first-order level is of particular relevance because it yields (to within order ϵ) the total amount of end products formed up to the critical time.

d. Polymerization. Polymerization reactions (which are, in a vague sense, the inverse of cracking ones), are of course a typical example of kinetics in a multicomponent mixture, with the different components being the molecules of different molecular weight. Because the latter may in principle reach infinity (as indeed it does in the case of cross-linked polymers), a continuous description where the label is the (appropriately normalized) molecular weight suggests itself naturally.

The literature on the kinetics of polymerization is so ample that it is hopeless to review it here; two recent works of general interest are by Dotson *et al.* (1993) and by Gupta and Kumar (1987). Here we limit our discussion to only two very recent works in the area where a continuous description has been used (Aris, 1993; McCoy, 1993).

McCoy (1993) has considered polymerization reactions that are reversible (i.e., oligomers may depolymerize), and may be either random or proportioned fissions. If the bimolecular kinetic constant and the molar concentration based equilibrium constant K^* are the same for all steps, and if the cracking reactions only break an *x*-mer into two equal (*x*/2)-mers, it is rather easy to write down the continuous equations describing the kinetics, which, after the appropriate normalizations, take up the following form (Aris, 1993):

$$c_t(x,t) = \int_0^x c(y,t)c(x-y,t)dy \qquad (129)$$
$$- c(x,t)<c(y,t)> + K^*[4c(2x,t) - c(x,t)],$$

where the first term on the right represents formation from lower molecular weight components of the *x*-mer, the second term represents its rate of polymerization to higher molecular weight components, the third term $4K^*c(2x,t)$ represents the rate of formation by cracking of the (2*x*)-mer, and the last term represents cracking. Equation (129) is easy to write down, but it is not at all easy to do anything with it (Aris, 1994). McCoy (1993) chooses to work with the moments of the distribution, and this leads to some moderately strong results.[18] Aris (1993) has limited attention to the case where the reactions are irreversible ($K^* = 0$) and where the mathematics are more tractable: If $c(x,0) = \phi(\alpha,x)$, a formal expression for $c(x,t)$ can be obtained. (It is rather nasty looking, but we encounter no problems when calculating it numerically.) As Aris states, "The final state is of course nothing of everything, though a little more of some things than of others. . . . another possible steady state is $c(x,t) = \delta(x)$, which is everything of nothing. But this is unstable—or, at any rate, nugatory."

e. Conclusions on Reaction Networks. The preceding discussion shows that a few simple cases where the topology of the reaction network is nontrivial can be analyzed, and formal solutions are (within bounds) possible. It also shows that as soon as some structure is assigned to the topology, the mathematics become quite complex, and therefore one concludes that there is little hope of dealing with complex topologies other than numerically. Robust numerical codes capable of dealing with such problems have still to be developed. A possibility that comes to mind is that of *lumping the reaction network;* that is, to substitute for the actual network a simpler one that, in some sense to be made precise, is either exactly or approximately equivalent to the original one. Some tentative steps in that direction have been taken (Allen, 1991; Frenklach, 1985; Froment, 1987; Vynckler and Froment, 1991), and perhaps a rigorous mathematical structure within which such problems should be couched can be found in a series of extremely interesting papers by Feinberg (1980, 1987, 1988, 1989, 1991a,b). A very concise review of some of the contents of these papers is given in Appendix C; further discussion of the Feinberg approach is given in the next section.

---

[18] Krambeck (1994b) has observed that the McCoy reversible polymerization kinetics are rather artificial, and that they do not satisfy detailed balance.

5. The Number Density Function

It is useful to introduce the concept of a number density function $s(x)$ by beginning with an important point that emerges from a very recent paper by Ho and White (1994). It may be taken for granted that a continuous description of chemical kinetics is only an approximation: Real mixtures are invariably discrete. Now the approximation is certainly going to be appropriate provided the number of components in the mixture is very large. However, as time progresses, the more reactive components tend to disappear from the mixture, which essentially contains a progressively smaller number of reactants. At very long times, only the most refractory components survive, and these may well be small enough in number that the continuous approximation breaks down. Ho and White have presented a detailed mathematical analysis of this point for the special case of hydrodesulfurization, which at long times is adequately described by intrinsic first-order kinetics. They define a *spacing*, δ, along the label axis x, which essentially measures the (average) thickness of a bar of the hystogram, which is being approximated by a distribution function in the continuous description, and they show that the underlying discreteness of the real mixture becomes significant (in the sense that the continuous approximation breaks down) at dimensionless times t where $t\delta$ exceeds unity. It follows, perhaps in enough generality, that if one is interested in the very long time kinetic behavior of the system, a discrete description based on the (few) very refractory components is preferable to a continuous one.

The Ho and White (1994) analysis is based on the idea of a constant spacing δ; $1/\delta$ then represents the (very large) number of individual components that are in reality present in the mixture over a unit segment of the label axis. In a strict interpretation of the continuous description, $\delta = 0$ (this is, essentially, De Donder's 1931 approach); but of course reality is a different matter. One can define a number density function $s(x)$ such that the number of individual compounds in any interval $[x_1, x_2]$ is

$$\int_{x_1}^{x_2} s(x)dx. \tag{130}$$

One way to interpret $s(x)$ is as a stochastic frequency of occurrence of compounds along the x line. This ameliorates the concern that the integral in Eq. (130) may well not be an integer. It also gives us the courage to apply the concept of a continuous description even when the number of compounds is not all that large, often with excellent results.

Sometimes it is useful to redefine $c(x)$ to $c'(x)$, where

$$c'(x) = c(x)/s(x), \quad \text{where } c(x) \text{ is continuous;} \tag{131}$$

$$c'(x) = c_I, \quad \text{where } c(x) = c_I\delta(x - x_I). \tag{132}$$

This has the advantage of being invariant under any distortion of the x scale, and it makes life easier when the mixture contains, in addition to the distributed components, a small number of discrete components, and it is particularly useful in some thermodynamic analysis, where a *semicontinuous* description is often used. It is also useful to note that, since $c'(x)$ is invariant with respect to rescaling the label x, the conceptual problem discussed in Appendix B is alleviated if one uses the logarithm of $c'(x)$.

6. Conclusion

The discussion in this section illustrates the wealth of results available for the analysis of the overall kinetic behavior of complex mixtures. The variety of available analytical results should not, however, mask the fact that there is a large area where very little has been done, that is, the area of nonlinear kinetics which are not of the cooperative uniform type. Powerful as the assumption of uniformity has turned out to be, remember that it is a strong assumption; in Aris's (1988) words, "your assumption of uniformity . . . also takes a bit of swallowing. That you can get results, and such striking ones, makes the medicine go down." We might not always have Mary Poppins around to help us out, and, at this stage, when we don't we really don't know what to do. See Appendix C.

C. Overall Reaction Engineering

In this section, we still restrict ourselves to the consideration of systems where only the overall behavior is of interest, but we extend the analysis to actual chemical reactors. Indeed, the discussion in the previous section was limited to the overall *kinetics* of multicomponent mixtures; seen from the viewpoint of chemical reaction engineering, the discussion was in essence limited to the behavior in isothermal batch reactors, or, equivalently, in isothermal plug flow reactors. In this section, we present a discussion of reactors other than these two equivalent basic ones. The fundamental problem in this area is concisely discussed next for a very simple example.

Suppose one has performed experiments with the mixture under consideration in a batch reactor, and one has obtained experimentally the overall kinetics—the $R(\ )$ function such that $dC/dt = -R(C)$. For instance, one could obtain $R(C) = C^2$ if the intrinsic kinetics are in fact first order and the initial concentration distribution is $\phi(1,x) = \exp(-x)$. If one were to regard $R(C)$ as a *true* (rather than an apparent) kinetic law, one would conclude that in a CSTR with dimensionless residence time T the exit overall concentration is delivered by the (positive) solution of $TC^2 + C = 1$. The correct value is in fact $C = <\exp(-y)/(1 + Ty)>$, and the difference is not a minor one. (To see that easily, consider the long time asymptote, $T \gg 1$. The positive solution of $TC^2 + C = 1$ yields $C = T^{-1/2}$, while the actual behavior is $C = \ln T/T$.)[19] What this implies is that, once an apparent overall kinetic function $R(C)$ has been determined experimentally, one needs to perform the following steps.

First, one needs to deconvolute $R(C)$ into some appropriate intrinsic kinetic law. What form the latter might take be suggested by experiments performed with single components.

Next, one needs to use the intrinsic kinetics as the basis for the analysis of a reactor other than a batch or plug flow reactor.

In this section, we restrict our analysis to the second step of this procedure. The nomenclature we use is as follows: $c_F(x)$ is the concentration distribution in the feed to any given reactor, normalized so that $\langle c_F(y) \rangle = \langle yc_F(y) \rangle = 1$, and $c_E(x)$ is the corresponding distribution in the product stream. Let $C = \langle C_E(y) \rangle$ is the overall residue. Finally, let T be the dimensionless residence time in the reactor, that is, the actual residence time times the average value of the frequency factor in the feed mixture.

1. The CSTR

The simplest reactor which is not a batch or a plug flow one is obviously the CSTR. The mass balance for component x is, if the kinetics are of the cooperative uniform type (Astarita and Nigam, 1989),

$$c_F(x) - c_E(x) = Txc_E(x)\mathbf{F}[c_E(z)]. \tag{133}$$

A warped time technique is again useful. Let the warped dimensionless residence time W be defined as

$$W = T\mathbf{F}[c_F(z)/(1 + Wz)]. \tag{134}$$

The solution of Eq. (133) is, as can be checked by substitution,

$$c_E(x) = c_F(x)/(1 + Wx), \tag{135}$$

$$C = \langle c_F(y)/(1 + Wy) \rangle. \tag{136}$$

The whole problem has thus been reduced to the solution of Eq. (134), which is a functional equation for the single scalar W. It is, however, a nasty functional equation, not so much for the possible nonlinearity of the functional $\mathbf{F}[\ ]$ itself, but because the argument function is nonlinear in the unknown W. The warped time technique is again useful, but, contrary to what happens in the single-component case, the solution for the CSTR is more difficult than the one for the plug flow reactor or batch reactor.

---

[19] This should be contrasted with the result discussed in the second paragraph after Eq. (101): Overall kinetics are far from being exact lumping, and they do not carry over to different reactor types.

a. First-Order Intrinsic Kinetics. For first-order intrinsic kinetics, the solution of Eq. (134) is trivial, since $F[\ ] = 1$ and $W = T$, so that the problem reduces to the evaluation of the integral on the right side of Eq. (136), which can of course always be obtained numerically with simple codes that are guaranteed to converge. Also, notice that, when $c_F(x) = \delta(x - 1)$ (the single-component case), Eq. (136) yields $C = 1/(1 + T)$, which is known to be the correct result. Now, more generally, suppose that $c_F(x) = \phi(\alpha, x)$. The integral on the right side of Eq. (136) can be expressed formally in terms of known functions if α is an integer (Astarita and Nigam, 1989). More generally, an expansion is possible for both low and high values of W. Remembering that W is known for first-order kinetics, one obtains

$$W \ll 1, \qquad C/(1 - C) = 1/W, \tag{137}$$

$$W \gg 1, \quad \begin{cases} C/(1 - C) = \alpha/(\alpha - 1)W, & \alpha > 1. \\ C/(1 - C) = \ln W/W, & \alpha = 1 \end{cases} \tag{138}$$

For a single component ($\alpha = \infty$), one has for all W's $C/(1 - C) = 1/W$ or, equivalently, $C = 1/(1 + W)$, which is known to be the correct result.

Aris (1991a) has analyzed the case of M CSTRs in series, each one endowed with the same residence time T/M. This is one of the homotopies spanning the range between a PFR and a CSTR: When $M = 1$, one has a CSTR, and when M approaches ∞ one has a PFR. The result is

$$C = <\phi(\alpha, y)/(1 + Ty/M)^M>, \tag{139}$$

where the right side is $(\alpha M/T)^\alpha$ times the Tricomi form (Abramowitz and Stegun, 1965; Tricomi, 1954) of the confluent hypergeometric function of α, $1 + \alpha - M$, $\alpha M/T$. It is easy to convince oneself that, when M approaches ∞, one indeed recovers the result of a PFR, and, quite obviously, one recovers Eq. (136) when $M = 1$.

b. Cooperative Uniform Kinetics. When the intrinsic kinetics are of the cooperative uniform type, the equations given earlier in terms of W still hold, but the solution is not completed until Eq. (134) has been solved for w. Astarita and Nigam (1989) have presented a general technique for solving that is based on the use of easily constructed robust numerical codes for the appropriate subproblems, and a graphical technique for the final solution. The case of M CSTRs in series has not been considered in the literature; the problem is not a trivial one, because even if the residence time in each reactor is T/M, the value of W is not the same in all reactors, and Eq. (134) would need to be solved for every reactor in the chain.

Astarita and Nigam (1989) have calculated the $C(T)$ curves for CSTRs for a few specific nonlinear kinetic functions. Their results clearly show that the difference between the value of T needed to achieve an assigned conversion in a CSTR and a PFR is, in the case of mixtures, very significantly larger than in the case of a single component with parameters having the average value of the feed mixture. The effect is particularly striking in the case of bimolecular reactions.

2. Maximum Segregation Reactors

We now move to the consideration of reactors with an assigned residence time distribution (RTD) $f(t)$, where t is the dimensionless residence time (i.e., the dimensional one times the average frequency factor in the feed). In this section, we indicate with curly braces integrals over t ranging from 0 to ∞. Then $\{f(t)\} = 1$ and $T = \{tf(t)\}$. We also make use of the complementary cumulative RTD, $F(t)$, which is defined as

$$F(t) = \int_t^\infty f(t')dt'; \quad F(0) = 1. \tag{140}$$

Now consider a maximum segregation reactor (Danckwerts, 1958; Zwietering, 1959). Let $C = p(t)$ be the overall concentration in a batch reactor at time t. (This can be calculated for linear and for cooperative uniform kinetics.) Then the product overall concentration in a maximum segregation reactor is easily calculated as (Aris, 1989)

$$C = \{f(t)p(t)\}. \tag{141}$$

Because $p(t) = <c(y,t)>$, the right side of Eq. (141) is a double integral over both x and t, and this can be reconducted to a double-label formalism if the second label is regarded as distributing over residence times (Aris, 1989). Of course, one expects Eq. (141) to hold for *any* degree of micromixing if the intrinsic kinetics are first order, but this remains to be demonstrated.

3. Maximum Mixedness Reactors

In the case of a maximum mixedness reactor, one works best with the life expectancy b. The life expectancy distribution in the feed stream, $f(b)$, is exactly the same as the residence time distribution in the product stream, $f(t)$. One can generalize the Zwietering (1959) equation for a maximum mixedness reactor to the case of continuous mixtures (Astarita and Ocone, 1990) to obtain the following functional differential equation for $c(x,b)$:

$$c_b(x,b) = xc(x,b)\mathbf{F}[c(z,b)] + f(b)[c(x,b) \\ - c_F(x)]/F(b), \tag{142}$$

where the quantity $f(b)/F(b)$ is the intensity function discussed by Shinnar (1993), that is, the residence time distributions of those elements that are still in the reactor.

Solving Eq. (142) poses some difficulties even for the case of a single component; indeed, recent literature on the subject deals with this (Astarita, 1990; Astarita and Ocone, 1993; Cicarelli and Astarita, 1992; Glasser et al., 1987; Glasser and Jackson, 1984; Guida et al., 1994a; Jackson and Glasser, 1986). The solution procedure usually recommended in the literature (see, e.g., Froment and Bischoff, 1990) is in fact applicable only if the limit of the intensity function $f(b)/F(b)$ for b approaching ∞ exists—which may well not be the case, as the counterexample of a plug flow reactor with recycle shows. In spite of these difficulties, Eq. (142) can be solved in general, if only formally.

The warping procedure is again useful. In this case we apply it to the life expectancy b by defining a warped life expectancy β, which is delivered by the following differential equation:

$$d\beta/db = \mathbf{F}[c(z,b)]; \qquad \beta(0) = 0. \tag{143}$$

Since the value of $\mathbf{F}[\ ]$ is nonnegative, $\beta(b)$ is invertible, $b = b(\beta)$, and one may define a warped life expectancy distribution:

$$q(\beta) = f(b(\beta))db/d\beta. \tag{144}$$

With this, a formal solution for $c(x,b)$ can be obtained:

$$c(x,b) = \frac{c_F(x)e^{x\beta(b)}}{F(b)} \int_\beta^\infty q(b')e^{-xb'} db'. \tag{145}$$

In the case of first-order kinetics, $\mathbf{F}[\ ] = 1$, $\beta = b$, $q(\ ) = f(\ )$, and Eq. (145) reduces to

$$c(x,b) = \frac{c_F(x)}{F(b)} \int_b^\infty f(b')e^{-xb'} db'. \tag{146}$$

The exit stream corresponds to $b = 0$, and hence one obtains

$$c_E(x) = c_F(x)\{f(b)e^{-xb}\}, \tag{147}$$

$$C = <c_F(y)\{f(b)e^{-yb}\}>$$
$$= \{f(b)<c_F(y)e^{-yb}>\} = \{f(b)p(b)\}; \tag{148}$$

that is, as expected, the result is the same as for a maximum segregation reactor.

Going back to the nonlinear case governed by Eq. (145), the distribution in the product stream and the residual are

$$c_E(x) = c_F(x)\{q(b)e^{-xb}\}, \tag{149}$$

$$C = \{q(b)<c_F(y)e^{-yb}>\}; \tag{150}$$

that is, the product stream has the same composition one would have in a maximum segregation reactor should the intrinsic kinetics be linear and the RTD be

$q(t)$: All nonlinearities have been buried in the function $q(\ )$. The result is deceivingly simple, because the function $q(b)$ is delivered by the solution of a formidable nonlinear functional differential equation.

Astarita and Ocone (1988) have analyzed a few special cases, and they have shown that for some kinetic schemes the maximum mixedness reactor at large total residence times may give a better overall conversion than the maximum segregation one with the same RTD, even for intrinsic kinetics that would never yield this result for a single component. This shows that the overall behavior of mixtures responds differently than single reactants to changes in the degree of micromixing.

4. Nonisothermal Reactors

So far, temperature has never entered the kinetic or reaction engineering equations, since we have essentially restricted our attention to isothermal systems. Of course, real reactors are seldom, if ever, isothermal. The problem of nonisothermality in multicomponent mixtures was considered in 1966 by Aris and Gavalas, and a discussion based on a discrete description was presented in 1972 by Golikeri and Luss, but after that nothing new seems to have appeared in the literature until the 1991 paper by Aris, who considered the rather special case of an adiabatic batch reactor. Aris (1991a) has identified the fundamental problem for nonisothermal reactors: Individual species cannot be simply identified with the frequency factor (or, in the nonisothermal terminology, the preexponential factor); one also needs to assign the activation energy. Hence, a two-label formalism is forced on the problem, with, say, x being proportional to the preexponential factor and x' to the activation energy.

This makes things complicated enough that even in the case of first-order kinetics there is more mathematical structure than one may wish for, and only that case has been considered in the literature. The following discussion is based on a paper by Cicarelli and Aris (1994), which provides a preliminary approach to the problem of a continuous mixture reacting with intrinsic first-order kinetics in a nonisothermal CSTR. As has been known since the pioneering work of Van Heerden (1958), this problem is far from being trivial even for a single reactant; in the case of mixtures, one has to deal with two distributions, one over x and one over x', which are justified on physical grounds, for example, for problems of coal gasification (Anthony and Howard, 1976; Pitt, 1962) and of heterogeneous catalysis (Brundege and Parravano, 1963).

Again making use of the powerful gamma distribution $\phi(\ )$, one could in general consider the case where the feed stream concentration distribution is given by

$$c_F(x,x') = \phi(\alpha,x)\phi(\alpha',x'). \tag{151}$$

However, Cicarelli and Aris (1994) limit attention to the case where only one of the variables is distributed (i.e., to the cases where either α or α' is ∞), so that, after the appropriate normalizations, we obtain

$$c_F(x,x') = \phi(\alpha,x)\delta(x' - 1), \quad \text{distributed over the frequency factors, DFF,} \quad (152)$$

$$c_F(x,x') = \phi(\alpha',x')\delta(x - 1), \quad \text{distributed over the activation energies, DAE.} \quad (153)$$

The mathematics involved are somewhat cumbersome, and numerical techniques are needed. The conclusions reached are the following ones:

1. *DAE case.* Even at $\alpha' = 20$, the steady-state multiplicity curves (Regenass and Aris, 1965) are not very close to those of a single component ($\alpha' = \infty$). The region of multiplicity seems to increase as α' increases, that is, in parameter space the single component case is the most likely to result in multiplicity.
2. *DFF case.* The curves are less sensitive to the value of α than they are to the value of α' in the DAE case; the qualitative behavior is, however, the same.
3. An infinity of multiple steady states (which in principle could exist in a mixture containing infinitely many reactants) is in fact attainable only for degenerate cases. For the DAE case the probability of getting an infinity of multiple steady states increases when α' decreases, but at the same time the region where they may occur shrinks until in the limit it becomes a single point in parameter space.

5. Multiplicity of Steady States

Multiple steady states as discussed in the previous subsection are related to the nonisothermicity of the CSTR. However, even in the isothermal case, a CSTR is known to be able to exhibit multiple steady states, periodic orbits, and chaotic behavior for sufficiently complex reaction network structures (see, e.g., Gray and Scott, 1990). When the number of reactions is very large, the problem becomes a formidable one. In a series of papers (Feinberg 1987, 1988, and the literature quoted therein), Feinberg and his coworkers have developed a procedure for CSTRs that can be applied to systems with arbitrarily large numbers of reactants and reactions. The procedure is based on the deficiency concept discussed in Appendix C.

First, one constructs an "augmented" reaction network by adding reactions of the type $0 \to A$, which represent addition of A to the CSTR via the feed stream, and reactions of the type $A \to 0$ representing removal of A in the product stream. Next, the deficiency of the augmented network is calculated (with "0" a complex which, by definition, is compatible with A). If the deficiency is zero, the strong deficiency zero theorem (Feinberg, 1987) applies: Provided the kinetics are of the mass action type, no matter what (positive) values the kinetic constants may have, the CSTR cannot exhibit multiple steady states, unstable steady states, or periodic orbits. The result is, in a sense, very strong because the governing differential equations may well constitute a gigantic nonlinear system. In another sense the result is not so strong, because "augmented" networks for CSTRs are almost never of deficiency zero, unless all the "true" chemical reactions are first-order ones.

If the deficiency of the augmented network is one, the results are less strong, but still very important. (It is important to realize that CSTRs where heterogeneous catalysis reactions occur, so that many species are not present in the product stream, often turn out to have deficiency one.) For a subclass of such systems, multiple steady states are impossible, but a single unstable steady state and a periodic orbit are possible. Outside of this subclass, the theory becomes delicate, and one has to make calculations based on an algorithm for the analysis of the topology of the reaction network. The algorithm, however, is now available on diskette (Feinberg, 1994).

It turns out that if a substantial fraction of the reactions involve adsorbed species on a catalyst (these are compounds that do not exit the reactor with the product stream), the deficiency is likely to be zero or one (but see the discussion in Appendix C). For homogeneous CSTRs, where all components are carried out by the product stream, it is easy to have high deficiencies, and the deficiency-oriented Feinberg method offers little advantage.

However, recent work by Schlosser and Feinberg (1994), which is specifically tailored for the analysis of homogeneous CSTRs, takes a completely different approach, where the possibly high deficiency of the augmented reaction network is no problem at all. The Schlosser–Feinberg method is based on the construction of a diagram called the species-complex-linkage (SCL) graph. The method is restricted to homogeneous CSTRs (where all species are present in the product stream), and in this sense it is less powerful than the deficiency-based theory. On the other side, for homogeneous CSTRs the method works for any value of the deficiency of the augmented network (if it works at all): It can tell whether multiple steady states are possible.

The Schlosser and Feinberg approach is based on the analysis of the geometrical topology of the true (not augmented) reaction network; all individual reactions are assumed to be governed by mass action kinetics. (This is a strong, if realistic, assumption: for instance, it implies that the reaction 2A = 2B is not the same thing as the reaction A = B.) The 1994 paper does not contain formal (or even informal) proofs. It gives elementary examples and counterexamples for almost all of the stated results, and the best we could do here is to essentially reword their presentation; hence we simply refer the reader back to that paper; proofs of the theorems can be found in Schlosser's thesis (1988). Their results are of three types:

1. For certain reaction networks (which can easily be identified from an analysis of the geometry) multiple steady states are impossible. For a comple-

mentary class multiple steady states (for *some* values of the residence time and the kinetic constants) are not excluded, but they are by no means guaranteed to be possible.
2. For those cases where only one stable steady state is possible, sustained oscillations cannot be excluded (the single steady state may be unstable). For a restricted class of reaction networks, stability (or lack thereof) of the single steady state can be demonstrated.
3. For some specific geometries of the reaction network, the possibility of multiple steady states for some appropriate set of values of the parameters is guaranteed.

An alternate approach to the analysis of the network geometry from the viewpoint of multiplicity and stability is due to Beretta and his coworkers (1979, 1981); the latter approach is based on knot theory. The Schlosser–Feinberg theory reproduces some of the Beretta-type theory results concerning stability from a somewhat different viewpoint.

There are many (otherwise unremarkable) instances of reaction networks where the Schlosser and Feinberg method gives no answer. Multiple steady states are not excluded but are not guaranteed to be possible either. In actual fact, multiplicity of steady states in isothermal homogeneous CSTRs has seldom been observed. This suggests, as Schlosser and Feinberg conclude, "the existence of as yet unknown theorems which, like theorem 4.1, deny the capacity for multiple steady states but which have an even wider range."

6. Homotopies between the CSTR and the PFR

Aris (1991a), in addition to the case of M CSTRs in series, has also analyzed two other homotopies: the plug flow reactor with recycle ratio R, and a PFR with axial diffusivity and Peclet number P, but only for first-order intrinsic kinetics. The values $M = 1(\infty)$, $R = \infty(0)$, and $P = 0(\infty)$ yield the CSTR (PFR). The M CSTRs in series were discussed earlier in Section IV,C,1. The solutions are expressed in terms of the Lerch function for the PFR with recycle, and in terms of the Niemand function for the PFR with dispersion. The latter case is the only one that has been attacked for the case of nonlinear intrinsic kinetics, as discussed below in Section IV,C,7,b. Guida *et al.* (1994a) have recently discussed a different homotopy, which is in some sense a basically different one; no work has been done on multicomponent mixture systems in such a homotopy.

7. Interference with Diffusion

The discussion in Section IV,B, as well as many of the arguments just given, show how powerful the procedure of warping the time scale is. Unfortunately, the procedure cannot be extended to problems where diffusion interferes with the reaction, since in the latter case there are two time scales (the reaction time and the diffusion time), and one may warp but not the other. There are two other ways of looking at the same problem. First, the warping procedure is used to linearize the kinetic equation; the procedure does not carry over to second-order differential equations such as those that arise when diffusion is taken into account (Aris and Astarita, 1989a,b). Second, as long as the governing equations are, for their differential part, first-order ones, the problems are hyperbolic and the method of characteristics can be used (Guida, 1994); when diffusion is included, the equations become parabolic and the technique does not work anymore. In the context of a discrete description, the problem had been identified in 1971 by Luss and Hutchinson.

a. Porous Catalysts. The (apparently—but there are subtleties involved) simplest problem where diffusion plays a role is the classical problem of the effectiveness factor in a porous catalyst. This problem was analyzed by Golikeri and Luss (1971), who restricted their attention to independent kinetics, which were shown to be a rather special and possibly unrealistic case in Section IV,B. A discrete description was considered as early as 1962 by Wei (1962a,b); this is restricted to systems of first-order reactions, and the analysis is couched in terms of a lumping scheme. The same approach has also been discussed by several authors, and a recent analysis is included in the paper by Li and Rabitz (1991c). The continuous description for the case of cooperative kinetics was analyzed by Ocone and Astarita (1993), and the following discussion is based on that paper.

Because a warped time technique does not carry over to second-order differential equations such as the ones arising when diffusion plays a role, the assumption of uniformity simplifies the mathematics involved only marginally. Let Eq. (102) be assumed as correctly describing the kinetics, with the parameter t being the dimensionless distance into the flat slab catalyst (complications arising for more realistic geometries of the catalyst pellet are easily dealt with). Furthermore, assume that the diffusivity of all reactants is the same (again, this assumption is easily relaxed), and let Φ be the overall Thiele modulus; that is, its square equals the average frequency factor in the external concentration distribution $c_F(x)$ times the square of the catalyst half-width and divided by the diffusivity. The differential equation and boundary conditions of the problem are

$$c_{tt}(x,t) = q\ \Phi^2 x c(x,t) \mathbf{F}[c(z,t)], \tag{154}$$

$$c(x,0) = c_F(x), \tag{155}$$

$$c_t(x,1) = 0. \tag{156}$$

The case of intrinsic first-order kinetics, $\mathbf{F}[\ ] = 1$, is easily solved. Let $h(x)$ be the effectiveness factor for reactant x; one simply obtains the obvious extension of the single reactant result:

$$h(x) = tgh(\Phi\sqrt{x})/\Phi\sqrt{x}. \tag{157}$$

The overall effectiveness factor H [which is *not* equal to $<h(y)>$] is given by

$$H = <c_F(y)\sqrt{y}tgh(\Phi\sqrt{y})>/\Phi, \tag{158}$$

which is not equal to the single-component effectiveness factor. For small values of Φ, one obtains

$$H = 1 - \Phi^2<y^2c_F(y)>/3 + O(\Phi^4), \tag{159}$$

which shows that one recovers the single-component result $H = 1$ only within $O(\Phi^2)$, since in general $<y^2c_F(y)>$ will be different from unity except in the single-reactant limit where $c_F(x) = \delta(x - 1)$. This can easily be seen if $c_F(x) = \phi(\alpha,x)$, where one gets $H\Phi = \Gamma(\alpha + 1/2)/\Gamma(\alpha)\sqrt{\alpha}$, which reduces to $H\Phi = 1$ only when α approaches ∞.

The case of cooperative uniform kinetics cannot be approached by warping the axial position t. A perturbation expansion is, however, useful for small values of Φ. The perturbation is a regular one, and deviations form the zero-order result $H = 1$ can be obtained. It turns out that if the apparent overall order of reaction (Astarita, 1989) at the distribution $c_F(x)$ is negative (which it may well be even if the intrinsic order is positive), the first-order correction for H is positive. That is, at a moderately small Thiele modulus the overall effectiveness factor is larger than unity. The perturbation expansion in the other limit of small values of $1/\Phi$ is a singular one (the perturbation parameter multiplies the highest order derivative). The outer solution is trivial but useless: c is simply zero at all levels of perturbation, reflecting the physical fact that at large Thiele moduli most of the catalyst is not reached by the reacting mixture. The inner (boundary layer) solution reproduces the entire problem, in the form that would apply to an infinitely deep catalyst. While this tells us that $H\Phi$ is going to approach some constant value, the procedure does not yield that value. A formal solution has been provided by Ocone and Astarita (1993), but it requires the solution of a highly nonlinear integrodifferential equation, and it is therefore of little practical use.

A different approach to the estimation of the overall effectiveness factor in porous catalysts was recently presented by Ho *et al.* (1994). These authors analyze the case of parallel bimolecular reactions, a case that is in general not one of uniform kinetics. Rather than trying to solve the coupled set of differential equations, Ho *et al.* chose to search directly for upper and lower bounds to the overall effectiveness factor, which are found by reducing the problem to that of finding the effectiveness factor for a single second-order reaction. The bounds can be estimated if information is available on the average and the spread of the individual Thiele moduli.

b. Flow Reactor with Axial Diffusion. For this problem, in addition to diffusion and reaction (which are present also in the porous catalyst problem), there is also convection; this, somewhat surprisingly, makes the analysis easier (Guida et al., 1994b). Let P be the ratio of the diffusivity (again assumed to be the same for all reactants) to the product of velocity and reactor length, and let t again be the dimensionless distance from the inlet. The differential equations and boundary conditions are, if Eq. (140) is used for the kinetics:

$$Pc_{tt}(x,t) = c_t(x,t) + xc(x,t)\mathbf{F}[c(z,t)], \tag{160}$$

$$c_F(x) = c(x,0) - Pc_t(x,0), \tag{161}$$

$$c_t(x,1) = 0. \tag{162}$$

This is a rather nasty problem to solve numerically, because boundary conditions over the whole range of x are assigned at $t = 0$ and $t = 1$. A perturbation expansion around $1/P = 0$ yields, as expected, the CSTR at the zero-order level; at all higher orders, one has a nested series of second-order linear nonhomogeneous differential equations that can be solved analytically if the lower order solution is available. The whole problem thus reduces to the solution of Eq. (133), which has been discussed before. This is, of course, the high-diffusivity limit that corresponds to a small Thiele modulus in the porous catalyst problem.

The limit $P = 0$ is again singular, but the perturbation expansion can be carried out usefully. The outer solution yields, at the zero-order level, the PFR, for which the solution is available. At every higher order level, one obtains a linear nonhomogeneous first-order integrodifferential equation, which can be reduced to a complete Volterra integral equation; numerical techniques that are guaranteed to converge can easily be set up for the solution. The inner (boundary layer) solution, which is a minor correction near the reactor exit, is an easy problem, since within the boundary layer convection and diffusion balance each other and the nonlinear kinetic term disappears from the equations.

8. Conclusion

Astarita (1991) presented a number of problems in chemical reaction engineering from the overall viewpoint in a continuous description, and this was discussed again in 1992 by Astarita and Ocone. Some moderate progress has been made since then in the solution of this kind of problem, but much still needs to be done. In particular, published analyses make use of old, traditional concepts of chemical reaction engineering, such as the RTD, micromixing versus macromixing, and so on; this is not the most modern approach in chemical reaction engineering (see, e.g., the ample literature on the significance of the RTD concept: Evangelista *et al.*, 1969; Glasser *et al.*, 1973; Hildebrand *et al.*, 1990; Leib *et al.*, 1988; Naor and Shinnar, 1963; Shinnar et al. 1972, Shinnar and Rumschintzki, 1989; Silverstein and Shinnar, 1975, 1982; Weinstein and Adler, 1967), but the science of chemical reaction engineering of complex mixtures is still in its infancy, and therefore it is not surprising that so far it has been developed on the basis of rather old concepts of reactor analysis.

APPENDIX A: ORTHOGONAL COMPLEMENT

Consider the three-component system formed by the three xylene isomers. This is clearly a merization system, and all values of **m** are stoichiometrically accessible from any initial **M**. There is only one vector (to within multiplication by an arbitrary scalar) to which all reactor vectors $\mathbf{m}_1 - \mathbf{m}_2$ are orthogonal, and that is $\mathbf{a}_1 = \mathbf{I}$; hence, in this case $C = 1$.

A slightly more complex case is that of a mixture of olefins that can only undergo reaction with ethylene:

$$C_{xH}2x + C_2H_4 = C_{x+2}H_{2x+4}, \qquad x = 0, 1, \ldots \qquad (163)$$

Again, **I** is a vector to which all reaction vectors $\mathbf{m}_1 - \mathbf{m}_2$ are orthogonal. However, in this case the total mass of odd-valued x-olefins is also constant, and hence one has also a second basis vector in the orthogonal complement:

$$\mathbf{a}_2 = \{0, 1/3, 0, 1/5, \ldots\}. \qquad (164)$$

Notice that x has no upper bound: $C = 2$ even when the number of different olefins in the system approaches infinity. Also notice that the balance of hydrogen does not yield any new linear constraint, in addition to the carbon balance. Finally, note that even if only one disproportionation reaction is allowed, leading from an odd- to an even-numbered olefin, $C = 1$ and the system is a merization one.

APPENDIX B

Consider the discrete equivalent of Eq. (21) so that $G^{ID}/RT = \Sigma X_I \ln X_I < 0$. First of all, it is interesting to determine what is the logical status of this equation (which, in most textbooks on thermodynamics, is derived either from an argument in statistical mechanics, or from consideration of the so-called Van't Hoff box with ideal gases and semipermeable membranes involved in the derivation). Let g^{MIX} be the total dimensionless free energy of mixing, $g^{MIX} = n_T G^{MIX}/RT$, where n_T is the total number of moles, and let μ_I be the dimensionless chemical potential of component I. We have, to within an irrelevant linear term,

$$g^{MIX} = \Sigma n_I \mu_I. \tag{165}$$

We want to write as simple a form as possible for the dependence of the chemical potentials on composition, and we want to define as "ideal" a mixture that satisfies this condition (hence the superscript ID). We impose the following requirement: The chemical potential of some component J depends only on the mole fraction of J itself, $\mu_J = f(X_J)$, with the function $f(\ )$ being the same for all components. This is clearly the simplest possible form. We obtain

$$g^{ID} = \Sigma n_I f(X_I). \tag{166}$$

Now the Gibbs–Duhem equation must be satisfied (recall that the Gibbs–Duhem equation is a simple consequence of the fact that the function, or functional, that delivers the free energy is homogeneous to the first degree; see Section V). For arbitrary δn_I resulting in some $\delta \mu_I$, one must have

$$\Sigma n_I \delta \mu_I = 0. \tag{167}$$

If the δn_I are of the form $x_I \delta n_T$, the Gibbs–Duhem equation would be trivially satisfied since $\delta \mu_I = 0$ for all I's. Now suppose the δn_I sum to some nonzero δn_T. One can define new δn_I's as the original ones minus $x_I \delta n_T$, and these would result in the same $\delta \mu_I$'s, but their sum would be zero. It follows that one only needs to make sure that the Gibbs–Duhem equation is satisfied for all sets of δn_I that satisfy $\Sigma \delta n_I = 0$. If Eq. (166) is to hold, one has

$$\delta \mu_I = f'(X_I) \delta n_I / n_T. \tag{168}$$

When this is substituted into Eq. (167), one obtains:

$$\Sigma X_I f'(X_I) \delta n_I = 0. \tag{169}$$

Equation (169) must hold for arbitrary δn_I satisfying $\Sigma \delta n_I = 0$, and clearly the only possibility is that $f'(X_I) = 1/X_I$, that is, $f(X_I) = \ln X_I$. The logarithmic form is the only one that satisfies the stated conditions, and hence the discrete equivalent of Eq. (21) simply follows from the definitions. Clearly, the argument can be generalized to a continuous description [one only needs to apply the Gibbs–Duhem equation in its continuous formulation to a $\delta n(x)$ satisfying $<\delta n(y)> = 0$], and hence Eq. (21) is identified as simply the *definition* of a continuous ideal mixture.

However, going back to the discrete description, consider the special case where all the mole fractions are equal to each other, $X_I = 1/N$; this composition maximizes the absolute value of G^{ID}/RT. One obtains $G^{ID}/RT = \ln(1/N)$, and one seems to have a problem with N approaches infinity: G^{ID}/RT approaches $-\infty$, if only logarithmically.[20]

The continuous generalization of the case $X_I = 1/N$ would seem to be one where $X^F(x)$ is constant, and this seems to create a problem, as discussed later. Now consider the case where $X^F(x)$ equals the gamma distribution $\phi(\alpha,x)$. One calculates

$$G^{ID}/RT = \ln[\alpha^\alpha/\Gamma(\alpha)] \\ + (\alpha - 1)<\phi(\alpha,y) \ln y> - \alpha. \tag{170}$$

When $\alpha = 1$ (an exponential distribution) one gets $G^{ID}/RT = -1$. When $\alpha > 1$, $<\phi(\alpha,y) \ln y>$ is guaranteed to converge and there is no problem. However, when $0 < \alpha < 1$, $<\phi(\alpha,y) \ln y>$ does not converge. This implies that a mole fraction distribution, if representable as a gamma function, is only legitimate for $\alpha \geq 1$. More generally, any mole fraction distribution is subject to the condition that G^{ID}/RT as given by Eq. (21) is finite.

The conceptual point is as follows. The discrete case $X_I = 1/N$ requires the mole fractions of all components to be equal to each other. That does *not*, however, correspond to a constant $X^F(x)$ in the continuous description, because $X^F(x)dx$ is the mole fraction of species between x and $x + dx$, and one would need to require $X^F(x)dx$ to be constant. But this can only be done if one has chosen a specific scaling for the label x: Any label x^* that is given by a monotonous function $x^*(x) = x^*$ would be legitimate, and of course $X^F(x)dx$ could be taken as constant for only one such scale. In other words, in a continuous description one has chosen some label x. The form of the mole fraction distribution must then satisfy certain constraints such as the one discussed earlier for the gamma distribution. The problem is related to the more general problem of the correct generalization to a continuous description of nonlinear formulas.

Indeed, consider the following (largely heuristic) argument. The generalization discussed earlier of the definition of an ideal mixture to the continuous description makes use of the implicit idea that $<n(y)>$ is finite. This seems so trivial that it does not need any discussion, but there is a subtlety involved, which is reminiscent of the so-called "thermodynamic limit" in classical statistical mechanics. Consider a box of volume V at some given temperature and pressure, such that it contains exactly 1 mol of an ideal gas (the problem discussed arises already with ideal gases); that is, it contains **N** molecules, where **N** is Avogadro's number. If we wish the mole fractions of all components to be the same, the largest number of components that can be squeezed into the box is **N** (there is just

---

[20] The problem arises also if the mole fractions are not all equal to each other. Let a_I be defined as NX_I. One obtains $G^{ID}/RT = \Sigma a_I \ln a_I + \ln(1/N)$, where the first term is always finite and in fact small, because $\Sigma a_I = N$. The offending term $\ln(1/N)$ cancels in all phase equilibria calculations.

one molecule of every component). In other words, the number of components in any given system cannot exceed the number of molecules in it. This in turn implies that the number of components may really approach infinity only if the total mass of the system also approaches infinity and, hence, that, at least in principle, one cannot take $<n(y)>$ to be finite. The continuous description can only be seen as an approximation to a discrete description—if based on a number of components, which may be as large as Avogadro's number (which is very large indeed).

APPENDIX C: THE FEINBERG APPROACH TO NETWORK TOPOLOGY

Feinberg bases his approach on the concept of *complex,* which is originally due (in its explicit form) to Horn and Jackson (1972), and which can be traced back in a more implicit form to the work of Krambeck (1970). Complexes are groupings of components that can appear in any given reaction; for instance, in the (molar-based described) reaction

$$A_1 = A_2 + A_3, \tag{171}$$

A_1 and $A_2 + A_3$ are complexes. Let N' be the total number of complexes in any given reaction network.

Next, Feinberg introduces the concept of *linkage class.* A linkage class is a group of complexes that can be formed from one another. Say, for example, in the classical xylene isomers example discussed in Appendix A, there is only one linkage class, since all three components (and complexes) can be formed from every other one. Let N'' be the number of linkage classes. Of course, all complexes within any given linkage class have to be "compatible": They all have to have the same number of atoms of each type. However, some compatible complexes may not belong to some linkage class, as later examples show. (In the part of the theory dedicated to the analysis of CSTRs, Feinberg introduces some *imaginary* complexes, which can be regarded as compatible only in a very formal sense; see Section IV,C,6. This does not concern us here.)

Finally, Feinberg considers the rank of the stoichiometric matrix, R. He then defines a "deficiency" of the reaction network, δ, as $\delta = N' - N'' - R$. In his series of papers, Feinberg obtains a number of strong results for systems of deficiency 0 and deficiency 1; these are not discussed here, and only very concisely in the next section. For the whole strength of the theory the reader is referred to the original Feinberg papers.

Feinberg also keeps track of how the "arrows" connecting any two complexes in a linkage class are oriented; for instance, the case of the three xylene isomers could be regarded as $A \rightarrow B \rightarrow C \rightarrow A$, or as $A \leftrightarrow B \leftrightarrow C \leftrightarrow A$, or any other intermediate case. Again, at this stage we do not need to be concerned with that part of the topology of the reaction network. The reason why, at this stage, we need not worry about the orientation of arrows within a given network is that the deficiency of a network is insensitive to the details of the arrow structure within the linkage classes: The precise nature of the arrows affects the deficiency only to the extent that the arrows determine the linkage classes (Feinberg, 1987, Section 2).

The deficiency of a reaction network depends very strongly on the topology one assigns to it, as the following two examples show. First, consider the system discussed in Appendix B, [see Eq. (165)]. Let N be the largest carbon number taken into consideration. Clearly one can write $N - 3$ reactions of the type of Eq. (165), and each one is a linkage class, so that $N'' = N - 3$. [The equals sign in Eq. (165) can still be interpreted as singly or doubly oriented.] Notice that many compatible complexes do not belong to their possible linkage class: All complexes with more than two compounds, and all complexes consisting of a pair where one of the members is not ethylene, do not belong to any linkage class, in spite of the fact that many such complexes are compatible with all linkage classes except the low-order ones. All of the reactions written down are independent, and hence $R = N - 3$. For each reaction, the quantities written on the right and on the left are complexes, and there are no other complexes (since olefins have been assumed to be able to react only with ethylene). It follows that $N' = 2(n - 3)$; for this system, the deficiency is zero, and the strong results developed by Feinberg for such systems apply (see Section IV,C,6).

However, consider the case where one allows for any two olefins to react with each other, say, with I and J, the carbon numbers of generic olefins:

$$A_I + A_J = A_{I+J}. \tag{172}$$

For $N = 4$, R is obviously 1; for $N = 5$, $R = 2$; and for $N = 6$, $R = 4$, because the reaction $2A_3 = A_6$ cannot be written as a linear combination of reactions of the type of Eq. (165). However, after that adding a new olefin only increases the number of independent reactions by one, so for $N > 5$ one has $R = N - 2$. For every component A_I, $I > 3$, there is one linkage class: all pairs which, as a complex, have the same carbon number (this still leaves out a large number of compatible complexes, because compatible triplets, etc., are excluded). However, the number of complexes and the deficiency now grow very rapidly with N, since they are delivered by a modified Fibonacci series:

N	4	5	6	7	8	9	10	11	12
R	1	2	4	5	6	7	8	9	10
N'	2	4	7	10	14	18	23	28	34
δ	0	0	0	0	1	3	5	8	11	15

One may ask why any compatible complexes should be excluded from any given linkage class. First of all, the Feinberg theory results are strong if the kinetics are of the mass action type, and hence all reactions that are written down are supposed to be elementary steps in the real chemical mechanism. Realistic elementary steps probably never include more than two molecules, and hence one tends to exclude complexes formed by more than a pair of compounds. Other exclusions may be suggested by our understanding of the chemistry involved; for instance, there may be situations where for some reason or another we come to the conclusion that olefins react to a significant extent only with ethylene.

The same kind of problem (rapid growth of N' and δ with growing N) arises with cracking and with polymerization reactions (of which olefin oligomerization is an example). For instance, in the case of cracking, if one assumes that only the end product (e.g., methane in the cracking of hydrocarbons) is cracked away at each reaction step, the deficiency is zero. [Notice that this assumption is exactly the opposite of assumption 2 just before Eq. (126), i.e., that all bonds are equally cleavable. Here one assumes that only the terminal bonds can be cleaved.] If, on the other side, one assumes the topology to be that considered in Section IV,B,4,c, the deficiency grows very rapidly with N. Notice that, in the first case, there would be no induction time, contrary to the result in Section IV,B,4,c.

However, polymerization and cracking may be the freak cases. Consider, for instance, hydrodesulfurization of an oil cut. Let RS represent any one of the N sulfurated compounds in the mixture; all the reactions are of the following type:

$$RS + H_2 = R + H_2S. \tag{173}$$

These are all independent, so $R = N$; each one represents a linkage class, so $N'' = N$; and each one yields two complexes, so $N' = 2N$, and $\delta = 0$. The deficiency tends to become very large in systems where most of the components have brute chemical formulas that are multiples of each other, which give rise to strongly interconnected networks (very many compatible complexes exist, even if complexes with more than two members are excluded); it stays at 0 or 1 in most systems of the hydrodesulfurization type, where parallel reactions with (almost) no interconnection take place.

APPENDIX D: MATHEMATICAL CONCEPTS

Consider a physical property (such as the total Gibbs free energy G) of a continuous mixture, the value of which depends on the composition of the mixture. Because the latter is a function of, say, the mole distribution $n(x)$, one has a mapping from a function to (in this case) a scalar quantity G, which is expressed by saying that G is given by a *functional* of $n(x)$. [One could equally well consider the mass distribution function $m(x)$, and consequently one would have partial mass properties rather than partial molar ones.] We use z for the label x when intended as a dummy variable inside a functional, and we identify functionals by writing them in the boldface type corresponding to the value, and their arguments within square brackets, say for the case at hand:

$$G = \mathbf{G}[n(z)]. \tag{174}$$

One now needs to extend the concept of a derivative from ordinary functions to functionals. This is done by writing, with $s(z)$ being a displacement function,

$$\mathbf{G}[n(z) + s(z)] = \mathbf{G}[n(z)] + \delta\mathbf{G}[n(z)|s(z)] + \mathbf{R}, \tag{175}$$

where $\delta\mathbf{G}[\ ]$ (called the Frechet differential of $\mathbf{G}[\ ]$) is a functional of both $n(z)$ and $s(z)$ which has the property of being linear in $s(z)$.

The *residual* **R** has the property:

$$\lim_{||s(x)||\,=\,0} \mathbf{R}/||s(x)|| = 0 \tag{176}$$

and $||s(x)||$ is the "norm" of $s(x)$.

The definition of the norm $||s(x)||$ [which is a nonnegative scalar assigned to any $s(x)$, which should satisfy the usual requirements for a norm, such as the triangular inequality and the fact that $||s(x)|| = 0$ only if $s(x)$ is identically zero] is a delicate question. First of all, in order to give an unequivocal meaning to Eq. (176), one does not need, in fact, to assign a norm, but only a topology. One need only know when the norm becomes vanishingly small, without knowing what the norm is when it is finite. In other words, assigning a norm establishes a "metric" in the space of functions [finite distances between $n(x)$ and $n(x) + s(x)$ are measured by the norm], but different metrics could yield the same topology. However, in practice the easiest way to assign a topology is to assign a metric.

The most common norm is the so-called Euclidean one:

$$||s(x)|| = <s^2(y)>. \tag{177}$$

A Euclidean norm declares two functions that differ only on a countable infinity of isolated points as being "close." This is not too much of a difficulty for the problems we consider, but there is another difficulty. If we want to consider distributions that include one or more discrete components (a semicontinuous distribution), $s(x)$ may well contain some delta functions. This implies, first, that all integrals have to be interpreted as a Stjieltjies ones; but even so one has a problem with the right-hand side of Eq. (177), because the delta function is not Stjieltjies square-integrable. One could be a bit cavalier here and say that we agree that $\delta^2(x) = \delta(x)$, but it is perhaps preferable to keep continuous and discrete components separate. Let, for instance, the mole distribution be $n_1, n_2, \ldots, n_N, n(x)$ in a mixture with N discrete components and a distributed spectrum. One can now define the scalar product as the ordinary one over the discrete components, plus the continuous one over the distribution, and therefore avoid the problem with delta functions.

Because $\delta \mathbf{G}$ is linear in its second argument, it must be expressible as a weighted integral of $s(z)$; the weighting function, however, may still depend on the particular "point" $n(z)$ in Hilbert space where the Frechet differential is evaluated. Hence, the weighting function is given by a functional $\mathbf{G}'[;]$ of $n(z)$, which depends parametrically on x; $\mathbf{G}'[;]$ is called the functional derivative of $\mathbf{G}[\,]$. One has

$$\delta \mathbf{G}[n(z)|s(z)] = <\mathbf{G}'[n(z);y]s(y)>. \qquad (178)$$

This formalism can now be applied, as an example, to our specific physical example where G is the Gibbs free energy and $n(x)$ is the mole distribution. The usual statement in thermodynamics that G is an extensive property can be formalized by requiring the functional $\mathbf{G}[\,]$ to be homogeneous of the first degree. Say for any positive scalar Ω one has

$$\mathbf{G}[\Omega n(z)] = \Omega \mathbf{G}[n(z)]. \qquad (179)$$

Subtracting $\mathbf{G}[n(z)]$ from both sides of Eq. (179), taking the limit as Ω approaches unity, and using the definitions given earlier, one obtains

$$G = <\mathbf{G}'[n(z);y]n(y)>. \qquad (180)$$

This leads to the obvious identification of the chemical potential distribution $\mu(x)$:

$$\mu(x) = \mathbf{G}'[n(z);x] \qquad (181)$$

so that one obtains the obvious extension of the classical result $G = <\mu, \mathbf{n}>$:

$$G = <\mu(y)n(y)>. \qquad (182)$$

Now consider any infinitesimal variation $\delta n(x)$ from some base distribution $n(x)$. Using the definitions of the functional derivative of $\mathbf{G}[\,]$ and of the chemical potential distribution $\mu(x)$, one would calculate the corresponding infinitesimal variation of G, δG, as $<\mu(y)\delta n(y)>$; differentiating Eq. (181) one would calculate $\delta G = <\mu(y)\delta n(y)> + <n(y)\delta \mu(y)>$. It follows that

$$<n(y)\delta\mu>(y)> = 0, \qquad (183)$$

which is the obvious continuous description equivalent of the Gibbs–Duhem equation.

Now consider again Eq. (182) and the requirement that the right-hand side of it be homogeneous of the first degree. It follows trivially that the functional giving $\mu(x)$, Eq. (181), is homogeneous of degree zero. Say for any positive scalar Ω,

$$\mathbf{G}'[\Omega n(z);x] = \mathbf{G}'[n(z);x]. \qquad (184)$$

If we now choose Ω as $1/<n(y)>$ (i.e., as the inverse of the total number of moles), so that $\Omega n(x)$ equals the mole fraction distribution $X(x)$, one obtains

$$\mathbf{G}'[X(z);x] = \mathbf{G}'[n(z);x]; \qquad (185)$$

that is, the chemical potential distribution depends only on the mole fraction distribution and not on the total number of moles. It follows that any variation $\delta n(x)$ which results in the same mole fraction distribution [i.e., any $\delta n(x)$ expressible as $n(x)\delta<n(y)>/<n(y)>$] results in a zero variation of $\mu(x)$. It follows that the Gibbs–Duhem equation, Eq. (183), is trivially satisfied for such variations, and hence that one only needs to ascertain, when writing down constitutive equations, that Eq. (183) is satisfied for those $\delta n(x)$ that satisfy $<\delta n(y)> = 0$. [It is useful to note that this argument applies provided $<n(y)>$ is finite—see Appendix B.]

The functional derivative (and the chemical potential distribution) are examples of a *function* that depends on another one. Consider the case of the chemical potential distribution:

$$\mu(x) = \mu[n(z);x]. \qquad (186)$$

If the argument function $[n(z)$ in Eq. (186)] and the value function $[\mu(x)$ in Eq. (186)] are defined over the *same* variable [as is the case in Eq. (186); z is the dummy variable form of x], it is legitimate to ask oneself whether a functional such as $\mathbf{G}'[;]$ is *invertible* at some $n(x)$. Whether the definition is over the same variable is easily kept track of in one-phase kinetics and thermodynamics: We use x, y, z for the component labels, and u, v, w for the reaction labels. The usual theory of equilibrium in Section III,A is based on the assumption that the functional $\mathbf{r}[q(w);u]$ is invertible at $\mathbf{q}^*(v)$. Also, in homogeneous systems where the isothermal free energy hypersurface is strictly convex, $\boldsymbol{\mu}[n(z);x]$ is invertible at every $n(x)$, and hence $n(x)$ is expressible as $\mathbf{n}[\mu(z);x]$.

Acknowledgments

We are indebted to a number of people for their help in preparing this article. We had many useful discussions on mathematical details with Vincenzo Guida. Professor R. Aris of the University of Minnesota, Professor M. Feinberg of the University of Rochester, Dr. T. C. Ho of Exxon Research and Engineering, and Professor S. I. Sandler of the University of Delaware helped us extract the key conceptual contents of their numerous contributions to the area, and they commented on our wording of the summary of that conceptual content.

Finally, a special debt of gratitude to Dr. F. J. Krambeck of Mobil Research must be acknowledged: Fred was involved in this project from the beginning, and if he is not a coauthor it is simply because his industrial commitments did not allow him to participate more actively than he has. He contributed much to this article, and he commented and criticized early drafts of many parts of it.

Although we acknowledge with gratitude the very significant help that we have received, we are solely responsible for any faults in this article.

References

Abbaian, M. J., and Weil, S. A., Phase equilibria of continuous fossil fuel process oils. *AIChE J.* **34,** 574 (1988).

Abramowitz, M., and Stegun, I. A., "Handbook of Mathematical Functions," Dover, New York, 1965.

Acrivos, A., and Amundson, N. R., On the steady state fractionation of multicomponent and complex mixtures in an ideal cascade. *Chem. Eng. Sci.* **4,** 29, 68, 141, 159, 206, 249 (1955).

Alberty, R. A., Kinetics and equilibrium of the polymerization of benzene series aromatic hydrocarbons in a flame. *In* "Kinetic and Thermodynamic Lumping of Multicomponent Mixtures." (Astarita, G., and Sandler, R. I., eds.), Elsevier, Amsterdam, 1991, p. 277.

Alberty, R. A., Use of an ensemble intermediate between the generalized ensemble and the isothermal-isobaric ensemble to calculate the equilibrium distribution of hydrocarbons in homologous series, *J. Chem. Phys.* **84,** 2890 (1986).

Alberty, R. A., Extrapolation of standard chemical thermodynamic properties of alkene isomers groups to higher carbon numbers, *J. Phys. Chem* **87,** 4999 (1983).

Alberty, R. A., and Oppenheim, I., Analytic expressions for the equilibrium distributions of isomer groups in homologous series. *J. Chem. Phys.* **84,** 917 (1986).

Alberty, R. A., and Oppenheim, I., A continuous thermodynamics approach to chemical equilibrium within an isomer group. *J. Chem. Phys.* **81,** 4603 (1984).

Allen, D. T., Structural models of catalytic cracking chemistry. *In* "Kinetic and Thermodynamic Lumping of Multicomponent Mixtures" (G. Astarita and R. I. Sandler, eds.). Elsevier, Amsterdam, 1991, p. 163.

Annesini, M. C., Gironi, F., and Marrelli, L., Multicomponent adsorption of continuous mixtures. *Ind. Eng. Chem. Res.* **27,** 1212 (1988).

Anthony, D. B., and Howard, J. B., Coal devolatilization and hydrogasification. *AIChE J.* **22,** 625 (1976).

Aris, R., Ends and beginnings in the mathematical modeling of chemical engineering systems. *Chem. Eng. Sci.* **48,** 2507 (1993).

Aris, R., Comments on mitigation of backmixing via catalyst dilution. *Chem. Eng. Sci.* **47,** 507 (1992).

Aris, R., The mathematics of continuous mixtures. *In* "Kinetic and Thermodynamic Lumping of Multicomponent Mixtures" (G. Astarita and R. I. Sandler, eds.). Elsevier, Amsterdam, 1991a, p. 23.

Aris, R., Multiple indices, simple lumps and duplicitous kinetics. *In* "Chemical Reactions in Complex Systems" (A. V. Sapre and F. J. Krambeck, eds.). Van Nostrand Reinhold, New York, 1991b, p. 25.

Aris, R., Manners makyth modellers. *Chem. Eng. Sci.* **46,** 1535 (1991c).

Aris, R., "Elementary Chemical Reactor Analysis." Butterworths, New York, 1989a.

Aris, R., Reactions in continuous mixtures. *AIChE J.* **35,** 539 (1989b).

Aris, R. Personal Communication, 1988.

Aris, R., How to get the most out of an equation without really trying. *Chem. Eng. Educ.* **10,** 114 (1976).

Aris, R. Prolegomena to the rational analysis of systems of chemical reactions. II, Some addenda. *Arch. Ratl. Mech. Anal.* **22,** 356 (1968).

Aris, R., Prolegomena to the rational analysis of systems of chemical reactions. *Arch. Ratl. Mech. Anal.* **19,** 81 (1965a).

Aris, R., "Introduction to the Analysis of Chemical Reactors." Prentice Hall, Englewood Cliffs, NJ, 1965b.

Aris, R., The algebra of systems of second-order reactions. *Ind. Eng. Chem. Fundament.* **3,** 28 (1964).

Aris, R., and Astarita, G., On aliases of differential equations. *Rend. Acc. Lincei* **LXXXIII,** (1989a).

Aris, R., and Astarita, G., Continuous lumping of nonlinear chemical kinetics. *Chem. Eng. Proc.* **26,** 63 (1989b).

Aris, R., and Gavalas, G., On the theory of reactions in continuous mixtures. *Proc. Roy. Soc. London* **260,** 351 (1966).

Astarita, G., Chemical reaction engineering of multicomponent mixtures: Open problems. *In* "Chemical Reactions in Complex Systems" (A. V. Sapre and F. J. Krambeck, eds.). Van Nostrand Reinhold, New York, 1991, p. 3.

Astarita, G., Multiple steady states in maximum mixedness reactors. *Chem. Eng. Comm.* **93,** 111 (1990).

Astarita, G., "Thermodynamics. An Advanced Textbook for Chemical Engineers." Plenum Press, New York, 1989a.

Astarita, G., Lumping nonlinear kinetics: Apparent overall order of reaction. *AIChE J.* **35,** 529 (1989b).

Astarita, G., A note on homogeneous and heterogeneous chemical equilibria. *Chem. Eng. Sci.* **31,** 1224 (1976).

Astarita, G., and Nigam, A., Lumping nonlinear kinetics in a CSTR. *AIChE J.* **35,** 1927 (1989).

Astarita, G., and Ocone, R., A perturbation analysis of the Aris puzzle. *Chem. Eng. Sci.* **48,** 823 (1993).

Astarita, G., and Ocone, R., Chemical reaction engineering of complex mixtures. *Chem. Eng. Sci.* **47,** 2135 (1992).

Astarita, G., and Ocone, R., Lumping parallel reactions. *In* "Kinetic and Thermodynamic Lumping of Multicomponent Mixtures" (G. Astarita and R. I. Sandler, eds.). Elsevier, Amsterdam, 1991, p. 207.

Astarita, G., and Ocone, R., Continuous lumping in a maximum mixedness reactor. *Chem. Eng. Sci.* **45,** 3399 (1990).

Astarita, G., and Ocone, R., Heterogeneous chemical equilibria in multicomponent mixtures. *Chem. Eng. Sci.* **44,** 2323 (1989).

Astarita, G., and Ocone, R., Lumping nonlinear kinetics. *AIChE J.* **34,** 1299 (1988).

Astarita, G., and Sandler, R. I., eds., "Kinetic and Thermodynamic Lumping of Multicomponent Mixtures." Elsevier, Amsterdam, 1991.

Avidan, A. A., and Shinnar, R., Development of catalytic cracking technology. A lesson in chemical reactor design. *Ind. Eng. Chem. Res.* **29,** 931 (1990).

Bailey, J. E., Diffusion of grouped multicomponent mixtures in uniform and nonuniform media. *AIChE J.* **21,** 192 (1975).

Bailey, J. E., Lumping analysis of reactions in continuous mixtures. *Chem. Eng. J.* **3,** 52 (1972).

Beretta, E., Stability problems in chemical networks. *In* "Nonlinear Differential Equations: Invariance, Stability and Bifurcations" (F. de Mottoni and L. Salvadori, eds.). Academic Press, New York, 1981, p. 11.

Beretta, E., Veltrano, F., Solimano, F. and Lazzari, C., Some results about nonlinear chemical systems represented by trees and cycles. *Bull. Math. Biol.* **41,** 641 (1979).

Bhore, N. A., Klein, M. T., and Bischoff, K. B., The delplot technique: A new method for reaction pathway analysis. *Ind. Eng. Chem. Res.* **29,** 313 (1990).

Bischoff, K. B., Nigam, A., and Klein, M. T., Lumping of discrete kinetic systems. *In* "Kinetic and Thermodynamic Lumping of Multicomponent Mixtures" (G. Astarita and R. I. Sandler, eds.). Elsevier, Amsterdam, 1991, p. 33.

Bowen, R. M., On the stoichiometry of chemically reacting materials. *Arch. Ratl. Mech. Anal.* **29**, 114 (1968a).

Bowen, R. M., Thermochemistry of reacting materials. *J. Chem. Phys.* **49**, 1625 (1968b).

Bowman, J. R., Distillation on an indefinite number of components. *IEC* **41**, 2004 (1949).

Briano, J. G., and Glandt, E. D., Molecular thermodynamics of continuous mixtures. *Fluid Phase Eq.* **14**, 91 (1983).

Brundege, J. A., and Parravano, G., The distribution of reaction rates and activatione energies on catalytic surfaces: Exchange reactions between gaseous benzene and benzene adsorbed on platinum. *J. Catal.* **2**, 380 (1963).

Cao, G., Viola, A., Baratti, R., Morbidelli, M., Sanseverino, L., and Cruccu, M., Lumped kinetic model for propene–butene mixtures oligomerization on a supported phosphoric acid catalyst. *Adv. Catal.* **41**, 301 (1988).

Caram, H. S., and Scriven, L. E., Nonunique reaction equilibria in nonideal systems. *Chem. Eng. Sci.* **31**, 163 (1976).

Chou, G. F., and Prausnitz, J. M., Supercritical fluid extraction calculations for high-boiling petroleum fractions using propane. Application of continuous thermodynamics. *Ber. Bunsenges. Phys. Chim.* **88**, 796 (1984).

Chou, M. Y., and Ho, T. C., Lumping coupled nonlinear reactions in continuous mixtures. *AIChE J.* **35**, 533 (1989).

Chou, M. Y., and Ho, T. C. Continuum theory for lumping nonlinear reactions. *AIChE J.* **34**, 1519 (1988).

Cicarelli, P., and Aris R., Continuous reactions in a non-isothermal CSTR. I, Multiplicity of steady states. *Chem. Eng. Sci.* **49**, 621 (1994).

Cicarelli, P., and Astarita, G., Micromixing and Shinnar paradoxes. *Chem. Eng. Sci.* **47**, 1007 (1992).

Cicarelli, P., Astarita, G., and Gallifuoco, A., Continuous kinetic lumping of catalytic cracking processes. *AIChE J.* **38**, 1038 (1992).

Coleman, B. D., and Gurtin, M. E., Thermodynamics with internal state variables. *J. Chem. Phys.* **47**, 597 (1967).

Cotterman, R. L., Bender, R., and Prausnitz, J. M., Phase equilibria for mixtures containing very many components: Development and application of continuous thermodynamics for chemical process design. *Ind. Eng. Chem. Proc. Des. Dev.* **24**, 194 (1985).

Cotterman, R. L., Chou, G. F., and Prausnitz, J. M., Comments on flash calculations for continuous and semi-continuous mixtures using an equation of state. *Ind. Chem. Eng. Proc. Des. Dev.* **25**, 840 (1986).

Cotterman, R. L., Dimitrelis, D., and Prausnitz, J. M., Supercritical fluid extraction calculations for high-boiling petroleum fractions using propane. Application of continuous thermodynamics. *Ber. Buns. Phys. Chem.* **88**, 796 (1984).

Cotterman, R. L., and Prausnitz, J. M., Continuous thermodynamics for phase-equilibrium calculations in chemical process design. *In* "Kinetic and Thermodynamic Lumping of Multicomponent Mixtures" (G. Astarita and R. I. Sandler, eds.). Elsevier, Amsterdam, 1991, p. 229.

Cotterman, R. L., and Prausnitz, J. M., Flash calculations for continuous or semicontinuous mixtures using an equation of state. *Ind. Eng. Chem. Proc. Des. Dev.* **24**, 434 (1985).

Coxson, P. G., Lumpability and observability of linear systems. *J. Math. Anal. Appl.* **99**, 435 (1984).

Coxson, P. G., and Bishoff, K. B., Lumping strategy. 2, A system theoretic approach. *Ind. Eng. Chem. Res.* **26**, 2151 (1987a).

Coxson, P. G., and Bishoff, K. B., Lumping strategy. 1, Introductory techniques and applications of cluster analysis. *Ind. Eng. Chem. Res.* **26**, 1239 (1987b).

Danckwerts, P. V. The effect of incomplete mixing on homogeneous reactions. *Chem. Eng. Sci.* **10,** 93 (1958).
De Donder, Th., "L'Affinite'. Seconde Partie." Gauthier-Villars, Paris, 1931.
Dotson, N. A., Galvàn, R., Laurence, R. L., and Tirrell, M., "Modeling of Polymerization Processes." VCH Press, New York, 1993.
Du, P. C., and Mansoori, G. A., Phase equilibrium computational algorithms of continuous thermodynamics. *Fluid Phase Eq.* **30,** 57 (1986).
Edminister, W. C., Improved intergral techniques for petroleum distillation curves. *Ind. Eng. Chem.* **47,** 1685 (1955).
Edminster, W. C., and Buchanan, D. H., Applications of integral distillation calculation method. *Chem. Eng. Prog. Symp. Ser.* **6,** 69 (1953).
Evangelista, J. J., Katz, S., and Shinnar, R., Scale-up criteria for stirred tank reactors. *AIChE J.* **15,** 843 (1969).
Feinberg, M., The chemical reaction network toolbox. Diskette for IBM compatibles, 1994.
Feinberg, M., Some recent results in chemical reaction network theory. *In* "Patterns and Dynamics in Reactive Media," IMA Volumes on Mathematics and its Applications, Springer-Verlag, Berlin, 1991a.
Feinberg, M., Applications of chemical reaction network theory in heterogeneous catalysis. *In* "Chemical Reactions in Complex Systems" (A. V. Sapre and F. J. Krambeck, eds.). Van Nostrand Reinhold, New York, 1991b, p. 179.
Feinberg, M., Necessary and sufficient conditions for detailed balancing in mass action systems of arbitrary complexity. *Chem. Eng. Sci.* **44,** 1819 (1989).
Feinberg, M., Chemical reaction network structure and the stability of complex isothermal reactors. II, Multiple steady states for networks of deficiency one. *Chem. Eng. Sci.* **43,** 1 (1988).
Feinberg, M., Chemical reaction network structure and the stability of complex isothermal reactors. I, The deficiency zero and deficiency one theorems. *Chem. Eng. Sci.* **42,** 2229 (1987).
Feinberg, M., Chemical oscillations, multiple equilibria, and reaction network structure. *In* "Dynamics and Modeling of Reactive Systems" (W. E. Stewart, W. H. Ray, and C. C. Conley, eds.). Academic Press, New York, 1980.
Feinberg, M., Mathematical aspects of mass action kinetics. *In* "Chemical Reactor Theory: A Review" (N. Amundson and L. Lapidus, eds.). Prentice Hall, Englewood Cliffs, NJ, 1977.
Feinberg, M., Complex balancing in general kinetic systems. *Arch. Ratl. Mech. Anal.* **49,** 187 (1972a).
Feinberg, M., On chemical kinetics of a certain class. *Arch. Ratl. Mech. Anal.* **46,** 1 (1972b).
Feinberg, M., and Horn, F., Chemical mechanism structure and the coincidence of the stoichiometric and kinetic subspaces. *Arch. Ratl. Mech. Anal.* **66,** 83 (1977).
Feinberg, M., and Horn, F., Dynamics of open chemical systems and the algebraic structure of the underlying reaction network. *Chem. Eng. Sci.* **29,** 775 (1974).
Frenklach, M., Computer modeling of infinite reaction sequences. A chemical lumping. *Chem. Eng. Sci.* **40,** 1843 (1985).
Froment, G. F., The kinetics of complex catalytic reactions. *Chem. Eng. Sci.* **42,** 1073 (1987).
Froment, G. F., and Bischoff, K. B., "Chemical Reactor Analysis and Design," 2nd ed. Wiley, New York, 1990.
Gal-Or, B., Cullihan, H. T., and Galli, R., New thermodynamic transport theory for systems with continuous component density distributions. *Chem. Eng. Sci.* **30,** 1085 (1975).
Glasser, D., Crowe, C. M., and Jackson R., Zwietering's maximum-mixed reactor model and the existence of multiple steady states. *Chem. Eng. Comm.* **40,** 41 (1986).
Glasser, D., Hildebrand, D., and Crowe, C. M., A geometric approach to steady flow reactors: The attainable region and optimization in concentration space. *Ind. Eng. Chem. Res.* **26,** 1803 (1987).
Glasser, D., and Jackson, R., A general residence time distribution model for a chemical reactor. *AIChE Symp. Ser.* **87,** 535 (1984).

Glasser, D., Katz, S., and Shinnar, R., the measurement and interpretation of contact time distributions for catalytic reactor characterization. *Ind. Eng. Chem. Fundament.* **12,** 165 (1973).

Golikeri, S. V., and Luss, D., Aggregation of many coupled consecutive first order reactions. *Chem. Eng. Sci.* **29,** 845 (1974).

Golikeri, S. V., and Luss, D., Analysis of activation energy of grouped parallel reactions. *AIChE J.* **18,** 277 (1972).

Golikeri, S. V., and Luss, D. Diffusional effects in reacting mixtures. *Chem. Eng. Sci.* **26,** 237 (1971).

Gray, P., and Scott, S. P., "Chemical Oscillations and Instabilities. Nonlinear Chemical Kinetics." Oxford Science Publications, Oxford, 1990.

Gualtieri, J. A., Kincaid, J. M., and Morrison, G., Phase equilibria in polydisperse fluids. *J. Chem. Phys.* **77,** 521 (1982).

V. Guida, Personal Communication, 1994.

Guida, V., Ocone, R., and Astarita, G., Micromixing and memory in chemical reactors. *Chem. Eng. Sci.* **49,** 5215 (1994a).

Guida, V., Ocone, R., and Astarita, G., Diffusion-convection-reaction in multicomponent mixtures. *AIChE J.* **40,** 1665 (1994b).

Gupta, S. K., and Kumar, A., "Reaction Engineering of Step Growth Polymerization." Pergamon, New York, 1987.

Gurtin, M. E., On the thermodynamics of chemically reacting fluid mixtures. *Arch. Ratl. Mech. Anal.* **43,** 198 (1971).

Gurtin, M. E., and Vargas, A. S., On the classical theory of raecting fluid mixtures. *Arch. Ratl. Mech. Anal.* **43,** 179 (1971).

Gutsche, B., Phase equilibria in oleochemical industry—Application of continuous thermodynamics. *Fluid Phase Eq.* **30,** 65 (1986).

Hendricks, E. M., Simplified phase equilibrium equations for multicomponent systems. *Fluid Phase Eq.* **33,** 207 (1987).

Hildebrand, D., Glasser, D., and Crowe, C. M., Geometry of the attainable region generated by reaction and mixing: With and without constraints. *Ind. Eng. Chem. Res.* **29,** 49 (1990).

Ho, T. C., A general expression for the collective behavior of a large number of reactions. *Chem. Eng. Sci.* **46,** 281 (1991a).

Ho, T. C., Collective behavior of many langmuir–Hinshelwood reactions. *J. Catal.* **129,** 524 (1991b).

Ho, T. C., Hydrodenitrogenation catalysts. *Catal. Rev. Sci. Eng.* **30,** 117 (1988).

Ho, T. C., and Aris, R., On apparent second order kinetics. *AIChE J.* **33,** 1050 (1987).

Ho, T. C., Li, B. Z., and Wu, J. H., Estimation of lumped effectiveness factor for many bimolecular reactions. *Chem. Eng. Sci.* (1994).

Ho, T. C., and White, B. S., Experimental and theoretical investigation of the validity of asymptotic lumped kinetics. *AIChE J.* Submitted for publication (1994).

Ho, T. C., and White, B. S., Mitigation of backmixing via catalyst dilution. *Chem. Eng. Sci.* **46,** 1861 (1991).

Ho, T. C., White, B. S., and Hu, R., Lumped kinetics of many parallel nth-order reactions. *AIChE J.* **36,** 685 (1990).

Horn, F., On a connexion between stability and graphs in chemical kinetics. I, Stability and the reaction diagram; II, Stability and the complex graph. *Proc. Roy. Soc.* **A334,** 299 (1973).

Horn, F., Necessary and sufficient conditions for complex balancing in chemical kinetics. *Arch. Ratl. Mech. Anal.* **49,** 172 (1972).

Horn, F., and Jackson R., General mass action kinetics. *Arch. Ratl. Mech. Anal.* **47,** 81 (1972).

Jackson, R., and Glasser, D., A general mixing model for steady flow chemical reactors. *Chem. Eng. Comm.* **42,** 17 (1986).

Jacob, S. M., Gross, B., Voltz, S. E., and Weekman, V. W., A lumping and reaction scheme for catalytic cracking. *AIChE J.* **22,** 701 (1976).

Jaumann, G., Geschlossenen system physikalischer und chemischer differentialgesetzte. *Sitzgsber. Akad. Wiss. Wien* **63,** 63 (1911).

Kehlen, H., and Ratzsch, M. T., Complex multicomponent distillation calculations by continuous thermodynamics. *Chem. Eng. Sci.* **42,** 221 (1987).

Kirchhoff, G. R., Ueber die ausflosung der gleichungen, auf welche man bei der untersuchung der linearlen vertheilung galvanischer strome gefurht wird. *Ann. Phys. Chem.* **72,** 497 (1847).

Krambeck, F. J. Thermodynamics and kinetics of complex mixtures, *Chem. Eng. Sci.* (1994a).

Krambeck, F. J., Personal Communication, 1994b.

Krambeck, F. J., Continuous mixtures in fluid catalytic cracking and extensions. *In* "Chemical Reactions in Complex Systems" (A. V. Sapre and F. J. Krambeck, eds.). Van Nostrand Reinhold, New York, 1991a, p. 42.

Krambeck, F. J., An industrial viewpoint on lumping. *In* "Chemical Reactions in Complex Systems" (A. V. Sapre and F. J. Krambeck, eds.). Van Nostrand Reinhold, New York, 1991b, p. 111.

Krambeck, F. J., Letter to the editor. *AIChE J.* **34,** 877 (1988).

Krambeck, F. J., Computers and modern analysis in reactor design. *Inst. Chem. Eng. Symp. Ser.* **87,** 733 (1984a).

Krambeck, F. J., Accessible composition domains for monomolecular systems. *Chem. Eng. Sci.* **39,** 1181 (1984b).

Krambeck, F. J., The mathematical structure of chemical kinetics in homogeneous single-phase systems. *Arch. Ratl. Mech. Anal.* **38,** 317 (1970).

Krambeck, F. J., Avidan, A. A., Lee, C. K., and Lo, M. N., Predicting fluid-bed reactor efficiency using adsorbing gas tracers. *AIChE J.* **33,** 1727 (1987).

Krambeck, F. J., Katz, S., and Shinnar, R., A stochastic model for fluidized beds, *Chem. Eng. Sci.* **24,** 1497 (1969).

Kuo, J., and Wei, J., A lumping analysis in monomolecular reaction systems. *Ind. Eng. Chem. Fundament.* **8,** 124 (1969).

Leib, T. M., Rumschitzki, D., and Feinberg, M., Multiple steady states in complex isothermal CFSTR's: I, General considerations. *Chem. Eng. Sci.* **43,** 321 (1988).

Lewis, G. N., A new principle of equilibrium. *Proc. Natl. Acad. Sci.* **11,** 179 (1925).

Li, B. Z., and Ho, T. C., Lumping weakly nonuniform bimolecular reactions. *Chem. Eng. Sci.* **46,** 273 (1991a).

Li, B. Z., and Ho, T. C., An analysis of lumping of bimolecular reactions. *In* "Kinetic and Thermodynamic Lumping of Multicomponent Mixtures" (G. Astarita and R. I. Sandler, eds.). Elsevier, Amsterdam, 1991.

Li, G., A lumping analysis in mono- or/and bimolecular reaction systems. *Chem. Eng. Sci.* **29,** 1261 (1984).

Li, G., and Rabitz, H., Determination of constrained lumping schemes for non-isothermal first order reaction systems. *Chem. Eng. Sci.* **46,** 583 (1991a).

Li, G., and Rabitz, H., New approaches to determination of constrained lumping schemes for a reaction system in the whole composition space. *Chem. Eng. Sci.* **46,** 95 (1991b).

Li, G., and Rabitz, H., A general analysis of lumping in chemical kinetics. *In* "Kinetic and Thermodynamic Lumping of Multicomponent Mixtures" (G. Astarita and R. I. Sandler, eds.). Elsevier, Amsterdam, 1991c, p. 63.

Li, G., and Rabitz, H., A general analysis of approximate lumping in chemical kinetics. *Chem. Eng. Sci.* **45,** 977 (1990).

Li, G., and Rabitz, H., A general analysis of exact lumping in chemical kinetics. *Chem. Eng. Sci.* **44,** 1413 (1989).

Liou, J. S., Balakotaiah, V., and Luss, D., Dispersion and diffusion influences on yield in complex reaction networks. *AIChE J.* **35,** 1509 (1989).

Liu, Y. A., and Lapidus, L., Observer theory for lumping analysis of monomolecular reactions systems. *AIChE J.* **19**, 467 (1973).
Luks, K. D., Turek, E. A., and Krogas, T. K., Asymptotic effects using semicontinuous vis-a-vis discrete descriptions in phase equilibria calculations. *Ind. Eng. Chem. Res.* **29**, 2101 (1990).
Luss, D., and Golikeri, S. V., Grouping of many species each consumed by two parallel first-order reactions, *AIChE J.* **21**, 865 (1975).
Luss, D., and Hutchinson P., Lumping of mixtures with many parallel N-th order reactions. *Chem. Eng. J.* **2**, 172 (1971).
Luss, D., and Hutchinson, P., Lumping of mixtures with many parallel first order reactions. *Chem. Eng. J.* **1**, 129 (1970).
Matthews, M. A., Mani, K. C., and Haynes, H. W., Continuous phase equilibrium for sequential operations. *In* "Kinetic and Thermodynamic Lumping of Multicomponent Mixtures" (G. Astarita and R. I. Sandler, eds.). Elsevier, Amsterdam, 1991, p. 307.
McCoy, B. J., Continuous mixture kinetics and equilibrium for reversible oligomerization reactions. *AIChE J.* **39**, 1827 (1993).
Moore, P. K., and Anthony, R. G., The continuous-lumping method for vapor–liquid equilibrium calculations. *AIChE J.* **35**, 1115 (1989).
Nace, D. M., Voltz, S. E., and Weekman, V. W., Application of a kinetic model for catalytic cracking: Effects of charge stocks. *Ind. Eng. Chem. Proc. Des. Dev.* **10**, 530 (1971).
Naor, P., and Shinnar, R., Representation and evaluation of residence time distributions. *Ind. Eng. Chem. Fundament.* **2**, 278 (1963).
Ocone, R., and Astarita, G., Lumping nonlinear kinetics in porous catalysts: Diffusion-reaction lumping strategy. *AIChE J.* **39**, 288 (1993).
Oliver, M. L. On balanced interactions in mixtures. *Arch. Ratl. Mech. Anal.* **49**, 195 (1972).
Othmer, G. H., Nonuniqueness of equilibria in closed reacting systems. *Chem. Eng. Sci.* **31**, 993 (1976).
Ozawa, Y., The structure of a lumpable monomolecular system for reversible chemical reactions. *Ind. Eng. Chem. Fundament.* **12**, 191 (1973).
Paynter, J. D., and Schuette, W. L., Development of a model for kinetics of olefin codimerization. *Ind. Eng. Chem. Proc. Des. Dev.* **10**, 250 (1971).
Pitt, G. J., The kinetics of the evolution of volatile products from coal. *Fuel* **42**, 267 (1962).
Prasad, G. N., Agnew, J. B., and Sridhar, T., Continuous reaction mixture model for coal liquefaction. *AIChE J.* **32**, 1277 (1986).
Prater, C. D., Silvestri, A. J., and Wei, J., On the structure and analysis of complex systems of first order chemical reactions containing irreversible steps. I, General properties. *Chem. Eng. Sci.* **22**, 1587 (1967).
Quann, R. J., Green, L. A., Tabak, S. A., and Krambeck, F. J., Chemistry of olefin oligomerization over ZSM-5 catalyst. *Ind. Eng. Chem. Res.* **27**, 565 (1988).
Quann, R. J., and Krambeck, F. J., Olefin oligomerization kinetics over ZSM-5. *In* "Chemical Reactions in Complex Systems" (A. V. Sapre and F. J. Krambeck, eds.). Van Nostrand Reinhold, New York, 1991, p. 143.
Ramage, M. P., Graziani, K. R., Schipper, P. H., Krambeck, F. J., and Choi, B. C., KINPtR (Mobil's kinetic reforming model): A review of Mobil's industrial process modeling philosophy. *Adv. Chem. Eng.* **13**, 193 (1987).
Ratzsch, M. T., Continuous thermodynamics. *Pure Appl. Chem.* **61**, 1105 (1989).
Ratzsch, M. T., and Kehlen, H., Continuous thermodynamics of complex mixtures. *Fluid Phase Eq.* **14**, 225 (1983).
Regenass, W., and Aris, R., Stability estimates for the stirred tank reactor. *Chem. Eng. Sci.* **20**, 60 (1965).

Rumschitzki, D., and Feinberg, M., Multiple steady states in complex isothermal CFSTR's. II, Homogeneous reactors. *Chem. Eng. Sci.* **43**, 329 (1988).

Salacuse, J. J., and Stell, G., Polydisperse sysems: Statistical thermodynamics, with applications to several models including hard and permeable spheres. *J. Chem. Phys.* **77**, 3714 (1982).

Sandler, S. I., Lumping or pseudocomponent identification in phase equilibrium calculations. In "Chemical Reactions in Complex Systems" (A. V. Sapre and F. J. Krambeck, eds.). Van Nostrand Reinhold, New York, 1991, p. 60.

Sandler, S. I., and Libby, M. C., A note on the method of moments in the thermodynamics of continuous mixtures. In "Kinetic and Thermodynamic Lumping of Multicomponent Mixtures" (G. Astarita and R. I. Sandler, eds.). Elsevier, Amsterdam, 1991, p. 341.

Sapre, A. V., and Krambeck, F. J., eds., "Chemical Reactions in Complex Systems." Van Nostrand Reinhold, New York, 1991.

Scaramella, R., Cicarelli, P., and Astarita, G., Continuous kinetics of bimolecular systems. In "Kinetic and Thermodynamic Lumping of Multicomponent Mixtures" (G. Astarita and R. I. Sandler, eds.). Elsevier, Amsterdam, 1991, p. 87.

Schlijper, A. G., Flash calculations for polydisperse fluids: A variational approach. *Fluid Phase Eq.* **34**, 149 (1987).

Schlijper, A. G., and Van Berge, A. R. D. A free energy criterion for the selection of pseudocomponents for vapour/liquid equilibrium calculations. In "Kinetic and Thermodynamic Lumping of Multicomponent Mixtures" (G. Astarita and R. I. Sandler, eds.). Elsevier, Amsterdam, 1991.

Schlosser, P. M., A graphical determination of the possibility of multiple steady states in complex isothermal CFSTR's. Ph.D. Thesis, University of Rochester (1988).

Schlosser, P. M., and Feinberg, M., A theory of multiple steady states in isothermal homogeneous CFSTR's with many reactions. *Chem. Eng. Sci.* **49**, 1749 (1994).

Schlosser, P. M., and Feinberg, M., A graphical determination of the possibility of multiple steady states in complex isothermal CFSTR's. In "Complex Chemical Reaction Systems: Mathematical Modeling and Simulation" (J. Warnatz, W. Jaeger, eds.). Springer-Verlag, Heidelberg, 1987.

Sellers, P., Algebraic complexes which characterize chemical networks. *SIAM J. Appl. Math.* **15**, 13 (1967).

Shapiro, N. Z., and Shapley, L. S., Mass action laws and the Gibbs free energy function. *SIAM J. Appl. Math.* **13**, 353 (1965).

Shibata, S. K., Sandler, S. I., and Behrens, R. A., Phase equilibrium calculations for continuous and semicontinuous mixtures. *Chem. Eng. Sci.* **42**, 1977 (1987).

Shinnar, R., Residence-time distributions and tracer experiments in chemical reactor design: The power and usefulness of a 'wrong' concept. *Rev. Chem. Eng.* **9**, 97 (1993).

Shinnar, R., Thermodynamic analysis in chemical processes and reactor design. *Chem. Eng. Sci.* **43**, 2303 (1988).

Shinnar, R., and Feng, C. A., Structure of complex catalytic reactions: Thermodynamic constraints in kinetic modeling and catalyst evaluation. *Ind. Eng. Chem. Fundament.* **24**, 153 (1985).

Shinnar, R., Glasser, D., and Katz, S., First order kinetics in continuous reactors. *Chem. Eng. Sci.* **28**, 617 (1973).

Shinnar, R., Naor, P., and Katz, S., Interpretation and evaluation of multiple tracer experiments. *Chem. Eng. Sci.* **27**, 1627 (1972).

Shinnar, R., and Rumschitzki, D., Tracer experiments and RTD's in heterogeneous reactor analysis and design. *AIChE J.* **35**, 1651 (1989).

Silverstein, J. L., and Shinnar, R., Effect of design on the stability and control of fixed bed catalytic reactors. 1, Concepts. *Ind. Eng. Chem. Proc. Des. Dev.* **21**, 241 (1982).

Silverstein, J. L., and Shinnar, R., Design of fixed bed catalytic microreactors. *Ind. Eng. Chem. Proc. Des. Dev.* **14**, 127 (1975).

Silvestri, A. J., Prater, C. D., and Wei, J., On the structure and analysis of complex systems of first order chemical reactions containing irreversible steps. III, Determination of the rate constants. *Chem. Eng. Sci.* **25,** 407 (1970).

Silvestri, A. J., Prater, C. D., and Wei, J., On the structure and analysis of complex systems of first order chemical reactions containing irreversible steps. II, Projection properties of the characteristic vectors. *Chem. Eng. Sci.* **23,** 1191 (1968).

Slaughter, D. W., and Doherty, M. F., Calculation of solid–liquid equilibrium and crystallization paths for melt crystallization processes. Submitted for publication (1994).

Smith, W. R., Chemical and phase equilibria calculations for complex mixtures. *In* "Chemical Reactions in Complex Systems" (A. V. Sapre and F. J. Krambeck, eds.). Van Nostrand Reinhold, New York, 1991, p. 298.

Smith, W. R., and Missen, R. W., "Chemical Reaction Equilibrium Analysis." Wiley, New York, 1982.

Stroud, A. J., and Secrest, D., "Gaussian Quadrature Formulas." Prentice-Hall, Englewood Cliffs, NJ, 1966.

Tricomi, F. G., "Funzioni Ipergeometriche Confluenti." Ediz. Cremonese, Rome, 1954.

Truesdell, C. A., "Rational Thermodynamics," 2nd ed. McGraw-Hill, New York, 1984.

Truesdell, C. A., "Rational Thermodynamics." McGraw-Hill, New York, 1969.

Van Heerden, C., The character of the stationary state for exothermic processes. *Chem. Eng. Sci.* **10,** 133 (1958).

Vynckler, E., and Froment, G. F., Modeling of the kinetics of complex processes based upon elementary steps. *In* "Kinetic and Thermodynamic Lumping of Multicomponent Mixtures" (G. Astarita and R. I. Sandler, eds.). Elsevier, Amsterdam, 1991, p. 131.

Weekman, V. W., Lumps, models and kinetics in practice. *Chem. Eng. Prog. Monog. Ser.* **75**(11), 3 (1979).

Weekman, V. W., and Nace, D. M., Kinetics of catalytic cracking selectivity in fixed, moving and fluid-bed reactors. *AIChE J.* **16,** 397 (1970).

Wegsheider, R., Uber simultane gleichgewichte und die beziehungen zwischen thermodynamik und reactionskinetik homogener systeme. *Z. Phys. Chem.* **39,** 257 (1902).

Wei, J., Structure of complex reaction systems. *Ind. Eng. Chem. Fundament.* **4,** 161 (1965).

Wei, J. An axiomatic treatment of chemical reaction systems. *J. Chem. Phys.* **36,** 1578 (1962a).

Wei, J., Intraparticle diffusion effects in complex systems of first order reactions. II, The influence of diffusion on the performance of chemical reactors. *J. Catal.* **1,** 526 (1962b).

Wei, J., Intraparticle diffusion effects in complex systems of first order reactions. I, The effects in single particles. *J. Catal.* **1,** 538 (1962c).

Wei, J., and Kuo, C. W., A lumping analysis in monomolecular reaction systems. *Ind. Eng. Chem. Fundament.* **8,** 114 (1969).

Wei, J., and Prater, C., The structure and analysis of complex reaction systems. *Adv. Catal.* **16,** 203 (1962).

Weinhold, F., Metric geometry of equilibrium thermodynamics. *J. Chem. Phys.* **63,** 2479 (1975).

Weinstein, H., and Adler, R. J., Micromixing effects in continuous chemical reactors. *Chem. Eng. Sci.* **22,** 67 (1967).

White, B. S., Ho, T. C., and Li, H. Y., Lumped kinetics of many irreversible bimolecular reactions. *Chem. Eng. Sci.* **49,** 781 (1994).

Ying, X., Ye, R., and Hu, Y., Phase equilibria for complex mixtures. Continuous thermodynamics method based on spline fit. *Fluid Phase Eq.* **53,** 407 (1989).

Zwietering, Th. N., The degree of mixing in continuous flow systems. *Chem. Eng. Sci.* **11,** 1 (1959).

COMBUSTION SYNTHESIS OF ADVANCED MATERIALS: PRINCIPLES AND APPLICATIONS

Arvind Varma, Alexander S. Rogachev[1], Alexander S. Mukasyan, and Stephen Hwang

Department of Chemical Engineering
University of Notre Dame
Notre Dame, Indiana 46556

I. Introduction	81
II. Methods for Laboratory and Large-Scale Synthesis	84
A. Laboratory Techniques	84
B. Production Technologies	87
III. Classes and Properties of Synthesized Materials	96
A. Gasless Combustion Synthesis From Elements	96
B. Combustion Synthesis in Gas–Solid Systems	107
C. Products of Thermite-Type SHS	115
D. Commercial Aspects	117
IV. Theoretical Considerations	120
A. Combustion Wave Propagation Theory in Gasless Systems	120
B. Microstructural Models	127
C. Cellular Models	130
D. Stability of Gasless Combustion	135
E. Filtration Combustion Theory	138
F. Other Aspects	149
V. Phenomenology of Combustion Synthesis	151
A. Thermodynamic Considerations	152
B. Dilution	158
C. Green Mixture Density for Gasless Systems	162
D. Green Mixture Density and Initial Gas Pressure for Gas–Solid Systems	165
E. Particle Size	169
F. Other Effects of Combustion Conditions	173
VI. Methods and Mechanisms for Structure Formation	180
A. Major Physicochemical Processes Occurring during Combustion Synthesis	182
B. Quenching of the Combustion Wave	183
C. Model Systems for Simulation of Reactant Interaction	190

[1] Permanent address: Institute of Structural Macrokinetics, Russian Academy of Sciences, Chernogolovka, 142 432 Russia.

D. Time-Resolved X-ray Diffraction (TRXRD) 195
E. Microstructure of Combustion Wave 197
F. Concluding Remarks 203
Nomenclature 206
References 209

Combustion synthesis is an attractive technique to synthesize a wide variety of advanced materials including powders and near-net shape products of ceramics, intermetallics, composites, and functionally graded materials. This method was discovered in the former Soviet Union by Merzhanov et al. (1971). The development of this technique by Merzhanov and coworkers led to the appearance of a new scientific direction that incorporates both aspects of combustion and materials science. At about the same time, some work concerning the combustion aspects of this method was also done in the United States (Booth, 1953; Walton and Poulos, 1959; Hardt and Phung, 1973). However, the full potential of combustion synthesis in the production of advanced materials was not utilized. The scientific and technological activity in the field picked up in the United States during the 1980s. The significant results of combustion synthesis have been described in a number of review articles (e.g., Munir and Anselmi-Tamburini, 1989; Merzhanov, 1990a; Holt and Dunmead, 1991; Rice, 1991; Varma and Lebrat, 1992; Merzhanov, 1993b; Moore and Feng, 1995). At the present time, scientists and engineers in many other countries are also involved in research and further development of combustion synthesis, and interesting theoretical, experimental, and technological results have been reported from various parts of the world (see SHS Bibliography, 1996).

This review article summarizes the state of the art in combustion synthesis, from both the scientific and technological points of view. In this context, we discuss wide-ranging topics including theory, phenomenology, and mechanisms of product structure formation, as well as types and properties of product synthesized, and methods for large-scale materials production by combustion synthesis technique.

I. Introduction

Combustion synthesis (CS) can occur by two modes: *self-propagating high-temperature synthesis* (SHS) and *volume combustion synthesis* (VCS). A schematic diagram of these modes is shown in Fig. 1. In both cases, reactants may be pressed into a pellet, typically cylindrical in shape. The samples are then heated by an external source (e.g., tungsten coil, laser) either locally (SHS) or uniformly (VCS) to initiate an exothermic reaction.

The characteristic feature of the SHS mode is, after initiation locally, the self-sustained propagation of a reaction wave through the heterogeneous mixture of reactants. The temperature of the wavefront can reach quite high values (2000–4000 K). In principle, if the physicochemical parameters of the medium, along with the instantaneous spatial distributions of temperature and concentration are known, we can calculate the combustion velocity and reaction rate throughout the mixture. Thus, the SHS mode of reaction can be considered to be a well-organized wavelike propagation of the exothermic chemical reaction through a heterogeneous medium, followed by the synthesis of desired condensed products.

During volume combustion synthesis, the entire sample is heated uniformly in a controlled manner until the reaction occurs essentially simultaneously through-

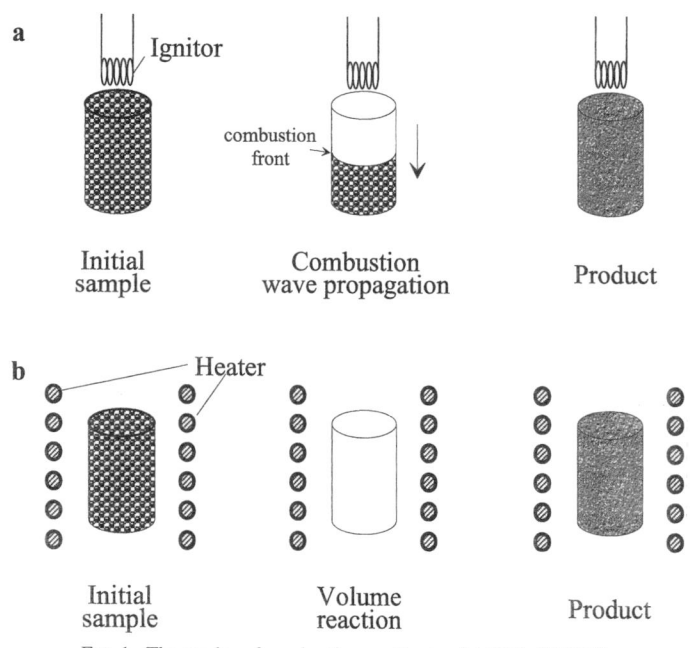

FIG. 1. The modes of combustion synthesis : (a) SHS; (b) VCS.

out the volume. This mode of synthesis is more appropriate for weakly exothermic reactions that require preheating prior to ignition, and is sometimes referred to as the *thermal explosion* mode. However, the term *explosion* used in this context refers to the rapid rise in temperature after the reaction has been initiated, and not the destructive process usually associated with detonation or shock waves. For this reason, volume combustion synthesis is perhaps a more appropriate name for this mode of synthesis.

From the viewpoint of chemical nature, three main types of CS processes can be distinguished. The first, *gasless combustion synthesis from elements,* is described by the equation

$$\sum_{i=1}^{n} X_i^{(s)} = \sum_{j=1}^{m} P_j^{(s,l)} + Q, \tag{1}$$

where $X_i^{(s)}$ are elemental reactant powders (metals or nonmetals), $P_j^{(s,l)}$ are products, Q is the heat of reaction, and the superscripts (s) and (l) indicate solid and liquid states, respectively. Perhaps the most popular example of this type of reaction is carbidization of titanium:

$$\text{Ti} + \text{C} = \text{TiC} + 230 \text{ kJ/mol}.$$

The second type, called *gas–solid combustion synthesis,* involves at least one gaseous reagent in the main combustion reaction:

$$\sum_{i=1}^{n-p} X_i^{(s)} + \sum_{i=1}^{p} Y_i^{(g)} = \sum_{j=1}^{m} P_j^{(s,l)} + Q, \tag{2}$$

where $Y_i^{(g)}$ represents the gaseous reactants (e.g., N_2, O_2, H_2, CO), which, in some cases, penetrate the sample by infiltration through its pores. This type of CS is also called *infiltration* (or *filtration*) *combustion synthesis.* Nitridation of titanium and silicon are common examples:

$$\text{Ti}^{(s)} + 0.5\text{N}_2^{(g)} = \text{TiN}^{(s)} + 335 \text{ kJ/mol},$$

$$3\text{Si}^{(s)} + 2\text{N}_2^{(g)} = \text{Si}_3\text{N}_4^{(s)} + 750 \text{ kJ/mol}.$$

The third main type of CS is *reduction combustion synthesis,* described by the formula

$$\sum_{i=1}^{n-q-r} (MO_x)_i^{(s)} + \sum_{i=1}^{r} Z_i^{(s)} + \sum_{i=1}^{q} X_i^{(s)} = \sum_{j=1}^{m-k} P_j^{(s,l)} + \sum_{j=1}^{k} (ZO_y)_j^{(s,l)} + Q, \tag{3}$$

where $(MO_x)_i^{(s)}$ is an oxide that reacts with a reducing metal $Z_i^{(s)}$ (e.g., Al, Mg, Zr, Ti), resulting in the appearance of another, more stable oxide $(ZO_y)_j^{(s,l)}$, and reduced metal $M_i^{(s,l)}$. This reaction may be followed by the interaction of $M_i^{(s,l)}$ with other elemental reactants $X_i^{(s)}$ to produce desired products $P_j^{(s,l)}$. Thus, in general, the reduction combustion synthesis can be considered to be a two-step process, where the first step is a *thermite* reaction:

$$\sum_{i=1}^{n-q-r} (MO_x)_i^{(s)} + \sum_{i=1}^{r} Z_i^{(s)} = \sum_{j=1}^{k} (ZO_y)_j^{(s,l)} + \sum_{j=1}^{l} M_j^{(s,l)} + Q_1, \quad (4a)$$

while the second step is the synthesis from elements similar to scheme (1):

$$\sum_{j=1}^{l} M_j^{(s,l)} + \sum_{i=1}^{q} X_i^{(s)} = \sum_{j=1}^{m-1} P_j^{(s,l)} + Q_2, \quad (4b)$$

with the total heat release, $Q = Q_1 + Q_2$. An example of this type of CS is

$$B_2O_3^{(s)} + 2Al^{(s)} + Ti^{(s)} = Al_2O_3^{(l)} + TiB_2^{(s,l)} + 700 \text{ kJ/mol},$$

where TiB_2 is the desired product and Al_2O_3 can be removed (e.g., by centrifugal separation) and used separately or a ceramic composite material ($Al_2O_3 + TiB_2$) can be produced. In some cases, the reducing reactant (Z_i) is the same as that used for the synthesis (X_i), for example,

$$2B_2O_3^{(s)} + 5Ti^{(s)} = 3TiO_2^{(s)} + 2TiB_2^{(s)} + 140 \text{ kJ/mol}.$$

Historically, combustion synthesis (both SHS and VCS) is a direct descendant of classic works on combustion and thermal explosion (e.g., Mallard and Le Chatelier, 1883; Semenov, 1929; Zeldovich and Frank-Kamenetskii, 1938; Williams, 1965; Glassman, 1977); see Hlavacek (1991) and Merzhanov (1995) for additional comments in this regard. We discuss later in Section IV how the theory of SHS grew directly from these works. The progress in combustion science made it possible to organize self-sustained exothermic reactions in powder mixtures that were controllable and predictable, hence avoiding the uncontrollable evolution of the reaction that is commonly associated with the terms *combustion, fire,* and *explosion.*

The number of products synthesized by CS increased rapidly during the 1970s and 1980s, and currently exceeds 400 different compounds (see Section III). Specifically, these materials include carbides (TiC, ZrC, SiC, B$_4$C, etc.), borides (TiB$_2$, ZrB$_2$, MoB$_2$, etc.), silicides (Ti$_5$Si$_3$, TiSi$_2$, MoSi$_2$, etc.), nitrides (TiN, ZrN, Si$_3$N$_4$, BN, AlN), and intermetallics (NiAl, Ni$_3$Al, TiNi, TiAl, CoAl, etc.). The methods used for production are described in Section II and the major products are presented in Section III. The theory of CS is discussed in Section IV, where we emphasize the physical basis of the theoretical approaches. The simplifying assumptions in the existing theories are also discussed in order to illustrate their limitations. The effects of different experimental conditions, including the role of gravity, for controlling the synthesis process to obtain desired products are presented in Section V. The unique aspects of CS, such as extremely high temperatures, fast heating rates, and short reaction times, cause certain problems for control of product microstructure. Thus it is necessary to develop ways of tailoring the microstructure of the product, based on the study of synthesis mechanisms and structure-forming processes in the combustion wave. This aspect of the problem is considered in Section VI, where two important features of the synthesis, product structure formation and combustion wave microstructure, are discussed.

Throughout this work, more emphasis is placed on the SHS mode of synthesis rather than the VCS mode because more information is available for SHS. Also, note that in this review, we do not consider production of powders by gas-phase combustion synthesis processes (e.g., Calcote *et al.*, 1990; Davis *et al.*, 1991).

II. Methods for Laboratory and Large-Scale Synthesis

In this section, we discuss the laboratory techniques and production technologies used for the combustion synthesis process. The laboratory studies reveal details of the CS process itself, while the technologies may also include other processing, such as densification of the product by external forces. In both cases, it is necessary to control the green mixture characteristics as well as the reaction conditions. For the production technologies, however, optimization of parameters related to external postcombustion treatment is also necessary in order to produce materials with desired properties.

The main characteristics of the green mixture used to control the CS process include mean reactant particle sizes, d_i; morphologies and size distribution of the reactant particles; reactant stoichiometry, v_i; initial density, ρ_0; size of the sample, D; initial temperature, T_0; dilution, b, that is, fraction of the inert diluent in the initial mixture; and reactant or inert gas pressure, p. In general, the combustion front propagation velocity, U, and the temperature–time profile of the synthesis process, $T(t)$, depend on all of these parameters. The most commonly used characteristic of the temperature history is the maximum combustion temperature, T_c. In the case of negligible heat losses and complete conversion of reactants, this temperature equals the thermodynamically determined adiabatic temperature T_c^{ad} (see also Section V,A). However, heat losses can be significant and the reaction may be incomplete. In these cases, the maximum combustion temperature also depends on the experimental parameters noted earlier.

A. LABORATORY TECHNIQUES

Laboratory studies in combustion synthesis are generally designed to determine the dependencies of U, $T(t)$, and T_c on the process parameters. A schematic diagram of the apparatus commonly used is shown Fig. 2 (cf. Lebrat *et al.*, 1992, for de-

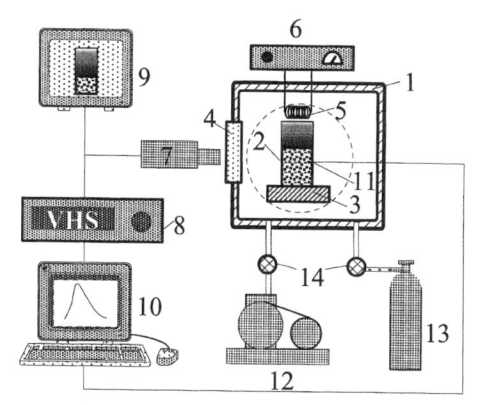

Fig. 2. Laboratory setup for combustion synthesis. 1–reaction chamber; 2–sample; 3–base; 4–quartz window; 5–tungsten coil; 6–power supply; 7–video camera; 8–video cassette recorder; 9–video monitor; 10–computer with data acquisition board; 11–thermocouple; 12–vacuum pump; 13–inert or reactant gas; 14–valve.

tails). The operating pressure in the reaction chamber varies from 150 atm to a vacuum of 10^{-2} torr. The reactant powders are dried, mixed in the appropriate amounts, and pressed to the desired green (i.e., initial) density. The sample is ignited typically by an electrically heated tungsten coil, and once initiated, the combustion wave self-propagates through the sample. The temperature is measured by a pyrometer or by thermocouples imbedded in the sample, while a video or movie camera is used to monitor the propagation of the combustion wave. The main features of the laboratory apparatus have remained essentially unchanged during the last 20 years, with the exception of significant improvements in data acquisition, control, and analysis by the use of computers and imaging by video techniques.

Owing to the large number of experimental parameters, it is not easy to determine the functional dependencies of U, $T(t)$ and T_c. Thus, a full description of these relationships has not been obtained for any system produced by combustion synthesis. Generally, experimental investigations have identified the combustion velocity, U, and temperature, T_c, dependencies on a single parameter (e.g., density, reactant particle size, dilution), while maintaining other parameters fixed. Merzhanov (1983) and later Rice (1991) generalized a large number of experimental results in order to extract some trends of combustion velocity and temperature variations on different experimental parameters.

Based on their analyses, and incorporation of additional details, we have outlined some general relationships for gasless combustion synthesis of materials from elements (type 1), as shown schematically in Fig. 3. Both characteristic features of the process, the combustion wave propagation velocity and maximum temperature, have maximum values when the composition of the green mixture corresponds to the most exothermic reaction for a given system (Fig. 3a). In gen-

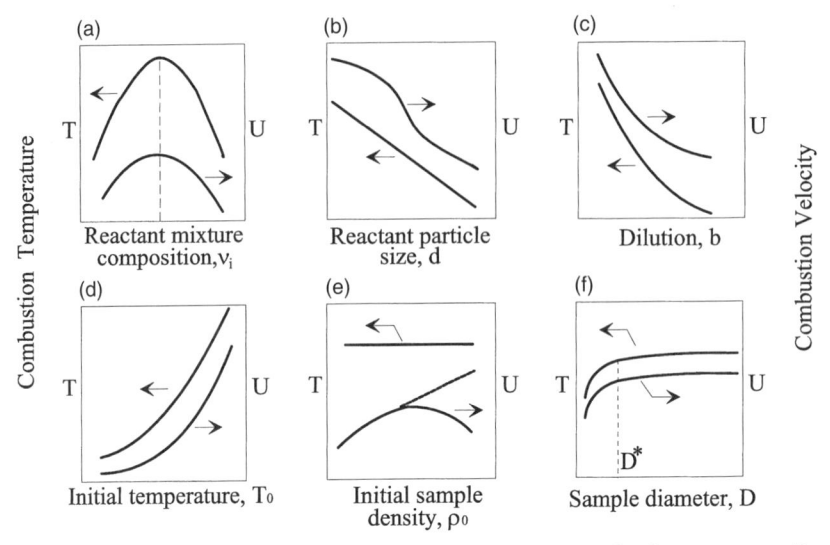

FIG. 3. Dependencies of combustion velocity, U, and maximum combustion temperature T_c on various CS parameters.

eral, U and T_c decrease with increasing initial reactant particle size, and with addition of an inert (nonreactive) diluent to the green mixture (Figs. 3b and c), while increasing significantly with increasing initial sample temperature (Fig. 3d). Different trends have been observed when the initial sample density is varied. With increasing ρ_0, the combustion front velocity either increases monotonically or goes through a maximum, while the combustion temperature generally remains constant (Fig. 3e). A decrease in the sample size (e.g., sample diameter, D) does not influence U and T_c when the size is larger than a critical value D^*, since heat losses are negligible compared to heat release from the chemical reaction. Below the critical sample size, both the combustion velocity and temperature decrease due to significant heat losses (Fig. 3f). Note that many exceptions to the dependencies discussed have been observed, even for the simplest case of gasless combustion synthesis from elements. The combustion wave behavior becomes more complicated in gas–solid and reduction type reactions. All of these effects are discussed in greater detail in Section V.

Within the region of optimal experimental parameters, the combustion wave velocity remains constant and the temperature profile $T(t)$ has the same form at each point of the reaction medium. This regime is called *steady propagation* of the combustion synthesis wave, or *steady SHS process*. As the reaction conditions move away from the optimum, where the heat evolution decreases and/or heat losses increase, different types of *unsteady propagation* regimes have been observed. These include the appearance of an *oscillating combustion synthesis* regime, where macroscopic oscillations of the combustion velocity and temperature occur. The reaction may also propagate in the form of a hot-spot, which, for example, may move along a spiral pattern in cylindrical samples, and is called the *spin combustion* regime of CS. The combustion regime has great importance in the production of materials, because it influences the product microstructure and properties.

B. Production Technologies

In general, methods for the large-scale production of advanced materials by combustion synthesis consist of three main steps: (1) preparation of the green mixture, (2) high-temperature synthesis, and (3) postsynthesis treatment. A schematic diagram of these steps is presented in Fig. 4. The first step is similar to

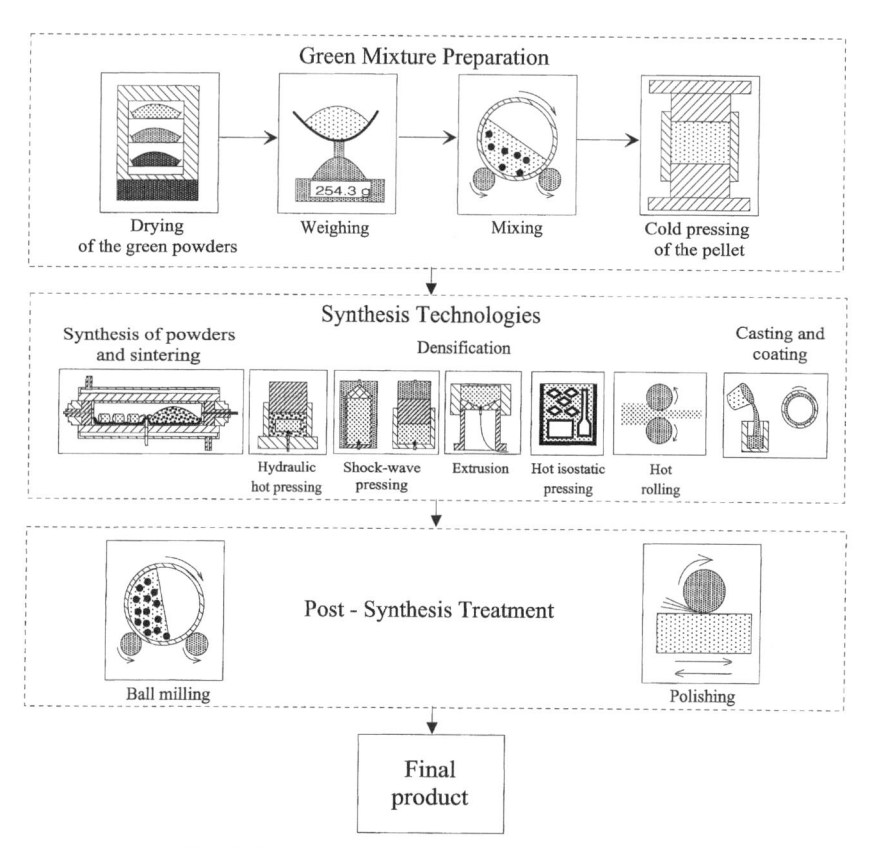

Fig. 4. Generalized schematic diagram of CS technologies.

the procedures commonly used in powder metallurgy, where the reactant powders are dried (e.g., under vacuum at 80–100°C), weighed into the appropriate amounts, and mixed (e.g., by ball mixing). For some applications, cold pressing of the green mixture is necessary, especially for the production of low porosity or poreless materials. Typically, no plasticizer is used, and the porosity of the cold-pressed compacts varies from 40 to 80% of the theoretical density for metal–nonmetal mixtures, and up to 90% for metal–metal mixtures. The final procedure in sample preparation determines the type of product to be synthesized: a powder product results from uncompacted powder reactants, while sintered products are yielded from cold-pressed compacts. Pressing the green mixture into special molds or machining pressed initial compacts yields complex-shaped articles.

The main production technologies of combustion synthesis are presented in the second block of Fig. 4. Following Merzhanov (1990a), they may be classified into several major types: powder production and sintering, densification, and casting and coating.

The volume combustion synthesis mode is used primarily for the synthesis of weakly exothermic systems. Various types of heaters, mostly commercially available furnaces, in addition to spiral coil and foil heaters, are used to preheat the sample up to the ignition point. To date, VCS synthesized materials have been produced only in laboratories, and no industrial or pilot production by this mode of synthesis has been reported.

The third main step of combustion synthesis technologies is postsynthesis treatment. This step is optional, since not all products require additional processing after synthesis. Powder milling and sieving are used to yield powders with a desired particle size distribution. Annealing at elevated temperatures (800–1200°C) removes residual thermal stress in brittle products. The synthesized materials and articles may also be machined into specified shapes and surface finishes.

1. Powder Production and Sintering

The design of a typical commercial reactor for large-scale production of materials is similar to the laboratory setup, except that the capacity of the former is larger, up to 30 liters. Since the synthesis of materials produced commercially is well understood, most reactors are not equipped with optical windows to monitor the process. A schematic diagram of such a reactor is shown in Fig. 5. Typically, it is a thick-walled stainless steel cylinder that can be water cooled (Borovinskaya *et al.,* 1991). The green mixture or pressed compacts are loaded inside the vessel, which is then sealed and evacuated by a vacuum pump. After this, the reactor is filled with inert or reactive gas (Ar, He, N_2, O_2, CO, CO_2). Alternatively, a constant flow of gas can also be supplied at a rate such that it permeates the porous reactant mixture. The inner surface of the reactor is lined with an inert material to

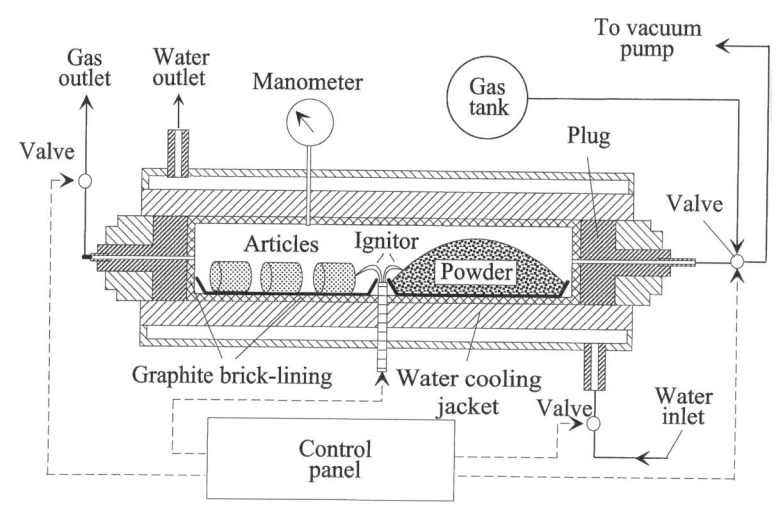

FIG. 5. Schematic diagram of an SHS reactor.

protect the vessel from the extreme reaction temperatures. Graphite is typically used for lining during carbide, boride, or silicide synthesis, while boron nitride and silicon nitride provide protection during nitride synthesis.

Two different types of reactors are used depending on the product synthesized. The first type can maintain pressures up to 150 atm, and is widely used for production of powders in gasless and gas–solid systems. Carbides, borides, silicides, intermetallics, chalcogenides, phosphides, and nitrides are usually produced in this type of reactor. The second type, a high-pressure reactor (up to 2000 atm), is used for the production of nitride-based articles and materials, since higher initial sample densities require elevated reactant gas pressures for full conversion. For example, well-sintered pure BN ceramic with a porosity of about 20–35% was synthesized at 100 to 5000-atm nitrogen pressure (Merzhanov, 1992). Additional examples are discussed in Section III.

2. *SHS with Densification*

The application of an external force during or after combustion is generally required to produce a fully dense (i.e., poreless) material. A variety of techniques for applying the external force, such as static, dynamic, and shock-wave loading have been investigated.

The oldest method uses relatively slow (static) loading provided by a hydraulic press along with a specially designed die (Borovinskaya *et al.*, 1975b; Miyamoto *et al.*, 1984; Adachi *et al.*, 1989; Zavitsanos *et al.*, 1990; Dunmead *et al.*, 1990a). Owing to the high temperature of the combustion products being densified, several approaches have been used to isolate the reacting sample from the die. One possible solution is the use of steel dies lined with graphite or BN ceramics (Nishida and Urabe, 1992). Another possibility is to use a pressure-transmitting medium such as SiO_2 (sand) or Al_2O_3, as shown in Fig. 6a (Merzhanov et al., 1981).

In this technique, pressure is applied immediately after the combustion process, or after a short time delay (typically a few seconds). The duration of this time delay is extremely important for achieving maximum density. The period of time before application of pressure should be long enough to allow the expulsion of gases from the sample, but shorter than the cooling time. An example of this critical time is given in Fig. 7a, which shows the dependence of residual porosity on the pressing delay time for TiC-Cr_2C_3-Ni cemented carbide synthesis. The effect of applied pressure on residual porosity is clearly important, and is illustrated for TiC-(Ni,Al) material in Fig. 7b. Essentially poreless (porosity less

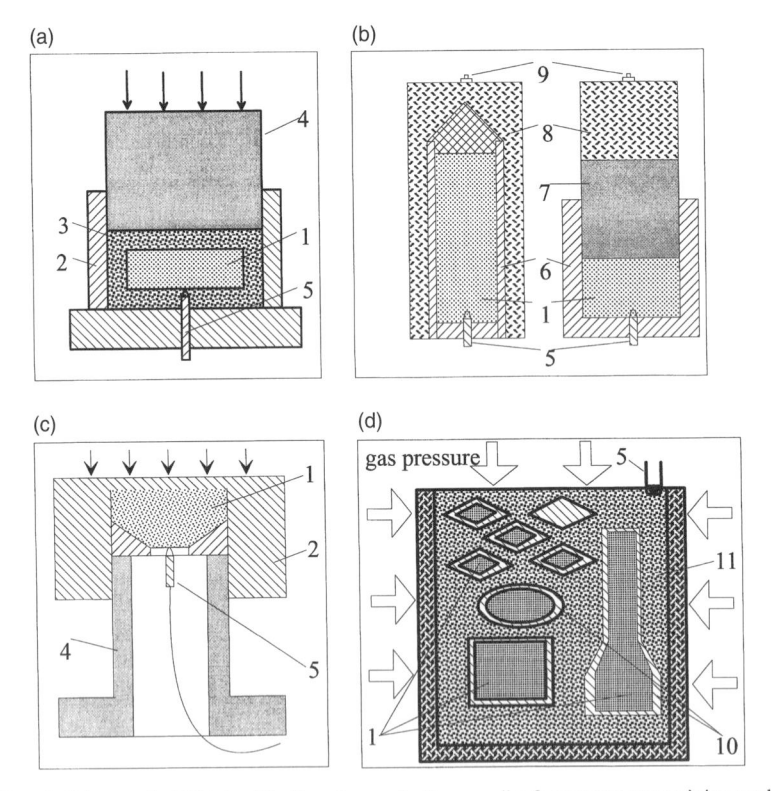

FIG. 6. Schemes for SHS densification. 1–sample; 2–press die; 3–pressure-transmitting medium; 4–pressing body; 5–ignitor; 6–metal container; 7–massive piston; 8–explosive; 9–electric fuse; 10–glass containers; 11–"chemical furnace" mixture.

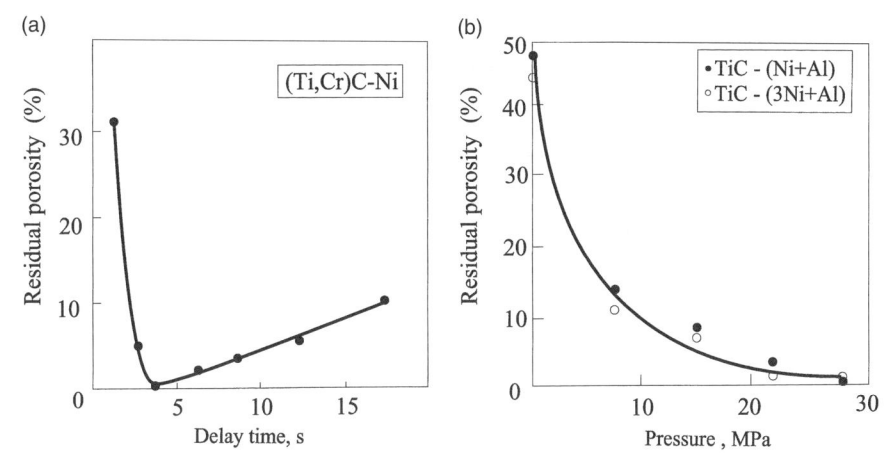

FIG. 7. Residual porosity for different samples as a function of (a) pressing delay time (Merzhanov, 1990a), (b) applied pressure (Adapted from Dunmead et al., 1990a).

than 1%) and large-scale (up to 1 m in diameter) ceramic and metal–ceramic composite materials can be produced by SHS using the static pressing method (Kvanin et al., 1993).

Another method of SHS densification involves high-speed loading by an explosion shock wave or by a fast-moving solid body. Two possible schemes for high-speed loading by shock waves are presented in Fig. 6b; the first method provides axial and the other radial densification. The dynamic axial densification method has been used to synthesize dense materials including TiC, TiB$_2$, and HfC (Kecskes et al., 1990; Grebe et al., 1992; Rabin et al., 1992a; Adadurov et al., 1992). The radial shock-wave densification technique was applied to produce several materials, including dense high-temperature ceramic superconductor YBa$_2$Cu$_3$O$_{7-x}$ (Gordopolov and Merzhanov, 1991; Fedorov et al., 1992).

The pressing body can also be accelerated by gas pressure and mechanical or electromagnetic forces. High gas pressures driving a mechanical piston have been applied in a number of SHS systems (Hoke et al., 1992; LaSalvia et al., 1994). In this case, the sample under combustion is impacted by a massive piston moving at a speed of ~10 m/s. Based on the magnitude of this speed, this method of compaction is between static pressing (~0.1 m/s) and shock-wave pressing (~1000 m/s). Materials produced by this method include ceramics (TiC, TiB$_2$), ceramic composites (TiB$_2$+Al$_2$O$_3$, TiB$_2$+BN, TiB$_2$+SiC), ceramic–metal composites (TiC+Ni, TiB$_2$+Ni), and intermetallics (TiNi, Ti$_3$Al). Electromagnetic forced pressing was designed for the pulse densification of cermets with a uniform or gradient distribution of the binder metal (Matsuzaki et al., 1990).

In the techniques just described, combustion synthesis was first initiated, and then dynamic loading was applied, to the hot product immediately after the combustion front propagated through the sample. Alternatively, shock compression applied to the reaction mixture may result in a rapid increase in temperature, initiating the chemical reaction with supersonic propagation rates. This approach, called *shock-induced synthesis,* has recently been developed (Work *et al.*, 1990; Vecchio *et al.*, 1994; Meyers *et al.*, 1994). A variety of dense silicides (e.g., $MoSi_2$, $NbSi_2$, Ti_5Si_3, $TiSi_2$) and aluminides (NiAl, $NiAl_3$) have been synthesized using this method. Although the boundary between combustion and shock-induced synthesis is not well defined, the difference between them is addressed in a review article that describes different shock-induced methods to synthesize materials (Thadani, 1993).

Another method used for the production of fully dense materials by combustion synthesis is SHS with extrusion, shown schematically in Fig. 6c (Podlesov *et al.*, 1992a,b). In this case, a powder compact of the reaction mixture is placed in the mold, and the process is initiated locally by a heated tungsten wire. After the combustion wave has propagated through the sample, relatively low pressure (<1000 MPa) is applied to the plunger, extruding the products through the hole of the conic die. A high plasticity at the elevated reaction temperature allows the formation of long rods of refractory materials. The form and size of the die hole determine the extruded product configuration. The most developed application of SHS extrusion to date is the production of TiC-based cermet electrodes used for electric spark alloying (Podlesov *et al.*, 1992b).

A promising method for densification of combustion-synthesized products is a combination of SHS with hot isostatic pressing (HIP). This idea was first applied to the synthesis of TiB_2 ceramics under a pressure of 3 GPa, which was provided by a cubic anvil press, resulting in 95% dense material (Miyamoto *et al.*, 1984; Yamada *et al.*, 1987). The relatively low exothermic reaction of SiC from elemental powders was also carried out under these conditions (Yamada *et al.*, 1985), and 96% conversion to β-SiC was achieved as compared to 36% conversion when the reaction was initiated locally.

Another approach is to use high gas pressure for densification of the product *simultaneously* with combustion synthesis. The method of SHS+HIP with pressing by gas has been developed by using a so-called *chemical furnace.* A schematic drawing of this *gas pressure combustion sintering* method is shown in Fig. 6d (Miyamoto *et al.*, 1984; Koizumi and Miyamoto, 1990). The green mixture is placed in evacuated glass containers, which are surrounded by a highly combustible mixture (e.g., Ti+C) that is enveloped by high-pressure gas (e.g., Ar at 100 MPa). After ignition, the combustible mixture acts as a chemical furnace, which heats the samples in the containers up to their ignition point. The heat evolved from the chemical furnace also heats the glass to its softening temperature, where it becomes plastic and easily deformable. Thus the material synthesized inside the containers is then pressed isostatically by the surrounding high gas pressure to zero porosity. A variety of ceramics, cermets, and functionally graded materials (FGMs) have been produced using this method, including TiC, TiB$_2$, TiB$_2$-Ni FGM (Miyamoto et al., 1990a,b), MoSi$_2$-SiC/TiAl FGM (Matsuzaki et al., 1990), and Cr$_3$C$_2$-Ni (Tanihata et al., 1992).

Along with the various methods of combustion product densification, the hot rolling technique has been investigated (Rice et al., 1986; Osipov et al., 1992). It was shown that simultaneous synthesis and hot rolling of intermetallic and ceramic materials (e.g., TiAl, TiC$_{0.47}$, TiC$_x$N$_{1-x}$-Ni) under vacuum yields articles with porosity in the range of 5 to 50% (Osipov et al., 1992). We expect that with further development of this technique, thin sheets and foils of combustion synthesized materials could be produced in the future.

3. SHS with Casting

During combustion synthesis, highly exothermic reactions (typically reduction-type) result in completely molten products, which may be processed using common metallurgical methods. Casting of CS products under inert gas pressure or centrifugal casting has been used to synthesize cermet ingots, corrosion- and wear-resistant coatings, and ceramic-lined pipes.

Casting under gas pressure is similar to conventional SHS production (see Section II,B,1). The reduction-type initial mixture (e.g., CrO$_3$+Al+C, WO$_3$+Al+C) is placed in a casting die and the reaction initiated under an inert gas pressure (0.1–5 MPa) to prevent product sputtering by gas evolution from the thermite reaction (Merzhanov et al., 1980; Yukhvid, 1992). Increased gas pressure was shown to decrease sputtering significantly, and subsequent product loss (see Fig. 8).

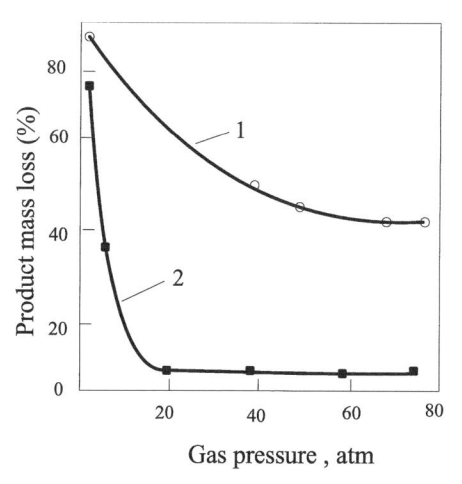

FIG. 8. The product mass loss as a function of Ar gas pressure. 1–CrO$_3$-Al-C; 2–WO$_3$-CoO-Al-C (Adapted from Yukhvid et al., 1983).

In some cases, the molten product may consist of two immiscible phases, either molten oxides (e.g., Al_2O_3, MgO) with dispersed nonoxides (e.g., metals carbides or borides) or molten metals with dispersed ceramic droplets. Owing to the difference in densities, the two phases may separate. By controlling the characteristic times of cooling (t_{cool}) and phase separation (t_{ps}), different distributions of phases may result. Thus, it is possible to obtain cermets with uniform ($t_{cool} \ll t_{ps}$), gradient ($t_{cool} \sim t_{ps}$), and layered ($t_{cool} \gg t_{ps}$) products.

Centrifugal casting allows for greater control of the distribution of phases, by controlling the time of separation. The two types of centrifugal casting equipment are shown in Fig. 9. For *radial centrifuges* (Fig. 9a), the sample is placed at a

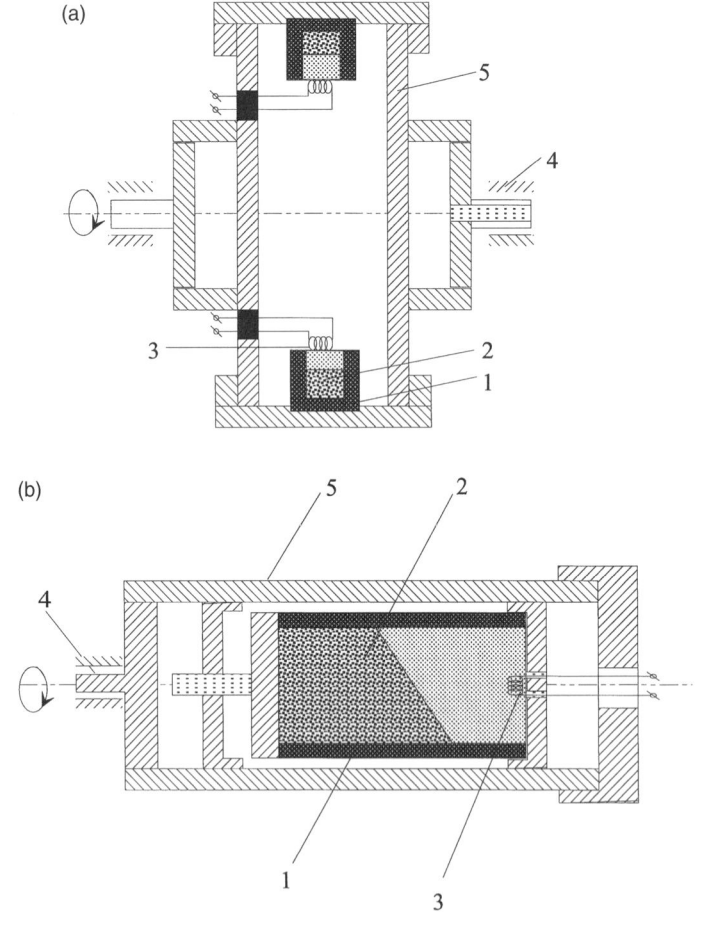

FIG. 9. Scheme of SHS+centrifugal casting: (a) radial centrifuge; (b) axial centrifuge. 1–sample container; 2–reactant mixture; 3–ignitor; 4–axle; 5–reactor. (Adapted from Merzhanov and Yukhvid, 1990)

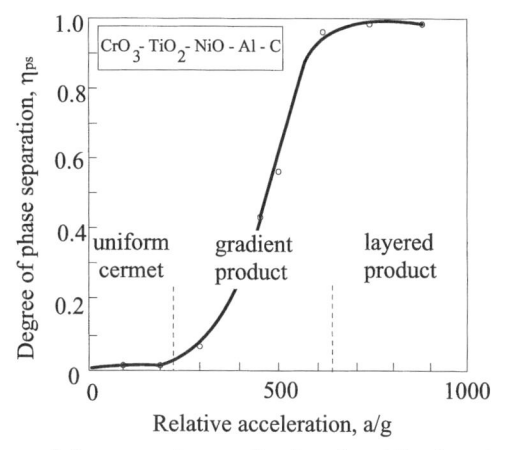

FIG. 10. The degree of phase separation as a function of centrifugal acceleration, a, where g is acceleration due to gravity (Adapted from Merzhanov and Yukhvid, 1990).

fixed radial position from the axis of rotation, and the applied centrifugal force is parallel to the direction of propagation. The influence of centrifugal acceleration on the degree of phase separation, η_{ps}, is shown in Fig. 10. This parameter characterizes the phase distribution in the final product:

$$\eta_{ps} = m_s/m_t,$$

where m_s is mass of the nonoxide phase (e.g., metal, carbide, boride) fully separated from the oxide matrix, and m_t is the total mass of this phase produced by the reaction. Thus, $\eta_{ps} = 0$ for the cermet product with uniformly distributed oxide and nonoxide phases, while $\eta_{ps} = 1$ for multilayer materials or when the metal-type ingot is totally separated from the oxide slag layer. Finally, the ingots have a gradient distribution of phases when $0 < \eta_{ps} < 1$ (see Fig. 10).

The second type of centrifugal casting apparatus, called an *axial centrifuge* (Fig. 9b) is used for production of ceramic, cermet, or ceramic-lined pipes. The axial centrifuge casting method was developed further for production of long pipes with multilayer ceramic inner coatings (Odawara and Ikeuchi, 1986; Odawara, 1992). In this process, a thermite-type SHS mixture (e.g., $Fe_2O_3/2Al$) is placed inside a rotating pipe, and ignited locally. A reduction-type combustion reaction propagates through the mixture, and the centrifugal force results in separate layers of metal and ceramic oxide, with the latter forming the innermost layer. The process is carried out in air under normal pressure, and pipes up to 5.5 m long have been obtained.

III. Classes and Properties of Synthesized Materials

The synthesis of several hundred materials, by both the SHS and VCS modes, has been reported in the literature. The types of compounds produced include carbides, borides, intermetallics, silicides, aluminides, composites, nitrides, hydrides, and oxides. The purpose of this section is to provide a description of the synthesized compounds, as well as the materials (i.e., powders, poreless materials, and functionally graded materials) and articles produced. The practical applications of experimental and technological methods are also described.

A. GASLESS COMBUSTION SYNTHESIS FROM ELEMENTS

1. Carbides

a. Group IV Metal Carbides. Group IVa transition metal carbides (TiC, ZrC, HfC) were among the first compounds synthesized by the SHS method (Merzhanov and Borovinskaya, 1972). With high melting points, hardness, and chemical stability, these materials can be used as abrasives (economical substitutes for diamond powders and pastes) or as components of cermet and ceramic materials (Storms, 1967; Samsonov, 1964).

These compounds have relatively high heats of formation (e.g., -185 kJ/mol for TiC and -198 kJ/mol for ZrC). The heat evolved upon initiation of the reaction allows the materials to be synthesized in the SHS mode, with a high degree of conversion. For example, the amount of unreacted carbon has been reported to be as low as 0.09 wt % for TiC and 0.01 wt % for ZrC and HfC (Merzhanov and Borovinskaya, 1972). Emission spectrochemical analysis of the powders (Holt and Munir, 1986) confirmed the high purity of the combustion synthesized TiC, where the largest impurity constituents were Al (up to 2000 ppm), Ar (300 ppm), Si (100 ppm), Fe (100 ppm), B (80 ppm), and Ca (30 ppm). With the exception of Fe, impurities in the product were two to seven times *lower* than in the initial mixture. Owing to the high temperature of combustion, the material *self-purifies* by purging any volatile impurities from the reaction mixture.

In particular, the Ti-C system has been used commonly in numerous investigations to understand the influence of various process parameters (see Section V), as well as the mechanisms of combustion and product structure formation (see Section VI). In Table I, some properties of TiC powders are presented. While many of the problems associated with producing TiC powders with tailored phase composition and purity have been addressed, this system continues to be investigated further. Recent work confirms and augments earlier results exploring the synthesis of pure TiC during combustion (Chang *et al.,* 1991; Lee and Chung, 1992; Chang *et al.,* 1995).

TABLE I
Some Characteristics of SHS Powders[a]

Powder	Chemical Composition (wt %)	Main Impurities (wt %)	Average Particle Size (μm) or Specific Surface Area (m^2/g)
TiC	C_{fixed} = 19.2–19.5	C_{free} < 0.8	3–5 μm
TiC (titanium carbide abrasive)	Ti = 77.0–79.5 C_{fixed} = 18.0–19.8	C_{free} < 0.2–2.0 Mg = 0.06 Fe = 0.1	polydispersed 1–200 μm
NbC	C_{total} = 11.0	C_{free} < 0.1 O < 0.2	1–100 μm polydispersed
SiC (β-phase)	C_{total} = 27.0–27.8	C_{free} < 0.3–0.8 O < 1.0 Fe < 0.2	8–10 m^2/g
BC$_4$	B_{total} = 74 C_{total} = 20.0	B_2O_3 < 0.5 Mg < 0.5	3.9 m^2/g
TiB$_2$ (polydispersed)	Ti = 69.6 B_{total} = 29.2	O < 0.5	1–200
TiB$_2$ (finely dispersed)	Ti = 66.7 B_{total} = 30.5	B_2O_3 < 0.2 Mg < 0.3	1–10
MoSi$_2$	Mo = 61.9–62.5 Si_{total} = 35.5–36.5	Si_{free} < 0.3 O < 0.3 Fe < 0.1	1–50
TiSi$_2$	Ti = 44.3 Si_{total} = 52.0	O < 0.1 Fe < 0.1	1–100

[a]Data from Borovinskaya et al., (1991) and Merzhanov (1992).

In addition, the high heat formation of titanium carbide and specific features of the Ti-C phase diagram make it possible to produce, by CS, *nonstoichiometric* carbides, TiC$_x$ (x=0.38–1.0) (Shkiro and Borovinskaya, 1975); *self-binding* materials, which consist of nonstoichiometric titanium carbide and titanium located along the carbide grain boundaries; and *cemented* carbides (Borovinskaya et al., 1992b). A wide variety of metal additives have been investigated to produce dense cemented carbides, where combustion-synthesized TiC grains are joined by metal (e.g., up to 60 wt % of Ni or Fe) or intermetallic binders.

Static pressing of hot TiC$_x$ immediately after synthesis yields dense ceramic or cermet (self-binded carbide) materials, with product densities as high as 99% of theoretical (Merzhanov et al., 1981; Holt and Munir, 1986; Borovinskaya et al., 1992b). Dense self-binded TiC$_x$-Ti cermets with residual porosity ~1% were also produced by pseudo HIP (Borovinskaya, 1992; Raman et al., 1995). Applying SHS with isostatic pressing to a Ti+0.42C green mixture, dense (4.54 g/cm^3) cermets with high hardness (86–87 HRA) and bending strength (652 MPa) were synthesized.

The production of TiC-based materials with different binders has been reported: TiC+Ni (Rogachev et al., 1987; Fu et al., 1993b), TiC+(Ni-Al) (Dunmead et al., 1990b; Tabachenko and Kryuchkova, 1993; Mei et al., 1994), TiC+(Ni-Mo) (Borovinskaya et al., 1992b; LaSalvia and Meyers, 1995), TiC+(Ni-Mo-Cu) and (TiC-Cr_3C_2)+Ni (Borovinskaya et al., 1992b), (TiC-Cr_3C_2)+steel and (TiC-TiN)+(Ni-Mo) (Merzhanov, 1992), TiC+(Ti-Al) (Maupin and Rawers, 1993), and TiC-Al (without pressing) (Choi and Rhee, 1993). Some properties of combustion-synthesized and static pressed cemented carbides are listed in Table II.

Dynamic densification has also been used to obtain dense TiC ceramics (Niiler et al., 1990; Meyers et al., 1991; Vecchio et al., 1992; Grebe et al., 1992; Kecskes et al., 1993; Wang et al., 1994a), TiC-Ni (LaSalvia et al., 1994), and (TiC-Cr_3C_2)+Ni cermets (Adadurov et al., 1992). Other densification methods applied to TiC and TiC-based cermets include SHS+HIP (Koizumi and Miyamoto, 1990; Choi et al., 1992), SHS+extrusion (Shishkina et al., 1992) and hot rolling (Rice et al., 1986). The synthesis of TiC and TiC-Fe in the volume combustion mode was also reported (Saidi et al., 1994). Dense ceramic materials of Cr_3C_2 and Cr_3C_2-TiC composites were obtained by the gas-pressure combustion sintering method, with a density of more than 99% of theoretical and flexural strength 625 MPa (Xiangfeng et al., 1992).

TABLE II
Some Properties of SHS Cermet Alloys[a]

Cemented Carbide Alloy	Basic Composition	Density (g/cm^3)	Average Grain Size (μm)	Hardness (HRA)	Ultimate Flex Strength (kg/mm^2)	Application
STIM-1B/3	TiC/TiB$_2$+Cu	4.94	5–7	93.5	70–80	Cutting edges, targets
STIM-2	TiC+Ni	5.50	5–7	90.0	100–120	Wear-resistant coatings
STIM-2A	TiC+(Ni-Mo)	6.40	1–2	87.0	160–180	Press tools
STIM 3B/3	TiC-Cr$_3$C$_2$+Ni	5.37	3–4	92.5	80–100	Cutting edges
STIM-3B	TiC-Cr$_3$C$_2$+steel	5.40	2–4	92.5	100–120	Corrosion-resistant articles
STIM-4	TiB+Ti	4.20	1–2	86.0	100–120	Thermal shock-resistant articles
STIM-5	TiC-TiN+Ni-Mo	5.0	1–2	91.0	120–140	Cutting edges

[a]Data from Borovinskaya et al. (1991).

b. Group Va Carbides. To a lesser extent, combustion synthesis from elements of transition (Group Va) metal carbides has also been investigated, especially for TaC (Shkiro *et al.*, 1978, 1979) and NbC (Martynenko and Borovinskaya, 1975). Some properties of reaction-synthesized NbC powders are presented in Table I.

c. Nonmetal Carbides. Among nonmetallic carbides, SiC synthesized from elemental Si and C powders has attracted the most attention. Silicon carbide has a relatively low heat of formation (69 kJ/mol), and additional heat input is required for combustion to occur. For example, locally igniting pressed Si+C samples, placed on a ribbon heater, yielded ~64% conversion to β-SiC, while increasing the area of ignition with a specially designed carbon-sleeve heater improved the conversion to ~99% (Yamada *et al.*, 1986). Dense SiC ceramics were produced by using the carbon-sleeve heater, combined with hot pressing (Yamada *et al.*, 1985). Another method of heating and ignition is to pass an electric current through the sample, yielding fine stoichiometric SiC powders (Yamada *et al.*, 1987). Fine-grained powders (0.2–0.5 μm) with high specific surface area (6.2 m^2/g) were synthesized in the VCS mode by inductively preheating the Si+C mixture to 1300°C (Pampuch *et al.*, 1987, 1989). Solid SiC and SiC-C blocks were also produced by infiltration of arc-melted silicon into porous carbon, during which the SHS reaction occurred (Ikeda *et al.*, 1990). Some properties of SiC powder synthesized by the SHS method are presented in Table I.

An interesting variation of passing electrical current through the reactant mixture has been reported (Feng and Munir, 1994). In this study, the gradient of the applied electric field was normal to the direction of the combustion front propagation. Since the reaction zone (containing molten silicon) has high electrical conductivity, the orientation of the electric field led to more intense Joule heating in this zone. It was demonstrated that the application of relatively low voltage (9.5–25 V) across samples of height 1.4 cm resulted in about a 10-fold increase in the combustion velocity (from 0.1 to ~1 cm/s).

Kharatyan and Nersisyan (1994) reported an alternative method of activating the reaction between Si and C without preheating. Potassium nitrate (KNO$_3$) is used as an activating agent. With this method, fine (<1-μm) SiC powders containing 28–29 wt % bound and 0.5 wt % free carbon were produced.

2. Borides

It has been demonstrated that most of the transition metal borides can be synthesized in the SHS mode (Holt *et al.*, 1985; Rice *et al.*, 1987; Mei *et al.*, 1992; Fu *et al.*, 1993a; Li, 1995). Early works reported synthesis of relatively pure (with residual unreacted boron less than 0.3 wt %) ZrB$_2$, TiB$_2$, HfB$_2$, and MoB (Merzhanov and Borovinskaya, 1972; Hardt and Holsinger, 1973). Combustion-synthesized borides also include TiB, TaB$_2$, LaB$_6$, and NbB$_2$ (Holt and Dunmead, 1991). Similar to TiC for carbides, the combustion synthesis of transition metal borides from elements has focused primarily on TiB$_2$. Some properties of TiB$_2$ powders obtained by the SHS method are listed in Table I.

The high heat of formation for these borides (e.g., -279 kJ/mol for TiB$_2$ and -160 kJ/mol for TiB) makes it possible to add chemically inert metals and alloys (or weakly exothermic mixtures) to the metal–boron system. The subsequent plasticity of the products at the combustion temperature allows for various methods of densification to be applied, yielding poreless boride ceramics or boride-based cermets. For example, dense TiB-Ti self-binding borides (Borovinskaya et al., 1992b; Shcherbakov et al., 1992) and TiB$_2$ ceramics with 95% theoretical densities (Zavitsanos et al., 1990) have been formed by hydraulic pseudo isostatic pressing and uniaxial pressing in a graphite-lined die, respectively. Dynamic compaction is considered to be another promising method for obtaining TiB$_2$-based ceramics (Niiler et al., 1990; Meyers et al., 1991; Wang et al., 1994a). Other dense boride-based materials were produced by hydraulic pressing of TiB$_2$-Fe immediately after synthesis (Fu et al., 1993a) and by reaction hot pressing of TiB$_2$-Al and TiB$_2$-Ni (DeAngelis, 1990). The hydrostatic pressing method was also applied to synthesize TiB$_2$-TiNi composites with 93% relative density and molar content of TiNi ranging from 0 to 95% (Yanagisawa et al., 1990). Dense TiB$_2$ ceramic materials were formed by high-pressure combustion sintering under 3 GPa in a cubic anvil press (Miyamoto et al., 1984), as well as by gas HIP at 100 MPa using a chemical furnace (Urabe et al., 1990). Synthesis of composite products, such as TiB$_2$-Al (Taneoka et al., 1989) and TiB$_2$-Fe (Fu et al., 1993a), has been reported. Mechanical testing of combustion-synthesized TiB$_2$-Fe composites demonstrates promising properties at elevated temperatures (Andrievski and Baiman, 1992).

3. TiC-TiB$_2$ Composite Ceramics

In some cases, several refractory compounds can result from two or more parallel reactions occurring simultaneously in the combustion wave. A typical example of this type is the Ti-C-B system, where both the Ti+C and Ti+2B reactions affect the combustion synthesis and structure formation processes (Shcherbakov and Pityulin, 1983). By adjusting the contents of carbon and boron powders in the reactant mixture, either carbide- or boride-based ceramics can be obtained.

The Ti-C-B system was used for a series of ceramic poreless composites produced by the SHS+HIP method (Borovinskaya et al., 1992b). Near-eutectic 40TiC-60TiB$_2$ (wt %) compositions were primarily investigated, while various metal additives were tested (e.g., Co, Cu) as the binder phase, with contents varying from 0 to 12 wt %. The materials demonstrated good mechanical properties (Table II) and may be used for cutting tools. Note that material with higher TiC content (80 wt %) possess higher bending strength.

4. Silicides

Compounds of metals with silicon were synthesized from elements by both the SHS (Sarkisyan *et al.*, 1978; Azatyan *et al.*, 1979; Deevi, 1991, 1992) and VCS modes (Trambukis and Munir, 1990; Subrahmanyam, 1994). Powders and sintered samples of various chemical compositions were obtained, including Ti_5Si_3, $TiSi$, $TiSi_2$, Mo_3Si, Mo_5Si_3, $MoSi_2$, Zr_2Si, Zr_5Si_3, $ZrSi$, and $ZrSi_2$. Among them, titanium and molybdenum silicides have received the most attention, due to their high melting points, excellent oxidation resistance, relatively low density, and high electrical and thermal conductivities (Samsonov and Vinitskii, 1980; Rozenkranz *et al.*, 1992; Shah *et al.*, 1992). The properties of some silicide powders are shown in Table I.

Combustion-synthesized $MoSi_2$ powders have been used to produce high-temperature heating elements (Merzhanov, 1990a). The application of the SHS+extrusion method for one-step production of $MoSi_2$ heaters has also been reported (Podlesov *et al.*, 1992a). Further, combustion synthesis of molybdenum alumosilicide (Mo-Al-Si) was demonstrated (Hakobian and Dolukhanyan, 1994).

5. Aluminides

Combustion synthesis of nickel aluminides in the VCS mode was first demonstrated during sintering of a Ni-Al mixture (Naiborodenko *et al.*, 1968). Shortly thereafter, the SHS mode of combustion synthesis was reported for a variety of aluminides, including those of Ni, Zr, Ti, Cr, Co, Mo, and Cu (Naiborodenko *et al.*, 1970, 1982; Naiborodenko and Itin, 1975a,b; Maslov *et al.*, 1976, 1979; Itin *et al.*, 1980).

In this class, most attention has been paid to nickel aluminides (Philpot *et al.*, 1987; Rabin *et al.*, 1990; Lebrat and Varma, 1992a; Lebrat *et al.*, 1992, 1994; Rogachev *et al.*, 1993; Wenning *et al.*, 1994; Alman, 1994) and titanium aluminides (Kuroki and Yamaguchi, 1990; Kaieda *et al.*, 1990a; Ho-Yi *et al.*, 1992; Kachelmyer *et al.*, 1993; Lee *et al.*, 1995; Hahn and Song, 1995). However, synthesis of other aluminides has also been investigated, including aluminides of iron (Rabin and Wright, 1991, 1992; Rabin *et al.*, 1992b), niobium (Maslov *et al.*, 1979; Kachelmyer and Varma, 1994; Kachelmyer *et al.*, 1995), and copper (Wang *et al.*, 1990).

As discussed earlier, aluminides have been used as binders for carbide- and boride-based cermets, for example, by adding Ni and Al powders to exothermic mixtures of Ti with C or B. On the other hand, some intermetallic compounds (e.g., NiAl, Ni_3Al, TiAl) possess high enough heats of formation so that composites with intermetallic matrices can be produced either in the VCS or SHS regimes. The ceramic components are added either in the green mixture or synthesized *in situ* during the reaction.

For example, Ni₃Al matrix composites reinforced by Al₂O₃, SiC, or B₄C whiskers have been formed in the SHS mode (Lebrat et al., 1992, 1994). Also, the reinforcement of combustion-synthesized Ni₃Al with TiC particles was investigated for two types of particles: commercially available TiC powders added in the reaction mixture and *in situ* synthesized TiC grains (Mei et al., 1993). The NiAl matrix composite reinforced with TiB₂ particles (10 and 30 wt %) was synthesized using SHS+hot pressing (Wang et al., 1993). The TiAl matrix was formed in the SHS mode, while reinforcing SiC particles were added in the initial Ti+Al mixture (Rawers et al., 1990). Also, Al, and TiAl₃ matrix composites, reinforced by TiC, TiB₂, and TiC+TiB₂ particles synthesized *in situ*, have been fabricated in the SHS regime (Gotman et al., 1994).

Owing to their high plasticity at the combustion temperature, poreless aluminides can be produced using various methods of densification. Dense titanium aluminide materials were obtained in the VCS regime with pseudo HIP (Shingu et al., 1990), while hydraulic hot pressing combined with SHS was used to form NiAl matrix composites with porosities of 1% or less (Wang et al., 1993). Hot rolling under vacuum was applied to form combustion-synthesized TiAl (Osipov et al., 1992). Other materials were synthesized by the extrusion of Ni-rich NiAl (Alman, 1994). Shock-induced synthesis with dynamic densification was reported for titanium, niobium, and nickel aluminides (Work et al., 1990; Ferreira et al., 1992; Strutt et al., 1994). Densification of iron aluminides and Fe₃Al matrix composites reinforced with Al₂O₃ particles was performed by pressing in graphite-lined dies at 10–70 MPa (Rabin and Wright, 1991), as well as by HIP and pseudo HIP techniques (Rabin et al., 1992b). It was shown recently that a 99.7% dense Fe₃Al-Nb alloy billet could be prepared by SHS, followed by superplastic consolidation (Dutta, 1995).

6. Compounds of Ti with Ni, Co, and Fe

While intermetallic Ti-(Ni, Co, Fe) compounds cannot be synthesized from elemental powders without preheating, a relatively low increase above room temperature is generally sufficient to initiate a self-sustained reaction. Almost all compounds of the Ti-Ni, Ti-Co, and Ti-Fe systems have been produced by this method (Itin et al., 1977, 1983; Bratchikov et al., 1980; Itin and Naiborodenko, 1989; Yi and Moore, 1989, 1990; Kaieda et al., 1990b; Zhang et al., 1995). Intermetallic TiNi compounds are especially attractive for their shape memory characteristics. Combustion-synthesized Ti-Ni powders and materials have been used in the industrial production of sheets, tubes, and wires with shape memory in Russia (Itin and Naiborodenko, 1989) and in Japan (Kaieda et al., 1990b). Cold-deformed binary alloys have retained up to 90% of their original shape, while 99.5% of the original shape may be restored for ternary alloys (e.g., Ti-Ni-Fe and Ti-Ni-Al). A 70-kg coil of hot-rolled TiNi wire produced by the CS+HIP technique is shown in Fig. 11.

FIG. 11. Hot-rolled coil of TiNi wire after forging. The weight of the coil is 70 kg (Adapted from Kaieda et al., 1990b).

7. Diamond Composites

At normal pressures, diamond is a metastable phase of carbon, which converts to graphite at temperatures higher than 1300 K (either in vacuum or inert gas). The rate of conversion is extremely rapid at temperatures higher than 2000 K (Howers, 1962; Evans and James, 1964). Thus, graphitization during sintering of diamond composites makes it difficult (if not impossible) to produce refractory matrix composite with diamond particles. Due to the short time of interaction, SHS is an attractive method for the production of diamond-containing composite materials with refractory and wear-resistant matrices. SHS combined with pressing (pseudo HIP) was used to obtain TiB-diamond composites from elemental Ti+B with up to 20 wt % synthetic diamond (Padyukov et al., 1992a,c). By decreasing the combustion temperature from 2200 K to 1900 K, the extent of graphitization was reported to decrease from 6% to 0.5% (Padyukov et al., 1992b). Further investigation of the combustion synthesis of TiB- and NiAl-matrix diamond-containing materials has shown that under certain conditions, diamond particles retain their shape, surface quality and mechanical properties after synthesis (Padyukov and Levashov, 1993; Levashov et al., 1993). The characteristic microstructure of the diamond-containing portion of an Ni-Al/diamond composite synthesized by the SHS method is shown in Fig. 12.

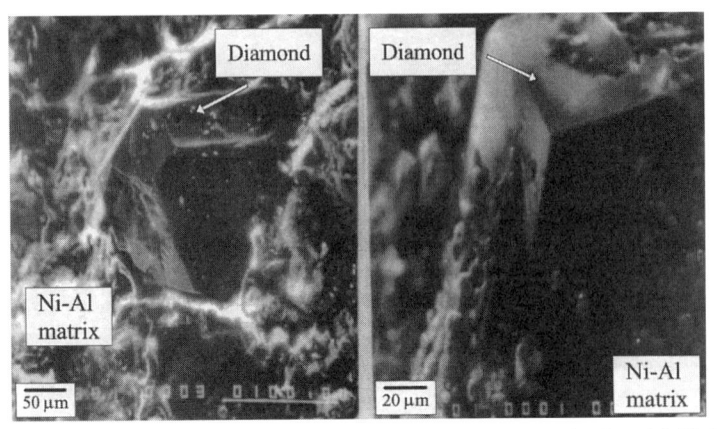

FIG. 12. Characteristic microstructure of the diamond-containing region in a Ni-Al diamond composite produced by the SHS method (Adapted from Levashov et al., 1993).

8. Functionally Graded Materials

The concept of functionally graded materials (FGMs) is to tailor nonuniform distribution of components and phases in materials, and hence combine mechanical, thermal, electrical, chemical, and other properties that cannot be realized in uniform materials. For example, the material structure may have a smooth transition from a metal phase with good mechanical strength on one side, to a ceramic phase with high thermal resistance on the other side (see Fig. 13). With a gradual variation in composition, FGMs do not have the intermaterial boundaries found in multilayer materials, and hence they exhibit better resistance to thermal stress.

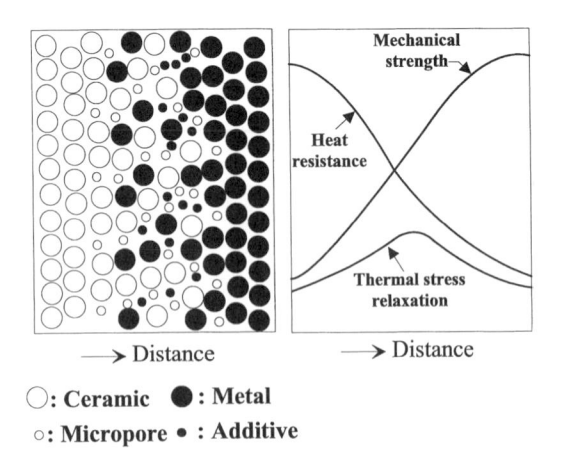

FIG. 13. The concept of functionally graded materials (Adapted from Sata et al., 1990b).

Combustion synthesis was applied successfully in the production of TiB$_2$-Cu FGMs with the amount of TiB$_2$ varying from 0 (metal side) to 100% (ceramic side) for cylindrical samples, 30 mm in diameter and 5 mm thick (Sata et al., 1990a,b; Sata, 1992). A predetermined distribution of elements in the green mixture was created using an automatic powder stacking apparatus. During synthesis, simultaneous densification by hydrostatic pressing was also applied. Similar methods were used to produce TiB$_2$-Cu FGM samples from cylindrical green compacts of 17 mm in diameter and 10 mm in height (Wang et al., 1994b). However, other phases (Ti$_3$Cu$_4$, Cu$_3$Ti, TiCu$_4$, TiB, Ti$_2$B$_5$) appeared when the Cu content was increased. The method of gas-pressure combustion sintering was applied to produce TiC-Ni FGM (Miyamoto et al., 1990a). The Vickers microhardness of this material gradually changed from 2 GPa at the nickel side to 23 GPa at the TiC side, where the TiC grain size decreased with increasing local Ni content. The microstructure of a Cr$_3$C$_2$/Ni FGM, consolidated from a green compact consisting of powder layers with different Cr-C-Ni compositions by the SHS+HIP method, is shown Fig. 14.

Another approach is to prepare a green mixture with relatively few (typically two or three) layers, while desired gradients in the product appear during the SHS process due to infiltration and migration of melts, diffusion, and other transport phenomena (Pityulin et al., 1992, 1994). This approach was used to produce poreless functionally gradient cermets, including TiC-Ni and (Ti,Cr)C-Ni, by combining SHS with pseudo HIP. An example of a multilayer material synthesized by the CS+hot pressing method is the TiB$_2$-Al$_2$O$_3$-ZTA (zirconia toughened alumina) composite. The microstructure of this layered material is shown in Fig. 15 (DeAngelis and Weiss, 1990). We can see that the layers can have different thicknesses and are well bonded to one another. Recently, the production of diamond-containing FGMs was also reported (Levashov et al., 1994), where properties

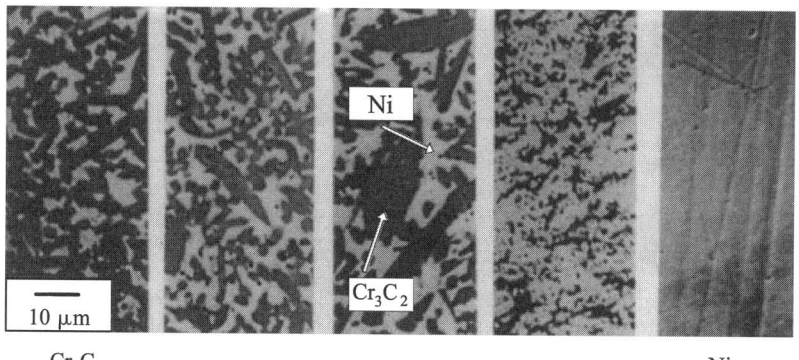

FIG. 14. SEM image of Cr$_3$C$_2$/Ni FGM microstructure (Adapted from Miyamoto et al., 1992).

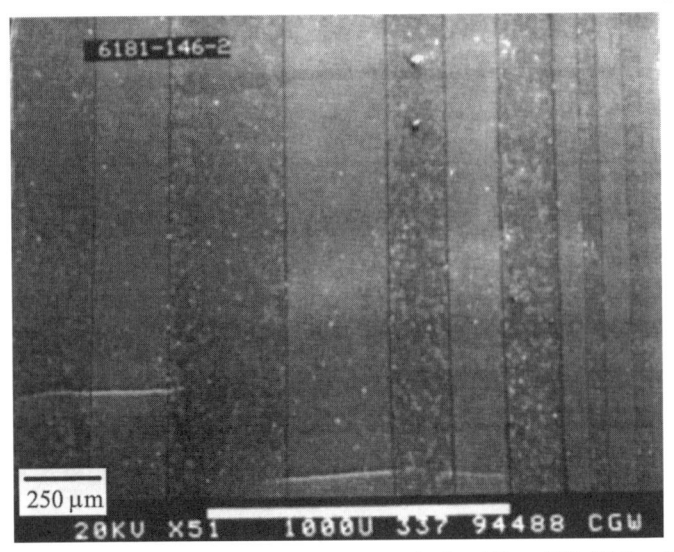

FIG. 15. Scanning electron microscope photomicrograph of multilayer TiB$_2$/Al$_2$O$_3$ with zirconia toughened alumina (Adapted from DeAngelis and Weiss, 1990).

of materials are determined by the gradient distribution of diamond in the combustion-synthesized matrix.

9. Other Classes of Products

Chalcogenides, including sulfides and selenides of Mg, Ti, Zr, Mo, and W, represent another group of compounds that can be synthesized from elements during the combustion process. Experimental synthesis of luminescent ZnS samples was also carried out (Molodetskaya et al., 1992). Combustion reactions between chalcogenides and alkali metal compounds were applied to produce GaAs and GaP from solid-state precursors (Wiley and Kaner, 1992). In addition, compounds of metals (Ni, Cu, Zr, etc.) with phosphorus were obtained (Muchnik, 1984; Muchnik et al., 1993). Also, phosphides of Al, Ga, and In were formed by direct combustion synthesis from elemental powders (Kanamaru and Odawara, 1994). An interesting example of the combustion synthesis of aluminum iodide Al$_2$I$_6$ from a mixture of Al powder with crystalline iodine was reported (Shtessel et al., 1986). Note, however, that synthesis of chalcogenides, phosphides, and iodides represents a process intermediate between gasless and solid–gas processes due to intense vaporization of the nonmetal reactant during combustion.

Among intermetallic compounds not mentioned previously, Mg$_2$Ni is a promising material for storing hydrogen. It has been shown that the combustion synthesis of this material is feasible, in spite of a relatively low heat of reaction.

For example, high-purity Mg$_2$Ni was obtained in the SHS mode (Akiyama et al., 1995).

B. COMBUSTION SYNTHESIS IN GAS–SOLID SYSTEMS

1. Nitrides and Nitride-Based Ceramics

a. Metallic Nitrides. A wide variety of metallic nitrides including TiN, ZrN, NbN, HfN, TaN, and VN have been produced by combustion synthesis. Among these, titanium and zirconium nitrides have been investigated more thoroughly. These nitride powders can be used to form refractory corrosion-resistant coatings and conducting materials in electronic applications (Borovinskaya et al., 1991). Depending on the nitrogen pressure, titanium and zirconium nitrides with a wide range of nitrogen compositions have been synthesized, ranging from solid solutions with a minimum amount of nitrogen (MeN$_{0.1}$) to single-phase supersaturated solutions (MeN$_{0.34}$-MeN$_{0.45}$), as well as stoichiometric MeN (Borovinskaya and Loryan, 1976, 1978). Even when the porosity of the green mixture is optimal (40–45%), the degree of conversion from Ti to TiN may not exceed 60–70% (Eslamloo-Grami and Munir, 1990a,b). In a recent work, Brezinsky et al. (1996) achieved conversions up to 75% during nitridation of titanium in supercritical nitrogen. However, full conversion can be achieved by dilution of the initial Ti powder with 40–50 wt % of TiN (Borovinskaya and Loryan, 1978).

Other metallic nitrides have also been synthesized by combustion synthesis. Vanadium nitride was first produced by Maksimov et al. (1979), and is the basis of ferrovanadium (VN+Fe), a commonly used alloying agent for the manufacture of steels with low-temperature applications (Maksimov et al., 1984). Also, a unique cubic structure of TaN, which has the highest microhardness among all transition metal nitrides (3200 kg/mm^2) and can be used as a hard alloy component, was formed during combustion of tantalum powder in liquid nitrogen (Borovinskaya et al., 1970). Niobium nitride powders, with a transition temperature to superconductivity at 14.3 K, were synthesized by SHS. In addition to powders, thin superconducting plates and wires of NbN were obtained using a Nb+NbN+N$_2$ mixture, as a chemical furnace (Ohyanagi et al., 1993). Recently, both the SHS and VCS modes were utilized for the synthesis of magnesium nitride, Mg$_3$N$_2$ (Li et al., 1994). It was found that addition of Mg$_3$N$_2$ as diluent to the initial Mg powder is necessary for carrying out the SHS process, in order to provide high permeability of nitrogen in the sample after melting of magnesium.

It is worth noting that a large number of publications (see Sections V and VI) deal with fundamental studies of combustion in metal–nitrogen systems. However, only a relatively few describe the properties of these powders and materials.

TABLE III
SOME CHARACTERISTICS AND PROPERTIES OF METALLIC SHS NITRIDES[a]

Products	Chemical Composition (wt %)	Impurities (wt %)	Particle Size and Specific Surface Area	Other Properties: Microhardness, H, and Transition Temperature to Superconducting State, T_{sc}
Titanium nitride (cubic, $a = 4.24$)	Ti = 77.0–78.5 N = 20.3–21.5	O = 0.3–0.5 Fe < 0.2	1–200 μm, easily grindable	$H = 1800$ kg/mm^2
Hafnium nitride (cubic, $a = 4.510$)	Hf = 92.0–94.0 N = 5.9–6.0	O = 0.3–0.5	1–50 μm	$T_{sc} = 6.6$ K
Tantalum nitride (cubic, $a = 4.323$)	Ta = 91.7–92.2 N = 7.3–7.6	O = 0.1–0.3 C = 0.05–0.1	1–10 μm 0.06 m^2/g	$H = 3200$ kg/mm^2
Zirconium nitride (cubic, $a = 4.576$)	Zr = 87.3 N = 12.6	O = 0.3–0.5 Fe < 0.2	1–200 μm	$H = 1570$ kg/mm^2 $T_{sc} = 9.6$ K
Niobium nitride (cubic, $a = 4.385$)	Nb = 86.9 N = 6.1	O = 0.1–0.3 C = 0.05–0.1	1–50 μm	$H = 1670$ kg/mm^2 $T_{sc} = 14.3$ K

[a]Data from Merzhanov and Borovinskaya (1972) and Borovinskaya et al. (1991).

The characteristics and properties of some combustion-synthesized transition metal nitrides are presented in Table III.

b. Nonmetallic Nitrides. Nonmetallic nitrides, including Si_3N_4, BN, and AlN, have important applications as high-temperature structural ceramics. As discussed earlier (see Section I), two methods can be used to produce nitrides by CS: elemental ($Si+N_2$, $B+N_2$, $Al+N_2$) and reduction (e.g., $B_2O_3+Mg+N_2$) reactions. In Tables IV, V, and VI, a comparison of the chemical composition and spe-

TABLE IV
SOME CHARACTERISTICS AND PROPERTIES OF Si_3N_4 POWDERS[a]

Content (wt %)	SHS β-phase	SHS α+β-phase, (α ≥ 80%)
Basic compound, Si_3N_4	98–99	98–99
Nitrogen, N	38.5–39.0	38.5–39.0
Oxygen, O	0.5–1.0	0.5–1.0
Iron, Fe	0.1–0.3	<0.01–0.3
Carbon, C	<0.1	≤0.1
Silicon free, Si_{free}	<0.1	<0.1
Specific Surface, m^2/g	1–3	3–11

[a]Data from Merzhanov (1992).

TABLE V
SOME CHARACTERISTICS AND PROPERTIES OF BN POWDERS[a]

Content (wt %)	SHS Product Ultrapure	SHS Product Technically Pure	Furnace Product ORPAC GRADE 99	Furnace Product Denka (Japan)
Basic compound, BN	>99.5	97.3	98–99	>98
Nitrogen, N	>55.7	54.9	54–55	54.5
Oxygen, O	<0.5	1.5	1.5	1.5
Carbon, C	<0.01	0.3	—	—
Metal impurities (Mg, Fe)	<0.2	0.3	—	—
Specific Surface, m^2/g	11.0	8–14	10	—

[a]Data from Merzhanov (1992).

cific surface area of powders produced by CS and by conventional methods is presented. Powders synthesized by SHS have comparable, if not higher, nitrogen contents than those produced by conventional methods. The metal impurities (mainly Fe) are the consequence of either inappropriate mixing/grinding equipment or impure raw materials, both of which may be remedied.

The industrial ceramics produced by conventional sintering of β-phase Si_3N_4 powders synthesized by SHS (Petrovskii et al., 1981) can be used as high-temperature articles with attractive dielectric properties ($\tan\delta = 4.4 \cdot 10^{-3}$ at $f = 10^6$ Hz, and dielectric strength, $E_d = 9.2$ kV/mm). Also, silicon nitride powders with a relatively high α-phase content ($\geq 80\%$) have been used for the production of advanced structural ceramics with good mechanical properties ($\sigma\hat{b} = 900$ MPa, $K_{Ic} = 6.7–7.2$ mN/m$^{3/2}$, and $H_V = 13.3–14.2$ GPa).

The dielectric materials produced by hot pressing SHS AlN powders have comparable thermal conductivities, but higher bending strengths than those produced in conventional furnaces (Gogotsi et al., 1984). For this reason, they are suitable for fabrication of panels in the microelectronics industry. Also, aluminum

TABLE VI
SOME CHARACTERISTICS AND PROPERTIES OF AlN POWDERS[a]

Content (wt %)	SHS Product Ultrapure	SHS Product Technically Pure	Furnace Product ART USA A-100	Furnace Product Starck (Grade B) (Germany)
Basic compound, AlN	99.7	98.8	99.0	98.1
Nitrogen, N	33.9	32.7	33.0	33.3
Oxygen, O	0.3	0.6	1.0	2.3
Iron, Fe	0.07	0.12	0.005	100 ppm
Specific Surface, m^2/g	2.0–20	1.5	2.5–4.0	1.0–8.0

[a]Data from Merzhanov (1992).

nitride powders synthesized by SHS have been used as a heat-conducting component in silicon–organic adhesive sealants. Used successfully in Russia for electronic applications, this glue has a thermal conductivity about twice that of alternative adhesives (Merzhanov, 1992).

Combustion synthesis of boron nitride powder, BN, was reported in one of the earliest works on SHS (Merzhanov and Borovinskaya, 1972). The mechanisms of combustion and product structure formation from elements were later investigated (Mukasyan and Borovinskaya, 1992). More recently, finely dispersed hexagonal boron nitride powder has been obtained from reduction-type reactions (Borovinskaya *et al.*, 1991).

The high temperatures present in the CS process make it possible to not only produce powders, but also to sinter materials and form net-shape articles in the combustion wave (Merzhanov, 1993a). Several types of SHS nitrides and their applications are presented in Table VII.

Since their mechanical properties are maintained at high temperatures (up to 1700 K), silicon nitride-based materials are attractive for use in high-temperature structural applications in chemically aggressive environments. For example, several properties of a Si_3N_4-based ceramic (Si_3N_4+SiC+TiN), suitable for ceramic engines, are shown in Table VIII.

TABLE VII
SHS Ceramics Based on Nonmetallic Nitrides[a]

Composition	Product Type	Applications
Si_3N_4	Bricks, plates, cylinders	Corrosion-resistant plugs for chemical pumps
Si_3N_4 + SiC + TiN (black ceramic)	Plates, rods, plugs, sleeves	Cylinders, plugs, and other parts for combustion engines
Si_3N_4 + SiC + BN + C	Plates, rods, balls	Friction pairs, ball bearings
SiAlON	Articles of complicated shape: turbine blades, honeycomb structures, etc.	Internal combustion engine parts for turbo-superchargers; catalyst supports for incineration of exhaust gases
SiAlON + BN	Bricks, plates	Parts for metallurgy
SiAlON + SiC + BN	Bricks, nozzle, turbine blades	Internal combustion engine parts; metallurgical tools
AlN	Cylinders, plates	Heat-conducting substrates
AlN + TiB_2	Bricks, plates, cylinders	Cathodes for coating by vacuum spraying
BN	Rods, bricks, plate, tubes, crucibles, rings, sleeves, bricks	Insulating sleeves; crucibles for pouring and tubes for flow of melted alloys
B + BN	Plugs	Nuclear safety
BN + SiO_2	Plates	Lining of MHD generator

[a]Data from Merzhanov (1993a).

TABLE VIII
SOME PROPERTIES OF SHS SILICON
NITRIDE-BASED CERAMICS[a]

Property	Value
Density, g/m^3	3–3.4
Porosity, %	<1–15
Elastic modulus, Gpa	180–250
Rockwell hardness, HRA	85–93
Vickers hardness, GPa	6–14.5
Bend strength, MPa, T = 1750 K	270–650
Critical stress intensity factor, MN/m$^{3/2}$	2.5–5
Thermal conductivity (873–1371 K), W/m K	15–20

[a]Data from Merzhanov (1990a).

Aluminum nitride is most commonly used for its high thermal conductivity. Recently, a poreless composite material, TiAl-TiB$_2$-AlN, was obtained by reacting a Ti+(0.7–0.95)Al+(0.05–0.50)B mixture at 30- to 100-atm nitrogen pressure (Yamada, 1994). The use of high-pressure nitrogen gas was found to be effective for simultaneous synthesis and consolidation of nitride ceramics with dispersed intermetallic compounds (e.g., TiAl). Dense, crack-free products with uniform grains (approximately 10 mm in size) were obtained.

The main applications for boron nitride materials take advantage of their dielectric properties, which are maintained at high temperatures (even above 2000 K) in nonoxidizing atmospheres. Some properties of boron nitride-based materials are listed in Table IX, and some examples of the

TABLE IX
SOME PROPERTIES OF SHS BN-BASED CERAMICS[a]

Properties	BN	BN + SiO$_2$
Chemical composition, wt %	N = 55; O < 0.5; B$_{free}$ < 0.5; B$_2$O$_3$ < 0.3; C < 0.5	BN = 74 SiO$_2$ = 26
Density, g/cm^3	1.5	1.85
Dielectric strength, kV/mm	25	19
Dielectric permeability	3.1	6.0
Tangent of dielectrical loss at 1 MHz	0.0034	0.036
Resistivity, Ω	2 × 10^8	1.3 × 10^8
Thermal conductivity, W/mK		
T = 400 K	20	
T = 900 K	8	8
Bend strength, MPa	25	50

[a]Data from Mukasyan et al. (1989).

various shapes and forms that can be produced from BN-based materials are shown in Fig. 16.

2. Hydrides

The heats of formation for transition metal hydrides are much lower than those for nitrides or oxides. However, the high diffusivity of hydrogen within the reaction medium makes it possible to organize self-sustained combustion (Dolukhanyan et al., 1976, 1978). More than 30 hydrides and deuterides of scandium, titanium, vanadium, and the major lanthanide metals have been obtained by gas–solid combustion synthesis (Dolukhanyan et al., 1992). Hydrides of the transition metal alloys and intermetallic compounds (e.g., $ZrNiH_3$, Ti_2CoH_3) were also obtained (Dolukhanyan et al., 1981). The production of complex hydrocarbides and hydronitrides is of special interest (Agadzhanyan and Dolukhanyan, 1990). For example, multicomponent single-phase hydronitrides of the following compositions $Zr_{0.9}Nb_{0.1}C_xN_{1-x}H_{0.19}$, $Zr_{0.4}Nb_{0.6}C_xN_{1-x}H_{0.26}$ can be synthesized.

3. Oxides

Combustion synthesis has been used to produce complex oxide materials, such as ferrites and ferroelectric, piezoelectric, and superconducting materials. In these systems, the heat required for SHS is supplied by the heat released from the reac-

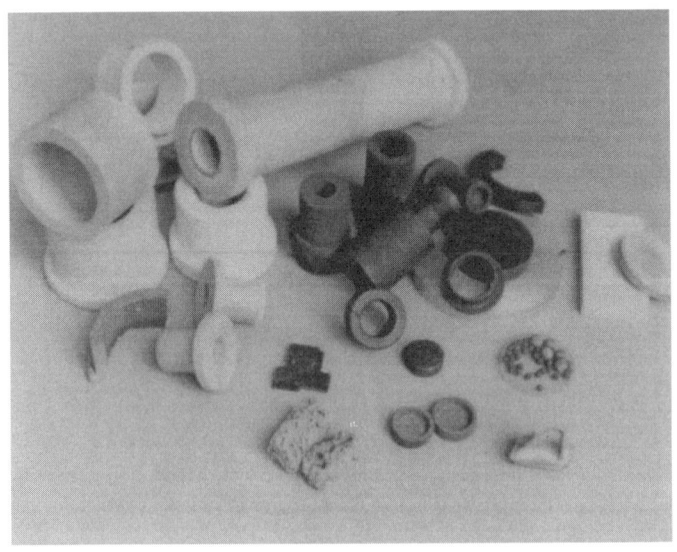

FIG. 16. Ceramic materials and net-shape articles produced by combustion synthesis. (Adapted from Mukasyan, 1986)

tion between oxygen and metal. Oxygen can be supplied to the reaction by either an external (oxygen gas) or an internal (e.g., reduction of oxides in the starting reaction mixture) source or by a combination of the two.

a. High-Temperature Superconductors. One of the promising oxide materials produced by SHS is the $YBa_2Cu_3O_{7-x}$ (Y_{123}) ceramic superconductor (Merzhanov, 1990a). This material has a number of applications, including targets for magnetron spraying, superconducting shields, wires, and cables. The compound is produced by SHS using the following reaction:

$$3Cu + 2BaO_2 + 1/2Y_2O_3 + (1.5-x)/2\ O_2 \rightarrow YBa_2Cu_3O_{7-x}\ (Y_{123}).$$

The effects of processing conditions, including green density, oxygen pressure, Cu, BaO_2, and Y_2O_3 particle sizes, as well as the use of Cu_2O and CuO as starting materials, have been investigated (Merzhanov, 1990b; Lebrat and Varma, 1992b; Lin et al., 1994). The reaction mechanism has also been identified by the quenching technique (Merzhanov, 1990b; Lebrat and Varma, 1991). It was shown that oxidation of Cu metal by gas-phase oxygen initiates the reaction, while the subsequent decomposition of barium peroxide provides an internal source of oxygen.

A comparison of Y_{123} powders produced by SHS (SHS-J1 and SHS-J2) with those produced by furnace synthesis is shown in Table X. The oxygen content, transition temperature, orthorhombic phase Y_{123} content, basic impurities, and mean particle sizes of SHS powders compare favorably with those produced by furnace synthesis. However, the time of Y_{123} production for a conventional fur-

TABLE X

Some Characteristics and Properties of Superconductive Powders ($YBA_2CU_3O_{7-x}$) Produced by Conventional and SHS Methods[a]

Powder	Oxygen Content (7-x)	Transition Temperature, T_{sc} (K)	Percent Orthorhombic (Y_{123}) Phase	Basic Impurities	Average Particle Size (μm)
SC5-S[b]	6.92	93.5	98	CuO	40.0
SC5-P[b]	6.87	93.5	~99	CuO	8.2
SC5-6.5[b]	6.88	90.5	97	$BaCuO_2$	6.8
SC5-7[b]	6.85	92.5	~99	CuO	5.7
CPS A-1203[b]	6.85	92.0	~99	$BaCuO_2$	3.0
SSC 03-0065[b]	6.89	92.0	~99	—	6.5
SHS-J1[c]	6.90	92.0	97	CuO	9.0
SHS-J2[c]	6.92	93.5	~99	—	8.0

[a]Data from Merzhanov (1991).
[b]Conventional methods.
[c]Combustion synthesis.

nace is an order of magnitude longer than that for SHS. Recently, relatively high-density (~90%) items with superconducting transition temperature, $T_{sc}=92$ K, have been fabricated by SuperConco Co. (U.S.) by sintering SHS-J2 powders (Avakyan et al., 1996). Also, thick films (1 mm) made from this powder exhibit a critical current density up to 10^6 A/cm^2. The materials of Bi$_2$Sr$_2$CaCu$_2$Oy and Tl$_2$Ba$_2$Ca$_2$Cu$_3$O$_y$ compositions with $T_{sc}=125$ and 135 K, respectively, have also been synthesized by SHS (Merzhanov, 1990b).

b. Ferroelectric Materials. The properties of various ferroelectric materials produced by SHS have recently been reported (Avakyan et al., 1996). It was noted that the properties of electronic ceramics produced by SHS, including chemical and phase compositions as well as electromagnetic properties, are strongly affected by the maximum combustion temperature. This feature suggests that the properties of the final product can be tailored for a specific application by changing the combustion temperature, which can be achieved by adjusting the processing conditions (see Section V).

Ferroelectric laminated bismuth compounds, which have application as high-temperature piezoelectrics and semiconductors, can be synthesized by SHS. In Table XI, the properties of several ferroelectric materials along with potential applications are presented. In addition, LiNbO$_3$ and LiTaO$_3$ can be obtained by SHS with densities two times higher than those produced by conventional methods, by varying the Li$_2$O/Me$_2$O$_5$ ratio in the product (Me is Nb or Ta). The formation of rare-earth (lithium and terbium) molybdate materials can also be achieved by SHS. All of the materials mentioned have electronic applications. During conven-

TABLE XI
SOME CHARACTERISTICS AND PROPERTIES OF SHS-PRODUCED FERROELECTRIC MATERIALS[a]

Composition	Resistivity, ρ ($\Omega \cdot$ cm)	Dielectric Constant, ϵ	Dielectric Dispersion, tan δ	Applications
Bi$_4$V$_2$O$_{11}$	4.7×10^9	63	0.089	Gas sensor (ethanol, acetone)
BiFeO$_3$	4.2×10^9	46	0.169	Gas sensor (ethanol, acetone, gasoline)
Bi$_2$Fe$_4$O$_9$	6.0×10^9	10	0.004	Gas sensor (ethanol, acetone)
Bi$_4$Fe$_2$O$_9$	2.8×10^9	40	0.109	Temperature-sensitive resistance, gas sensor (ethanol, acetone, gasoline)
BaBi$_2$Ta$_2$O$_9$	4.0×10^{11}	49	0.0022	Gas sensor (ethanol, acetone, natural gas, H$_2$O
Bi$_3$TiNbO$_9$	1.9×10^{10}	70	0.0047	Moisture sensor
BaBi$_2$Ni$_2$O$_9$	3.5×10^{11}	92	0.0081	Moisture sensor
Na$_{0.5}$Bi$_{4.5}$Ti$_4$O$_{15}$	1.45×10^9	82	0.0059	Moisture sensor

[a]Data from Avakyan et al. (1996).

tional synthesis, difficulties arise when forming single crystals of lithium and terbium molybdates due to the high volatility of molybdenum oxide. This difficulty has been overcome by using SHS-produced stoichiometric powders, which are characterized by minimal losses during heating, melting, and crystallization.

C. Products of Thermite-Type SHS

It is impossible to describe, even briefly, all of the products obtained by thermite-type self-propagating reactions. Some examples were given previously in Section I, and some additional examples involving Mg metal are presented later in Table XX. The use of Mg leads to MgO product that can be easily leached, leaving behind the desired product composition in powder form. In this section, we consider some selected products reported in the literature, specifically those whose phase distribution and properties are controlled by gravitational or centrifugal forces. Owing to high heats of formation, the products are completely molten at the combustion temperature. As a result, gravity-induced processes of convection and buoyancy may affect phase separation and crystallization, as well as combustion wave propagation.

Centrifugal SHS casting (Section II,B,3) was used to produce oxide-carbide ceramics by the following reaction: $MO_x + Al + C \rightarrow MC_y + Al_2O_3$ (Yukhvid *et al.*, 1994). In general, titanium, chromium, molybdenum, and tungsten carbides are synthesized with an alumina melt in the combustion wave, for example, Cr_3C_2-Al_2O_3, WC-Al_2O_3, (Ti,Cr)C-Al_2O_3, and MoC-Al_2O_3. Also, addition of metal (Fe group) oxides (NiO, CoO, Fe_2O_3) results in formation of cermet composites containing a metal binder [e.g., (Ti,Cr)C-Ni] in addition to the carbide phase. The lighter oxide phase can be fully separated from the cermet by centrifugal acceleration. A centrifugal overload of 150 g is sufficient to separate MoC from Al_2O_3, while 300 and 1000 g are required to separate WC-Co cermet and WC, respectively, from the alumina melt (Merzhanov and Yukhvid, 1990). Lower centrifugal overloads result in incomplete separation and the formation of graded materials. This phenomenon has been used to obtain refractory and chemically stable protective coatings on various substrates, which possess relatively complex composition and structure (Grigor'ev and Merzhanov, 1992; Yukhvid, 1992).

An interesting application of a thermite-type SHS reaction with centrifugal overload is the fabrication of multilayered composite pipes (Odawara, 1982, 1990; Odawara and Ikeuchi, 1986; Kachin and Yukhvid, 1992). A simple and well-known composition, $Fe_2O_3 + 2Al$, was used to produce a two-layer lining inside the steel pipe; see Fig. 17 (Odawara, 1992). The iron outer layer provides good adhesion of the coating to the inner surface of the steel pipe, while the inner layer (Al_2O_3) provides protection from corrosion and abrasion. Separation of the layers was achieved by rotating the pipe (centrifugal acceleration between 50 and

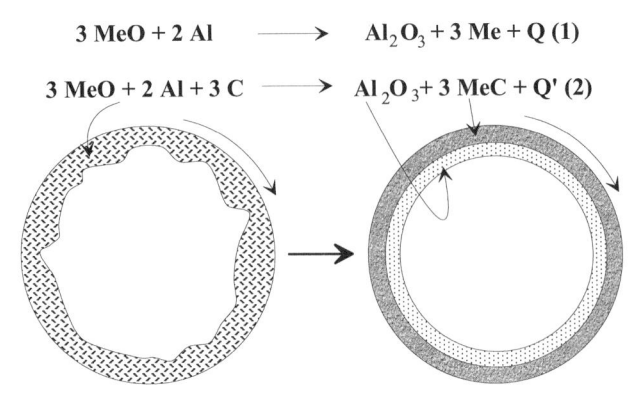

$$3\,MeO + 2\,Al \longrightarrow Al_2O_3 + 3\,Me + Q \quad (1)$$
$$3\,MeO + 2\,Al + 3\,C \longrightarrow Al_2O_3 + 3\,MeC + Q' \quad (2)$$

FIG. 17. The concept of centrifugal process for production of ceramic-lined steel pipes (Adapted from Odawara and Ikeuchi, 1986).

200 g). Some properties of the ceramic layer are presented in Table XII. The effects of adding SiO_2 and Al_2O_3 to the $Fe_2O_3 + 2Al$ mixture, on the structure of pipe lining layer have been investigated (Orru' et al., 1995). More complex structures were formed using multicomponent mixtures (Merzhanov and Yukhvid, 1990). For example, the Ti-C-Ni-Al-O reaction system yielded multiphase gradient layers, where the composition gradually changed from TiC-Ni cermet to the $Al_2O_3 + TiO_2$ ceramic. The local density changed from 6.8 g/cm^3 at the cermet side, down to 3.9 g/cm^3 at the ceramic side. The production of steel pipes lined with a porous alumina–zirconia ceramic layer by reaction of $Fe_2O_3 + Al + Zr +$ additives has also been described (Li et al., 1995).

TABLE XII
PROPERTIES OF THE ALUMINA-BASED CERAMIC LAYER
PRODUCED BY THE CENTRIFUGAL THERMITE PROCESS[a]

Property	Value
Porosity	5.6%
Bulk density	3.86 g/cm^3
Linear expansion (293–1273 K)	8.57×10^{-6} K^{-1}
Thermal conductivity (at 293 K)	13.9 W m^{-1} K^{-1}
Resistivity (at 293 K)	2×10^7 Ω
Vickers hardness	1200 MPa
Compressive strength	>850 MPa
Bending strength	150 MPa
Bulk modulus	2.2×10^5 MPa

[a]Data from Odawara (1992).

D. COMMERCIAL ASPECTS

As the extent of fundamental research in the field increases, so does the feasibility of using CS technology either to produce novel materials or to replace existing commercial synthesis processes. In this section, we consider the technological and economic advantages of using CS to produce advanced materials, as well as describe some current commercial projects.

1. Technological Advantages

As noted in Section I, the CS process is characterized by high temperatures, fast heating rates, and short reaction times. These features make CS an attractive method for commercial synthesis, with the potential for new materials and lower costs, compared to conventional methods of furnace synthesis.

The typical production technologies for combustion synthesis were described in Section II. It is evident that apparatus needed for synthesis is relatively simple, especially since no additional equipment is needed for bulk heating of the material (e.g., furnaces, plasma generators, etc.). In addition, it has been reported that replacing conventional furnaces with a CS apparatus leads to significant reduction in workspace requirements (Merzhanov, 1992). Finally, owing to low power requirements, combustion synthesis technology is one of the few methods of materials synthesis that is feasible in outer space (Hunter and Moore, 1994).

In terms of operation, CS offers several advantages of production. The foremost is that the energy for synthesis is supplied solely by the heat of chemical reaction, instead of an external source. Further, when layer-by-layer combustion is organized, the heating of the reactant charge is uniform. On the other hand, heating materials by external sources results in temperature gradients during synthesis, which can lead to undesired nonuniformities in the product. This effect may be especially pronounced for large or irregular-shaped bodies. In some cases, the synthesis temperatures during CS may be higher than those attainable in conventional furnaces.

The time of production by CS methods is much shorter than that for conventional methods of powder metallurgy. Also, materials and articles can be produced directly in the combustion wave. Furthermore, a continuous mode of production may be realized, where reactants are fed at a rate equal to the combustion wave velocity.

The quality of the materials produced by CS is another important consideration. One of the attractive aspects of the process is its ability to produce materials of high purity, since the high temperatures purge the powders of any volatile impurities adsorbed on the reactants. Furthermore, the temperature gradients, combined with rapid cooling in the combustion wave, may form metastable phases and unique structures not possible by conventional methods.

TABLE XIII
COST BREAKDOWN FOR Si_3N_4 POWDER PRODUCTION[a]

	Conventional		SHS-based	
Category	Cost ($/kg)	% of Total	Cost ($/kg)	% of Total
Raw materials	9.4	27.8	9.4	33.6
Fixed costs	7.7	22.9	4.4	15.6
Utilities	0.8	2.2	0.7	2.4
Expendable materials	13.5	44.6	11.2	40.2
Labor	2.3	2.3	2.3	8.3
Total	33.7	100.0	28.0	100.0

[a]Data from Golubjatnikov et al. (1993).

2. Economic Considerations

Some comparisons of production costs by SHS and conventional processes have been reported. Analysis for Si_3N_4, AlN, and SiC materials has shown that combustion synthesis is economically favorable to existing technologies, as shown in Tables XIII, XIV, and XV (Golubjatnikov et al., 1993). For all materials, the total cost per kilogram of material produced is lower for SHS-based processing than for conventional methods, with a major contribution to savings from reduced fixed costs (i.e., capital equipment costs). The significant difference in the raw material cost for SiC tiles occurs due to the use of Si+C powder mixtures in the SHS case, compared with SiC powder for conventional methods.

3. Current Commercial Projects

As noted recently, the scale-up of SHS reactors is nontrivial due to the extreme reaction conditions (Hlavacek and Puszynski, 1996). However, the existence of numerous manufacturing plants in the former Soviet Union demonstrates that the

TABLE XIV
COST BREAKDOWN FOR AlN TILE PRODUCTION[a]

	Conventional		SHS-based	
Category	Cost ($/kg)	% of Total	Cost ($/kg)	% of Total
Raw materials	8.2	32.0	8.2	34.9
Fixed costs	6.7	26.2	4.6	19.7
Utilities	1.9	7.2	1.8	7.7
Expendable materials	7.4	28.8	7.4	31.4
Labor	1.5	5.7	1.4	6.2
Total	25.5	100.0	23.4	100.0

[a]Data from Golubjatnikov et al. (1993).

TABLE XV
COST BREAKDOWN FOR SiC TILE PRODUCTION[a]

Category	Conventional Cost ($/kg)	% of Total	SHS-based Cost ($/kg)	% of Total
Raw materials	16.2	75.7	5.7	58.7
Fixed costs	2.1	10.0	0.8	8.5
Utilities	1.6	7.5	0.5	5.6
Expendable materials	0.3	1.1	0.4	4.4
Labor	1.2	5.7	2.2	22.9
Total	21.4	100.0	9.7	100.0

[a]Data from Golubjatnikov et al. (1993).

production of materials by SHS is feasible. To date, the following products have been manufactured commercially: powders of TiC, TiN, TiC-TiN, TiB_2, Cr_3C_2, TiC-Cr_3C_2, TiH_2, B, BN, AlN, β-Si_3N_4, nitrided ferovanadium, $MoSi_2$, ceramic insulators, lithium niobate, and nickel–zinc ferrites (Merzhanov, 1992).

The first commercial application of SHS technology was the production of $MoSi_2$ at the Kirovakan Plant (Armenia) of High-Temperature Heaters in 1979. Replacing induction furnaces with SHS technology resulted in improved productivity and product quality for this oxidation-resistant material used for high-temperature heating elements. Another example is the manufacture of abrasive TiC pastes (commercial-grade KT), which can be used for one-step grinding and polishing. The larger agglomerated TiC grains carry out the grinding, and then break up into smaller particles which polish the sample. In addition, large-scale production of articles and net-shape materials has been reported in Russia. Examples include BN-based ceramic bushings and STIM cutting inserts (Merzhanov, 1992).

In Japan, several commercial projects have been reported in the literature. For example, at the National Research Institute for Metals, the NiTi shape-memory alloy is produced by combustion synthesis from elemental powder for use as wires, tubes, and sheets. The mechanical properties and the shape-memory effect of the wires are similar to those produced conventionally (Kaieda et al., 1990b). Also, the production of metal–ceramic composite pipes from the centrifugal-thermite process has been reported (Odawara, 1990; see also Section III,C,1).

In Spain, SHS Prometeus Espana AIE was created in 1993 as a joint venture between Tecnologia y Gestion de la Innovation (Spain), Empresa Nacional Santa Barbara (Spain), ISMAN (Russia), and United Technologies (USA). They have reported successful production of Si_3N_4, SiAlON, and AlN powders, as well as carbide-based composites (Cr_3C_2, Cr_3C_2-TiC) with or without a metal matrix (Borovinskaya et al., 1995).

In the United States, a Georgia-based company has begun commercial production of improved high-performance titanium diboride-based materials using a combustion synthesis process (*Chemical Engineering Progress,* 1995). This technology was developed by researchers at Georgia Institute of Technology and licensed to Advanced Engineered Materials (AEM), Woodstock, Georgia. Commercialization of the process has begun at AEM's plant, with a capacity of 1.6 million lb/yr of materials.

Currently, other U.S. companies have also acknowledged the economic benefits of combustion-synthesized materials and are involved in pilot-scale production. However, for proprietary reasons, the details are not available. It is likely that in the future, as materials produced by combustion synthesis demonstrate unique properties (e.g., nanophase and disordered materials) compared with conventional methods and its economical benefits become more evident, the technique will receive more attention.

IV. Theoretical Considerations

Progress is being made toward a theory of combustion synthesis. Much of the development to date has been based on classical combustion theory, where the emphasis is on the prediction of macroscopic characteristics, such as combustion front velocity and stability nature. Using this approach, a large number of impressive results have been obtained. However, the principal issues concerning the microstructural aspects of the process have not been resolved. In this section, we present a variety of theoretical results, as well as some remaining problems and possible ways to approach them.

A. COMBUSTION WAVE PROPAGATION THEORY IN GASLESS SYSTEMS

1. Combustion Wave Velocity

The application of combustion theory to gasless synthesis processes is based on a comparison of the mass and thermal diffusivities in the solid reactant mixture. In these types of systems, the Lewis number is very small:

$$\text{Le} = \frac{\mathcal{D}}{\alpha} \ll 1 \qquad (5)$$

where $\mathcal{D}$ and α are the mass and thermal diffusivities, respectively, indicating that heat conduction occurs much faster than mass diffusion. As a result, mass transfer by diffusion at the *macroscopic* scale may be neglected, and an average concentration of the reactants in any local region of the heterogeneous mixture may be used. Thus, the physical and thermal properties (e.g., density, thermal conductivity, heat capacity) are the average of reactant and product values. In this case, the combustion synthesis process is controlled only by heat evolution from the exothermic reaction and heat transfer from the reaction zone to the unreacted mixture. Thus, assuming a flat combustion wave, its propagation is described by the energy continuity equation with a heat source:

$$c_p \rho \frac{\partial T}{\partial t} = \frac{\partial}{\partial X}\left(\lambda \frac{\partial T}{\partial X}\right) + \sum_k \Phi_k(\eta, T), \tag{6}$$

where Φ_k represents a heat source or sink, and η denotes reactant conversion. For example, heat evolution from a single-stage chemical reaction, along with the chemical kinetic equation, would be presented as

$$\Phi_k = Q\rho\phi(\eta, T), \tag{7a}$$

$$\frac{\partial \eta}{\partial t} = \phi(T, \eta), \tag{7b}$$

where Q represents the heat of reaction.

Phase transformations with a significant thermal effect (e.g., melting, evaporation) are also sometimes included. For example, the effect of melting can be accounted for by introducing a negative heat source (Aldushin and Merzhanov, 1978):

$$\Phi_k = -\Delta H_{\text{fus}} \rho \delta(\Delta T) \zeta$$

$$\delta(\Delta T) = \delta(T - T_m) = \begin{cases} 0, & \text{if } T < T_m \text{ or } T > T_m \\ 1, & \text{if } T = T_m, \end{cases} \tag{8}$$

where ΔH_{fus} is the latent heat of fusion and ζ represents the fraction of reactant melted.

Similarly, heat losses from the combustion zone to the surroundings yield other types of negative sources. For convective heat losses,

$$\Phi_k = h\frac{S}{V}(T - T_a), \tag{9a}$$

where h is the convective heat transfer coefficient, S and V are the sample surface area and volume, respectively, and T_a is the temperature of the surroundings. Similarly, for heat losses by radiation,

$$\Phi_k = \epsilon_m \sigma \frac{S}{V}\left(T^4 - T_a^4\right), \tag{9b}$$

where ϵ_m is the emissivity of sample surface and σ is the Stefan–Boltzmann constant.

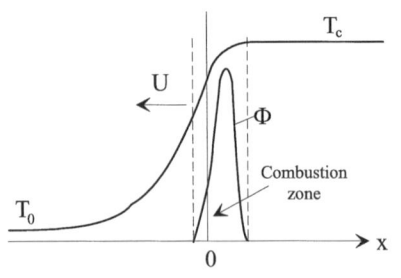

FIG. 18. Temperature and heat generation profiles in the combustion zone for a single reaction.

For a combustion wave that propagates through the reactant mixture with a velocity U, we can change to a coordinate system attached to the wave (see Fig. 18):

$$x = X + U \cdot t. \tag{10}$$

Substituting Eq. (10) into Eq. (6) yields

$$\rho c_p \frac{\partial T}{\partial t} + \rho c_p U \frac{\partial T}{\partial x} = \frac{\partial}{\partial x}\left(\lambda \frac{\partial T}{\partial x}\right) + \sum_k \Phi_k(\eta, T),$$

$$\frac{\partial \eta}{\partial t} + U \frac{\partial \eta}{\partial x} = \phi(\eta, T). \tag{11}$$

Assuming that the temperature and concentration profiles do not change with time and further that there are no heat losses (i.e., constant-pattern propagation), we can set $\partial T/\partial t$ and $\partial \eta/\partial t$ equal to zero. Then Eq. (11) takes the following form:

$$\rho c_p U \frac{dT}{dx} = \frac{d}{dx}\left(\lambda \frac{dT}{dx}\right) + \sum_k \Phi_k(\eta, T), \tag{12a}$$

$$U \frac{d\eta}{dx} = \phi(\eta, T), \tag{12b}$$

along with the boundary conditions (BCs):

$$T = T_0, \quad \frac{dT}{dx} = 0, \quad \eta = 0 \text{ at } x \to -\infty$$

$$T = T_c, \quad \frac{dT}{dx} = 0, \quad \eta = 1 \text{ at } x \to +\infty \tag{13}$$

The solution of Eqs. (12) and (13) was first obtained by Zeldovich and Frank-Kamenetskii (1938) in the context of gas flames. This solution was later adapted for solid mixtures by Novozilov (1961) and others (Khaikin and Merzhanov, 1966; Hardt and Phung, 1973; Margolis, 1983). In addition to those mentioned earlier, two other assumptions are made: (1) Heat evolution in the combustion wave is a one-stage process, due only to one chemical reaction, and (2) the rate of heat evolution may be neglected everywhere, except in a narrow zone near the maximum combustion temperature, T_c, called the *combustion* (or *reaction*) *zone*, due to the strong dependence of the reaction rate on temperature. In this section, we describe a method for estimating the combustion velocity from Eqs. (12) and (13).

Using Eq. (7), we can rewrite Eq. (12a) as

$$\frac{d}{dx}\left(\lambda \frac{dT}{dx}\right) - \rho c_p U \frac{dT}{dx} + Q\rho U \frac{d\eta}{dx} = 0. \tag{14}$$

Integrating Eq. (14) from x to $+\infty$, using BC [Eq. (13)] leads to

$$\frac{\lambda}{\rho U}\frac{dT}{dx} = Q(1 - \eta) - c_p(T_c - T). \tag{15}$$

Dividing Eq. (12b) by Eq. (15) yields

$$\frac{\rho U^2}{\lambda}\frac{d\eta}{dT} = \frac{\phi(\eta,T)}{Q(1-\eta) - c_p(T_c - T)} \tag{16a}$$

with the following BCs:

$$\eta = 0 \text{ at } T = T_0,$$
$$\eta = 1 \text{ at } T = T_c. \tag{16b}$$

From the narrow-zone approximation, the heat release function, $\phi(\eta,T)$, is negligible everywhere except near T_c. Thus replacing the value of T in the denominator of the right side of Eq. (16a) with the combustion temperature T_c yields the equation

$$\frac{\rho U^2}{\lambda}\frac{d\eta}{dT} = \frac{\phi(\eta,T)}{Q(1-\eta)}. \tag{17}$$

For most heterogeneous reactions, the kinetic function, $\phi(\eta,T)$ may be separated into two parts, each depending on one variable alone:

$$\phi(\eta,T) = \varphi(\eta)k_0 \exp(-E/RT).$$

In this case, Eq. (17) becomes

$$\frac{\rho U^2}{\lambda}\frac{d\eta}{dT} = \frac{\varphi(\eta)k_0 \exp(-E/RT)}{Q(1-\eta)}, \tag{18}$$

which may be integrated through a separation of variables with BCs [Eq. (16b)]:

$$\frac{\rho U^2 Q}{\lambda}\int_0^1 \frac{1-\eta}{\phi(\eta)}d\eta = k_0 \int_{T_0}^{T_c} \exp(-E/RT)dT. \tag{19}$$

Unfortunately, the right side of Eq. (19) cannot be integrated analytically owing to the nature of the exponential function. However, using the *Frank-Kamenetskii approximation*,

$$\exp(-E/RT) \cong \exp(-E/RT_c) \exp\left[\frac{-E(T - T_c)}{RT_c^2}\right]$$

so that

$$\int_{T_0}^{T_c} \exp(-E/RT)\,dT = \frac{RT_c^2}{E}\exp(-E/RT_c). \tag{20}$$

From Eqs. (19) and (20), we obtain the following expression:

$$U^2 = \frac{\lambda}{\rho Q} \frac{RT_c^2}{E} \frac{k_0 \exp(-E/RT_c)}{\int_0^1 \frac{1-\eta}{\varphi(\eta)}\,d\eta}. \tag{21}$$

This expression can be solved for various types of kinetics, $\varphi(\eta)$. A simple approach is to ignore the influence of conversion, η, on the reaction rate [i.e., $\varphi(\eta) = 1$], which is reasonable to an extent, since the temperature dependence of the reaction rate is generally much stronger than the concentration dependence. This, of course, corresponds to a zero-order reaction. Using this assumption, we can solve Eq. (4.17) to yield

$$U^2 = \frac{2\lambda}{\rho Q} \frac{RT_c^2}{E} k_0 \exp(-E/RT_c), \tag{22}$$

an expression first derived by Zeldovich and Frank-Kamenetskii (1938). This formula is widely used to determine effective kinetic constants from experimental data, since

$$\ln\left(\frac{U}{T_c}\right) = -\frac{E}{2RT_c} + \text{constant}. \tag{23}$$

Thus by varying T_c (by dilution or by changing T_0) and plotting $\ln(U/T_c)$ vs. $1/T_c$, the activation energy can be obtained readily. In this manner, the effective values of E have been measured for various SHS systems (cf Munir and Anselmi-Tamburini, 1989). However, it is not clear if these reported values correspond to actual elementary processes or whether they are the result of a complex interaction between transport and reaction phenomena. The simplifying assumptions made to derive Eq. (22) were stated earlier, and they all need to be satisfied under the experimental conditions in order to obtain *intrinsic* kinetic data.

In some systems, the effect of conversion on the heterogeneous reaction rate is significant. For example, substituting n'th order type kinetics [i.e., $\varphi(\eta)=(1-\eta)^n$] into Eq. (21) yields

$$U^2 = \frac{(2-n)\lambda}{\rho Q} \frac{RT_c^2}{E} k_0 \exp(-E/RT_c). \tag{24}$$

Note that Eq. (24) is not valid for $n \geq 2$, since it leads to zero ($n=2$) and imaginary ($n>2$) values of velocity. The origin of this discrepancy is that when the dependence of reaction rate on conversion is sufficiently strong (e.g., for higher-order reactions), the reaction rate away from the adiabatic combustion temperature may also be significant (Merzhanov and Khaikin, 1988). Utilizing this idea, while maintaining the thin-zone approximation, Khaikin and Merzhanov (1966) derived an expression for the case of n'th order kinetics:

$$U^2 = \left\{ 2\left[\Gamma\left(\frac{n}{2}+1\right)\right]^{1-n/2} \left(\frac{n}{2e}\right)^{n/4} \right\} \frac{\lambda}{\rho Q} \frac{RT_c^2}{E} k_0 \exp(-E/RT_c), \tag{25}$$

where $\Gamma(z) = \int_0^\infty e^{-t} t^{z-1} dt$ is the gamma function. For a zero-order reaction, the expression in brackets equals 2, and Eq. (25) simplifies to Eq. (22). For a first-order reaction, this expression equals 1.1, as compared to 1 in Eq. (24).

The theory discussed here applies mainly to kinetic-controlled reactions. However, combustion synthesis reactions involve many processes, including diffusion control, phase transitions, and multistage reactions, which result in a complex heat release function, $\Sigma\Phi_k(\eta,T)$. An analysis of the system of combustion equations for various types of heat sources was compared with experimental data, which led to the conclusion that in general, the combustion wave velocity can be represented as follows (Merzhanov, 1990a):

$$U^2 = A(\eta^*, T^*) \exp(-E/RT^*), \tag{26}$$

where T^* and η^* are the values of temperature and corresponding conversion, respectively, which control the combustion front propagation, and $A(T^*, \eta^*)$ is a function weaker than the exponential. Based on Eq. (26), four types of combustion reactions can be identified (Merzhanov, 1977):

1. *Combustion with a thin reaction zone* (discussed earlier), where $\eta^*=1$, $T^*=T_c \approx T_0+Q/c_p$, and for which the combustion velocity is determined by the adiabatic combustion temperature
2. *Combustion with wide zone,* which occurs when a product layer strongly inhibits the reaction, and $T^*=T_0+(Q/c_p)\eta^*$, where η^* is determined from

$$\frac{Q}{c_p} \frac{\partial \phi(\eta^*, T^*)}{\partial T} + \frac{\partial \phi(\eta^*, T^*)}{\partial \eta} = 0$$

3. *Combustion with phase transformations* (e.g., melting reactants), where $T^* = T_m$ (melting point) and $\eta^* = (T_m - T_0)/(Q/c_p)$
4. *Combustion with multistage spatially separated reactions*, where $\eta_1^* = 1$ and $T^* \approx T_0 + (Q_1/c_p)$, where the subscript 1 refers to the first low-temperature stage of the complex reaction.

2. Temperature Profiles

The temperature profile for the preheating zone of the combustion wave can be derived readily by introducing the variable $q = \lambda \, dT/dx$ for the heat flux, so that Eq. (14) rearranges as

$$q \frac{dq}{dT} - \rho c_p U q + \lambda Q \rho \phi(\eta, T) = 0, \qquad (27)$$

and BCs [Eq. (13)] take the following form:

$$\begin{aligned} q = 0, \quad \eta = 0 \text{ at } T = T_0, \\ q = 0, \quad \eta = 1 \text{ at } T = T_c. \end{aligned} \qquad (28)$$

Because there is no heat evolution in the preheating zone, we can neglect the third term of Eq. (27):

$$\frac{dq}{dT} - \rho c_p U = 0; \qquad x \leq 0. \qquad (29)$$

Direct integration of Eq. (29) with BCs [Eq. (28)] yields the Michelson (1930) solution for the temperature profile in the preheating zone:

$$T = T_0 + (T_c - T_0) \exp\left(\frac{U}{\alpha} x\right), \qquad x \leq 0,$$

where $\alpha = \lambda/(\rho c_p)$, and $x_T = \alpha/U$ denotes the characteristic length scale of the combustion wave. Classic combustion theory does not describe the temperature profile in the reaction zone, since it is assumed to be close to T_c. For the four types of combustion synthesis reactions just discussed, the corresponding temperature profiles are shown schematically in Fig. 19a–d. The actual temperature profile may be a combination of several elemental ones (cf Zenin *et al.*, 1980, 1981). Some computer simulations of gasless combustion synthesis exhibit the temperature profile shown in Fig. 19e, where the temperature of the combustion front may exceed T_c^{ad} (Huque and Kanury, 1993). It is interesting to note that this profile resembles early works in combustion of gases, where an ignition temperature model was used (Mallard and Le Chatelier, 1883; Mason and Wheeler, 1917; Daniel, 1930).

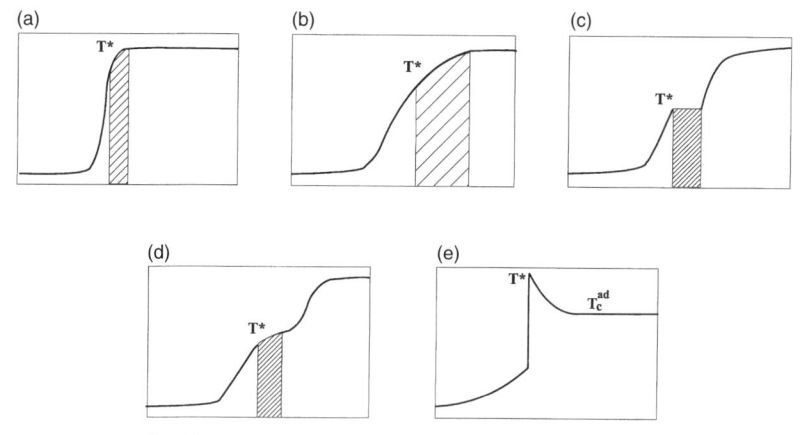

FIG. 19. Elementary temperature profiles of SHS propagation.

B. Microstructural Models

While providing a simple method for analyzing the redistribution of energy in the combustion wave, the models discussed in the previous section do not account for the local structural features of the reaction medium. *Microstructural models* account for details such as reactant particle size and distribution, product layer thickness, etc., and correlate them with the characteristics of combustion (e.g., U, T_c).

The equations used to describe the combustion wave propagation for microstructural models are similar to those in Section IV,A [see Eq. (6)]. However, the kinetics of heat release, Φ_k, may be controlled by phenomena other than reaction kinetics, such as diffusion through a product layer or melting and spreading of reactants. Since these phenomena often have Arrhenius-type dependences [e.g., for diffusion, $\mathscr{D} = \mathscr{D}_0 \exp(-E_D/RT)$], microstructural models have similar temperature dependences as those obtained in Section IV,A. Let us consider, for example, the dependence of velocity, U, on the reactant particle size, d, a parameter of medium heterogeneity:

$$U = f(d)F(T). \tag{30}$$

The first microstructural models were developed independently and essentially simultaneously (Aldushin *et al.*, 1972a,b; Hardt and Phung, 1973; Aldushin and Khaikin, 1974). For these models, the elementary reaction cell, which accounts for the details of the microstructure, consists of alternating lamellae of the two reactants (A and B), which diffuse through a product layer (C), to react (see Fig. 20a). Assuming that the particles are flat allows one to neglect the change in reac-

(a) A(s)+B(s) → C(s)
(b) A(l)+B(s) → C(s)
(c) A(l) + B(s) → C(l)
(d) A(l) + B(l) → C(l)

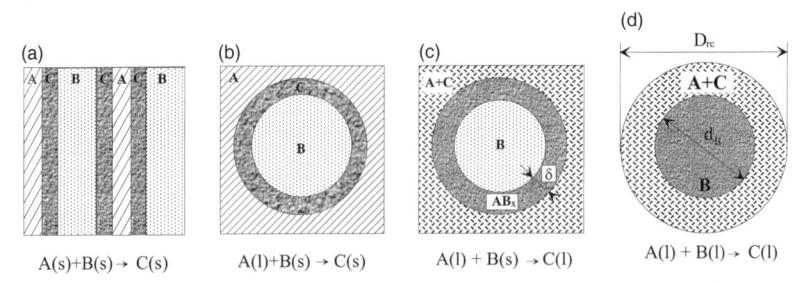

FIG. 20. Geometry of the reaction cells considered in the theoretical models.

tion surface area during synthesis. The characteristic particle size, d, is equivalent to the layer thickness, and the relative thicknesses of the initial reactant layers are determined by stoichiometry.

Hardt and Phung (1973) developed a simple analytical solution for the case of diffusion-controlled kinetics, and found that $f(d) \propto 1/d$. A more accurate expression, based on the same physical geometry, was developed by Aldushin et al. (1972b) using the following kinetic function:

$$\phi(\eta,T) = k_0 \exp(-E/RT) e^{-m\eta} \eta^{-n}, \qquad (31)$$

which describes linear ($m=0$, $n=0$), parabolic ($m=0$, $n=1$), cubic ($m=0$, $n=2$), and exponential ($m>0$, $n=0$) kinetic dependences. The solution of Eqs. (6) and (7) for power-law kinetics (i.e., $m=0$, $n \geq 0$) yields

$$U = \frac{(n+1)(n+2)}{d^{n+1/2}} F(T), \qquad (32)$$

and for exponential kinetics leads to:

$$U = \frac{m^2}{(e^m - 1 - m)d} F(T). \qquad (33)$$

The velocity can also be calculated for polydispersed systems with a particle size distribution function, $\chi(d)$ (Aldushin et al., 1976a). For a unimodal particle size distribution, an effective particle size, d_{eff}, defined as

$$d_{\text{eff}} = \left[\int_{d_{\min}}^{d_{\max}} d^2 \chi(d) dd \right]^{1/2},$$

where $d_{\max}$ and $d_{\min}$ are the upper and lower size limits, respectively, can be used in expressions such as Eqs. (32) and (33). For a bimodal distribution, the combustion wave has a two-stage structure similar to the multistage chemical reactions (Fig. 19d), where two combustion fronts propagate in sequence. Using effective particle sizes for fine (d_1) and coarse (d_2) particles, the velocity dependence for the *leading* combustion front is described by the following equation:

$$U = \frac{F(T)}{[d_1^2 m_1 + d_2^2(1 - m_1)]^{1/2}}, \tag{34}$$

where m_1 is the mass fraction of finer particles.

For the next two types of theoretical models, the elementary reaction cell consists of a spherical particle of one reactant surrounded by a melt of the other reactant. In the first case, the product layers (C) grow on the surface of the more refractory particles (B) due to diffusion of atoms from the melt phase (A) through the product layer (see Fig. 20b). At a given temperature, the concentrations at the interphase boundaries are determined from the phase diagram of the system. Numerical calculations by Nekrasov *et al.* (1981, 1993) have shown satisfactory agreement with experimental results for a variety of systems.

The second model (Fig. 20c) assumes that upon melting of reactant A, a layer of initial product forms on the solid reactant surface. The reaction proceeds by diffusion of reactant B through this layer, whose thickness is assumed to remain constant during the reaction (Aleksandrov *et al.*, 1987; Aleksandrov and Korchagin, 1988). The final product crystallizes (C) in the volume of the melt after saturation. Based on this model, Kanury (1992) has developed a kinetic expression for the diffusion-controlled rate. Using this rate equation, an analytical expression for the combustion wave velocity has been reported (Cao and Varma, 1994)

$$U = \left[\frac{1}{d}\left(\frac{2}{d} + \frac{1}{\delta}\right)\right]^{1/2} F(T), \tag{35}$$

where δ is the thickness of the initial product layer.

For the case where both reactants melt in the preheating zone and the liquid product forms in the reaction zone, a simple combustion model using the reaction cell geometry presented in Fig. 20d was developed by Okolovich *et al.* (1977). After both reactants melt, their interdiffusion and the formation of a liquid product occur simultaneously. Numerical and analytical solutions were obtained for both kinetic- and diffusion-controlled reactions. In the kinetic-limiting case, for a stoichiometric mixture of reactants (A and B), the propagation velocity does not depend on the initial reactant particle sizes. For diffusion-controlled reactions, the velocity can be written as

$$U = \frac{\overline{b}}{d} \cdot F(T), \tag{36}$$

where

$$\bar{b} = \begin{cases} 1 & , \text{ if } d < d_B^{st} \\ \dfrac{d^3}{D_{rc}^3 - d^3} \cdot \dfrac{\nu_B MW_B}{\nu_A MW_A}, & \text{ if } d > d_B^{st}. \end{cases}$$

Here d_B^{st} is the diameter of the refractory reactant (B) in a corresponding stoichiometric mixture, and D_{rc} is diameter of the reaction cell.

The microstructural models described here represent theoretical milestones in gasless combustion. Using similar approaches, other models have also been developed. For example, Makino and Law (1994) used the solid–liquid model (Fig. 20c) to determine the combustion velocity as a function of stoichiometry, degree of dilution, and initial particle size. Calculations for a variety of systems compared favorably with experimental data. In addition, an analytical solution was developed for diffusion-controlled reactions, which accounted for changes in λ, ρ, and c_p within the combustion wave, and led to the conclusion that $U \propto 1/d$ (Lakshmikantha and Sekhar, 1993).

C. Cellular Models

As discussed in previous sections, the reaction medium for the study of combustion wave propagation in powder mixtures is typically assumed to be quasihomogeneous. The implicit assumption is that the width of the heat release zone is much larger than the heterogeneous scale of the medium. In other words, a reacting particle can be considered as a point heat source within the combustion zone. Thus, we replace the heterogeneous system with a homogeneous medium, characterized by some effective thermal properties. As a result, the relations for heat release and transport in a homogeneous medium can be used, as in the classical theory of flame propagation.

Cellular models (Kottke and Niiler, 1988; Astapchik et al., 1993; Hwang et al., 1997) represent another direction for modeling the combustion synthesis process. For these models, the reaction medium is tessellated into a discrete matrix of cells, where the temperature and effective properties are assumed to be uniform throughout the cell. The interactions between the cells are based on various rules, often a discretized form of the heat conduction equation.

At this point, a distinction should be made between cellular and finite difference/element models. The latter are finite approximations of continuous equations [e.g., Eq. (11)], with the implicit assumption that the width of the reaction zone is larger than other pertinent length scales (diffusion, heterogeneity of the medium, etc.). However, no such assumptions need to be made for cellular models, and thus local transport processes in a heterogeneous medium can be accounted for (Hwang et al., 1997).

As with continuous models, the basis of cellular models in CS is the simulation of heat conduction and reaction [see Eq. (11)]. The discrete form of this equation is often given as

$$[\rho c_p]^{(i,j)} \frac{\partial T^{(i,j)}}{\partial t} = \left(\lambda_x \frac{\partial T}{\partial x}\bigg|^{(i+\frac{1}{2},j)} - \lambda_x \frac{\partial T}{\partial x}\bigg|^{(i-\frac{1}{2},j)}\right) + \left(\lambda_y \frac{\partial T}{\partial y}\bigg|^{(i,j+\frac{1}{2})} - \lambda_y \frac{\partial T}{\partial y}\bigg|^{(i,j-\frac{1}{2})}\right) + \left[\rho Q \frac{\partial \eta}{\partial t}\right]^{(i,j)}, \quad (37)$$

where (i,j) refers to the (x,y) position in the grid coordinates. The thermal conductivity and temperature gradient between cells, respectively, are:

$$\lambda_x\bigg|^{(i+\frac{1}{2},j)} = \frac{2\lambda_x^{(i+1,j)}\lambda_x^{(i,j)}}{\lambda_x^{(i+1,j)} + \lambda_x^{(i,j)}}, \quad \frac{\partial T}{\partial x}\bigg|^{(i+\frac{1}{2},j)} = \frac{T^{(i+1,j)} - T^{(i,j)}}{\Delta x}, \quad (38)$$

and similarly for λ_y and $\partial T/\partial y$.

In an early work by Kottke and Niiler (1988), a cellular model was used to simulate the combustion wave initiation and propagation for the Ti+C model system. The interactions between neighboring cells were described by the electrical circuit analogy to heat conduction. At the reaction initiation temperature (i.e., melting point of titanium), the cell is instantly converted to the product, TiC, at the adiabatic combustion temperature. The cell size was chosen to be twice as large as the Ti particles (44 μm). Experimentally determined values for the green mixture thermal conductivity as a function of density were used in the simulations. As a result, the effects of thermal conductivity of the reactant mixture on combustion wave velocity were determined (see Fig. 21). Advani et al. (1991) used the same model, and also computed the effects of adding TiC as a diluent on the combustion velocity.

In a similar manner, Bowen and Derby (1993) developed a cellular model for the reaction

$$3TiO_2 + 4Al + 3C \rightarrow 3TiC + 2Al_2O_3.$$

The one-dimensional form of Eq. (37) was used to describe the distribution of energy in the system. Also, the ignition temperature kinetics were used, where the cell instantly converts to TiC and Al_2O_3 when the ignition temperature (900°C) is reached (determined from DTA experiments). In addition, heat losses due to convection and radiation from the first cell were also taken into account.

The size of the cells was comparable to the largest powder size (i.e., Al). The thermal conductivity of each *phase* was determined using a model developed by Luikov et al. (1968), with the thermal conductivity of the *cell* determined using

132   A. VARMA ET AL.

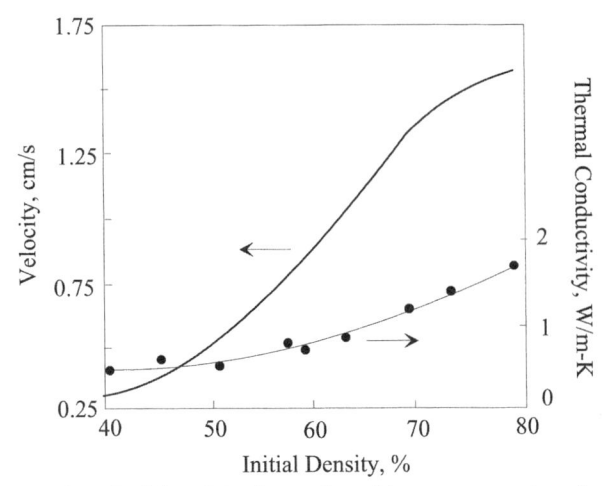

FIG. 21. Thermal conductivity of titanium and graphite compacts and predicted combustion velocity, as a function of sample density (Adapted from Kottke et al., 1990).

the three-phase rule (Brailsford and Major, 1964). The temperature–time profiles were analyzed (see Fig. 22) and compared to the mechanism of structural transformation (i.e., melting of aluminum, solidification of alumina). The effect of varying cell size (i.e., the diameter of the Ti particle) was also considered for the Ti+2B system, where the reaction velocity decreased with increasing cell size.

Another application of cellular models was based on a stochastic, rather than deterministic, approach (Astapchik et al., 1993). Based on the same equations for heat transfer as described earlier, the criterion for ignition was not when the tem-

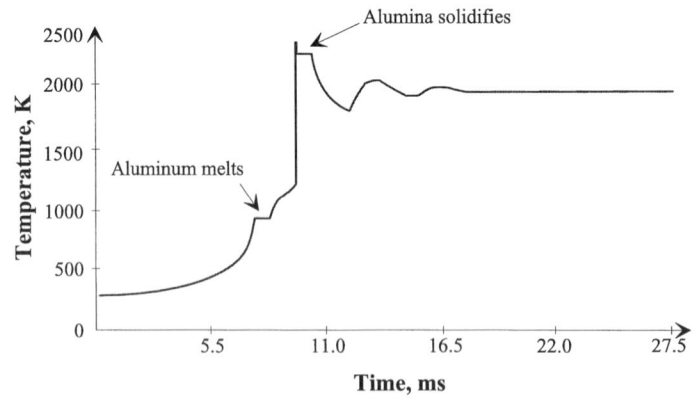

FIG. 22. Temperature–time profile of a single cell for the TiO_2-Al-C reaction (Adapted from Bowen and Derby, 1993).

perature of the cell reached an ignition temperature, but the *probability* that it would react (as a function of temperature of the cell):

$$p(T) = k_0\tau \cdot \exp\left(-\frac{E}{RT}\right), \tag{39}$$

where τ is the time step and k_0 is the preexponential factor for the reaction rate. The stochastic approach accounts for any nonuniform surface structure, shape, size, etc. The results of the simulations were compared to deterministic velocities. The occurrence of oscillatory unstable behavior and its dependence on cell size was also studied using this approach.

In all the cellular models described, the cells throughout the medium were initially uniform. A *microheterogeneous cell model* has recently been developed, where the cells have varying initial properties (Hwang et al., 1997). In such a model, the *structure* of the reaction medium is stochastic, and the heterogeneity of the reactant medium is considered explicitly. For example, the structure can be described as a regular two-dimensional matrix of circular cylinders (with nominally flat sides) in contact, with a number of cylinders randomly removed until the ratio of the voids to total volume equals the porosity (Fig. 23a). The cylinder diameter is assumed to be similar to the particle diameter.

The porous structure is divided into cells of dimension l (see Fig. 23b). An elastic contact model (Cooper et al., 1969) is used to describe the contact region between two particles, which is related to the loading pressure and mechanical

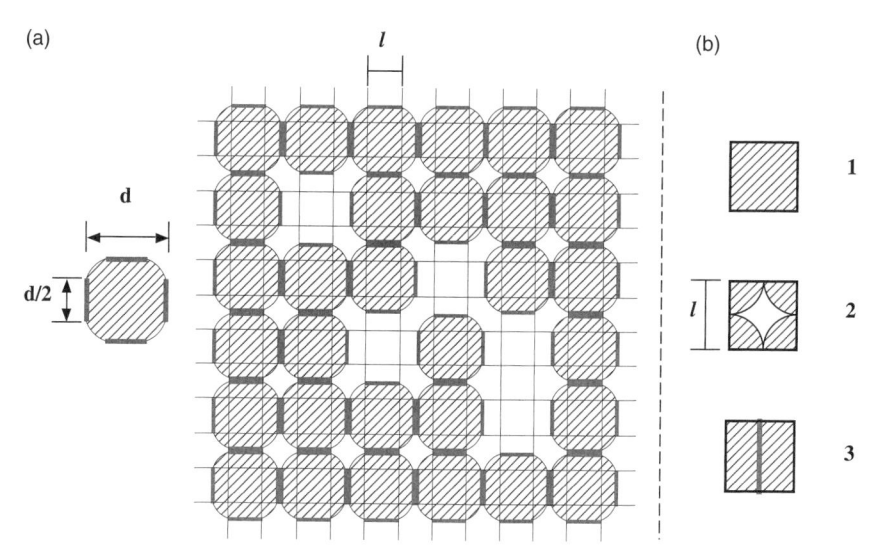

FIG. 23. Model structure of porous reaction medium: (a) Overall view, (b) schematic diagram of cell model structure.

properties of the particle. This gives rise to three basic cell types, which consist of "cores" (Fig. 23b, 1), "interstices" (Fig. 23b, 2), and "edges" (Fig. 23b, 3). The cylinder cores (Fig. 23b, 1) have the highest thermal conductivity corresponding to the bulk value for the solid, while the interstices (Fig23b, 2) have the lowest value owing to a large fraction of gas. The cylinder edges have an intermediate thermal conductivity, which corresponds to the thermal contact resistance between two particles. To determine the effective thermal conductivity of each cell, the work of Goel *et al.* (1992) was used, which accounts for the distribution of phases within the cell. The heat transfer between cells is described by Eq. (37). It is assumed that the reaction within the cell is initiated when the cell reaches a certain ignition temperature.

With this model, the microstructure of the combustion wave was studied, and compared with experimental results (Hwang *et al.*, 1997; Mukasyan *et al.*, 1996). For example, sequences of combustion front propagation at the microscopic level, obtained experimentally and by calculation, are shown in Fig. 24. In addition, it was demonstrated that fluctuations in combustion wave shape and propagation correlate with the heterogeneity of the reactant mixture (e.g., porosity and particle size).

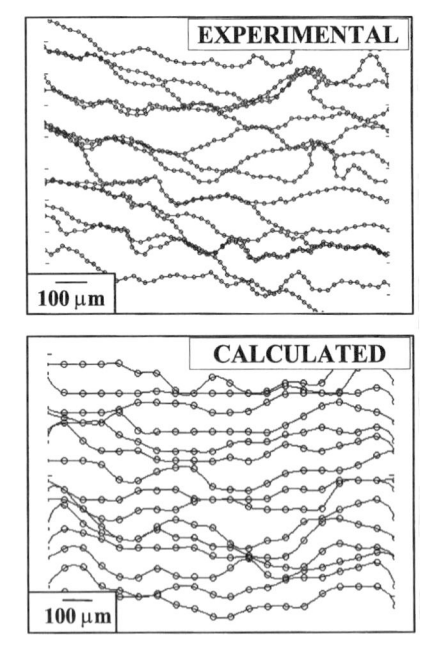

FIG. 24. Profiles of the experimental and calculated combustion wave microstructure.

D. STABILITY OF GASLESS COMBUSTION

The stability of combustion wave propagation is an important factor in determining the quality of materials produced by CS. To produce uniform product, a stable combustion regime is desired. Furthermore, it is also important to know the boundaries where combustion, stable or unstable, can propagate.

From classic combustion theory (Zeldovich et al., 1985), the following two conditions must be satisfied for a constant pattern combustion wave with thin reaction zone to be self-propagating:

$$\beta = \frac{RT_c}{E} \ll 1, \quad \gamma = \frac{c_p}{Q}\frac{RT_c^2}{E} = \frac{c_p T_c}{Q}\beta = \frac{RT_c^2}{E(T_c - T_0)} \ll 1, \quad (40)$$

so that the wave velocity can be computed using Eq. (21).

Numerical results indicate, however, that a condition weaker than Eq. (40) is sufficient to ensure constant-pattern propagation (Aldushin et al., 1978):

$$\Psi = \gamma - \beta = \frac{RT_0 T_c}{E(T_c - T_0)} \ll 1. \quad (41)$$

In essence, the preceding conditions require the reaction rate at the combustion temperature, T_c, to be much greater than that at the initial temperature, T_0. The ratio of reaction rates at the two temperatures is proportional to

$$\frac{\exp(-E/RT_c)}{\exp(-E/RT_0)} = \exp\left(-\frac{1}{\Psi}\right). \quad (42)$$

Puszynski et al. (1987) have suggested the use of the homogeneous explosion criterion for the existence of constant-pattern propagation:

$$\frac{2RT_c}{E}\left(\frac{2T_c}{T_c - T_0} + 1\right) < 1, \quad (43)$$

which may be rearranged as

$$2\gamma + \beta < \frac{1}{2}. \quad (44)$$

According to the theory of Zeldovich (1941) developed for premixed gas flames, the velocity of the combustion front just before the wave ceases to propagate (e.g., by addition of inert dilent of mole fraction, b) is lower, compared to the maximum value (i.e., with no diluent added), by a factor of $\sqrt{e}$. It was shown experimentally that this conclusion can also be applied to gasless systems with a narrow reaction zone (Maksimov et al., 1965). Based on this idea and using Eq.

(22), Munir and Sata (1992) have suggested a criterion to determine the SHS/non-SHS boundary:

$$\left(\frac{U^{max}}{U^{min}}\right)^2 = (1-b)\left(\frac{T^{max}}{T^{min}}\right)^2 \exp\left[-\frac{E}{R}\left(\frac{1}{T^{max}} - \frac{1}{T^{min}}\right)\right] = e, \qquad (45)$$

where the superscript "max" refers to the maximum combustion temperature (T_c) and "min" corresponds to the lowest combustion temperature where the reaction wave will still propagate. Based on Eq. (45), an interesting method to describe the regimes of combustion synthesis (called *SHS diagrams*) was proposed, and a good match with experimental data for a variety of systems was reported (Munir and Sata, 1992; Munir and Lai, 1992). An example of such a diagram for the TiB_2 synthesis is shown in Fig. 25.

A stability criterion for steady wave propagation was developed by numerical calculation, in the form of an approximate relation (Shkadinskii *et al.*, 1971):

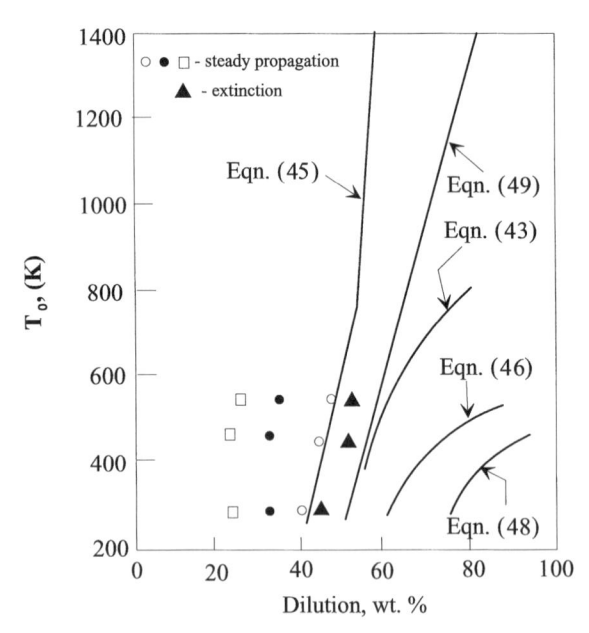

$$\alpha_{st} = \left(9.1\frac{T_c}{T_c - T_0} - 2.5\right)\frac{RT_c}{E}. \qquad (46)$$

FIG. 25. The SHS diagram for TiB_2 synthesis (Adapted from Munir and Sata, 1992). The curves are computed values; the symbols represent experimental data.

The boundary separating the steady and oscillating combustion modes is given by the following conditions:

$$\alpha_{st} > 1, \text{ stable steady-state combustion,}$$
$$\alpha_{st} < 1, \text{ oscillatory combustion.} \tag{47}$$

Formula (46) is in good agreement with an analytical criterion for stable propagation under the condition of a narrow reaction zone (Makhviladze and Novozilov, 1971; Matkowsky and Sivashinsky, 1978):

$$\frac{E(T_c - T_0)}{2RT_c^2} < (2 + \sqrt{5}). \tag{48}$$

For the case of melting reactants (Margolis, 1983, 1985), the stability criterion is modified as follows:

$$\frac{E(T_c - T_0)}{2RT_c^2 \left\{ 1 - \exp\left[\frac{E(T_c - T_m)}{RT_c^2} \right] \right\}} < (2 + \sqrt{5}). \tag{49}$$

Based on Eq. (49), it can be concluded that for gasless combustion processes, unstable combustion is more likely to occur for higher melting temperatures, T_m, in addition to higher activation energies and lower combustion temperatures (Margolis, 1992).

An interesting application of stability analysis was performed for the case of porous samples with melting and flow of reactants (Aldushin et al., 1994a). It was determined how various parameters of the system affect stability, and the pulsing instabilities of uniformly propagating solutions were found. These instabilities can result in nonuniform composition of the product.

Analysis of the theoretical expressions proposed for the boundaries between stable propagation and oscillations, as well as between combustion and noncombustion regions, leads one to the conclusion that in spite of good agreement of theory with experimental data reported in some cases, a direct comparison is generally not available for most SHS systems. In this context, we agree with Novozilov (1992), who concluded that current theoretical results cannot be compared quantitatively with experimental data. According to his analysis, this is due, first, to the fact that in analytical and numerical investigations, the mechanism chosen to simulate combustion process was oversimplified. Thus, for example, most theoretical studies assume an infinitely narrow reaction zone, and the majority of the computations were performed for one-step reactions. In addition, accurate thermophysical and kinetic data are not available for the extreme conditions associated with the combustion wave. Hence, new experimental results describing the detailed mechanisms of combustion synthesis must be obtained in order to develop the theory further and to bridge the gap between the often abstract mathematical approach and the experimental reality of the process. Experimental techniques that provide details about the reaction mechanism are described in Section VI.

E. FILTRATION COMBUSTION THEORY

In contrast with the theory for gasless reactions (Sections IV,A and IV,B), which was adapted from classical combustion theory, filtration combustion (FC) theory was developed specifically to describe gas–solid reactions in CS (Merzhanov et al., 1972; Aldushin et al., 1974).

A general schematic of FC is shown in Fig. 26. The initial reaction medium consists of a porous matrix of solid reactants and inert diluents, where the pores are filled with gas-phase reactants. The combustion front propagates through the sample with a velocity, U, due to the chemical interaction between the gas- and condensed-phase reactants. Behind the front, the final product is formed, which in some cases may approach poreless structure, since the volume of the final product grains is typically greater than that of reactant particles. The amount of reactant gas present in the pores decreases rapidly in the combustion front owing to the intense gas–solid reaction. The resulting pressure gradient is the driving force for infiltration of gas from the surroundings, which transports the gas-phase reactant and thermal energy (by convection). With this type of combustion, the important consideration is the availability and permeability of reactant gas to the reaction zone.

Depending on reaction and permeation conditions, two modes of FC can be defined: *natural* and *forced* filtration. In the first case, infiltration flow is a consequence of the natural pressure gradient between the atmosphere (typically constant) and the reaction zone. In the second case, forced gas flow is induced by

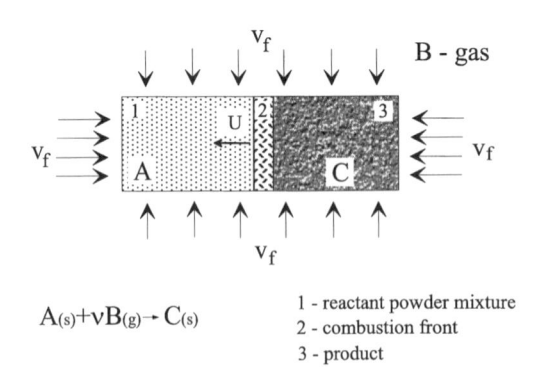

$A_{(s)} + \nu B_{(g)} \to C_{(s)}$

1 - reactant powder mixture
2 - combustion front
3 - product

FIG. 26. Schematic of filtration combustion.

maintaining a pressure gradient along the sample. In general, current SHS technologies are based on the natural filtration combustion mode.

For natural permeation combustion, the reactant gas may flow in either the same (cocurrent flow, Fig. 27a) or opposite (countercurrent flow, Fig. 27b) direction as that of combustion wave propagation, and in some cases even in both directions Fig. 27c). One-dimensional flow regimes can be achieved by placing the sample in a quartz tube, which in turn can be closed at one end (Pityulin *et al.*, 1979). In the general case (see Fig. 26), the reactant gas also enters from the side surface, and the process can be described by two- or three-dimensional numerical simulations (cf Ivleva and Shkadinskii, 1981; Dandekar *et al.*, 1990, 1993). In this section, results for the simpler one-dimensional case are analyzed and features of the higher dimensional studies are briefly discussed.

1. One-Dimensional Models with Natural Permeation of Gas

Let us consider the reaction front propagation for a binary system:

$$A(s) + \nu B(g) \rightarrow C(s,l).$$

In addition to equations for heat conduction and chemical kinetics used for gasless systems [cf Eq. (6)], gas–solid reactions require equations describing the

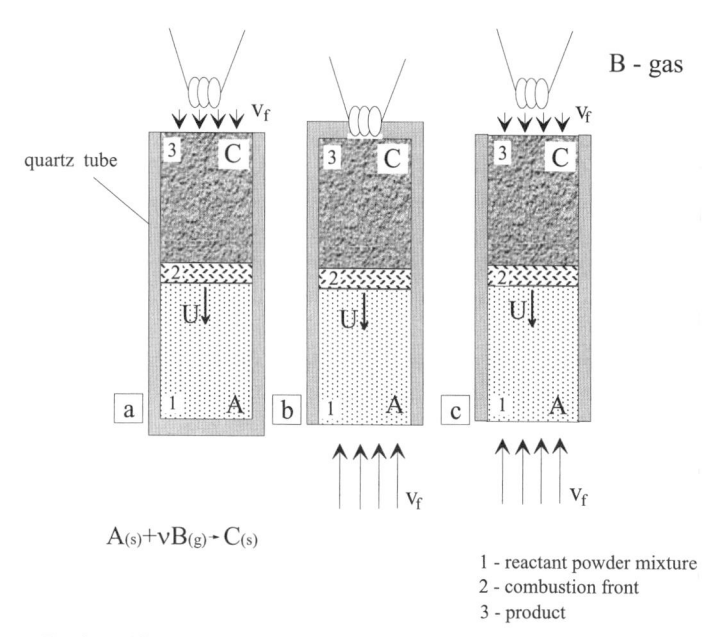

FIG. 27. Different types of filtration combustion (one-dimensional schemes).

1 - reactant powder mixture
2 - combustion front
3 - product

mass balance for the gas-phase reactant, gas flow, and equation of state. In this case, stable combustion wave propagation with velocity U, where the coordinates are attached to the moving front, is described by the following set of equations (Aldushin et al., 1976b):

$$U\frac{dT}{dx} = \alpha\frac{d^2T}{dx^2} + qU\frac{d\eta}{dx}, \quad (50)$$

$$\frac{d}{dx}(U\rho_g + v_f\rho_g) = -\nu\rho_s U\frac{d\eta}{dx}, \quad (51)$$

$$U\frac{d\eta}{dx} = W(\eta,T,p), \quad (52)$$

$$\rho_g = \frac{\epsilon p}{RT}, \quad v_f = -k_f\frac{dp}{dx}, \quad (53)$$

where

$$q = \frac{Q}{c_{p,s}(1 + \Delta \cdot \eta_c)}, \quad \alpha = \frac{\lambda}{c_{p,s}\rho_s(1 + \Delta \cdot \eta_c)}, \quad \Delta = \nu\frac{c_{p,g}}{c_{p,s}}$$

and p is the gas pressure, ϵ is the sample porosity, ρ_i is the amount of reactant i per unit sample volume, $c_{p,i}$ is the specific heat of reactant i, λ is the effective thermal conductivity of the porous medium, v_f is velocity of gas filtration, k_f is the permeability coefficient (k/μ), $W(\eta,T,P)$ is the chemical reaction rate, ν is the stoichiometric coefficient of the gas reactant, and the subscripts "s" and "g" refer to the solid and gas reactants, respectively.

Because it is difficult to account for changes in the properties of the reaction medium (e.g., permeability, thermal conductivity, specific heat) due to structural transformations in the combustion wave, the models typically assume that these parameters are constant (Aldushin et al., 1976b; Aldushin, 1988). In addition, the gas flow is generally described by Darcy's law. Convective heat transfer due to gas flow is accounted for by an effective thermal conductivity coefficient for the medium, that is, quasihomogeneous approximation. Finally, the reaction conditions typically associated with the SHS process ($T \geq 2000$ K and $p < 10^3$ MPa) allow the use of ideal gas law as the equation of state.

According to Aldushin et al. (1976b), there are two characteristic length scales that should be compared to examine the filtration of reactant gas from the atmosphere to the reaction front: the length of the sample, L, and the length of the filtration during combustion, l_f. If the characteristic filtration length is much shorter than that of the sample ($l_f \ll L$), then there is no filtration from the atmosphere to the combustion front, and the reaction propagation occurs only due to the gas present initially in the pores. On the other hand, when the characteristic length of filtration is much longer than the sample length ($l_f \gg L$), the reactant gas permeates from the atmosphere to the combustion front. In both cases, the reactant gas may infiltrate to the sample in the postcombustion zone *after* the combustion front has passed.

a. Combustion without Filtration to the Reaction Front from the Atmosphere. Because only the gas initially in the sample pores is available for propagation of the combustion wave, in practice, this FC regime can be realized at relatively high reactant gas pressures. For example, the minimum nitrogen pressures (p_0^{cr}) required to form nitrides from samples with initial porosities equal to 50% are presented in Table XVI. At such pressures, the amount of reactant gas in the sample pores is sufficient for full conversion to the final product. However, thermal expansion in the preheating zone and gas consumption in the reaction zone lead to pressure gradients in the medium. Thus, gas permeation within the sample may become an important factor for combustion front propagation.

Analysis of Eqs. (50)–(53), along with the BCs:

$$x \to -\infty, \quad T = T_0, \quad \eta = 0, \quad p = p_0,$$
$$x \to +\infty, \quad T = T_c, \quad \eta = \eta_c, \quad p = p_c, \quad (54)$$

demonstrates that another dimensionless parameter, L_f, can be defined (Aldushin et al., 1976b):

$$L_f = \frac{l_f}{x_T} = \frac{k_f p_0/2U}{\alpha/U} = \frac{k_f p_0}{2\alpha}, \quad (55)$$

which is the ratio of the filtration zone length, l_f, to the thermal length scale of combustion, x_T. Although analogous to the Lewis number, Le, the filtration parameter, L_f, is not assumed to be much smaller than unity [cf Eq. (5)], and can vary significantly for different conditions. Values for L_f are presented in Table XVII for

TABLE XVI
CRITICAL NITROGEN PRESSURES (p_0^{cr}) FOR COMPLETE CONVERSION (AT $T = 300$ K, $\epsilon = 0.5$)

System	Product	Critical Pressure (atm)
Nb–N$_2$	NbN	885
Zr–N$_2$	ZrN	900
Ta–N$_2$	TaN	1000
Ti–N$_2$	TiN	1170
Al–N$_2$	AlN	1250
Si–N$_2$	Si$_3$N$_4$	1375
B–N$_2$	BN	2700

TABLE XVII
VALUES OF FILTRATION PARAMETERS FOR VARIOUS SOLID-NITROGEN SYSTEMS

System	p_0 (N/m^2)	α (m^2/s)	k_f [m^4/(N·s)]	L_f	L_f^{cr}	L_f/L_f^{cr}
Zr–N$_2$	10^5–10^7	10^{-4}	10^{-8}	10–10^2	~7	~1–10^2
Ta–N$_2$	2×10^6–3×10^8	2×10^{-5}	4×10^{-11}	10^2	10	10
Si–N$_2$	10^{-7}	10^{-7}	10^{-15}	~10^{-1}	~7	10^{-2}
B–N$_2$	3×10^8	10^{-7}	10^{-18}	~10^{-3}	~8	10^{-4}

various gas–solid systems at different initial pressures. The filtration parameter has been shown to determine the pressure distribution in the combustion wave (cf Aldushin, 1988). If L_f is above some critical value ($L_f^{cr}=T_c/T_0$), then the pressure decreases monotonically to the reaction front (Fig. 28a). Otherwise, a maximum in calculated pressure is observed (Fig. 28b), due to gas expansion in the preheating zone when the permeability of the medium is low. From a practical point of view, elevated pressures in the combustion front may adversely affect product quality, since cracks may form in the material. However, these reaction conditions may also yield unique phase compositions, such as cubic boron nitride (Borovinskaya, 1977).

Using Eqs. (50)–(53), along with the narrow-zone approximation, an expression for the combustion velocity in gas–solid systems can be obtained that is sim-

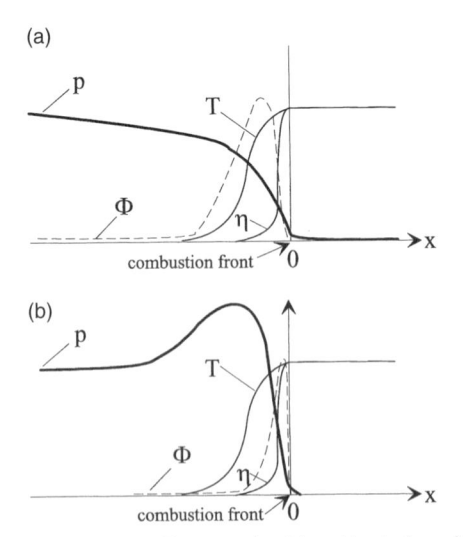

FIG. 28. Gas pressure (p), temperature (T), conversion (η), and heat release function (Φ) distributions in the combustion front: (a) $L_f/L_f^{cr} > 1$; (b) $L_f/L_f^{cr} < 1$ (Adapted from Aldushin et al., 1976b).

ilar to that derived for gasless systems [Eq. (21)]. For this, let us consider a reaction rate function of the form

$$W(\eta,T,p) = \frac{d\eta}{dt} = k_0 \exp(-E/RT) \cdot f(\eta) \cdot p^m, \qquad (56)$$

where $f(\eta)$ is the dependence of reaction rate on conversion of the solid reactant and m is the reaction order relative to the gaseous reactant. Using these kinetics and assuming an average gas pressure, $\bar{p}$, in the reaction zone, we obtain (Aldushin, 1988)

$$U^2 = \frac{\lambda \cdot F(\eta) \cdot k_0 \exp(-E/RT_c)}{Q(T_c)\rho_s} \cdot \frac{RT_c^2}{E} \cdot (\bar{p})^m, \qquad (57)$$

where the function $F(\eta)$ depends on $f(\eta)$. For kinetics $f(\eta)=\eta^{-n}$, describing reaction retardation by the product layer, the function $F(\eta)$ becomes

$$F(\eta) = (n+1)(n+2). \qquad (58)$$

As a result, the only convenient method for controlling the combustion velocity is by varying the average gas pressure, $\bar{p}$, in the reaction zone. For the case of excess gas reactant ($\nu_0 = \rho_{g0}/\rho_{s0} > \nu$), $\bar{p}$ is approximately equal to the gas pressure of the combustion products:

$$\bar{p} = p_0 \frac{T_c}{T_0}\left(1 - \frac{\nu}{\nu_0}\right) \qquad (59)$$

On the other hand, for the case where reactant gas is deficient (i.e., $\nu_0 < \nu$), using the approximate relation $p^2 = b(T_c - T)$, the following expression was obtained:

$$\bar{p} = \left(b \cdot \frac{RT_c^2}{E}\right)^{m/2} \Gamma(1 + m/2), \qquad (60)$$

where $\Gamma(x)$ is the Gamma function, and

$$b = \frac{2\alpha\rho_{g0}RT_c}{\epsilon k_f} \frac{p_0}{(T_c - T_0)}. \qquad (61)$$

It is interesting to note that the average gas pressure in the reaction zone (and consequently the combustion velocity) decreases with higher sample permeation coefficients, k_f, and porosity, ϵ, owing to higher reaction rates. In addition, the relationship between combustion temperature and initial pressure, for an excess of ($\nu_0 > \nu$) and lack ($\nu_0 > \nu$) of reactant gas, can be represented as

$$\begin{aligned}\text{for } \nu_0 > \nu, \quad & T_c = T_0 + \frac{Q \cdot \rho_{s0}}{c_{p,g}\rho_{g0} + c_{p,s}\rho_{s0}}, \\ \text{for } \nu_0 < \nu, \quad & T_c = T_0 + \frac{Q}{c_{p,g}\rho_{g0} + c_{p,s}\rho_{s0}} \cdot \frac{\rho_{g0}}{\nu}.\end{aligned} \qquad (62)$$

In general, the maximum combustion temperature and velocity are achieved for a stoichiometric mixture of reactants.

b. Combustion with Filtration to the Reaction Front from the Atmosphere. Let us now consider the case when there is not enough reactant gas initially in the pores for propagation to occur. In this case, layer-by-layer combustion with complete conversion can only be achieved with infiltration of gas from the surroundings. Two directions of gas flow relative to that of propagation can be considered: countercurrent (Fig. 27b) and cocurrent (Fig. 27a). In both cases, for samples of finite length, the distance between the reaction front and the exposed surface, $L(t)$ changes with time. When wave propagation results from filtration to the combustion front rather than the gas initially present in the pores [i.e., $\eta_c \gg (\rho_{g0}/\rho_s)\nu$], analysis has shown that a quasisteady approximation [i.e., $L(t)$ becomes a parameter] can be used to solve Eqs. (50)–(52). Furthermore, the reactant gas flow term, $U\rho_g$, attributed to front propagation is considered negligible when compared to filtration flow, and can be neglected in the gas balance equation (51).

(i) Countercurrent Infiltration. For gas flow in the direction opposite that of the combustion propagation, the boundary conditions may be written as (Aldushin et al., 1974):

$$\text{at } x = L(t), \quad T = T_0, \quad \eta = 0, \quad p = p_0, \tag{63}$$

$$\text{at } x = -L_1(t), \quad \frac{dT}{dx} = \frac{dp}{dx} = 0, \quad W(\eta_c, T_c, p_c) = 0, \tag{64}$$

where $L_1(t)$ is the length of the reacted portion. For gas–solid systems, the reaction can terminate [i.e., $W(\eta,T,p) = 0$] when either the solid reactant is completely converted or when there is not enough reactant gas. The solution of Eqs. (50)–(53) determines which of the two cases is realized.

Theoretical analysis of the problem (Aldushin et al., 1974, 1975) shows that combustion wave propagation can occur for incomplete conversion of the solid reactants. Whether full conversion is achieved can be estimated by the dimensionless parameter, $\omega = U_k/v_f$, the ratio of the kinetically controlled combustion velocity (U_k), and gas filtration (v_f) rates. The former can be calculated from Eq. (57), while the latter can be calculated as follows:

$$v_f = \frac{L_f \alpha}{L(t)} \cdot \frac{\rho_{g0}}{\rho_{s0} \nu} \cdot \frac{\Pi^{m/4}}{\sqrt{\Gamma(1 + m/2)}}, \tag{65}$$

where

$$\Pi = \frac{E(T_c - T_0)}{RT_c^2} \cdot \frac{T_0}{T_c} \cdot \frac{L_f(1 + \Delta)}{\nu}.$$

When $\omega<1$, there is no filtration limitation for the solid to react completely in the combustion front ($\eta_c=1$), and $T_c=T_c^{ad}$. In this case, the combustion velocity equals U_k. On the other hand, for $\omega>1$, full conversion is not achieved in the combustion front ($\eta_c<1$) and the combustion temperature is $T_c=T_0+Q/c_m(1+\eta_c\delta)$. For highly exothermic reactions (i.e., $\kappa=T_0 c_{p,s}/Q\ll 1$), the reaction conversion can be expressed as (Aldushin et al., 1974)

$$\eta_c = \frac{1 - \kappa(1+\Delta)\cdot\dfrac{RT_c^{ad}}{E}\cdot\ln\omega}{1+(1+\Delta)\cdot\dfrac{RT_c^{ad}}{E}\cdot\ln\omega}. \tag{66}$$

Thus, the degree of conversion depends significantly on the heat of reaction, and not much on the other parameters. It is interesting that decreasing Q (e.g., by dilution) results in increasing η_c. This occurs because the decrease of T_c reduces the reaction rate, so that the effect of filtration limitation is not as severe. Finally, in this case, the combustion velocity is given by $U=v_f/\eta_c$.

Let us consider the tantalum–nitrogen system to illustrate different combustion regimes. Since the adiabatic combustion temperature for this system ($T_c^{ad}=3000$ K) is lower than the melting point of tantalum (3300 K), no melting occurs in the combustion front, and the permeability of the reaction medium changes only slightly in the preheating zone. As a result, the Ta-N_2 system can be described by the theory discussed earlier. In addition, the thermal conductivity of the sample remains essentially unchanged during reaction. The reaction kinetics for 10-μm Ta powders are given by

$$\frac{d\eta}{dt}=\frac{k(T)}{\eta}$$

where $k(T)=1.2\exp(-77600/RT)\,10^6\,\text{s}^{-1}$ (Vadchenko et al., 1980). Substituting the values for λ (59 W/m K), $m=0$, and Q (6.7×10^5 J/kg) (Samsonov, 1964) along with green mixture density, $\rho_0=20\%$, yields $U_k=3$ cm/s, which is comparable to the experimentally measured value, $U_{exp}\sim 2$ cm/s (Pityulin et al., 1979). From Eq. (65), the filtration velocity can be estimated from

$$v_f = \frac{k_f p_0}{2L}\cdot\frac{\nu_0}{\nu}. \tag{67}$$

Using the k_f data from Table XVII, the calculated value for U_k, and the criterion $\omega=1$, we can estimate the maximum sample length, L_{max}, in which complete conversion can be achieved for different nitrogen pressures, p_0 (Table XVIII). Thus, for example, the external nitrogen pressure must be greater than 100 atm in order to achieve complete conversion in a 1-cm-long sample.

TABLE XVIII
CRITICAL SAMPLE
LENGTH (L_{max})
FOR COMPLETE SOLID
CONVERSION
IN TA-N$_2$ SYSTEM

p_0 (atm)	L_{max} (cm)
1	10^{-2}
10	10^{-1}
100	1
1000	10

(ii) Cocurrent Infiltration. In practice, cocurrent infiltration of reactant gas to the combustion front is rarely realized, because the permeability of the final product is generally lower than that of the initial mixture. Consequently, this type of combustion can be achieved experimentally only if the sample surface ahead of the combustion front is enclosed (Fig. 27a). However, the cocurrent regime is attractive from the applications viewpoint, because in contrast with the countercurrent mode, it can occur with *full* conversion over a wider range of reaction parameters.

The characteristics of this type of combustion were investigated first by Aldushin *et al.* (1980). It was shown that because the reactant gas flows over the product, the reaction is not limited by the concentration of reactant gas. However, since the filtration length increases progressively with time, the combustion velocity decreases ($U \sim t^{-1/2}$). On the other hand, the combustion wave maintains a constant temperature profile. Two possible cases can arise depending on the sample permeability. When it is low, the combustion temperature may be lower than the adiabatic value, due to heat carried away from the reaction front by the gas flow. For high sample permeabilities, the heat gained by the infiltration of reactant gas flowing through hot product may cause T_c to be greater than the adiabatic temperature.

A model of filtration combustion in a thin porous layer, immersed in a bath of gaseous reactant, has also been investigated (Shkadinskii *et al.*, 1992a). In this case, only filtration of gas from the surroundings to the sample, normal to the direction of combustion propagation (cross-flow filtration), should be considered. New pulsating instabilities associated with the gas–solid chemical reaction and mass transfer of gas to the porous medium were identified.

2. Two-Dimensional Models with Natural Permeation of Gas

Covering the side surfaces of the sample limits filtration to one dimension, which simplifies the experimental and theoretical study of gas–solid combustion systems. However, two- and three-dimensional infiltration of the gaseous reactant is more commonly used in practice. Ivleva and Shkadinskii (1981) were the first to model and numerically analyze two-dimensional FC for cylindrical samples reacted under constant pressure.

More recently, the cross-flow filtration combustion configuration (Fig. 29), commonly used for powder production, was investigated (Dandekar *et al.*, 1990). Within the reactor, a rectangular container holds the solid reactant powder, and the gaseous reactant is transported from the surroundings to the front not only by longitudinal flow from above, but also by countercurrent flow from ahead of the front. The assumptions used in the model were as follows:

- The gas pressure of the surroundings remains constant (i.e. an open reactor).
- The reaction rate is diffusion controlled at the single particle level.
- The reactant mixture is pseudohomogeneous.
- The physical properties of all reactants are constant.
- The permeability of the reaction medium is a function of porosity, given by the Kozeny–Carman equation.
- The thermal conductivity and heat capacity of the reaction medium are linear functions of porosity.
- No dissociation of the formed product occurs.

Also, changes in reaction mixture porosity, ϵ, during the process were accounted for by the volumetric expansion of solids due to chemical reaction.

It was shown that the reactant gas pressure strongly influences front propagation (Dandekar *et al.*, 1990). At high pressures ($\nu_0 \geq \nu$), the combustion front is essentially planar, although a hot-spot may appear under certain conditions. This effect may lead to melting or sintering of the solid, creating an undesired nonuniform structure. At low pressures ($\nu_0 < \nu$), the filtration resistance controls reaction front propagation and for this reason, the front is skewed forward at the top of the sample. Also, the conversion is complete at the surface, and decreases deeper into the sample. Thus, the depth of the container is also an important pa-

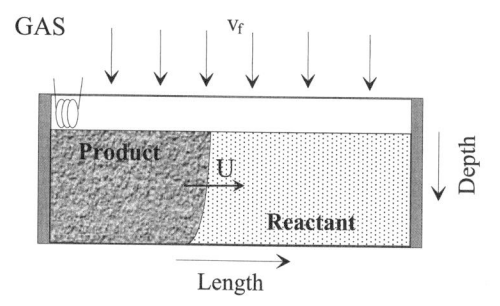

FIG. 29. Cross-flow combustion configuration, two-dimensional scheme.

TABLE XIX
Typical Values of Expansion Coefficient, Z

Solid Reactant	Nitride	Z
Hf	HfN	1.02
Zr	ZrN	1.06
Ti	Ti	1.11
Nb	NbN	1.19
Ta	TaN	1.23
Si	Si_3N_4	1.20
Al	AlN	1.25
B	BN	2.40

rameter. For a given pressure, it was shown that there exists a critical container depth above which the combustion front extinguishes in the bulk. In addition, the ratio of solid product and solid reactant volumes, Z, was used to analyze the process. For example, if Z>1, then the solid expands on reaction, decreasing the medium permeability. Typical values of Z for several nitrogen–solid systems are given in Table XIX.

Numerical simulations analyzed the effect of sample porosity on reaction. For $\epsilon_0=0.4$, $L_f=17.5$, and $Z=1.2$, the porosity decreased rapidly along with sample permeability, due to high conversion near the point of ignition (see Fig. 30a). The

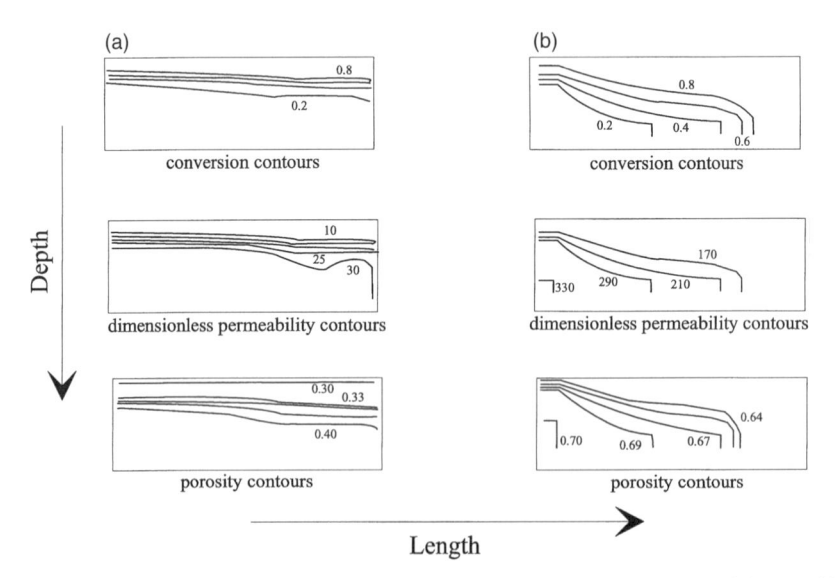

Fig. 30. Effect of initial porosity on reaction: (a) lower initial porosity, $\epsilon_0 = 0.4$, $\tau = 40$; (b) higher initial porosity, $\epsilon_0 = 0.7$, $\tau = 11$ (Adapted from Dandekar et al., 1990).

reaction front eventually stops when gas filtration becomes severely limited. For higher porosity ($\epsilon_0=0.7$), the filtration parameter L_f equals 165, and permeation of gas to the deeper layers is not restricted, so complete conversion can be achieved (see Fig. 30b).

Recently, the cross-flow configuration was also investigated for a closed reactor, that is, constant volume (Grachev et al., 1995). In this case, the amount of gas-phase reactant decreases due to chemical reaction, which would decrease the gas pressure. However, the gas is heated during reaction, which would increase the gas pressure. Depending on the reactor volume, for typical reaction conditions, the gas pressure was shown to have a maximum either at full or intermediate conversion. This conclusion is important, since in practice a decrease in gas pressure is usually associated with reaction completion.

The role of changing porosity and medium permeability due to volumetric expansion of the solids during chemical reaction was also studied for the titanium–nitrogen system (Borovinskaya et al., 1993). In this work, it was assumed that the reaction of titanium and nitrogen followed the equilibrium phase diagram. Thus, the feasibility of combustion due to formation of solid solution was demonstrated, and changes in the melting point of intermediate phases with addition of nitrogen could be described.

F. OTHER ASPECTS

A variety of results in FC for heterogeneous systems have recently been reported. Some of these are briefly discussed next.

1. Forced Gas Permeation

As mentioned previously, the gas–solid CS process can be conducted with forced infiltration of the reactant gas through the porous medium. In this case, two combustion modes can be realized, depending on the parameter (Aldushin and Sepliarskii, 1987a,b):

$$\pi = \frac{v_T}{U},$$

which is the ratio of the velocity of the convective heat exchange front, $v_T = c_{p,g} G/c_{p,v}$ (where G is the forced gas velocity, and $c_{p,v}$ is the effective specific heat capacity of the reaction medium) and the combustion velocity, U.

In the case of a relatively low forced gas flux ($\pi<1$), a *normal wave* structure is observed; that is, the cold reactants are ahead of the combustion front, while the hot products are behind. However, a so-called *superadiabatic* condition is achieved, where the temperature in the front exceeds the adiabatic combustion temperature, due to additional energy carried by filtration flow from the high-temperature products (Fig. 31a). This type of FC is promising, since weakly exothermic reactions (e.g., Cr-N$_2$, Si-C) can be conducted in the SHS mode.

In the case of strong forced gas flux ($\pi > 1$), the so-called *inverse wave* structure of FC occurs (Fig. 31b). The unreacted solids in zone I^* are heated by a gas flux to a high combustion temperature $T_c < T_c^{ad}$, but reaction does not occur since there is no reactant gas present. In some cases (e.g., Si-N$_2$, B-N systems), dissociation of the product occurs in the combustion front, limiting final conversion. Decreasing combustion temperature by strong gas flow alleviates this problem (Aldushin, 1993).

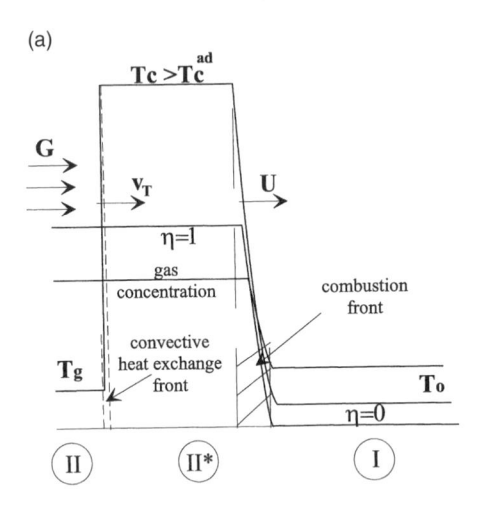

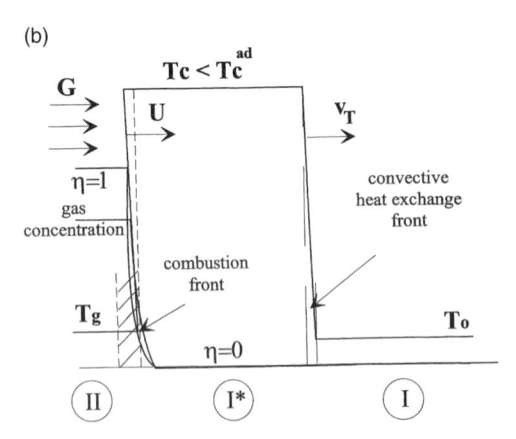

FIG. 31. Combustion wave structure in the case of forced gas filtration: (a) low gas flux ($\pi < 1$); (b) high gas flux ($\pi > 1$) (Adapted from Aldushin, 1993).

2. Effects of Other Forces

Different external (gravitational, centrifugal, etc.) or internal (e.g., the gas pressure gradient between the reaction zone and surroundings) forces may lead to densification or expansion of the reaction medium during FC (cf Shkadinskii *et al.*, 1992b). Product expansion and cracking due to gravity and impurity gasification have been investigated. In addition, the possibility of self-compaction during reaction has also been demonstrated. These works have relevance for the production of compact and foamed materials by the SHS method in gas–solid systems.

3. Interaction of Gasless and Filtration Combustion

The interaction of gasless and filtration combustion has also been examined (Aldushin *et al.*, 1994b). The competition between simultaneous gasless and gas–solid reactions

$$A(s)+B(g) \rightarrow D,$$
$$A(s)+C(s) \rightarrow E,$$

was considered. Assuming that both reactions have large activation energies, the combustion wave may propagate uniformly, with both reactions localized near the maximum temperature. The simultaneous occurrence of parallel reactions may also be used to intensify weakly exothermic reactions that otherwise would not occur. Also, in this case, the combustion temperature may be higher than the adiabatic value, due to heating by gas filtration from the high-temperature product region.

V. Phenomenology of Combustion Synthesis

As described in Section IV, the characteristic feature of combustion synthesis, as compared to conventional powder metallurgy, is that the process variables, such as combustion wave velocity, U, and temperature–time history, $T(t)$, are strongly related. For example, a small change in the combustion temperature may result in a large change in the combustion front velocity [see, for example, Eq. (23)] and hence the characteristic time of synthesis. The process parameters (e.g., green mixture composition, dilution, initial density, gas pressure, reactant particle characteristics) influence the combustion velocity and the temperature–time profile, and in turn can be used to control the synthesis process. In this chapter, the effects of reaction conditions on the characteristics of combustion are analyzed, leading to some general conclusions about controlling the CS process.

A. Thermodynamic Considerations

For any reaction system, the chemical and phase composition of the final product depends on the *green mixture composition*, gas pressure, reactive volume, and initial temperature. As shown in Section I, CS reactions can be represented in the following general form:

$$\sum_{i=1}^{N_X} \nu_i X_i^{(s)} + \sum_{i=1}^{N_Y} \nu_i Y_i^{(g)} + \sum_{i=1}^{N_{MO_X}} \nu_i (MO_X)_i^{(s)} + \sum_{i=1}^{N_Z} \nu_i Z_i^{(s)} = \sum_{j=1}^{N_P} \nu_j P_j^{(s)} + \sum_{j=1}^{N_M} \nu_j M_j^{(s)}, \quad (68)$$

where $X_i^{(s)}$ are the metal or nonmetal solid reactants, $Y_i^{(g)}$ the gas-phase reactants, $(MO_X)_i^{(s)}$ the oxide reactants, $Z_i^{(s)}$ the reducing metal reactants, $P_i^{(s)}$ the solid products (e.g., carbides, borides, nitrides), and $M_i^{(s)}$ the reduced metal products. In limiting cases, Eq. (68) leads to the three main classes of CS reactions discussed in Section I. Thus we have

$N_Y=0$, $N_{MO_X}=0$, and $N_M=0$,   gasless combustion [cf Eq. (1)]
$N_{MO_X}=0$, $N_Z=0$, and $N_M=0$,   gas-solid combustion [cf. Eq. (2)]
$N_{MO_X} \neq 0$ and $N_Z \neq 0$,   reduction-type combustion [cf. Eq. (3)]

Thermodynamic calculations can identify the adiabatic combustion temperature, as well as the equilibrium phases and compounds present at that temperature. The composition of the equilibrium final products is determined by minimizing the thermodynamic potential. For a system with $N^{(g)}$ gas and $N^{(s)}$ solid number of components, at constant pressure, this may be expressed as

$$F(\{n_k\},\{n_l\}) = \sum_{k=1}^{N^{(g)}} n_k \left(\ln \frac{p_k}{p} + G_k \right) + \sum_{l=1}^{N^{(s)}} n_l G_l, \quad (69)$$

where p_k is the partial pressure of the k'th gas-phase component, while n_i and G_i are the number of moles and molar Gibbs free energy of component i (cf Prausnitz et al., 1986). The adiabatic combustion temperature, T_c^{ad}, is determined by total energy balance:

$$\sum_{i=1}^{N_0} H_i(T_0) = \sum_{k=1}^{N(g)} n_k H_k(T_c^{ad}) + \sum_{l=1}^{N(s)} n_l H_l(T_c^{ad}), \quad (70)$$

where the enthalpy of each component is

$$H_i(T) = \Delta H_{f,i}^\circ + \int_{T_0}^{T} c_{p,i} dT + \sum \Delta H_{s,i} \quad (71)$$

and $\Delta H^\circ_{f,i}$ is the heat of formation at 1 atm and reference temperature T_0, $c_{p,i}$ is the heat capacity, and $\Delta H_{s,i}$ is the heat of s'th phase transition for component i.

Thermodynamic calculations of adiabatic combustion temperatures and their comparison with experimentally measured values have been made for a variety of systems (cf Holt and Munir, 1986; Calcote *et al.*, 1990; Glassman *et al.*, 1992). Under conditions that lead to full conversion, a good agreement between theoretical and experimental values has generally been obtained (Table XX).

TABLE XX
ADIABATIC AND MEASURED COMBUSTION TEMPERATURES FOR VARIOUS REACTION SYSTEMS

System	Adiabatic Combustion Temperature, T_c^{ad} (K)	Measured Combustion Temperature, T_c (K)	Lowest Melting Point On the Phase Diagram (K)
Carbides			
Ta + C	3290	2550	3295 (Ta)
Ti + C	1690	3070	1921 (eut)
Si + C	1690	2000[a]	1690 (Si)
Borides			
Ta + B	2728	2700	2365 (B)
Ti + 2B	3193	3190	1810 (eut)
Ti + B	2460	2500	1810 (eut)
Silicides			
Mo + 2Si	1925	1920	1673 (eut)
Ti + 2Si	1773	1770	1600 (eut)
5Ti + 3Si	2403	2350	1600 (eut)
Intermetallics			
Ni + Al	1912	1900	921 (eut)
3Ni + Al	1586	1600	921 (eut)
Ti + Al	1517	n/a	933 (Al)
Ti + Ni	1418	n/a	1215 (eut)
Ti + Fe	1042	n/a	1358 (eut)
Nitrides[b]			
2Ta + N_2	3165	2500	3000 (Ta)
2Nb + N_2	3322	2800	2673 (NbN)
2Ti + N_2	3446	2700	1943 (Ti)
2Al + N_2	3639	2300	933 (Al)
3Si + $2N_2$	2430	2250	1690 (Si)
2B + N_2	3437	2600	2350 (B)
Reduction type			
B_2O_3 + Mg	2530	2420	1415 (eut)
B_2O_3 + Mg + C	2400	2270	n/a
B_2O_3 + Mg + N_2	2830	2700	n/a
SiO_2 + Mg	2250	2200	1816 (eut)
SiO_2 + Mg + C	2400	2330	n/a
La_2O_3 + Mg + B_2O_3	n/a	2400	n/a

[a]With preheating.
[b]T_c^{ad} calculated for 1 atm.

Thermodynamic calculations have also been used to determine the equilibrium products (Mamyan and Vershinnikov, 1992; Shiryaev et al., 1993), and to illustrate new possibilities for controlling the synthesis process, even for complex multicomponent systems. Correlating these calculations with the equilibrium phase diagram for each system provides a basis for predicting possible chemical interactions and even limits of combustion during CS of materials. Some examples are discussed in the following subsections.

1. Gasless Systems

For the Ti-Si system, a comparison of the calculated T_c^{ad} and the phase diagram (Fig. 32) shows (Rogachev et al., 1995) that the combustion temperature of all silicide compounds exceeds the following:

- The melting point of both eutectic compositions (1330°C)
- The silicon melting point (but lower than the titanium melting point), for combustion of Ti+Si, 5Ti+4Si and 3Ti+Si mixtures
- The titanium melting point in the case of the 5Ti+3Si mixture (with up to 15 wt % of the final product molten).

Consequently, the two temperature regions of interest in the experimental study of the combustion mechanism are

Region I: 1330°C $< T <$ 1670°C

(interaction of solid Ti particles with melt of Si or eutectics)

Region II: 1670°C $< T <$ 2130°C

(melting of both Si and Ti)

These regions correspond to zones I and II, respectively, in the temperature profile, also shown schematically in Fig. 32. It was confirmed experimentally that the interaction in zone I occurs through a solid–liquid reaction, Ti(s)+Si(s,l)+L$_1$(eut.) → TiSi$_2$(s)+L$_2$, for 1330°C $< T <$ 1670°C. However, in zone II, a liquid–liquid interaction, Ti(l) + Si(l) → Ti$_5$Si$_3$ (l,s), takes place with the formation of Ti$_5$Si$_3$ for the temperature range 1670–2130°C:

$$Ti(l)+Si(l) \rightarrow Ti_5Si_3(s,l).$$

2. Reduction-Type Systems

The thermodynamic calculations for the equilibrium combustion temperature and product compositions for a three-component (Ni-Al-NiO) reduction-type system are shown in Figs. 33a and b, respectively (Filatov et al., 1988). In Fig. 33a, the curves correspond to constant adiabatic combustion temperature at $P=1$ atm. The maximum T_c^{ad} was calculated to be 3140 K for the 2Al+3NiO green mixture

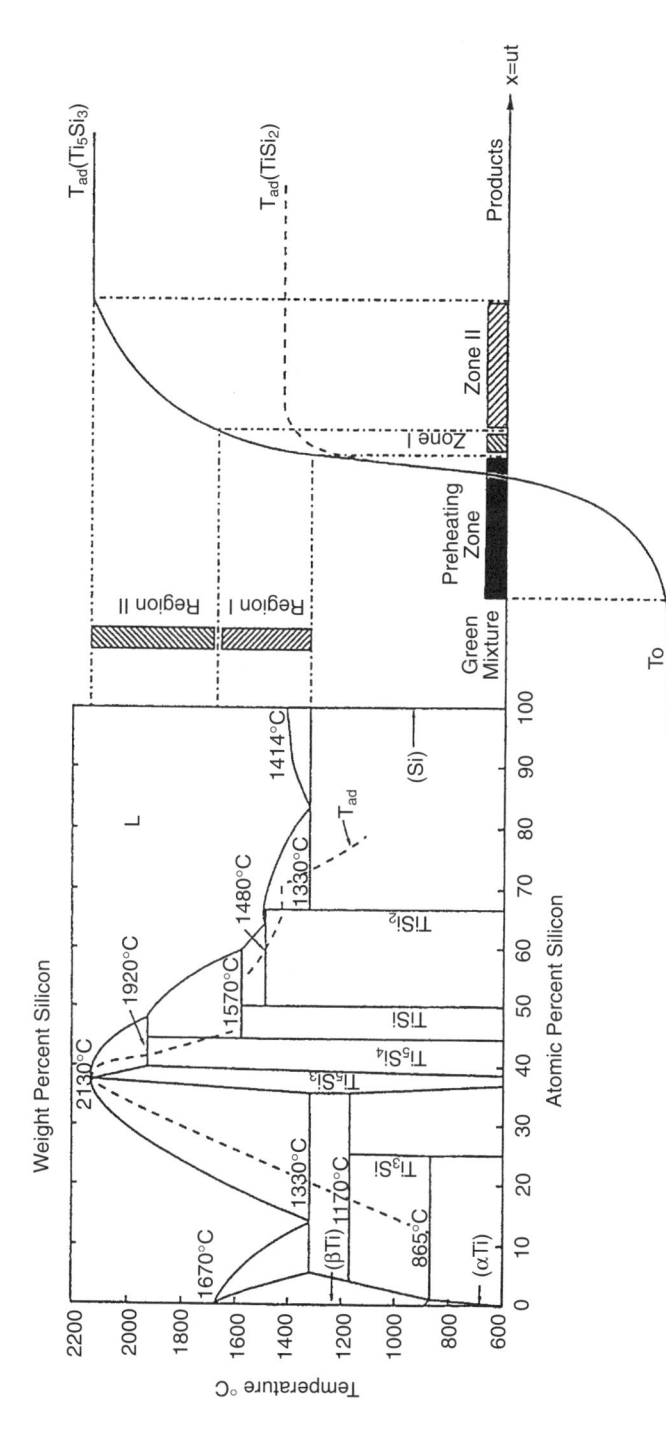

FIG. 32. Equilibrium combustion temperature and phase diagram of Ti-Si system (Adapted from Rogachev et al., 1995).

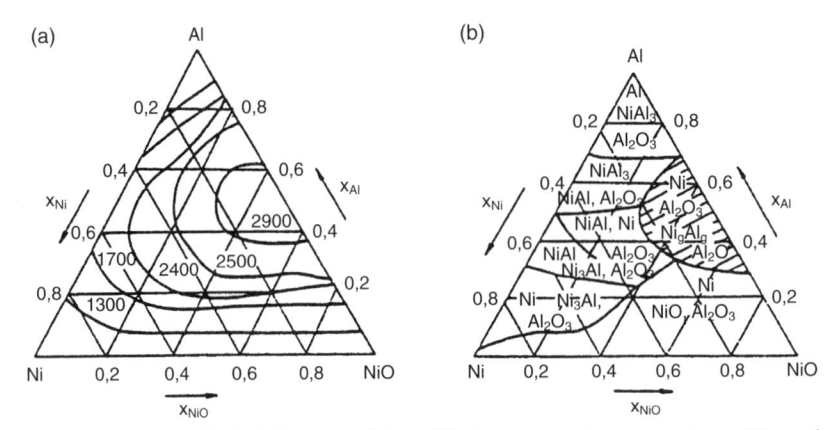

FIG. 33. For the Ni-Al-NiO system: (a) equilibrium combustion temperatures; (b) product composition (Adapted from Filatov et al., 1988).

($Q=810$ kcal/g), and is constrained by the nickel boiling point. Also, it was shown that the adiabatic combustion temperature for the 5Al+3NiO composition, with a higher heat of reaction ($Q=860$ kcal/g) was only 3000 K. The lower T_c^{ad} in the latter case can be attributed to the dissociation of formed nickel aluminides and the consequent reduction in overall heat generation. Note that the thermodynamic calculations reveal the presence of a gas-phase product for initial mole fractions of Al from 0.3 to 0.7 (hatched region of Fig. 33b). Increasing the amount of nickel in the green mixture reduces T_c^{ad}, and promotes the formation of nickel aluminides. With relatively low melting points, these compounds add another liquid phase to the system, allowing better contact for reaction and expanding the limits of combustion. These calculations have also been confirmed by experimental investigations (Filatov and Naiborodenko, 1992).

3. Gas–Solid Systems

The thermodynamic results discussed previously were calculated for constant gas pressure. However, it is also necessary to treat the case of constant reaction volume, especially for gas–solid systems where the initial reactant ratio may be varied by changing the reactant gas pressure.

Consider the combustion reaction between a solid reactant and a gas oxidizer present initially in the constant volume of a porous medium (see Section IV,D,1). In this case, thermodynamic calculations for the silicon–nitrogen system have been made for constant volumes (Skibska et al., 1993b). The calculations yield the adiabatic combustion temperature, as well as pressures and concentration, as functions of the silicon conversion. As shown in Fig. 34a, the reactant gas pressure (curve 3) increases even though conversion increases. This occurs because

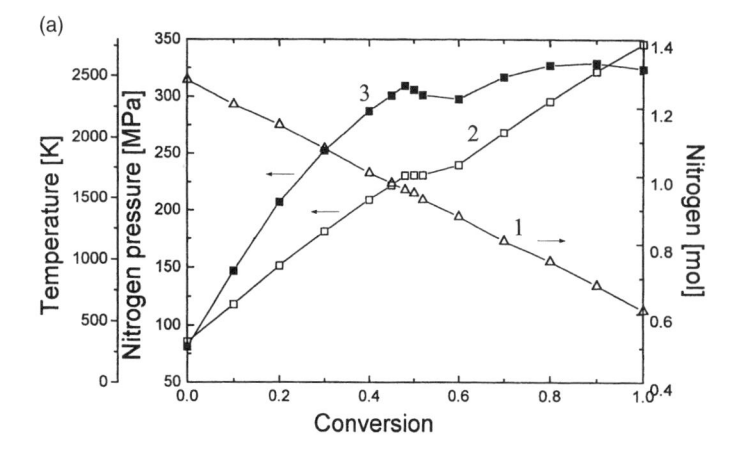

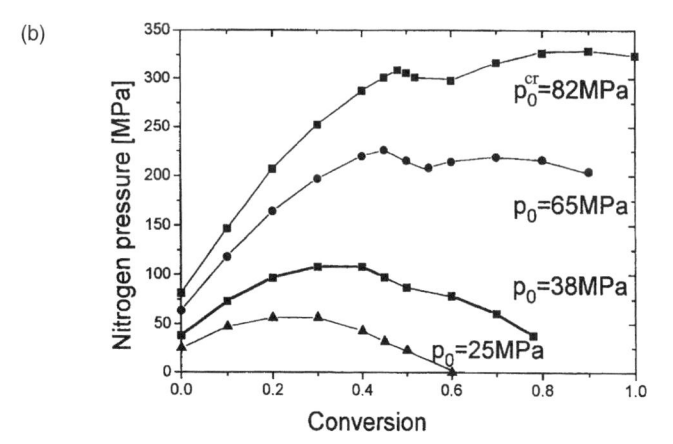

FIG. 34. Thermodynamic calculations of combustion characteristics as a function of degree of conversion in Si-N$_2$ system; $T_0 = 300$ K, $\epsilon = 0.6$ (Adapted from Skibska et al., 1993b).

the temperature (curve 2) also increases with conversion, while the amount of nitrogen gas (curve 1) in the system decreases. The discontinuity in slopes of the various curves occurs at the melting point of silicon (1414°C). In addition, the change in nitrogen pressure during reaction depends on its initial value, p_0 (see Fig. 34b). Above a critical initial value (p_0^{cr}), the nitrogen pressure increases with increasing conversion to full reaction. However, for lower initial nitrogen pressures, a maximum is observed. These results provide an explanation for the unusual dependence of U vs. p_{N_2}, and suggest methods for optimizing the synthesis process (Skibska et al., 1993a,b).

As shown by the preceding examples, changes in the initial composition of the green mixture can result in a wide variation of synthesis conditions and products.

Equilibrium thermodynamic calculations can help to elucidate these conditions and products, and provide useful information for controlling the process.

B. Dilution

One of the most effective and commonly used procedures to modify the synthesis conditions is to dilute the initial mixture with a chemically inert compound, often the final product. The typical dependencies of combustion temperature and velocity on the amount of dilution, b, are shown in Fig. 3e. In general, an increase in dilution decreases the overall heat evolution of the system, and therefore decreases the combustion temperature and velocity. Some examples of this type are presented in the review by Rice (1991), where experimental data for group IV metals with boron and carbon, as well as Mo-Si systems are summarized. A similar result was obtained for the Hf-B system shown in Fig. 35 (Andreev et al., 1980).

However, some exceptions to these trends have also been observed. In Fig. 36, the dependencies of combustion temperature and velocity on dilution for the Ni-Al system are shown (Maslov et al., 1976). As expected, the combustion velocity decreases with increasing dilution. Similarly, for lower ($b < 11.5\%$) and higher ($b > 35.5\%$) dilutions, T_c decreases with increasing dilution. However, for intermediate dilutions ($11.5\% < b < 35.5\%$), the combustion temperature is constant,

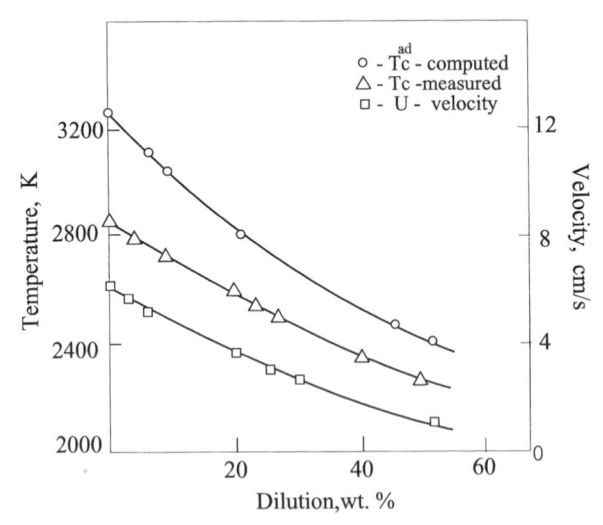

FIG. 35. Combustion temperature and velocity of hafnium–boron mixture as a function of dilution by final product (Adapted from Andreev et al., 1980).

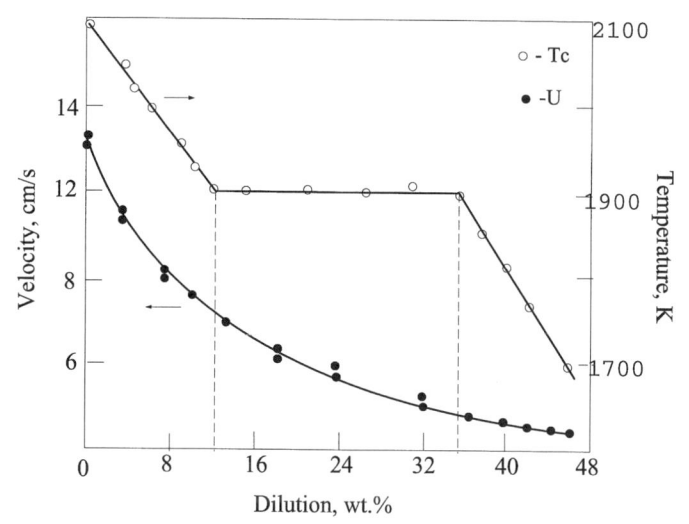

FIG. 36. Dependence of combustion velocity and temperature on dilution by final product in Ni-Al system; $T_0 = 740$ K (Adapted from Maslov et al., 1976).

and corresponds to the melting point of NiAl. This type of behavior has also been observed for other systems, when the T_c^{ad} may correspond to the melting point of the product (Shcherbakov and Pityulin, 1983; Subrahmanyam et al., 1989; Kottke et al., 1990).

The particle size of the inert diluent may also influence combustion characteristics. For example, the combustion velocity in the Ti-C system as a function of dilution for different TiC particle sizes is shown in Fig. 37 (Maslov et al., 1990). For the smaller particles ($d_{TiC} < 100$ μm), the combustion velocity decreased with increasing b (curve 1). However, for larger TiC particles ($d_{TiC} > 1000$ μm), owing

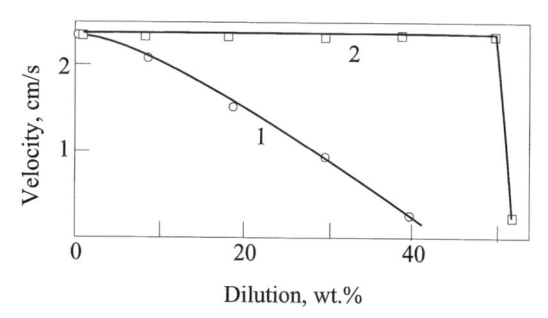

FIG. 37. Combustion velocity as a function of inert dilution: curve 1, $d_{TiC} = 50$–90 mm; curve 2, $d_{TiC} = 1000$–1500 mm (Adapted from Maslov et al., 1990).

to their incomplete heating in the combustion wave, U remained practically constant up to about 50 wt % dilution before eventually decreasing (curve 2).

The dependence of combustion temperature and velocity for the Si-N$_2$ system as a function of dilution with β-Si$_3$N$_4$ powder is shown in Fig. 38. In this case, at constant gas pressure, T_c remains constant and is equal to the dissociation temperature of silicon nitride, for dilutions up to 60 wt %. However, with increasing dilution, the combustion front velocity increases. Also, increasing the overall heat evolution, smaller dilution by nonmelted nitride powder promotes coalescence of liquid Si particles, leading to an increase in the average reactant particle size, as well as to formation of thin liquid Si films, which blocks nitrogen infiltration to the reaction zone (Mukasyan et al., 1986). The same effects were also observed for the Al-N$_2$ and B-N$_2$ systems (Mukasyan, 1994), which are characterized by dissociation of the final product in the combustion wave.

Moreover, dilution is an effective method of increasing the final extent of conversion for some gas–solid systems, especially when melting of the reactants or products occurs, and the sample permeability is reduced. Figure 39 shows the dependence of final conversion, η, on the extent of dilution for the Ti-N$_2$ system. In this case, the adiabatic and measured combustion temperatures (see Table XX) are higher than the melting point of titanium (1670°C) and the conversion in the combustion front, η^*, is less than unity. With an increase in solid-phase dilution, the combustion temperature decreases, which diminishes the effect of titanium melting and coalescence on sample permeability. However, excessive dilution leads to a decrease in conversion and eventual process extinction.

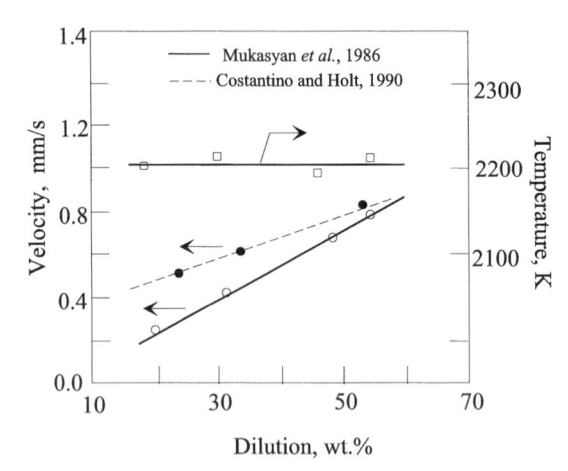

FIG. 38. Combustion velocity and temperature as a function of dilution by final product in Si-N$_2$ system.

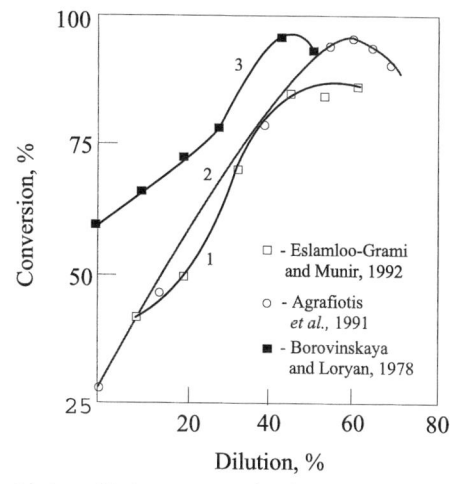

FIG. 39. Effect of solid phase dilution on conversion for Ti-N$_2$ system: curve 1, $P = 1$ atm; curve 2, $P = 10$ atm; 3, $P = 2000$ atm.

In the Ta-N$_2$ system, the measured combustion temperature is lower than the melting points of both the metal and final products. In this case, there is no significant influence of dilution on final conversion. However, an interesting effect of dilution on the product phase composition was found by Borovinskaya et al. (1975a). Increasing dilution results in a change, from δ-TaN (cubic) to the ε-TaN (hexagonal) phase (see Table XXI).

Finally, the following conclusions may be made about the effects of dilution:

- For a wide range of systems, the combustion front velocity and temperature decrease with increasing dilution.
- The atypical dependencies of T_c and U as a function of dilution are usually associated with phase transition processes occurring in the combustion wave (e.g., melting, dissociation, evaporation).

TABLE XXI
PHASE AND CHEMICAL COMPOSITIONS OF SHS TANTALUM NITRIDE POWDERS

Dilution (wt %)	Product Chemical Composition	Product Phase Composition
0	TaN$_{1.05}$	δ–TaN(cub)
6.25	TaN$_{1.03}$	δ–TaN(cub)
18.75	TaN$_{1.01}$	δ–TaN(cub)
31.25	TaN$_{0.97}$	TaN2+δ–TaN(cub)
43.75	TaN$_{0.94}$	TaN2+δ–TaN(cub)
56.25	TaN$_{0.91}$	ε–TaN(hex)

- For gas–solid systems with combustion temperatures higher than the melting points of the reactants or products, dilution is an effective method for increasing final conversion, by inhibiting the effects of melting, particle coalescence, and the subsequent reduction of sample permeability.

C. Green Mixture Density for Gasless Systems

As shown earlier in Fig. 2b, the dependence of combustion velocity on green mixture density for gasless systems is frequently characterized by a maximum at some intermediate value (Merzhanov, 1983; Rice, 1991). However, combustion theory [Eq. (20)] suggests that the velocity under adiabatic conditions should increase with increasing density, since higher densities correspond to higher thermal conductivities (Aleksandrov et al., 1985; Kottke et al., 1990). Different explanations have been suggested to explain this discrepancy.

It has been proposed that the effect can be attributed to two competing phenomena (Varma and Lebrat, 1992). As the density increases, intimate contact between the reactant particles is augmented, which enhances both reaction and front propagation. On the other hand, due to an increase in thermal conductivity at higher densities, greater heat is lost due to conduction from the reaction zone to the unreacted portion, inhibiting front propagation. However, this idea, which considers the thermal conductivity as a function of reactant particle contact and thermal losses, has not been investigated experimentally. Furthermore, most theoretical models do not take into account the contact resistance. Kottke et al. (1990) calculated the velocity dependence on initial density, based on experimental measurements of $\lambda = f(\rho_0)$, and showed that U increases monotonically with increasing ρ_0 (Fig. 21).

Another hypothesis was suggested by Kirdiyashkin et al. (1981) for the combustion synthesis systems characterized by melting of a reactant metal (e.g., Ti-C, Ti-B), where capillary spreading may control the combustion process (Shkiro and Borovinskaya, 1976). In these cases, it was suggested that an optimal density occurs where the volume of pores equals the volume of the molten metal. However, an analysis of the experimental data for the Ti-B system showed that this hypothesis may not be valid over the entire range of particle sizes investigated (Munir and Anselmi-Tamburini, 1989).

It is well known that powders with high surface areas usually have a large amount of adsorbed gas impurities. In Fig. 40, the dependence of combustion velocity on green density in the Ni-Al system is shown for combustion of mixtures with (curve 2) and without (curve 1) heat treatment of powders to remove adsorbed contaminants. While the untreated samples exhibit a maximum, for the pretreated samples, the velocity increases monotonically with increasing density (Kasat'skiy et al., 1991). The monotonic increase of combustion velocity and

COMBUSTION SYNTHESIS OF ADVANCED MATERIALS   163

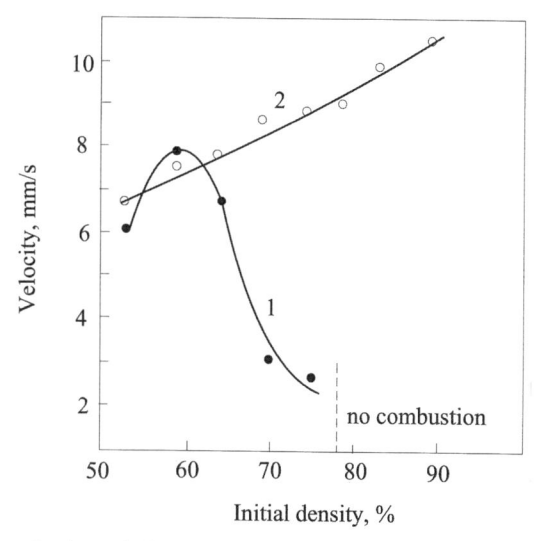

FIG. 40. The combustion velocity as a function of initial sample density for the Ni-Al system: curve 1, with no pretreatment; curve 2, with pretreatment at $T = 570$ K and $P = 13.3$ Pa (Adapted from Kasat'skiy et al., 1991).

temperature with increasing green density (see Fig. 41) was also observed for 3Ni+Al samples pressed from high-purity powders (Lebrat and Varma, 1992a). In addition, for Ti+C and 5Ti+3Si systems, without preliminary outgassing of adsorbed hydrogen from the surface of powders (mainly Ti), a maximum in the U dependence with density was observed at $\rho_0 \sim 0.56$. However, heat treatment (400°C) of the green mixture under vacuum ($\sim 10^{-5}$ atm) resulted in a monotonically increasing combustion velocity for densities up to $\rho_0 = 0.75$ (Vadchenko 1994, 1996).

In an early work on combustion in the Ti-C system (Shkiro and Borovinskaya, 1975), stable combustion propagation was observed with increasing velocity for $\rho_0 = 0.38$–0.56, while for higher densities, the oscillatory mode occurred and the average velocity decreased (Fig. 42). The oscillatory mode also occurred for the Nb-B and Ti-B systems (Filonenko, 1975) and was shown to be related to the formation of thin flaky layers due to stratification of the reactant medium during combustion. In constrained samples, stratification did not occur and the combustion wave propagated without oscillations. The formation of the flaky layers was associated with adsorbed gas escaping from the sample.

Based on these results, an explanation for the combustion velocity dependence on green mixture density has recently been suggested (Vadchenko et al., 1996). For relatively low sample densities ($\rho_0 < 0.6$), escape of desorbed gases in the combustion wave is not limited by permeation and occurs without destruction of the reactant medium structure. In this case, increasing the green density

164   A. VARMA *ET AL.*

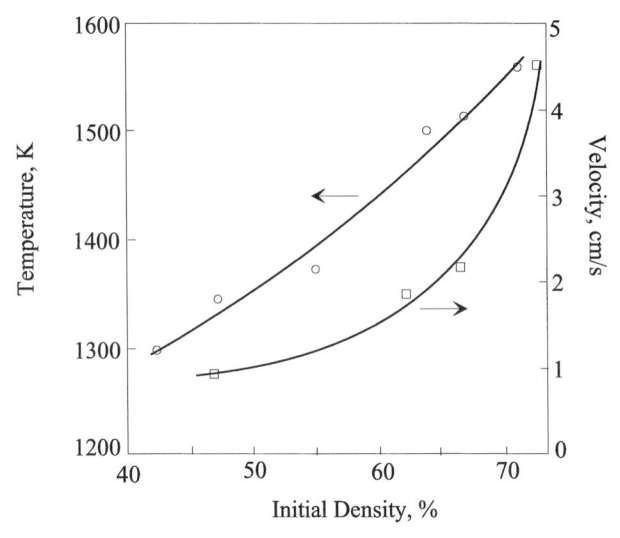

FIG. 41. Effect of initial density on combustion temperature and velocity in 3Ni-Al system (Adapted from Lebrat and Varma, 1992a).

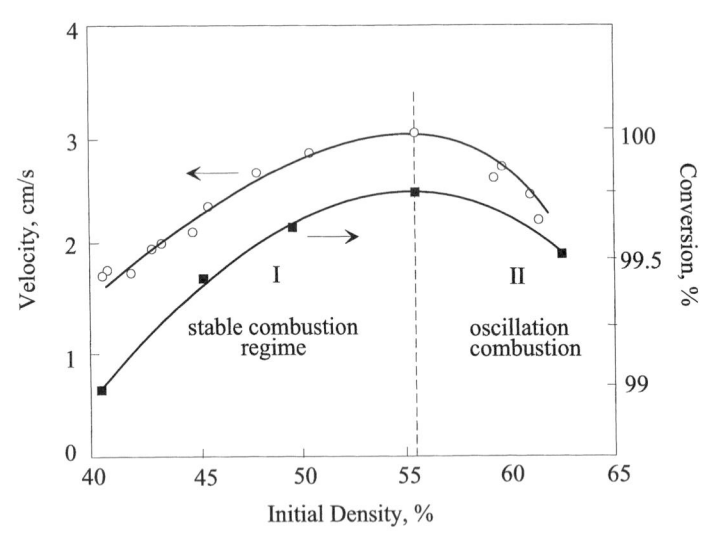

FIG. 42. The dependences of combustion velocity and conversion on initial sample density in Ti-C system (Adapted from Shkiro and Borovinskaya, 1975).

(ρ_0=0.45–0.6) results in increasing thermal diffusivity of the medium, and hence increasing velocity. At higher densities (ρ_0>0.6), adsorbed gases are unable to escape due to low sample permeability. As the temperature increases, the gas pressure inside the pores also increases, forming cracks in the sample, which lower the effective thermal conductivity of medium in the combustion zone. With higher green densities, ρ_0, the extent of crack formation is greater, which leads to lower combustion velocities.

The effect of the expansion of adsorbed gases on the sample final density was also found in the investigation of the Nb-3Al system, using volume combustion synthesis technique (Kachelmyer *et al.*, 1995). In pellets with higher initial densities, the passage of expanding contaminant gases during sample preheating was restricted due to lower permeability. This leads to a lower density of the final product. However, at lower ρ_0, the expanding gases are expelled with greater ease, resulting in higher final density.

D. GREEN MIXTURE DENSITY AND INITIAL GAS PRESSURE FOR GAS–SOLID SYSTEMS

For gas–solid systems, the green density affects not only the thermal conductivity, but also the availability and permeability of the reactant gas within the compact. For this reason, in gas–solid systems, it is necessary to analyze the effects of green density along with initial reactant gas pressure on the combustion synthesis process. As discussed in Section IV, different combustion regimes can occur in these systems, depending on the initial sample porosity and reactant gas pressure. Two types of reaction propagation may be distinguished: with and without reactant gas infiltration to the combustion front. In both cases, reactant gas permeation may take place after passage of the front (i.e., postcombustion reaction).

1. Without Gas Infiltration to the Combustion Front

This type of propagation may follow three different combustion modes, including *surface combustion* and *layer-by-layer combustion* with either incomplete or complete conversion. The specific mode followed depends on the initial amount of reactant gas present in the sample pores.

a. Surface Combustion. In this case, the initial amount of gas present in the pores is not enough to sustain chemical reaction within the bulk of the sample. Thus the combustion wave propagates *only* on the outer surface of the sample in a relatively thin layer (usually 0.5–2 mm). However, reaction in the postcombustion zone may play a significant role in the final product formation process. Surface combustion mode typically occurs with reactant gas pressures, P_g<10 MPa and initial sample density ρ_0>0.5, and has been investigated primarily for various nitrogen-metal systems: Ti-N_2 (Eslamloo-Grami and Munir, 1990a,b; Agrafiotis *et al.*, 1991; Borovinskaya *et al.*, 1992a), Zr-N_2 (Merzhanov *et al.*, 1972), Ta-N_2 (Agrafiotis *et al.*, 1991), Hf-N_2 (Borovinskaya and Pityulin, 1978), V-N_2 (Maksimov *et al.*, 1979), and Nb-N_2 (Mukasyan, 1994).

b. Layer-by-Layer Combustion (with Incomplete Conversion). Above a critical initial gas reactant pressure, the combustion wave can propagate in a layer-by-layer regime through the sample without additional gas infiltration. However, the amount of gaseous reactant in the pores may not be sufficient for complete conversion of the solid reactant. This mode is possible owing to the fact that in many gas–solid systems, the combustion wave can be sustained by *only* a relatively small degree of conversion (e.g., $\eta^* \sim 0.1$). For example, in the Ti-N_2 system, this type of combustion occurs when nitrogen pressure is in the range of 10–15 MPa and green density $0.6 > \rho_0 > 0.5$, where η^* achieved does not exceed 0.2 (Borovinskaya *et al.*, 1992a). Of course, complete conversion may be achieved during postcombustion reaction. This mode of combustion has been observed in several systems, including Ti-N_2 (Hirao *et al.*, 1987; Borovinskaya *et al.*, 1992a) and Ta-N_2 (Borovinskaya *et al.*, 1975a; Agrafiotis *et al.*, 1991).

c. Layer-by-Layer Combustion (with Complete Conversion). This mode occurs when the amount of reactant gas within the pores of the sample is enough for complete conversion of the solid reactant. In this case, the reaction is complete *immediately* after the combustion wave propagates through the sample. Calculations for the critical reactant gas pressure as a function of initial density to realize this mode of combustion have been reported (Munir and Holt, 1987; Merzhanov, 1993a). For example, in solid–nitrogen systems, for an initial green density of $\rho_0 \sim 0.5$, the critical nitrogen pressure ranges from about 100 MPa for Ta, up to 300 MPa for B (see Table XVI). These critical conditions have been studied experimentally for several systems: Zr-N_2 and Ti-N_2 (Borovinskaya and Loryan, 1978), Ta-N_2 (Borovinskaya and Pityulin, 1978), and B-N_2 (Mukasyan, 1986).

2. With Gas Infiltration to the Combustion Front

In this case, the combustion front propagates due to the interaction between the solid reactant and the gas infiltrated to the reaction zone from the surrounding atmosphere (see Section IV,D). For metal–nitrogen reactions, this type of combustion was observed for low initial sample density (~0.2) in the Ta–N_2 system, where the metal does not melt in the combustion wave (Pityulin *et al.*, 1979; Kumar, 1988). Additional examples were identified during the synthesis of non-metallic nitrides, including Si_3N_4 (Mukasyan *et al.,* 1986; Hirao *et al.,* 1987; Costantino and Holt, 1990), AlN (Dunmead *et al.,* 1990a), and BN (Mukasyan and Borovinskaya, 1992). In these systems, the rates of chemical reactions are relatively small, hence low gas permeation rates, achievable with gas pressures up to ~100 atm, are sufficient to sustain the infiltration mode of combustion.

The dependencies of combustion velocity on the sample green density for several gas–solid systems are presented in Fig. 43. In general, the velocity decreases with increasing density. Even more important from the practical point of view is the extent of final conversion as a function of initial density. For the Hf-N_2 (Borovinskaya and Pityulin, 1978) and B-N_2 (Mukasyan, 1986) systems, the conversion decreases with increasing density, owing to the lower availability and permeability of nitrogen in the sample (curves 1 and 2, Fig. 44). However, for the Ti-N_2 system at 1-atm nitrogen pressure, a maximum final conversion was observed at $\rho_0 = 0.58$ (curve 3, Fig. 44). At low densities, the high temperature in the reaction zone (~1800°C) partially melts the titanium, which reduces the permeability

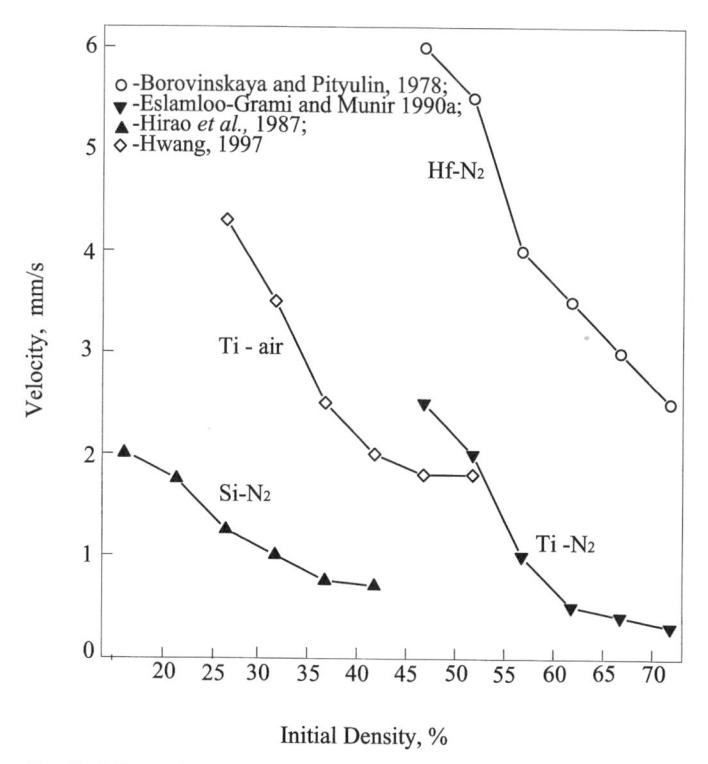

FIG. 43. Effects of initial density on combustion velocity in gas–solid systems.

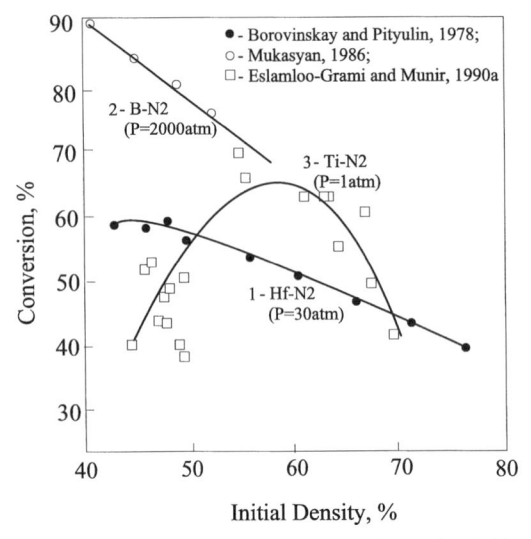

FIG. 44. Effect of initial density on conversion to the nitride.

of the sample surface. As a result, the amount of reactant gas available in the postcombustion zone is reduced, lowering the final conversion. Increasing the green density decreases the combustion temperature, until it becomes lower than the melting point of titanium (1760°C). Since Ti does not melt, the surface permeability does not change dramatically, and consequently a relatively wide postcombustion zone was observed, with higher conversions. For higher green densities (>60%), the initial sample permeability is lower, and it further decreases due to chemical reaction leading to less final conversion.

Nonmonotonic behavior of final conversion was also observed for the Zr-N$_2$ system as a function of initial nitrogen pressure (Merzhanov et al., 1972). Again, it was shown that this effect is related to the melting of zirconium in the combustion zone. The gravimetric curves of zirconium nitridation at different initial pressures (Fig. 45) demonstrate that for relatively low pressure ($P=5$ atm), there exists a wide postcombustion zone where conversion in the reaction front, η^* is about 0.20, and nearly 0.60 in the final product (curve 1). At higher pressure ($P=30$ atm), η^* is higher (~0.25), but no reaction occurs in the postcombustion zone. Increasing the initial nitrogen pressure even further (>30 atm) increases final conversion, which is due only to an increase in the amount of nitrogen reacted in combustion front (curve 4, $P=100$ atm). For the Nb-N$_2$ system, in which no melting of metal occurs at $P=30$ atm, a wide postreaction zone is obtained (curve 5). The reactant pressure can also influence the phase composition of the product, as was demonstrated for Ti-N$_2$ (Borovinskaya and Loryan, 1978) and Ta-N$_2$ (Borovinskaya et al., 1975a; Agrafiotis et al., 1991) systems.

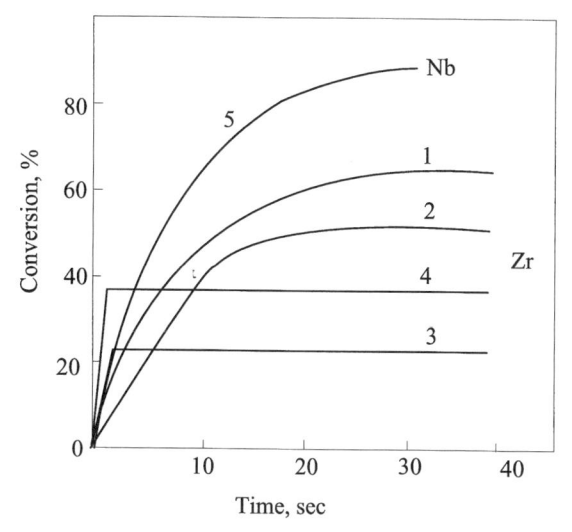

COMBUSTION SYNTHESIS OF ADVANCED MATERIALS 169

FIG. 45. Continuous gravimetric measurements for Zr (curves 1–4) and Nb (curve 5) samples reacted with nitrogen at different gas pressures: curve 1, $P = 5$ atm; curve 2, $P = 10$ atm; curves 3 and 5, $P = 30$ atm; curve 4, $P = 100$ atm. (Adapted from Merzhanov et al., 1972).

Finally, note that a relationship between the product density, ρ_f, with that of the initial mixture, ρ_0, can be obtained readily as follows:

$$\rho_f = \rho_0(1 + A\eta_f)\chi, \tag{72}$$

where A is constant for the investigated system, and $\chi = V_0/V_f$ is the ratio of the initial and final sample volumes. For net-shape production of articles, $\chi = 1$, although if a densifying load is applied during synthesis, then $\chi > 1$. Examples of relationship (72) for the B-N_2 (Mukasyan, 1986) and Al-N_2 (Dunmead et al., 1990a) systems are presented in Fig. 46. To obtain materials with high final density (i.e., low porosity), we need to increase both initial density and conversion η_f. However, as discussed earlier (Fig. 44), η_f generally decreases with increasing ρ_0. Thus optimization of these two parameters is the main problem when attempting to produce materials with a tailored composition and mechanical properties.

E. PARTICLE SIZE

To date, the dependencies of combustion velocity, temperature, and conversion as functions of reactant particle size have not been studied as extensively as the other parameters described earlier. This is somewhat surprising in light of the fact that particle diameter is an important parameter that can be conveniently

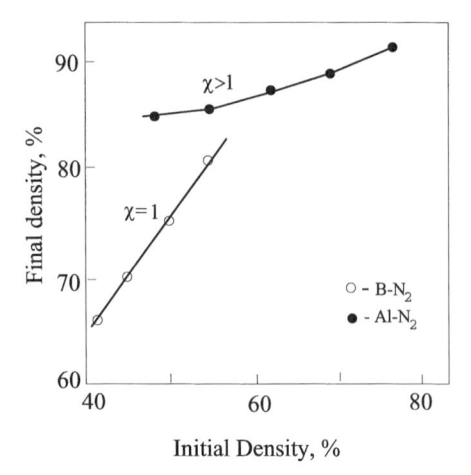

FIG. 46. Final sample density as a function of initial density for B-N$_2$ (Mukasyan, 1986) and Al-N$_2$ (Dunmead et al., 1990a) systems.

studied experimentally. Some investigations have shown that the particle size, d, influences the process, and typical relationships are shown in Fig. 3b. While the combustion velocity generally decreases with increasing particle size, the actual function may vary: $U \sim 1/d$, $U \sim 1/d^2$, $U \sim 1/d^{1/2}$ (Merzhanov, 1981). These relationships were developed theoretically (see Section IV), taking into account the mechanisms of chemical interaction in the combustion wave. Often, the differences can be explained by comparing the adiabatic combustion temperature to the melting points of the reactants (see Table XX).

1. "Solid" Flame Systems

Tantalum–carbon, tantalum–boron, and niobium–boron mixtures may be classified as so-called "solid flame" systems, where T_c^{ad} is lower than the melting points of both reactants and products. Consequently, the sizes of reactant particles are not expected to change in the combustion wave. The dependence of U on the average metal particle size for these systems is presented in Fig. 47.

The tantalum–nitrogen system is also characterized by a relatively low measured combustion temperature (see Table XX) compared to the melting points of Ta and TaN. However, it was shown that the propagation velocity and temperature in this system are essentially independent of the Ta particle size (Agrafiotis et al., 1991). This phenomenon was explained by accounting for the morphology of the fine and coarse powders. The coarser particles used were highly porous and had a large surface area, while the finer particles were nonporous. Thus, each coarse particle acted as a large number of discrete nonporous grains, and resulted in the same U as the fine powders.

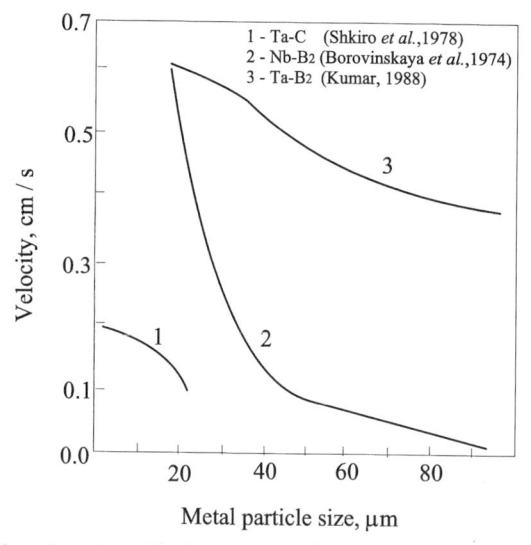

FIG. 47. Effect of metal particle size on combustion velocity in "solid flame" systems.

2. Systems with One Reactant Melting

In this case, the adiabatic combustion temperature is higher than the melting point of one reactant. The titanium–carbon system is the most widely investigated in this category. Data from different investigators for the $U=f(d)$ dependence for the TiC system are presented in Fig. 48. Only one study (Kumar et al., 1988a) demonstrated a constant velocity for the range of Ti particle sizes investigated (up to 100 μm), which one may expect if the metal were to melt in the preheating zone. The others show a general trend where U decreases with increasing Ti particle size in the range of 40–250 μm. A similar trend was observed when the Ti particle size was constant, while the carbon particle size (d_C) was varied from 2 to 70 μm (Deevi, 1991). For relatively small Ti particle sizes (<20 μm), Shkiro and Borovinskaya (1976) have shown that the combustion velocity is independent of titanium particle size.

Nekrasov et al. (1978) considered three possible situations, which can be used to explain these results. When rather small metal particles are used, that is,

$$d_{Ti}^2 < \sigma \lambda r_C \mu U^2 \ln\left(\frac{T_c - T_0}{T_m - T_0}\right), \tag{73}$$

where μ and σ are the viscosity and surface tension of the liquid, the interaction is controlled by the rate of chemical reaction between molten Ti and solid C, and the

172   A. VARMA ET AL.

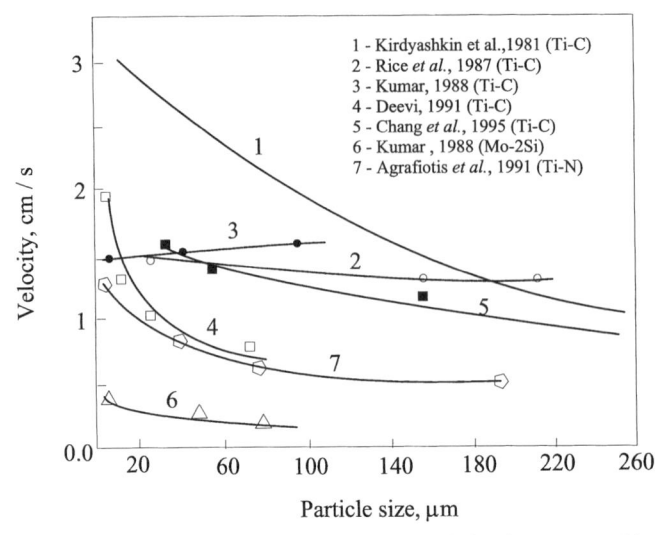

FIG. 48. Effect of metal particle size on combustion velocity in systems with one melting reactant.

combustion velocity is essentially independent of d_{Ti}. When relatively large Ti particles are used,

$$d_{Ti}^2 > \frac{\sigma \cdot r_c^3}{\mu D},\qquad(74)$$

where D is the diffusion coefficient of the molten reactant in the product, capillary flow becomes significant, and spreading of molten titanium through the solid carbon matrix dominates. Accordingly, the velocity decreases with increasing Ti particle size. For intermediate sizes, diffusion of titanium through the solid TiC product layer controls the interaction.

The molybdenum–silicon and titanium–nitrogen systems also belong to this group. For both, it was shown that U decreases when metal (Mo, Ti) particle size increases (curves 6 and 7, Fig. 48).

It has been shown that the ratio of reactant particle sizes influences the microstructure of the final product. In the 3Ni-Al system, the use of Ni (32–44 μm) and Al (<10 μm) led to the optimum microstructure (i.e., full conversion and dense final product) as d_{Ni}/d_{Al} approached 3.0. For example, using smaller Ni (<10 μm) and the same Al (<10 μm) particles, a more porous final product was obtained (Lebrat and Varma, 1992a).

3. Systems with Both Melting Reactants

Results for titanium–boron and several intermetallic systems (Ni-Al, Ti-Al, Ni-Ti, Co-Ti) are presented in Fig. 49. In all cases, the measured combustion tem-

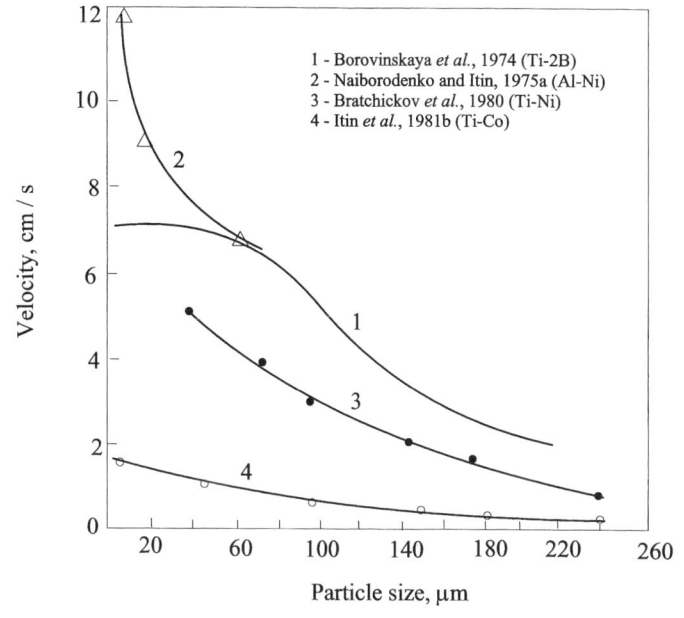

FIG. 49. Effect of metal (Ti, Al) particle size on combustion velocity in systems with both melting reactants.

peratures were higher than melting points of both reactants (see Table XX), and no exceptions to general trends (Fig. 3b) were obtained.

F. OTHER EFFECTS OF COMBUSTION CONDITIONS

In addition to the main variables already considered, other parameters also influence the combustion synthesis process. These include the ignition conditions, initial temperature of the green mixture, sample dimensions, and their effects and they are discussed next.

1. Initiation of SHS Process

A recent review by Barzykin (1992) summarizes experimental and theoretical studies on the initiation of CS systems. In this section, we focus only on the effects of initiation conditions on the reaction pathway during CS.

It has been shown theoretically that for the case of two competing parallel reactions, two stable combustion modes can exist (Khaikin and Khudiaev, 1979), where the reaction route depends on the ignition conditions. This effect was first observed experimentally by Martirosyan et al. (1983) for the Zr-C-H_2 system (Fig. 50). Using relatively low ignition temperatures (1300 K), the combustion

174   A. VARMA ET AL.

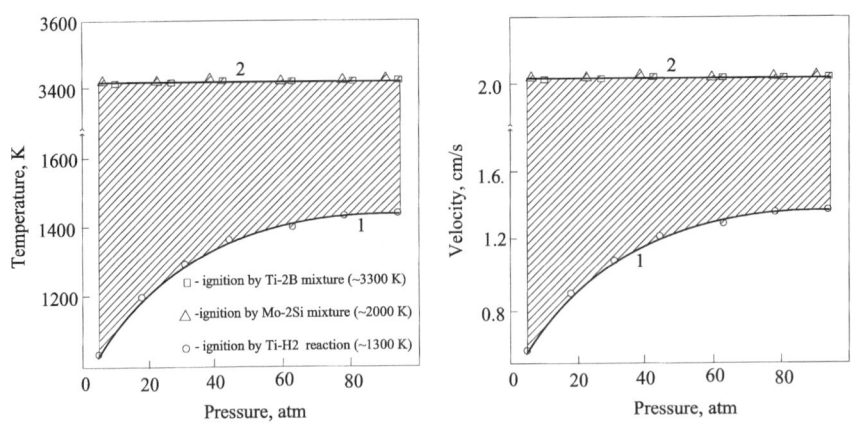

FIG. 50. Nonuniqueness of combustion modes in the Zr-C-H$_2$ system (Adapted from Martirosyan et al., 1983).

proceeded via the low-temperature mode (curves 1) forming zirconium hydrides as the final product. For higher ignition temperatures (2100–3300 K), the high-temperature mode was realized (curves 2), and zirconium carbides were formed.

Recently the nonuniqueness of combustion modes due to different ignition conditions was observed both for the Ti-N$_2$-O$_2$ (Merzhanov et al., 1995; see also Section VI,D,2) and Nb-B-O$_2$ (Mukasyan et al., 1997a) systems. In the latter case, for ignition temperatures lower than 2050 K, a stable low-temperature (~1000 K) combustion regime resulted in the formation of niobium oxide (Nb$_2$O$_5$), while for ignition temperatures higher than 2350 K, NbB$_2$ was formed in a high-temperature (~2700 K) combustion wave (Fig. 51). It is worth noting the main difference between phenomena observed in the Nb-B-O$_2$ and Zr-C-H$_2$ systems. While both Nb$_2$O$_5$ and NbB$_2$ products are stable in the range of ignition temperatures used, zirconium hydride dissociates at temperatures higher than 1300 K and ZrC is the only stable phase existing under these conditions.

2. Initial Temperature

Experimental data for the combustion velocity dependence on initial temperature for different systems are presented in Fig. 52. Examples of intermetallic systems (Al-Ni and Co-Ti) with melting of both reactants show a general trend in which U increases with increasing T_0. The same behavior was obtained for the Mo-Si (Kumar et al., 1988b), Ti-C (Kirdyashkin et al., 1981) and 3Ni-Al (Lebrat and Varma, 1992a) systems, which are characterized by melting of only one reactant.

For gas–solid systems, this effect has not been studied well. In the only example available, which involves combustion of silicon in nitrogen, it was observed

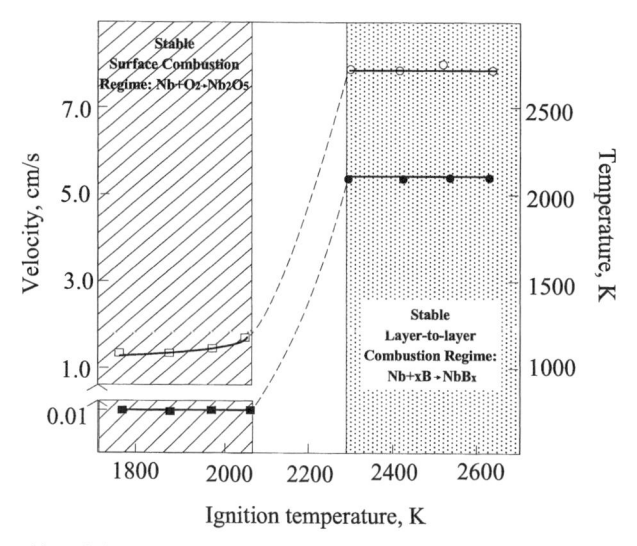

FIG. 51. Competition of chemical reactions in the Nb-B-O_2 system (Adapted from Mukasyan et al., 1997a).

that the combustion velocity does not always increase monotonically with increasing T_0 (Fig. 53). For low dilutions (≤30%), the combustion velocity has a maximum (curve 3). This effect can be attributed to greater melting and coalescence of silicon particles in the combustion zone for increasing T_0, which decreases the sample permeability.

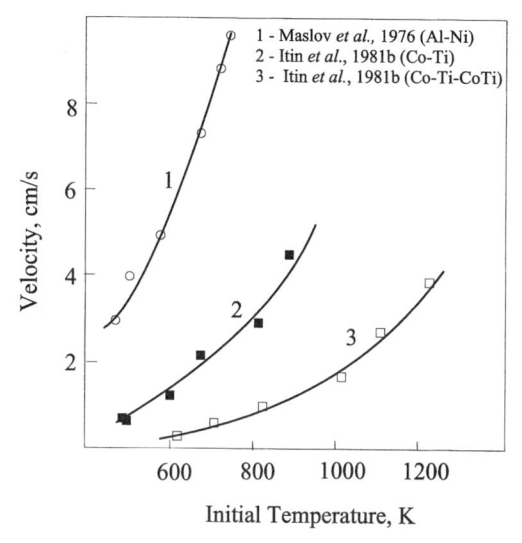

FIG. 52. Combustion velocity for intermetallic systems as a function of initial temperature.

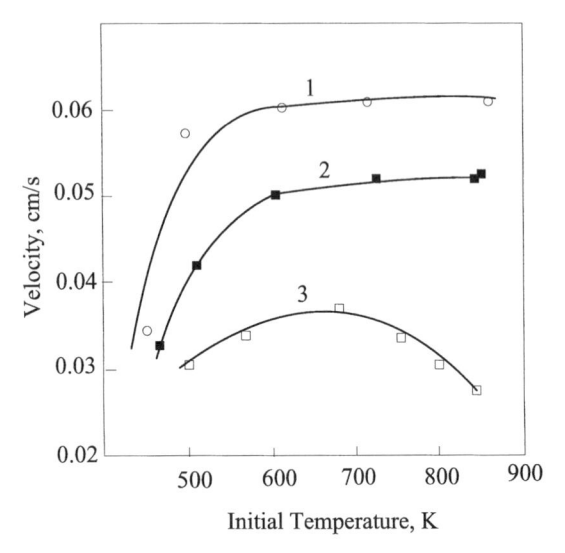

FIG. 53. Combustion velocity for Si-N$_2$ system as a function of initial temperature; $P = 120$ atm and different dilution levels: curve 1, 60 wt %; curve 2, 40 wt %; curve 3, 30 wt % (Adapted from Mukasyan, 1986).

3. Sample Dimensions

Increasing the sample diameter (D) increases the ratio of volumetric heat generation ($\propto D^3$) to surface heat loss ($\propto D^2$). The dependence of U on D, for different systems, is shown in Fig. 54. By increasing the diameter above a critical value, the combustion temperature approaches the adiabatic value, and U becomes constant. This behavior has been observed for the "solid flame" Ta-C system (curve 1), the Ti-C system (curve 2) where one reactant melts, and in Ni-Al (curve 3) where both reactants melt. The maximum measured temperature as a function of radial position in a cylindrical pellet with a diameter of 2 cm for the Mo+2Si system is presented in Fig. 55. The data show significant heat losses from the specimen. For this reason, incomplete combustion often occurs for samples with small diameter and may lead to an undesirable product phase composition (Martynenko and Borovinskaya, 1975; Bratchikov et al., 1980).

For gas–solid systems, adjusting the sample diameter may alter the combustion regime due to changes in the characteristic filtration length (Section IV,D,2). For small diameters, a layer-by-layer infiltration mode with a high degree of conversion is observed in the Ti-N$_2$ system (Borovinskaya et al., 1992a). However, to achieve high conversion for systems with a wide postcombustion zone, a slow rate of cooling is desired. In Fig. 56, the dependence of final conversion on sample diameter for the B-N$_2$ system is presented. A complete reaction was achieved only for samples with a relatively large diameter (≥ 8 cm).

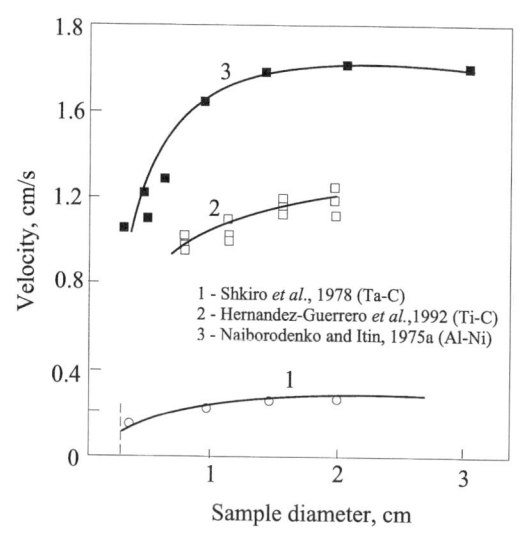

FIG. 54. Combustion velocity as a function of sample diameter.

4. *Effects of Gravity*

To control the product properties during the synthesis of advanced materials, the mechanism of product synthesis must be understood (see also Section VI). In other words, it is essential to develop a fundamental understanding of the physi-

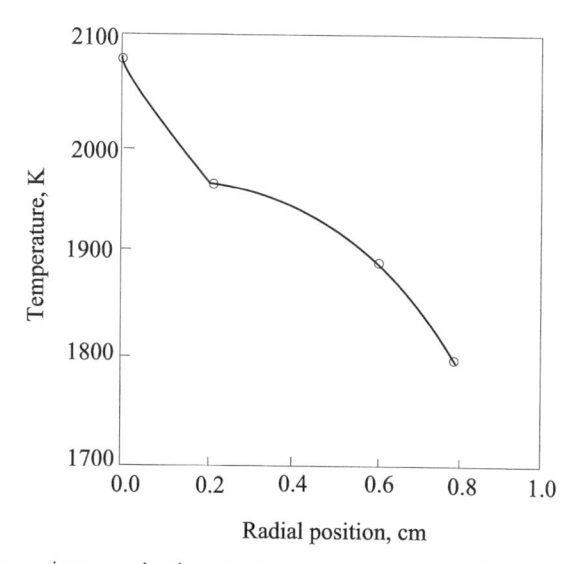

FIG. 55. The maximum combustion temperature as a function of radial position in the cylinder pellet; Mo-2Si system (Adapted from Kumar *et al.*, 1988b).

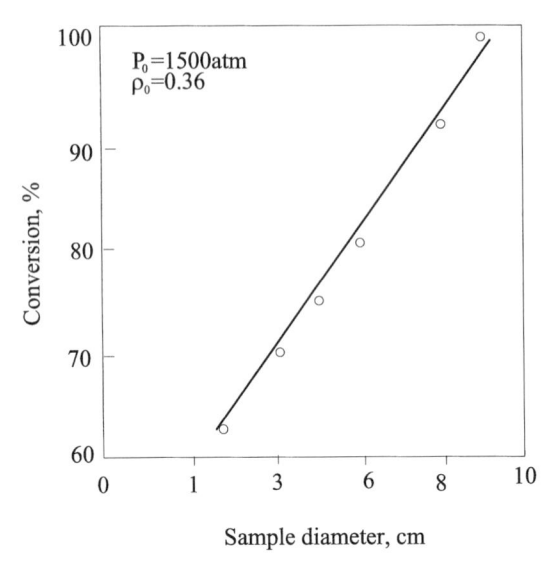

FIG. 56. Final conversion as a function of initial sample diameter for the B-N$_2$ system (Adapted from Mukasyan, 1986).

cochemical processes that occur during the extreme conditions of the combustion wave. A variety of reaction systems have been studied previously under normal gravity conditions, and results have shown that the mechanisms of combustion and structure formation involve several stages. These include melting of reactants and products, spreading of the melt, droplet coalescence, diffusion and convection in the molten phase, nucleation of solid products, crystal growth, buoyancy of solid particles and bubbles in the melt, and natural convection in the gas phase. Most of these processes are affected by gravity.

Relatively little research has been done to date on combustion synthesis of materials under microgravity. Aircraft experiments carried out in Russia (Shteinberg *et al.*, 1991) resulted in production of highly porous (up to 95%) TiC-based materials. The materials were produced using combustion synthesis from elements, with the addition of special gasifying compounds to the reaction mixture to increase pore formation. It was shown that the materials synthesized in microgravity (10^{-2} g) had higher degrees of expansion, with larger final porosities (see Fig. 57), than those produced in normal gravity conditions. Moreover, the distribution of porosity along the sample was more uniform under microgravity conditions. In the United States, foamed ceramic B$_4$C-Al$_2$O$_3$ composites were synthesized during parabolic flights on-board a Lear jet, using a thermite-type reaction (Moore *et al.*, 1992). Note, however, that the mechanical and physical properties of these foamed materials have not been well characterized. Nevertheless, they may provide interesting combinations of materials properties, especially if the

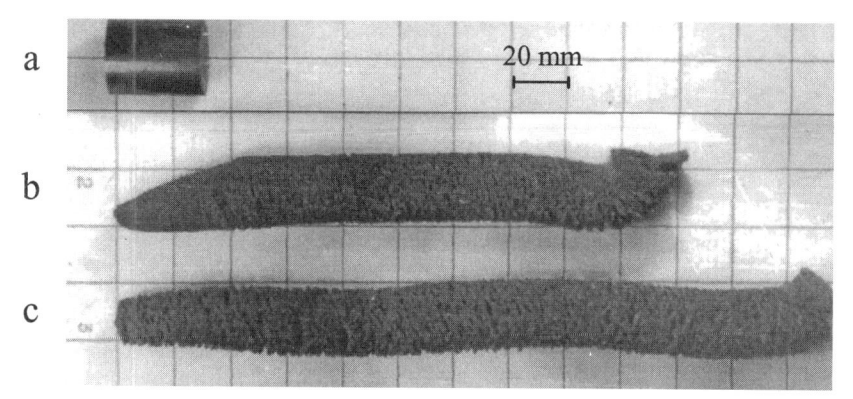

FIG. 57. Photographs of (a) initial sample and products (TiC) synthesized in (b) terrestrial and (c) microgravity conditions (Adapted from Shteinberg et al., 1991).

distribution and morphology of the porosity, as well as micro- and macrostructures, can be controlled.

Another promising direction is production of poreless materials in microgravity. Reducing the effects of gravity during synthesis can inhibit segregation of phases with different densities (Yi et al., 1996). Some experiments have been conducted in Lear jet planes (Hunter and Moore, 1994) to obtain dense ceramic–metal composites (e.g., TiC-Al$_2$O$_3$-Al, TiB$_2$-Al$_2$O$_3$-Al, ZrB$_2$-Al$_2$O$_3$-Al, B$_4$CAl$_2$O$_3$-Al) in thermite-type systems. These studies have shown that gravity significantly influences the CS of composite materials (Moore, 1995). For example, a more uniform product microstructure was obtained in μG as compared with normal gravity conditions. Also, experiments in microgravity for thermite systems (Zr-Al-Fe$_2$O$_3$) have been conducted in Japan (Odawara et al., 1993).

In a recent work by Mukasyan et al. (1997b), experiments were conducted in a 2.2-s drop tower (NASA Lewis Center, Cleveland, OH) providing microgravity conditions ($\sim 10^{-4}$ m/s^2) during 2.2 s of drop time. The role of gravity on the combustion process and product microstructure during synthesis of various materials was examined. For example, in the Ni$_3$Al-TiB$_2$ composite system, the Ni$_3$Al ($-\Delta H_f^\circ = 153.1$ kJ/mol) phase melts during reaction, allowing the buoyancy of the ceramic solid phase to be examined. The amount of liquid phase was controlled by varying the TiB$_2$ ($-\Delta H_f^\circ = 323.8$ kJ/mol) content, which generates the additional heat. It was observed that gravity influences not only propagation of the combustion wave, but the microstructure of the final product as well. Statistical analysis revealed that the average particle size of TiB$_2$ grains in the material produced in normal gravity was 1.2 μm, while in microgravity it was 0.9 μm. These results were confirmed with quenching experiments, where the rate of grain growth during combustion synthesis can be obtained (Fig. 58). These experiments indicate that the difference is related to the process of grain growth dur-

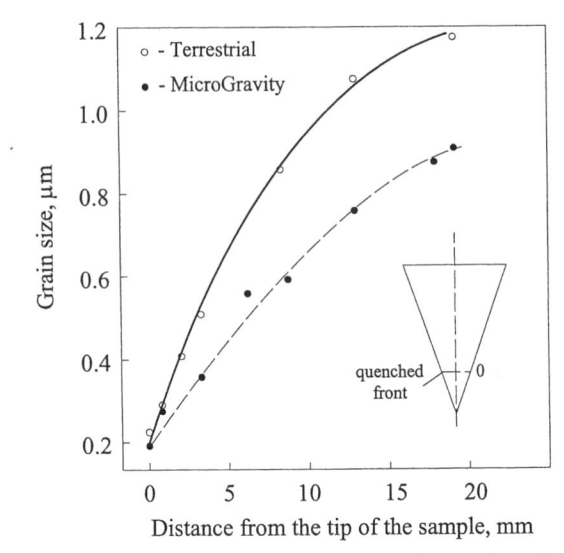

FIG. 58. Evolution of TiB$_2$ (dark phase) grain size in Ni$_3$Al (bright phase) matrix during combustion synthesis in the (Ti-2B)-(3Ni-Al) system (Adapted from Mukasyan et al., 1997b).

ing CS. Although this effect should be investigated in more detail, it appears that enhanced mass transfer, owing to melt convection and buoyancy in normal gravity, promotes grain growth. Thus, gravity appears to be a promising way for controlling the microstructure of composite materials produced by combustion synthesis, and opens the possibility of nanophase materials production in the combustion wave.

VI. Methods and Mechanisms for Structure Formation

The characteristics of the combustion wave, including velocity and combustion temperature, are determined by the processes occurring in the heating and reaction zones and in some cases, the postcombustion zone. These different synthesis zones were determined experimentally for many systems by an analysis of the temperature profiles measured using microthermocouples (cf. Zenin et al., 1980, 1981; Dunmead et al., 1992a,b,c), and are shown schematically in Fig. 59. The length of the preheating zone varies from 0.05 to 0.3 mm, while the heating time and rate are equal to 10^{-4}–1 s and 10^2–10^6 K/s, respectively. The total wavelength (including zones of preheating, combustion and post-combustion) is generally 1–2 mm and may be as wide as 20–30 mm for multistage reactions. The reaction time within the combustion zone does not exceed 10 s even in the case of

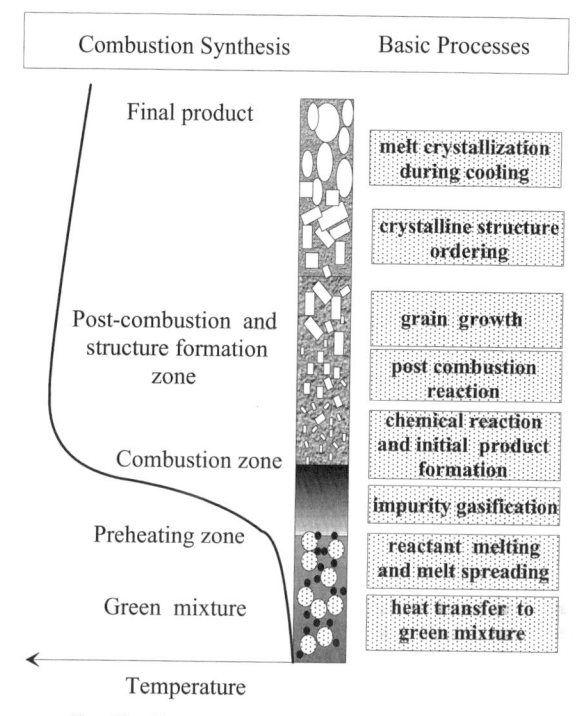

FIG. 59. Characteristic structure of combustion wave.

broad-zone waves and the temperature of most systems in this zone varies within the range of 2000–3000 K. Hence, the characteristic features of combustion synthesis are high temperatures, fast heating rates, and relatively short reaction times.

In addition to the specific features of the zones (i.e., temperature, length, heating rate, and time), the divisions between them can also be distinguished by the differences in physicochemical nature of the initial and final product structures. The initial stage of structure formation is concurrent with the chemical reaction, where the driving force of the process is the reduction of Gibbs free energy resulting from the formation of new chemical bonds, under nonequilibrium conditions (see Fig. 59). In the final structure formation process, physical effects are predominant where the free energy reduces further due to interfacial surface reduction, ordering of the crystal structure, and other related processes that occur without changes in the chemical composition under quasi-equilibrium conditions. In general, the initial product structure may be defined as that formed during a chemical reaction in the combustion zone, which becomes the starting point of the final structure formation step that yields the desired product in the postcombustion zone.

The boundary between the initial and final structure formation often coincides with the zone between combustion and postcombustion. However, in the case of considerable changes in the chemical composition taking place behind the combustion zone (e.g., during a multistage reaction), the boundary between the initial and final structure formation may coincide with the zones between postcombustion and cooling. If many intermediate products are formed in a multistage reaction, the initial structure of each should be considered.

As discussed earlier, analysis of temperature profiles obtained by microthermocouple measurements have elucidated the unique conditions associated with the combustion synthesis process. However, this approach does not directly identify the composition or microstructure of the phases formed. It is important to recognize that most published investigations in the field of combustion synthesis only address the *final* product structure. Considerably less has been reported about the structure formation processes leading to the final product. Most results that describe the evolution from the initial reactants to the final product are inferred by the effects of processing variables (e.g., density, dilution, particle size) on the final microstructure (see Section V). To date, only a few investigations have directly identified *initial* product structure. As discussed earlier, identification of this structure is important since the initial structure represents the starting point for all subsequent material structure formation processes. Thus, the focus of this section is on the initial stages of the structure formation mechanisms in combustion synthesis and novel methods developed especially for this purpose.

A. Major Physicochemical Processes Occurring during Combustion Synthesis

The following physicochemical processes affecting structure formation during combustion synthesis can be identified:

1. Heat transfer from the reaction zone to unreacted particles in the green mixture ahead of the reaction front
2. Phase transitions of solid reactants (e.g., $\alpha Ti \rightarrow \beta Ti$)
3. Formation of eutectic melts and contact melting
4. Melting of reactants
5. Spreading of a molten phase under the action of capillary forces and due to the reduction of surface tension
6. Coalescence of fused particles
7. Gasification of volatile impurities and reactants
8. Chemical reaction with initial product formation
9. Melting of intermediate products
10. Melt crystallization upon cooling

11. Crystal growth
12. Phase transitions in solid products during cooling
13. Ordering of the crystal structure

The first nine processes proceed during the rapid increase in temperature to a maximum at the combustion front and the last four take place behind the combustion front with a constant or gradual lowering of temperature. Therefore, dynamic methods of investigation providing *in situ* monitoring of fast processes with a short duration are necessary to describe adequately the first group of synthesis steps during rapid heating. On the other hand, methods that analyze quenched reaction waves or final products can be applied to the study of the processes occurring during cooling (processes 10–13).

B. QUENCHING OF THE COMBUSTION WAVE

Before the development of dynamic techniques discussed in later sections, the quenching technique, where there is rapid cooling of the combustion front, was the only method available to study mechanisms of structure formation during combustion synthesis. Following quenching, the phase composition and microstructure of the different zones in the combustion wave can be identified using a layer-by-layer analysis of the quenched regions by scanning electron microscopy (SEM) and X-ray diffraction (XRD). Several methods have been developed to quench the combustion wave during synthesis and are described next.

1. Quenching in Liquid

The first attempt to terminate the combustion wave and obtain quenched products was by immersing the sample in liquid argon (Merzhanov *et al.*, 1972). A schematic of this method is shown in Fig. 60a along with the measured cooling rate. However, the cooling rates were not high due to a gaseous layer that formed on the surface of the sample, which acted as a thermal insulator and decreased the quenching rate. The $Zr-N_2$ solid–gas system quenched in liquid argon revealed that the driving force for combustion wave propagation is heat evolution from a highly exothermic solid solution formed in the reaction zone (Borovinskaya and Loryan, 1976). The mechanism of nickel and cobalt aluminide formation in the reaction zone was also investigated by dropping the reacting mixtures in water (Naiborodenko and Itin, 1975a). In this work, the conclusion was made that the direct dissolution of Ni into molten Al is the controlling step at higher temperatures, while the interaction through a solid product layer limits the process at lower temperatures.

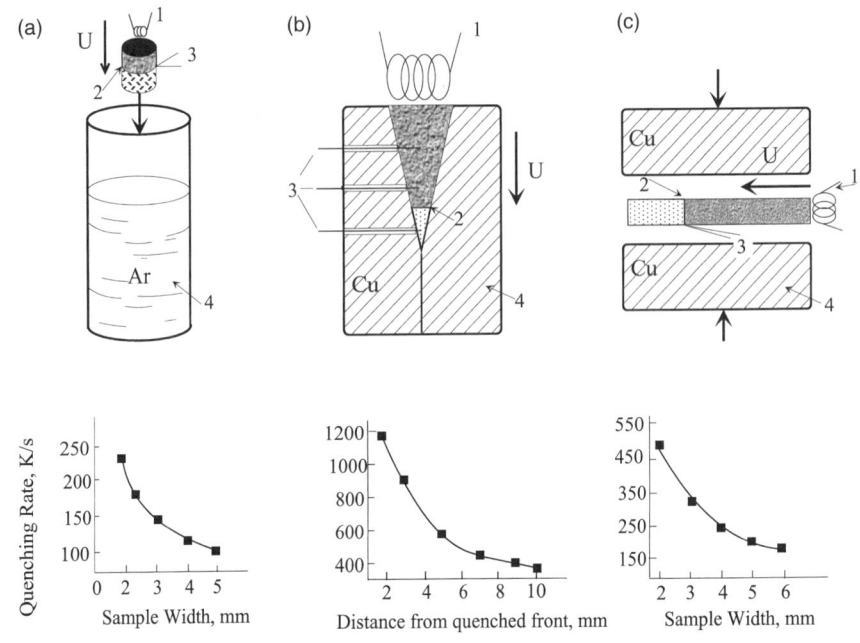

FIG. 60. Schematic drawings of quenching methods and cooling rates in (a) liquid argon; (b) a wedge-shaped cut in a copper block; (c) between two copper plates. 1, ignitor; 2, quenched front; 3, thermocouples; 4, quenching medium (Adapted from Mukasyan and Borovinskaya, 1992).

The highest rate of cooling (10^4–10^5 K/s) was reported for a relatively thin sample quenched with a high-speed water jet (Khusid et al., 1992). However, the high water velocity (up to 150 m/s) necessary to achieve high rates of quenching often destroyed the sample and interaction occurred between the water and reacting sample.

2. Quenching with Copper Block

The use of a massive copper block to remove heat rapidly was originally demonstrated during the combustion of heterogeneous solid fuels (Andreev, 1966). This technique was later applied to quench reacting pellets during combustion synthesis (Rogachev et al., 1987). In this method, a reactant mixture is pressed into a wedge-shaped cut in a copper block (Fig. 60b). The sample is ignited and the reacting mixture progressively quenched due to a decrease of the reactant mixture volume, which increases conductive heat losses to the walls of the copper block as the combustion wave travels to the apex of the wedge. A series of photodiodes and/or thermocouples can be used to measure the temperature at several locations of the reacting pellet. The continuous acquisition of temperature during the experiment where quenching rates and times can be determined directly from temperature measurements is clearly an advantage of this method over quenching in a liquid.

Another quenching method utilized in gas–solid reacting systems involves a thin rectangular sample and two copper plates (Mukasyan and Borovinskaya, 1992). At an appropriate instant during the progress of the combustion wave, the reacting sample is squeezed between the copper plates. Thus, the copper plates serve not only as heat sinks but also isolate the reacting solid compact from the gaseous reactant (Fig. 60c). However, it can be seen from Fig. 60 that the quenching rate with copper plates is not as high as with the wedge-shaped cut of a copper block.

3. Quenching with Shock Waves

The use of shock waves has also been investigated to quench the combustion synthesis reaction in some systems, including titanium and air (Molokov and Mukasyan, 1992). In this method, the titanium compact was placed in a steel tube, which permitted access of air to the combustion zone. The steel tube was surrounded by an explosive charge. The titanium sample was ignited by a tungsten coil at one end. When the combustion wave reached a specific location in the sample, it was detected by an embedded thermocouple, and the explosive charge was detonated at the other end. This explosive shock wave rapidly altered the heat release conditions in the reacting sample by forcing the sample to contact the steel liner, which was at a lower temperature. In addition, the detonation force increased the sample density, which in turn increased the thermal conductivity, and quenched the reacting mixture. A quenching rate of $\sim 10^4$ K/s for this method was estimated indirectly from microstructural features. Using this technique, it was observed for the first time that titanium reacted with air in a two-stage process. The first stage was the interaction between titanium and nitrogen, while oxygen was involved in the postcombustion zone. A similar result was obtained by time-resolved X-ray diffraction, a technique discussed later in this section.

4. Quenching Results

Most quenching results reported in the literature to identify structure formation mechanisms during combustion synthesis were obtained by a wedge-shaped sample pressed into a copper block. As discussed earlier, this method is relatively simple and provides high cooling rates. The stages of structure formation were identified using this technique in several systems, including Ti-C and Ti-B (Rogachev *et al.*, 1987), Ti-C-Ni (Rogachev *et al.*, 1988), Ti-Al-C (Volpe and

Evstigneev, 1992), Ni-Al (Lebrat *et al.*, 1994), and Ti-Si (Rogachev *et al.*, 1995). Some of these results are discussed here.

Figure 61 shows the velocity measurements and temperature profiles for a TiC sample pressed in a wedge-shaped cut of a copper block and quenched (Merzhanov *et al.*, 1990a). With the exception of regions near the point of ignition and the quenched combustion front, the velocity of the combustion front remained approximately constant (Fig. 61a), which implies that within the region of constant velocity, conditions of product formation in the combustion zone are not significantly influenced by heat losses. A relatively high velocity near the ignition of the sample was attributed to an excess of enthalpy provided by the ignition heat pulse and a lower velocity near the point of extinction due to significant heat loss to the copper. However, there is a gradual decrease in quenching time in the sample as the front propagates from the top (wide portion) to the bottom (apex), as shown in the temperature profiles at different locations (Fig. 61b). Thus, different regions in the sample have progressively less reaction completeness and represent

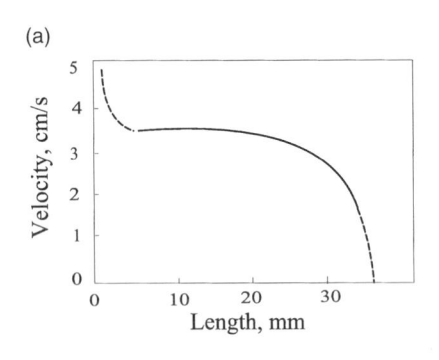

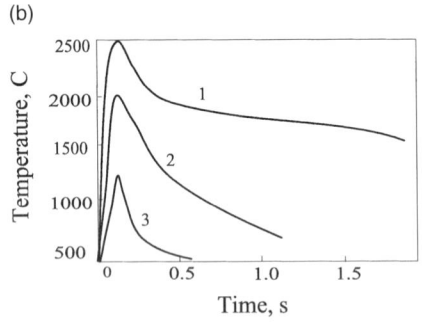

FIG. 61. Combustion front quenching in a wedge-shaped cut of a copper block for Ti-C system: (a) velocity as a function of distance from ignition surface; (b) temperature profiles at different locations relative to ignition surface: curve 1, 3 mm; curve 2, 26 mm; curve 3, 36 mm (Adapted from Rogachev *et al.*, 1987).

the sequence of steps in the structure formation process of the postcombustion zone. By combining the resulting phase formation with the measured quenching time, it is possible to evaluate the kinetic characteristics of the final structure grain growth. The initial structure formation processes can be identified by studying the extinguished area.

The microstructure of the initial titanium-graphite mixture is shown in Figure 62a. When the temperature in the combustion wave reaches 1660°C, titanium melts. It was determined from quenching results that a thin film (~0.1 μm) of the Ti melt spreads over the solid carbon surface with *simultaneous* formation of titanium carbide grains (Fig. 62b). Small rounded TiC particles were observed to appear *within* the liquid rather than in the form of a continuous product layer (Rogachev et al., 1987). To illustrate this fact further, the typical microstructure formed during combustion reaction of titanium melts with graphite whiskers (10 μm in diameter) is shown in Fig. 62c.

A similar mechanism of initial product formation within a molten film was also observed for TiB$_2$, with the difference that at a distance ~1 mm from the combustion front, the rounded boride particles acquired morphological anisotropy followed by crystallographic faceting. Also, in a study of various quenched Ti+C+Ni mixtures (Rogachev et al., 1988), it was determined that titanium carbide grains (~1 μm) were surrounded by a metallic melt composed of Ti and Ni. The product grains continued to grow in the postcombustion zone. The fraction of C in the product TiC$_x$ grains was greater than the initial C/Ti ratio, owing to the reaction of titanium with nickel.

The product grain growth rate was determined from measurements of the grain size in the different quenched regions obtained by SEM observations. For example, a sequence of microstructures taken from different zones of Ti-C-Ni quenched samples is shown in Fig. 63. The results of statistical analysis of such microstructures for Ti-C, TiB$_2$, and Ti-C-Ni systems are presented in Fig. 64, and illustrate an initial rapid increase in grain size that approaches a constant value in

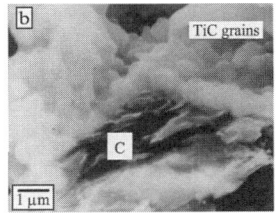

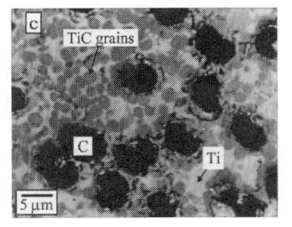

FIG. 62. Characteristic microstructures during synthesis of Ti-C system: (a) initial reactant mixture, Ti+graphite; (b) initial product formation; (c) formation of TiC grains in Ti melt around graphite whiskers.

188   A. VARMA ET AL.

FIG. 63. Sequence of microstructures taken from different zones of Ti-C-Ni (Rogachev et al., 1988): (a) 1 mm; (b) 10 mm; (c) 35 mm behind quenched front.

the postcombustion zone. The growth rate of the mean product grains ranges between 1 and 30 μm/s and demonstrates that the initial product grains increase by an order of magnitude in a few seconds after passage of the combustion front. A layer-by-layer X-ray diffraction (XRD) phase composition analysis of quenched Ti+C and Ti+2B mixtures confirms that the refractory carbide and boride crystal

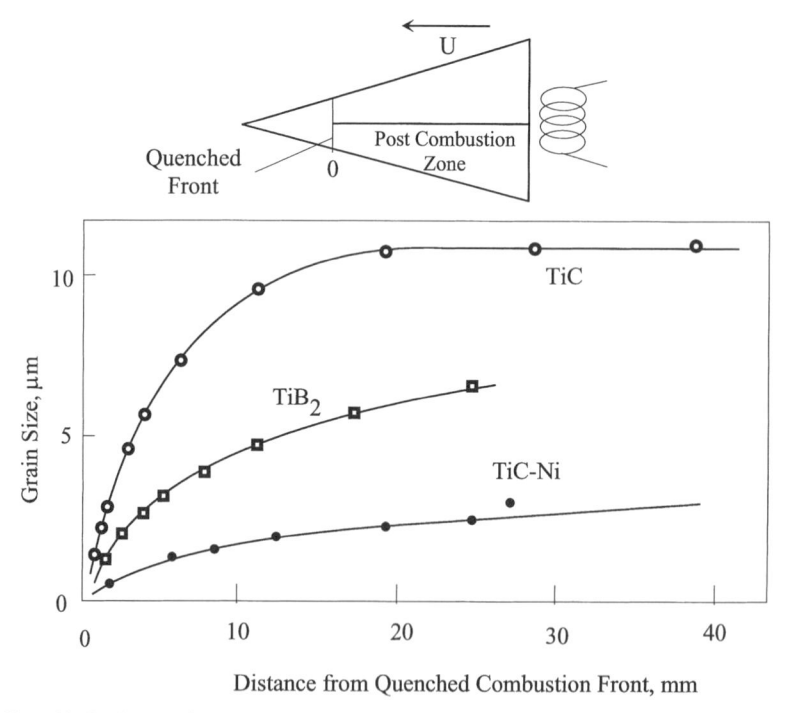

FIG. 64. Product grain growth in quenched samples of different systems (Adapted from Merzhanov and Rogachev, 1992).

structures form in the reaction zone, and the phase composition does not change behind the combustion front (Fig. 65).

The evolution of the microstructure in combustion synthesis of Ni$_3$Al, showed that the Ni$_2$Al$_3$ and NiAl layers initially appear on the surface of Ni, followed by formation of the final Ni$_3$Al product (Lebrat *et al.*, 1994). It was shown that the *dissolution* of Ni into the Al melt controls the process. On the other hand, an investigation of quenching Ni+Al mixtures (Rogachev *et al.*, 1993) led to the conclusion that Ni and Al particles melt simultaneously and coalesce in the reaction zone, resulting in the appearance of NiAl crystals in the volume of melt. Homogenization of the chemical composition occurred in the postcombustion zone. This mechanism of structure formation was observed for the first time and is referred to as the *reaction coalescence* mechanism (Rogachev *et al.*, 1993). The results from the two studies demonstrate that in the case of the 3Ni+Al system, the final product is preceded by the NiAl and Ni$_2$Al$_3$ intermediate phases, whereas in the Ni+Al reaction, the initial product has the same overall phase composition as the final product.

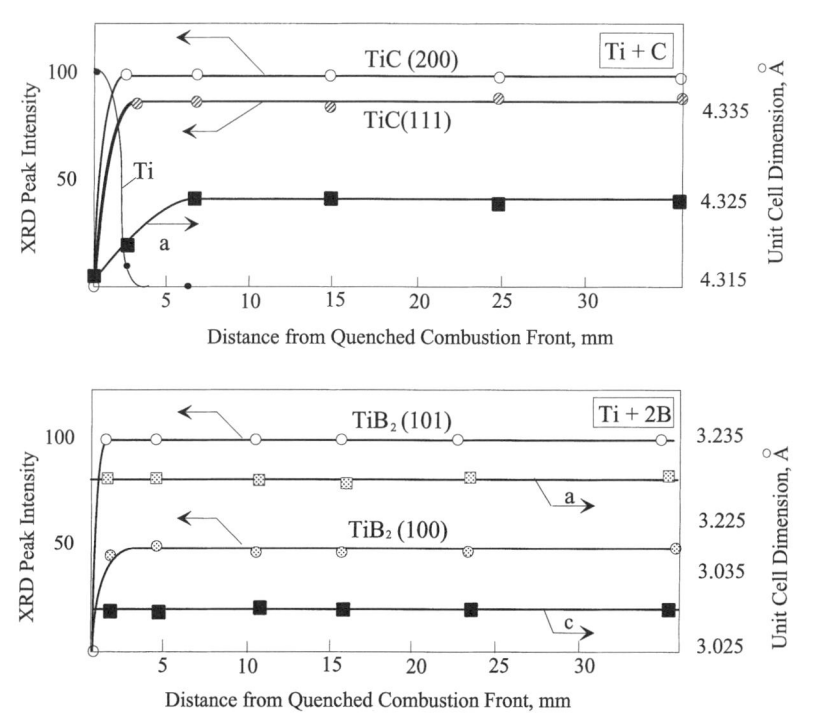

FIG. 65. Results of layer-by-layer X-ray diffraction phase composition analyses of quenched Ti-C and Ti-2B mixtures (Adapted from Rogachev, 1995).

C. Model Systems for Simulation of Reactant Interaction

The spatial structure of a heterogeneous reactant powder mixture with its random distribution of reactant particles is too complex for a study of the elemental stages of structure formation mechanisms. This problem has been overcome by the development of experimental techniques that utilize reactants with simpler geometries, such as wires and foils. These methods isolate the reactant interaction, thus yielding additional information.

1. Metal Wires

The *electrothermography* method, in which metal wires are heated by electric current in an oxygen gas atmosphere, was originally developed to study oxidation processes (Grigor'ev, 1975). This technique was also used to study the mechanism of ignition for Ti+C and Zr+C systems, by coating Ti and Zr wires with carbon black (Vadchenko et al., 1976). The interaction between reactants was quenched by simply shutting off the current to the wire. In this work, after the metal filament (Ti or Zr) was heated to the melting point, the liquid metal penetrated the pores of carbon under the action of capillary forces. This led to self-propagating reaction conditions in which chemical heat evolution was due to the dissolution of carbon in the liquid metal, forming carbide phases. The interaction mechanism was also investigated with titanium and aluminum wires electrodeposited with nickel (Vadchenko et al., 1987). In this study, cross-sections of the reacted wires were polished and the influence of initial temperature on intermetallic layer growth was investigated. The critical temperatures of the two systems were also determined. In titanium wires coated with nickel, the critical temperature of 1220 ± 20 K corresponded to the eutectic reaction $\beta Ti + Ti_2Ni$ (1128 K) below the melting points of Ti and Ni. For nickel-coated aluminum wires, the temperature of 910 ± 20 K was close to the eutectic reaction $Al + NiAl_3$ (913 K) and the aluminum melting point (933 K). The dependence of interaction on the Ti wire diameter and the nickel coating thickness was also identified.

2. Laminated Metallic Foils

To provide a well-defined boundary between reactants, experiments have been conducted using ensembles of laminated Ni and Al metallic foils (Anselmi-Tamburini and Munir, 1989). These experiments were motivated by modeling efforts that used a foil geometry to understand interactions between powder reactants (Hardt and Phung, 1973; see Section IV,B). In the experimental technique, nickel (12.5 and 125 μm) and aluminum (12.5 μm) foils were laminated in specific stoichiometric amounts corresponding to the $NiAl$ and $NiAl_3$ compounds. These ensembles were then ignited in a chemical oven consisting of Ni+Al powders. An inert layer (Al_2O_3), placed at the bottom of the laminated foils, quenched the wave propagation through the foils. In both the NiAl and $NiAl_3$ foil compositions, melting of aluminum occurred ahead of the wave and ignition was triggered at the nickel melting point.

The appearance of molten aluminum *ahead* of the combustion front was not observed in quenching studies for nickel aluminides (Lebrat *et al.*, 1994). However, the quenched foil result using a Ni+Al composition is in partial agreement with quenching experiments for NiAl (Rogachev *et al.*, 1993) where the presence of molten nickel and aluminum was observed *at* the combustion front. It is possible that the presence of an oxide film on the aluminum foils could have delayed ignition until the nickel melting point. In any case, it appears that the foil assembly does not adequately model the interaction between particulate reactants during combustion synthesis.

3. Particle-Foil Experiments

In both the electrothermographic and foil assembly methods, the rapid heating rates associated with combustion synthesis are reproduced. However, the powder reactant contact found in a compacted green mixture of particulate reactants is not adequately simulated. One way to overcome this is to investigate interactions of particles of one reactant placed on the surface of the coreactant in the form of a thin foil. The physical simulation corresponds to the reaction of a powder mixture where the particle size of one reactant is small while that of the coreactant is relatively large. Two methods have been used to initiate the interaction.

a. Reaction Initiated by Electron Beam. This method was proposed by Korchagin and Podergin (1979), where the electron beam not only initiates the reaction but also allows *in situ* observation of morphological and microstructural changes in a transmission electron microscope (TEM). The beam creates temperature conditions (i.e., rapid heating rates) similar to those found in a combustion wave propagating through a pressed compact of reactants. It was switched off at various stages in order to quench the interaction and the product formed was examined with the microscope.

The interactions of Fe_2O_3 particles over Al (Korchagin and Podergin, 1979) and Ti particles over C (Korchagin and Aleksandrov, 1981) were investigated using this technique. It was determined that the reaction between components is initiated after the appearance of a liquid phase that diffuses into the solid phase. This is followed by the formation of a primary product layer, which permits the solid to pass to the liquid phase by dissolution. It is interesting to note that the diffusion coefficients of the liquid component, estimated from the growth rate and width of the primary product layer, exceeded known values for bulk bimetallic samples by several orders of magnitude.

In similar TEM work by Sharipova and coworkers (1992), the interaction of Cr_2O_3 particles with a sputter coated Al film was observed. From this work, it was determined that Cr_2O_3 was reduced in stages: $Cr_2O_3 \rightarrow CrO \rightarrow Cr$. The chromium aluminides that formed ($CrAl_2$, Cr_3Al_2, Cr_9Al_{17}) were found to depend on the aluminum content of the sample and on the initial temperature, maintained using a tungsten coil prior to initiation of combustion with the electron beam.

b. Reaction Initiated by Electric Current. Based on microscopy and electrothermographic methods, another experimental technique (Shugaev *et al.*, 1992a,b) involving particle and foil reactants has been developed that allows a direct continuous observation of the process (see Fig. 66). In this approach, the foil substrate and particles are heated at rates up to 10^3–10^4 K/s by passing an electrical current through the foil, which allows better control in heating the two reactants and simulates heating rates found in the combustion wavefront. It is apparent that the upper

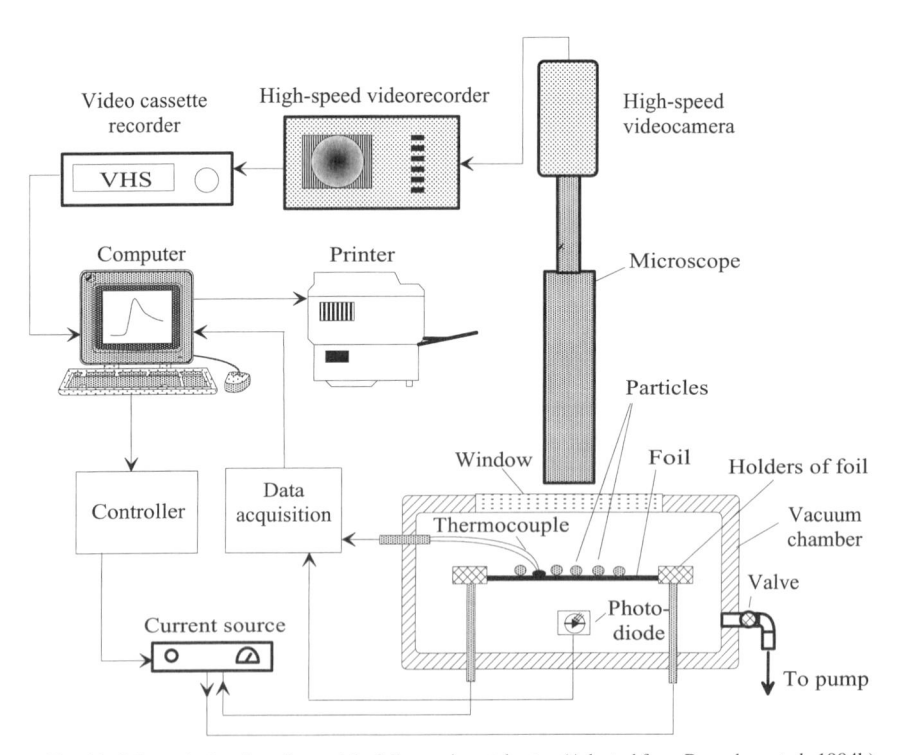

FIG. 66. Schematic drawing of a particle-foil experimental setup (Adapted from Rogachev *et al.*, 1994b).

temperature limit of this method is the melting point of the foil reactant. In the works cited earlier, the temperature of the heated foil was calculated from the emissivity registered with a photodiode, and the error in the temperature measurement was ±1%. The extent of interaction between the reactants was controlled by varying the current and quenching that resulted from switching off the current, making it possible to examine the initial stages of structure formation. A direct continuous observation of the interaction was obtained with a high-temperature microscope attached to a high-speed motion picture camera. The quenched intermediate structures were analyzed by means of SEM and electron probe microanalysis (EPMA).

Two groups of systems have been identified: (1) with rapid and broad spreading of the reaction area and (2) without considerable spreading (Rogachev *et al.*, 1994a). In the first case, the spreading rate is on the order of 1 mm/s where a thin, fine-grained product layer appears on the surface of a nonmelting foil substrate. In the second case, the spreading rate is less than 0.1 mm/s, where the molten particles penetrate into the substrate rather than spread over the surface. The first group includes most carbide, boride, and silicide systems (Ti-C, B-Nb, B-Ta, B-Mo, Si-Nb, Si-Ta, Si-Mo, etc.) while the second group consists mainly of intermetallic systems (Al-Ni, Al-Ti, Al-Nb, etc.). Some systems (e.g., Si-Ti) exhibit intermediate behavior.

It is important to note that the initial product microstructures observed in the model particle-foil experiments are similar to those detected in the quenched reaction zone of the combustion wave in bulk samples. For example, grain size distributions for both experiments in the TiC system are shown in Fig. 67. Rather fine grains form at the early stages of both experiments (curves 1 and 2), followed by significant coarsening of grains which takes place in a few seconds up to tens of seconds (curves 3 and 4). Their agreement suggests that the particle-foil experiment simulates particulate reactant interaction within a pressed compact.

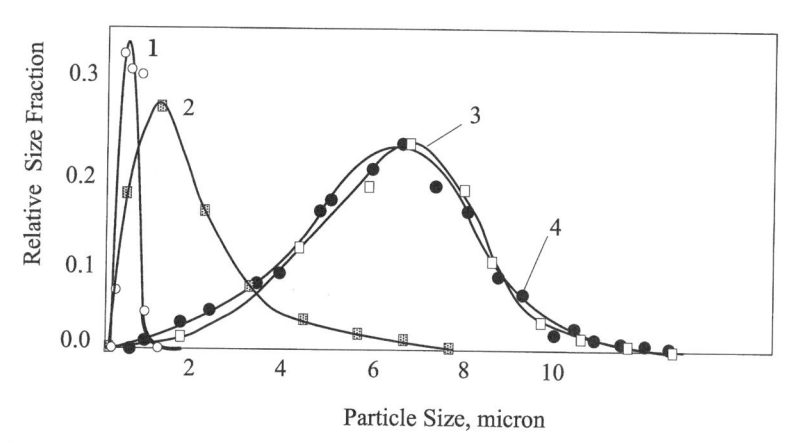

FIG. 67. A comparison of the TiC product grain size distributions in the initial (curves 1, 2) and final (curves 3, 4) stages of product formation obtained by quenching (curves 2, 4) and particle-foil techniques (curves 1, 3) (Adapted from Rogachev, 1995).

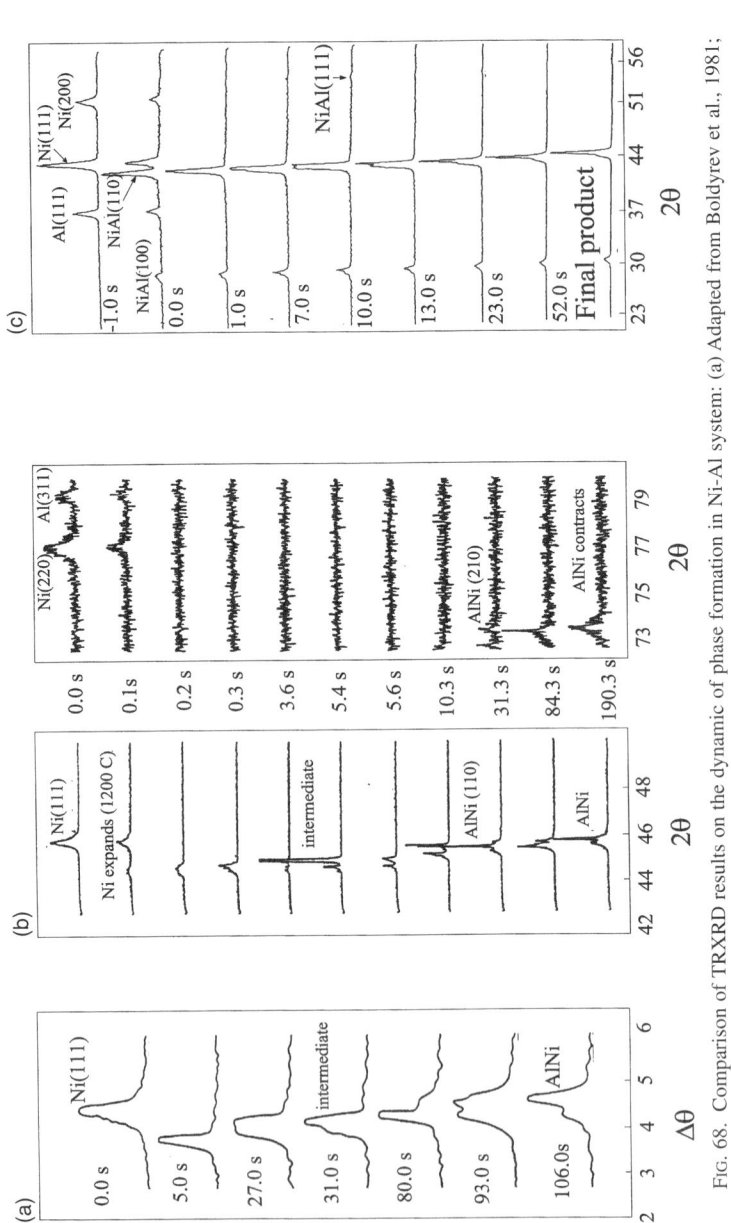

FIG. 68. Comparison of TRXRD results on the dynamic of phase formation in Ni-Al system: (a) Adapted from Boldyrev et al., 1981; (b) Adapted from Wong et al. 1990; (c) Adapted from Rogachev et al. 1994a.

D. TIME-RESOLVED X-RAY DIFFRACTION (TRXRD)

1. TRXRD Using Synchrotron Radiation

Continuous monitoring of crystal structure formation during combustion of a heterogeneous mixture of reactants is a powerful tool for understanding the structure-forming mechanisms. This idea was first realized using synchrotron radiation (Boldyrev *et al.,* 1981) and provided interesting results. A sequence of XRD patterns in the vicinity of 100% Ni and NiAl peaks, recorded during the combustion synthesis of a Ni+Al mixture, demonstrated that the diffraction line of nickel monoaluminide appeared 70–75 s after the combustion wave passage. The formation of the NiAl product was preceded by two lines shifted into a region of larger interplanar spacing (i.e., lower angles) than that of NiAl (Fig. 68a), which were assigned to unidentified intermediate phases.

The TRXRD synchrotron radiation method was developed further by Holt *et al.* (1990) where the scanning time of each XRD pattern was reduced to 0.01 s and two position-sensitive detectors allowed simultaneous monitoring of two regions of the X-ray diffraction spectrum. The combustion syntheses of NiAl, TiC, TiC-NiTi (Holt *et al.,* 1990; Wong *et al.,* 1990), TaC, and Ta_2C (Larson *et al.,* 1993) were investigated by this improved method. In an investigation of the Ni+Al reaction, the most intensive XRD peak of NiAl (110) appeared about 10 s after the passage of the combustion front and was preceded by several shifted and split peaks that were attributed to an unidentified intermediate product. The appearance of the NiAl (210) peak, 20 s after the NiAl (110) line, was attributed to the process of crystal lattice ordering after passage of the combustion front (Fig. 68b).

2. TRXRD Using an X-Ray Tube

A significant advance in this field was achieved by Merzhanov and coworkers (1993) with a new method of TRXRD using an ordinary X-ray tube (~2 kW). A schematic diagram of the apparatus is shown in Fig. 69. Using a one-dimensional position-sensitive detector, a wide range of XRD spectra (up to 30–40 degrees) can be monitored, which allows simultaneous observation of several nonoverlapping lines of each phase in complex systems. To calculate the XRD peak shift due to thermal expansion of the crystal lattice during combustion, the temperature of the sample area under investigation was measured by a thermocouple.

Several SHS systems have been investigated using this equipment, including Ti-air, $Ti-N_2$ and $Ti-O_2$ (Khomenko *et al.,* 1993), Ni-Al (Rogachev *et al.,* 1994b), Ti-Si (Rogachev *et al.,* 1995), Nb-B, Ti-C, Ta-C, and Ti-Cr-C (Merzhanov *et al.,* 1995). The results from these investigations demonstrate one or multistage mechanisms of product structure formation, where the sequence of stages was generally found to be different in the combustion wave relative to synthesis by conventional methods involving lower heating rate conditions.

The TRXRD patterns obtained during combustion of a Ni+Al mixture are shown in Fig. 68c (Rogachev *et al.*, 1994b). Corrections made taking into account thermal expansion led to the conclusion that partially disordered NiAl is the first phase to appear in the combustion synthesis wave rather than an unidentified phase as previously concluded by Boldyrev *et al.* (1981) and Wong *et al.* (1990). The splitting of peaks was determined to be due to the ordering process in NiAl that takes place in the postcombustion zone. In 2Ni+3Al mixtures, the disordered Ni$_2$Al$_3$ phase was found as a primary product during combustion. An ordering process of the crystal lattice was observed during the cooling of samples. These results suggest the possibility of obtaining disordered intermetallic phases by combustion synthesis.

Another example of the application of the TRXRD method involved combustion of Ti in air (Khomenko *et al.*, 1993). A complicated mechanism was observed involving five intermediate phases that preceded the formation of the final TiO$_2$ product. In general, the TRXRD results were similar to those found in quenching the reaction with shock waves (Molokov and Mukasyan, 1992) where the first stage involves the titanium–nitrogen reaction, followed by the interaction with oxygen in the postcombustion zone. The TRXRD experiment identified the formation of additional oxide phases that may have remained undetected during microprobe analysis of the quenched sample. The kinetics of the appearance and disappearance of each phase are shown qualitatively in Fig. 70.

An interesting TRXRD result during the combustion of titanium with air is the dependence of the reaction mechanism on temperature. At low heating rates (10°C/s), where the maximum combustion temperature was determined to be ~650°C, the Ti reacts directly with oxygen to form TiO$_2$ (Merzhanov *et al.*, 1995). At a considerably higher heating rate of ~10^3°C/s, the maximum combustion tem-

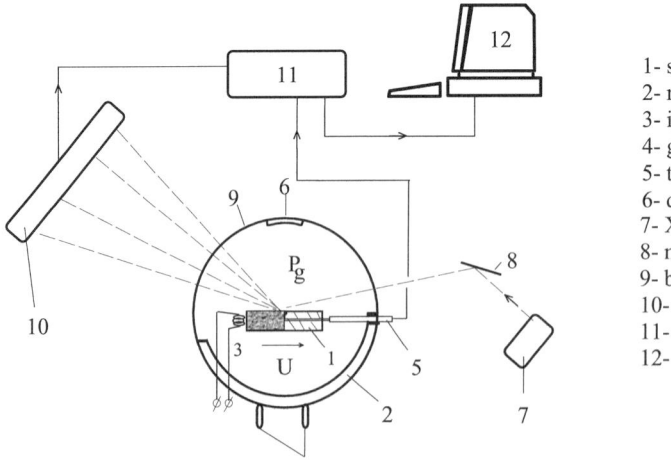

FIG. 69. Schematic diagram of the laboratory TRXRD apparatus (Adapted from Khomenko *et al.*, 1993).

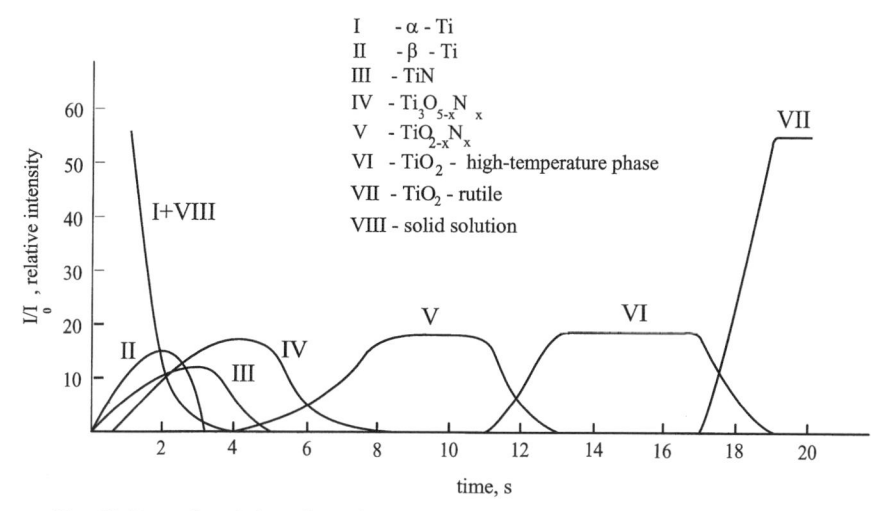

FIG. 70. Dynamics of phase formation during combustion reaction between titanium and air in high-temperature mode (Adapted from Khomenko et al., 1993).

perature was ~1700°C and a more complex mechanism occurs: $\alpha Ti \rightarrow \beta Ti \rightarrow TiN_x \rightarrow Ti_3O_{5-x}N_x \rightarrow Ti_6O_{11-x}N_x \rightarrow TiO_{2-x}N_x \rightarrow TiO_{2-x}N_x$ (rutile structure). These results demonstrate that the sequence of intermediate phases formed may depend strongly on the heating rate.

E. Microstructure of Combustion Wave

While the microscopic processes occurring in the combustion wave need to be understood, the microstructure of the combustion wave itself is also a significant consideration. In other words, it is important not only to understand the local structural transformations occurring during CS, but to link them with variations in local conditions of the combustion wave. The two approaches for investigating the combustion wave microstructure examine the temperature–time profiles and heterogeneity of the combustion wave shape and propagation, at the local level.

In the first approach, the temperature–time profiles of the combustion wave are measured using thin thermocouples (Zenin et al., 1980; Dunmead et al., 1992a,b,c). In the work of Zenin et al. (1980), 7-μm-diameter microthermocouples protected by a layer of boron nitride were used. The temperature–time profiles were analyzed using the one-dimensional heat conduction equation with heat generation (see Section IV,A,1) taking numerical derivatives of the spatial temperature distribution. An implicit assumption in the analysis is that the combustion wave is stable and planar at both the macro- and microscopic scales.

Several reaction systems (e.g., Nb-B, Zr-B, Ti-Si, Ti-C) have been studied using this approach. For example, the results for a Nb+2B mixture are shown in

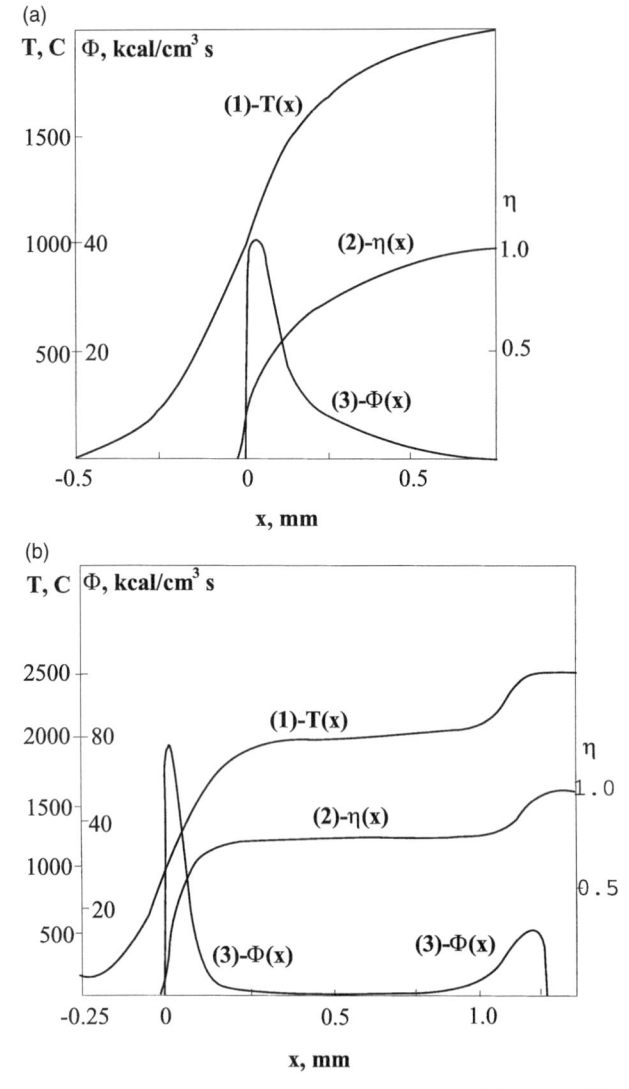

FIG. 71. Average temperature distribution over the combustion zones of the mixture Nb+xB, rate of heat release Φ, and degree of conversion, η: (a) the mixture Nb+2B; (b) Nb+B (Adapted from Zenin et al., 1980).

Fig. 71. Based on the measured temperature profile data (curve 1), the distribution of conversion along the combustion wave, $\eta(x)$ (curve 2), and the heat release function, $\phi(x)$ (curve 3), have been determined using Eq. (15). The characteristic length of the zones, L_r is given by the size of the domain where $\phi(x)$ is nonzero. The preheating zone, x_T, is defined as the sample length ahead of the front where the heat flux (i.e., $q=\lambda \cdot dT/dx$) varies linearly with temperature, T (see Section IV,A,2).

The second approach considers the heterogeneous structure of the compact particulate reaction mixture. Since the combustion wave propagates through a mixture of powder reactants with varying particle sizes (0.1–100 μm), nonuniformity of the combustion front and variations in propagation velocity can be expected at the microscopic scale.

Using a high-speed movie camera, the heterogeneity of the combustion front for Ti-C-Ni-based systems (reacted in air) was observed (Levashov et al., 1991). Recently, a new method using microscopic high-speed video imaging was developed (Rogachev et al., 1994b) to study the microstructure of the combustion wave for various systems including Ti-C, Ti-Si, Ti-N$_2$, and Ni-Al. A schematic of the experimental apparatus is shown in Fig. 72. The setup is equipped with a long-focus microscope (K-2, Infinity Photo-Optical Company) and a high-speed video camera (EktaPro 1000 Imager and Processor, Kodak). With this method, the combustion wave can be observed at high spatial (800×) and temporal (1-ms) resolutions.

Figure 73 shows the propagation of the combustion front (bright field) through the green mixture (dark field) for the gasless Ti-Si system. At conventional mag-

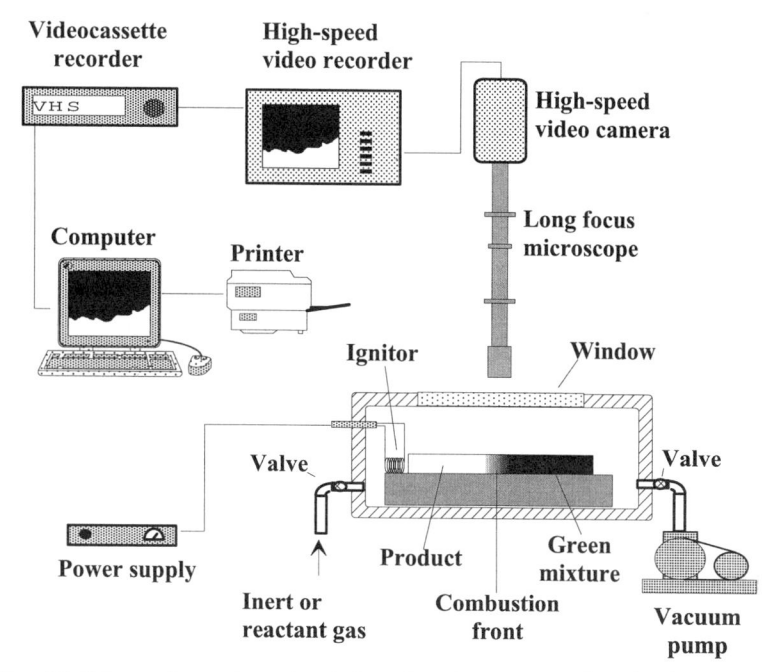

FIG. 72. Scheme of experimental technique for microstructure of combustion wave study (Adapted from Rogachev et al., 1994b).

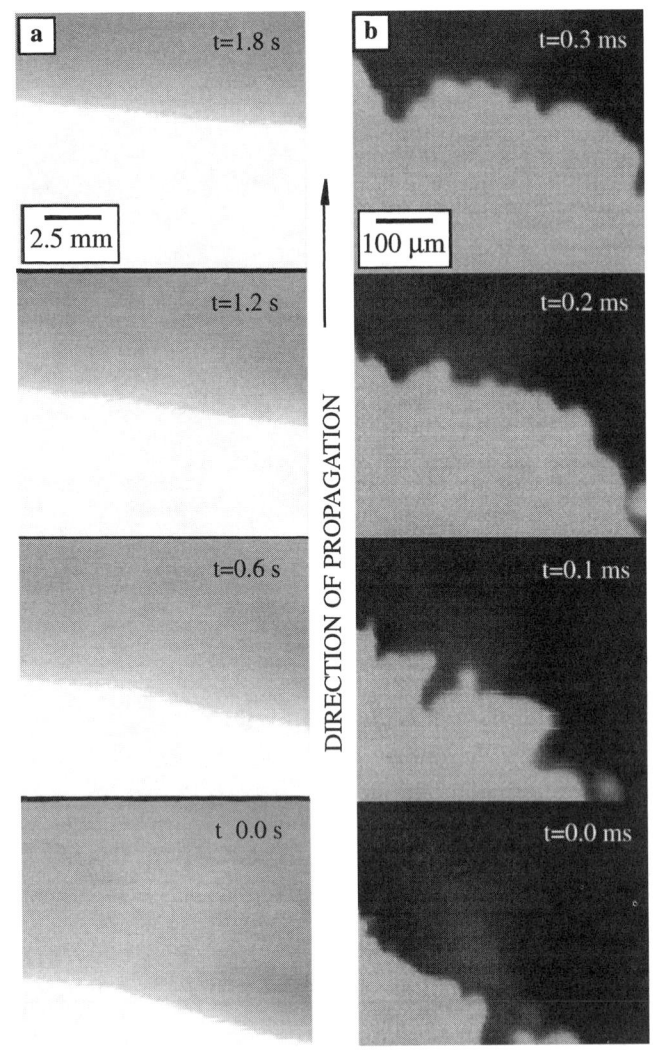

FIG. 73. Frames of combustion front propagation through the 5Ti-3Si system obtained with different magnifications and image rates: (a) 4× and 30 frames/s; (b) 100× and 1000 frames/s (Adapted from Mukasyan et al., 1996).

nification (4×) and imaging rate (30 frames/s), the combustion front profile appears flat, and moves steadily through the pellet (Fig. 73a). However, at high magnification (100×) and imaging rate (1000 frames/s), the shape of the combustion front is seen to be irregular with pronounced arc-shaped convexities and concavities (see Fig. 73b). Also, the combustion wave moves in an unsteady manner, often hesitating or moving rapidly. The results show that the microstructure of the combustion front is comparable to the titanium particle size and shape.

COMBUSTION SYNTHESIS OF ADVANCED MATERIALS 201

TABLE XXII
CHARACTERISTICS OF COMBUSTION FRONT PROPAGATION[a]

Characteristic	Parameter
Average position of combustion front at time t	$\overline{F}(t) = \dfrac{\int_0^{y_0} F(y,t) dy}{y_0}$
Dispersion of combustion front	$\sigma_F = \sqrt{\dfrac{\int_{t_i}^{t_f} \int_0^{y_0} [\overline{F}(t) - F(y,t)]^2 dy\, dt}{y_0(t_f - t_i)}}$
Instantaneous combustion front velocity	$U(y,t) = \dfrac{\partial F(y,t)}{\partial t} \approx \dfrac{F(y,t_i) - F(y,t_{i-1})}{t_i - t_{i-1}}$
Average combustion front velocity for duration of experiment	$\overline{\overline{U}} = \dfrac{\int_{t_i}^{t_f} \int_0^{y_0} U(y,t) dy\, dt}{y_0(t_f - t_i)}$
Dispersion of combustion velocities for duration of experiment	$\sigma_U = \sqrt{\dfrac{\int_{t_i}^{t_0} \int_0^{y_0} [\overline{U}(t) - U(y,t)]^2 dy\, dt}{y_0(t_f - t_i)}}$

[a]Data from Mukasyan et al. (1996).

The microscopic high-speed video recording method has recently been developed further and used to investigate the combustion wave microstructure in the Ti-Si and Ti-N_2 systems (Mukasyan et al., 1996; Hwang et al., 1997). The location of the combustion front was estimated using image analysis techniques, and its shape and propagation were characterized quantitatively.

Table XXII shows some of the parameters used to characterize the microstructure of the combustion wave. In general, the parameters can be divided into two groups. The first group characterizes the *shape* of the combustion front, and includes the local [$F(y,t)$] and average [$\overline{F}(t)$] front profiles, as well as the front dispersion, σ_F, which is a measure of "roughness" of the combustion front. The second group describes the combustion front *propagation* at the microscopic level. For this, the instantaneous, $U(y,t)$ and average, $\overline{\overline{U}}$, velocities of the combustion wave, as well as the dispersion of the instantaneous velocities, σ_U, are calculated.

The microstructure of the combustion wave was investigated as a function of two parameters characteristic of the heterogeneity of the reactant medium: the average particle size, d, of the more refractory titanium (note that the Ti-Si system initiates with the melting of silicon; Rogachev et al., 1995) and the initial sample porosity, ϵ. Similar results were observed for both the gasless Ti-Si and gas–solid Ti-N_2 systems. Either with increasing particle size or increasing porosity, the fluctuations of the combustion wave, both in terms of shape and propagation, increase monotonically (e.g., see Fig. 74).

Further, we can normalize three measures of the combustion wave microstructure: d, σ_F, and σ_U, with the appropriate length (x_T) or velocity (U_{macro}) scales (see

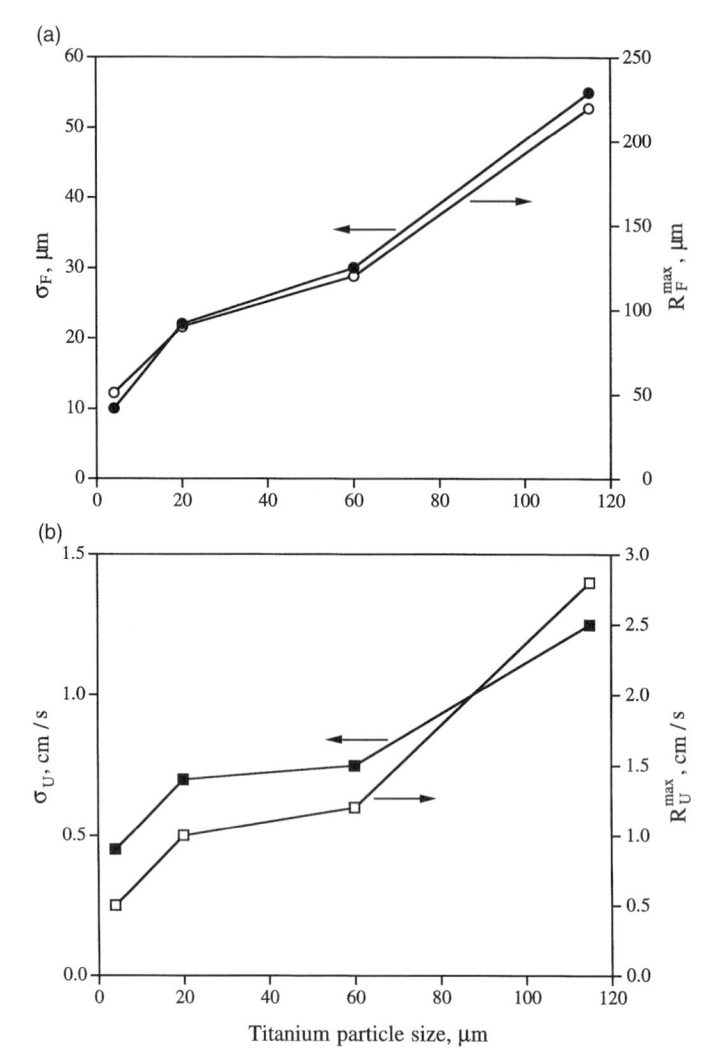

FIG. 74. Dependence of combustion front heterogeneity on Ti particle size in the 5Ti-3Si system: (a) shape; (b) propagation (Adapted from Mukasyan et al., 1996).

Table XXIII). The length scale of combustion, x_T, can be evaluated using the relation:

$$x_T = \alpha/U, \tag{75}$$

where α is the medium thermal diffusivity. In addition, an estimate of the length scale during combustion conditions can also be obtained from the temperature profiles (Zenin et al., 1980). Table XXIII shows that thermal length scales obtained from the two approaches are comparable in magnitude.

TABLE XXIII
COMPARISON OF EXPERIMENTAL AND CALCULATED THERMAL LENGTH SCALES, X_T, μm FOR THE 5Ti+3Si SYSTEM[a]

d (μm)	ϵ	$X_T = \alpha/U$	From Zenin et al. (1980)
4	0.5	20	—
20	0.5	35	60 ± 20
60	0.5	40	—
110	0.5	50	60 ± 20
115	0.5	50	70 ± 20

[a]Data from Mukasyan et al. (1996).

Several characteristic ratios of heterogeneity as functions of titanium particle size and sample porosity are shown in Figs. 75a and b, respectively. For the quasi-homogeneous approximation to be valid, the heterogeneity of the reactant medium should be small in comparison to the thermal length scale. As a result, the heterogeneity of the combustion front (shape and propagation) should also be small relative to the macroscopic behavior:

$$d/X_T \ll 1, \quad \sigma_F/X_T \ll 1, \quad \sigma_U/U_{macro} \ll 1. \tag{76}$$

The data in Fig. 75 indicate that the characteristic measures of heterogeneity increase with increasing reactant particle size (at fixed porosity) and sample porosity (at fixed particle sizes). Also, these ratios are small *only* for relatively small particle sizes and porosity values, limiting the regions where the quasi-homogeneous approximation may be expected to hold (region I). For larger particle sizes and sample porosities, the microscopic scales of heterogeneity are large (region II), and consequently the nature of combustion wave propagation is more complex.

Based on the results described earlier, the following mechanism of heterogeneous combustion was suggested (Mukasyan et al., 1996). This process consists of two sequential steps: rapid reaction of a single reactant particle, typically the most refractory, followed by ignition delay during which preheating of the neighboring particle occurs. Since in this case the particles react sequentially, one after another, this mechanism may be called a *relay-race* mechanism of heterogeneous combustion. The thermal resistance between particles is significant in heterogeneous systems such as powder mixtures, which leads to the delay between reactions of neighboring particles.

F. Concluding Remarks

To study the variety of structural transformation processes that occur simultaneously in the combustion wave, it is necessary to utilize a wide range of methods. The evolution and morphological features of the microstructure during com-

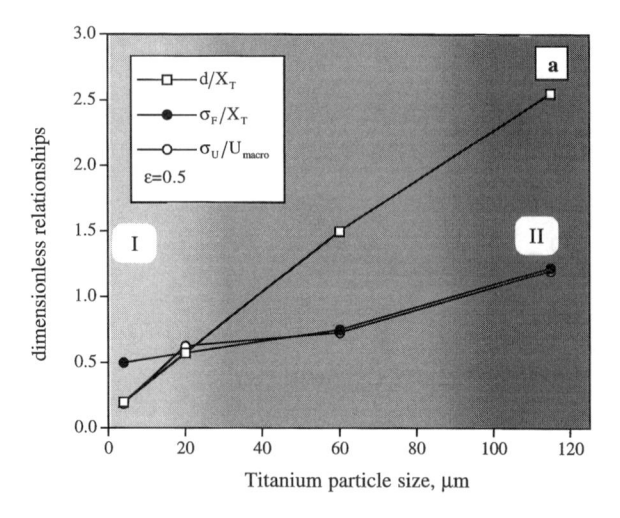

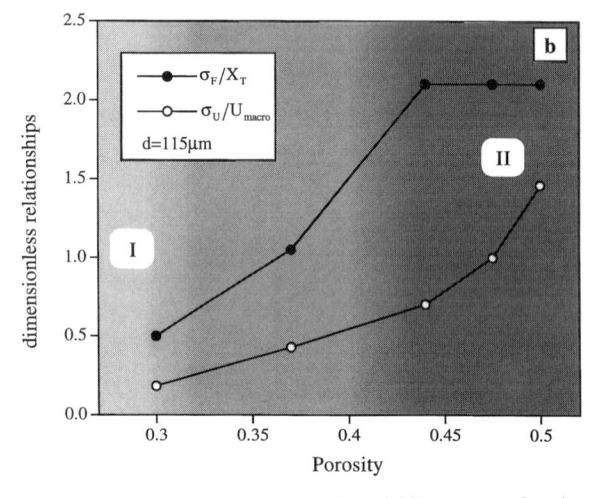

FIG. 75. Characteristic ratios of heterogeneity in the 5Ti-3Si system as functions of (a) titanium particle size; (b) initial sample porosity (Adapted from Mukasyan et al., 1996).

bustion synthesis can be identified using a layer-by-layer analysis of quenched samples. In particle-foil experiments, the initial interaction between molten and solid reactants can be isolated, and the spreading rate of the liquid phase as well as features of the resulting product layer (e.g., phase composition, product grain size, layer thickness) identified. The dynamics of the phase composition and crystal structure ordering can be monitored continuously by TRXRD. Finally, microstructural analysis of the combustion wave provides important information about the local conditions which affect the synthesis process.

Each method described in this chapter elucidates some specific aspects of the structure-forming processes during combustion synthesis. These techniques have been developed primarily for combustion synthesis, where extreme reaction conditions such as short reaction times and steep thermal gradients exist. However, it should be noted that each technique also has limitations associated with it. In the particle-foil experiment, the foil temperature can be measured accurately, but the precise temperature of the particle in contact with the foil is difficult to determine. If a system follows a multistage reaction, where intermediates form and decompose in the postcombustion zone, quenching may affect the course of the reaction, and products not found in bulk samples may be present in the quenched sample. Finally, the TRXRD method does not detect amorphous, molten phases or solid solutions that may form during combustion synthesis, nor does it provide a direct visualization of the phases formed. Despite their shortcomings, these methods complement each other, and if applied to a given system, can lead to a reasonably complete description of the structure-forming processes.

For example, four microstructural models of the product formation process in solid-solid systems can be distinguished based on the results of the experimental methods discussed in this chapter. Brief descriptions of the models are presented in Fig. 76.

The first mechanism, *reaction-diffusion,* is based on the assumption that a layer of the solid product (or products) appears on the boundary between the two

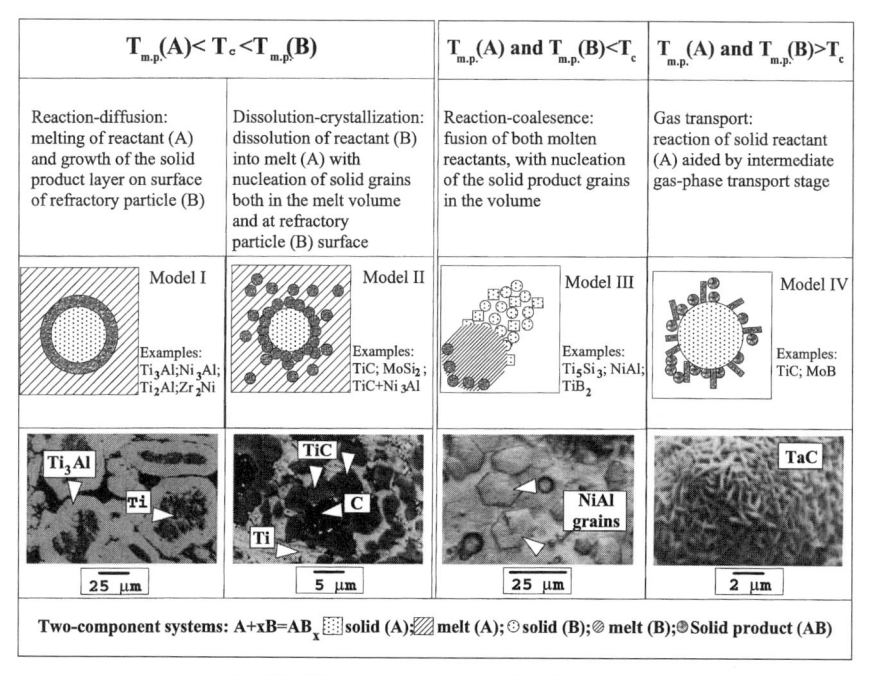

$T_{m.p.}(A) < T_c < T_{m.p.}(B)$		$T_{m.p.}(A)$ and $T_{m.p.}(B) < T_c$	$T_{m.p.}(A)$ and $T_{m.p.}(B) > T_c$
Reaction-diffusion: melting of reactant (A) and growth of the solid product layer on surface of refractory particle (B)	Dissolution-crystallization: dissolution of reactant (B) into melt (A) with nucleation of solid grains both in the melt volume and at refractory particle (B) surface	Reaction-coalescence: fusion of both molten reactants, with nucleation of the solid product grains in the volume	Gas transport: reaction of solid reactant (A) aided by intermediate gas-phase transport stage
Model I — Examples: Ti$_3$Al; Ni$_3$Al; Ti$_2$Al; Zr$_2$Ni	Model II — Examples: TiC; MoSi$_2$; TiC+Ni$_3$Al	Model III — Examples: Ti$_5$Si$_3$; NiAl; TiB$_2$	Model IV — Examples: TiC; MoB

Two-component systems: A+xB=AB$_x$ □ solid (A); ▨ melt (A); ☉ solid (B); ⊘ melt (B); ● Solid product (AB)

FIG. 76. Microstructural models of initial product.

reactants. This mechanism has been discussed thoroughly in the literature (e.g., Hardt and Phung, 1973; Aldushin and Khaikin, 1974; Kanury, 1992; Cao and Varma, 1994). It appears that this microstructural model applies mainly when intermetallic compounds are synthesized. In the synthesis of refractory ceramic materials, such as TiC, ZrC, etc., the carbides form separate grains of the initial product, which are difficult to pack into a continuous layer on the nonmelted reactant particle surface. The second model, *dissolution-crystallization,* represents the case where particle size of the nonmelted reactant is comparable to the initial product grain size (~0.1–1 μm), which is a common situation for most carbide and boride systems. This mechanism describes synthesis of polycrystalline ceramic materials or ceramic grain–metal matrix composites, depending on composition of the green mixture. When the combustion temperature is higher than melting points of both reactants, the *reaction coalescence* mechanism becomes dominant. Finally, when the temperature of combustion is lower than melting points of the reactants, *gas-transport* between solid particles plays a significant role in the initial product formation, as was shown for the Ta-C system (Merzhanov *et al.,* 1990b). In this case, an initial product with microstructure characteristic of gas phase deposition (lamellar and column crystals, whiskers) appears.

Note that even for the same reaction system, the combustion temperature T_c can be varied relative to the reactant melting points, by varying the initial temperature, dilution and stoichiometry. Thus different models may apply for the same system, depending on the experimental conditions. The examples listed in Fig. 76 correspond to the most common conditions, with no preheating or dilution.

Acknowledgments

We gratefully acknowledge financial support from the National Science Foundation (grants CTS 92-14009 and CTS 95-28941), including supplements for U.S.–Russia joint cooperative research, and NASA (NAG3-1644). Dr. Cynthia Kachelmyer contributed to early versions of this review, in particular, section VI.

Nomenclature

$A(T^*,\eta^*)$	Function; see Eq. (26)
b	Fraction of inert or product diluent
c_p	Heat capacity
d	Particle size

D	Characteristic length of the sample (e.g., pellet diameter)
D_{rc}	Diameter of the reaction cell; see Eq. (36)
D^*	Critical sample size
$\mathcal{D}$	Mass diffusivity
$f(d)$	Dependence of U on particle size (see Section IV,B)
$F(T)$	Dependence of U on temperature (see Section IV,B)
$F(y,t)$	Local combustion front profile
$\overline{F}(t)$	Average combustion front profile at time t (see Table XXII)
G	Forced gas velocity
G_i	Molar Gibbs free energy of component i
h	Convective heat transfer coefficient
H	Microhardness
ΔH_{fus}	Heat of fusion
$H_i(T)$	Enthalpy of component i
k_0	Preexponential constant for kinetics
k_f	Permeability coefficient, k/μ
L	Length of sample
Le	Lewis number, $\mathcal{D}/\alpha$
L_f	Filtration combustion parameter; see Eq. (55)
l_f	Length of filtration during combustion
MW_i	Molecular weight of component i
M_i	Reduced-metal intermediate or product i
$(MO_x)_i$	Metal oxide reactant
m	Mass
m_s	Mass of nonoxide phase (see Section I,B,3)
m_t	Total mass of the phase produced by the reaction (see Section I,B,3)
n_i	Number of moles of i
N_k	Number of components of type k
P_i	Product i
p	Gas pressure
$p(T)$	Probability of cell reaction; see Eq. (39)
p_0^{cr}	Critical gas pressure for full reaction
p_k	Partial pressure of gas phase component k
$\overline{p}$	Average gas pressure
q	Heat flux variable, $\lambda \cdot dT/dx$
Q	Heat of reaction
S	Surface area of sample
t	Time
T	Temperature
U	Combustion velocity
U_{macro}	Macroscopic combustion velocity (see Table XXII)
$\overline{\overline{U}}$	Average combustion velocity (see Table XXII)

V	Volume of sample
v_f	Velocity of gas filtration
v_T	Velocity of convective heat exchange front (see Section IV,3,a)
$W(\eta,T,p)$	Chemical reaction rate for gas–solid systems
x	Spatial coordinate in traveling wave frame; see Eq. (10)
x_T	Characteristic length of thermal relaxation in combustion wave, α/U
X	Spatial coordinate
X_i	Solid-phase reactant
y	Spatial coordinate
Y_i	Gas-phase reactant i
Z_i	Reducing-metal reactant
$(ZO_x)_i$	Oxide product i

Greek Letters

α_{st}	Stability parameter; see Eq. (46)
α	Thermal diffusivity
β	Stability parameter; see Eq. (40)
$\chi(d)$	Particle size distribution function (see Section IV,B)
δ	Thickness of product layer; see Eq. (35)
$\delta(x)$	Dirac delta function
ϵ_m	Emissivity
ϵ	Porosity
$\Phi_k(\eta,T)$	Heat release function for source or sink, k
$\phi(\eta,T)$	Reaction kinetic function
$\Gamma(z)$	Gamma function
γ	Stability parameter; see Eq. (40)
η	Reactant conversion
η^*	Reactant conversion controlling combustion front propagation
η_{ps}	Degree of phase separation (see Section I,B,3)
$\varphi(\eta)$	Kinetic dependence of reactant conversion
λ	Thermal conductivity
μ	Viscosity
ν_i	Stoichiometric coefficient of reactant i
π	Forced flux parameter, v_T/U (see Section IV,3,a)
ρ	Density
ρ_i	Amount of reactant i per unit sample volume
σ	Stefan–Boltzmann constant
σ_F	Dispersion of combustion front shape (see Table XXII)
σ_U	Dispersion of combustion front velocities (see Table XXII)
τ	Time step; see Eq. (39)

ω	Counter-current filtration parameter, U_k/v_f (see Section IV,1,b,1)
Ψ	Stability parameter; see Eq. (41)
ζ	Fraction of reactant melted; see Eq. (8)

Subscripts

0	Initial
a	Ambient value
c	Combustion
cool	Cooling
f	Filtration
i	Component i
k	Kinetic control
m	Melting point
max	Maximum
min	Minimum
ps	Phase separation
sc	Superconducting

Superscripts

ad	Adiabatic
cr	Critical
(g)	Gas-phase reactant or product
(l)	Liquid-phase reactant or product
(s)	Solid-phase reactant or product
*	Controlling combustion front propagation

References

Adachi, S., Wada, T., Mihara, T., Miyamoto, Y., Koizumi, M., and O. Yamada, Fabrication of titanium carbide ceramics by high-pressure self-combustion sintering of titanium powder and carbon fiber. *J. Am. Ceram. Soc.*, **72**, 805 (1989).

Adadurov, G. A., Borovinskaya, I. P., Gordopolov, Y. A., and Merzhanov, A. G., Technological fundamentals of SHS compacting. *J. Eng. Phys. Thermophys.*, **63**, 1075 (1992).

Advani, A. H., Thadhani, N. N., Grebe, H. A., Heaps, R., Coffin, C., and Kottke, T., Dynamic modeling of self propagating high temperature synthesis of titanium carbide ceramics. *Scripta Metall. Mater.*, **25**, 1447 (1991).

Agadzhanyan, N. N., and Dolukhanyan, S. K., Burning in the Zr+Nb+N+H system: Hydronitride synthesis. *Combust. Explos. Shock Waves*, **26**, 739 (1990).

Agrafiotis, C. C., Puszynski, J. A., and Hlavacek, V., Experimental study on the synthesis of titanium and tantalum nitrides in the self-propagating regime. *Combust. Sci. Tech.*, **76**, 187 (1991).

Akiyama, T., Isogai, H., and Yagi, J., Combustion synthesis of magnesium nickel. *Int. J. SHS,* **4,** 69 (1995).
Aldushin, A. P., New results in the theory of filtration combustion. *Combust. Flame,* **94,** 308 (1993).
Aldushin, A. P., Filtration combustion of metals. *In* "Thermal Wave Propagation in Heterogeneous Media (Russ.)." Nauka, Novosibirsk, 1988, p. 52.
Aldushin, A. P., and Khaikin, B. I., Combustion of mixtures forming condensed reaction products. *Combust. Explos. Shock Waves,* **10,** 273 (1974).
Aldushin, A. P., and Merzhanov, A. G., Filtration combustion theory: General approaches and results. *In* "Thermal Wave Propagation in Heterogeneous Media (Russ.)." Nauka, Novosibirsk, 1988, p. 9.
Aldushin, A. P., and Merzhanov, A. G., Gasless combustion with phase transformation. *Dokl. Phys. Chem.,* **236,** 973 (1978).
Aldushin, A. P., and Sepliarskii, B. S., Inversion of wave structure in porous medium during gas blow-through. *Phys. Dokl.,* **24,** 928 (1978a).
Aldushin, A. P., and Sepliarskii, B. S., Propagation of waves of exothermic reaction in porous medium during gas blow-through. *Phys. Dokl.,* **23,** 483 (1978b).
Aldushin, A. P., Matkowsky, B. J., Shkadinsky, K. G., Shkadinskaya, G. V., and Volpert, V. A., Combustion of porous samples with melting and flow of reactants. *Combust. Sci. Tech.,* **99,** 313 (1994a).
Aldushin, A. P., Matkowsky, B. J., and Volpert, V. A., Interaction of gasless and filtration combustion. *Combust. Sci. Tech.,* **99,** 75 (1994b).
Aldushin, A. P., Sepliarskii, B. S., and Shkadinskii, K. G., Theory of filtration combustion. *Combust. Explos. Shock Waves,* **16,** 33 (1980).
Aldushin, A. P., Lugovoi, V. D., Merzhanov, A. G., and Khaikin, B. I., Conditions of degeneration of steady combustion wave. *Phys. Dokl.,* **23,** 914 (1978).
Aldushin, A. P., Khaikin, B. I., and Shkadinskii, K. G., Effect of the inhomogeneity of the internal structure of the medium on the combustion of condensed mixtures, interacting through a layer of product. *Combust. Explos. Shock Waves,* **12,** 725 (1976a).
Aldushin, A. P., Merzhanov, A. G., and Sepliarskii, B. S., Theory of filtration combustion of metals. *Combust. Explos. Shock Waves,* **12,** 285 (1976b).
Aldushin, A. P., Ivleva, T. P., Merzhanov, A. G., Khaikin, B. I., and Shkadinskii, K. G., Combustion front propagation in porous exothermic metallic samples with oxidizer filtration. *In* Combustion Processes in Chemical Engineering and Metallurgy (Russ.) (A. G. Merzhanov, ed.). USSR Academy of Science, Chernogolovka, Russia, 1975, p. 245.
Aldushin, A. P., Merzhanov, A. G., and Khaikin, B. I., Conditions for the layer filtration combustion of porous metals. *Dokl. Phys. Chem.,* **215,** 295 (1974).
Aldushin, A. P., Martem'yanova, T. M., Merzhanov, A. G., Khaikin, B. I., and Shkadinskii, K. G., Propagation of the front of an exothermic reaction in condensed mixtures with the interaction of the components through a layer of high-melting product. *Combust. Explos. Shock Waves,* **8,** 159 (1972a).
Aldushin, A. P., Merzhanov, A. G., and Khaikin, B. I., Some features of combustion of condensed system with high-metal-point reaction products. *Dokl. Phys. Chem.,* **204,** 475 (1972b).
Aleksandrov, V. V., and Korchagin, M. A., Mechanism and macrokinetics of reactions accompanying the combustion of SHS systems. *Combust. Explos. Shock Waves,* **23,** 557 (1988).
Aleksandrov, V. V., Korchagin, M. A., and Boldyrev, V. V., Mechanism and macrokinetics of a reaction of components in a powder mixture. *Dokl. Phys. Chem.,* **292,** 114 (1987).
Aleksandrov, V. V., Gruzdev, V. A., and Kovalenko, Y. A., Thermal conductivity of certain aluminum-based SHS systems. *Combust. Explos. Shock Waves,* **21,** 93 (1985).
Alman, D. E., Reaction synthesis of Ni-36.8 wt% Al. *J. Mater. Sci. Lett.,* **13,** 483 (1994).
Andreev, K. K., "Thermal Decomposition and Combustion of Explosive Powders." Nauka, Moscow, 1966.

Andreev, V. A., Mal'tsev, N. M., and Seleznev, V. A., Study of the combustion of hafnium-boron mixtures by optical pyrometry. *Combust. Explos. Shock Waves,* **16,** 374 (1980).
Andrievski, R. A., and Baiman, I. F., Physical-mechanical properties of boride composites obtained by SHS compacting method. *Int. J. SHS,* **1,** 298 (1992).
Anselmi-Tamburini, U., and Munir, Z. A., The propagation of a solid-state combustion wave in Ni-Al foils. *J. Appl. Phys.,* **66,** 5039 (1989).
Astapchik, A. S., Podvoisky, E. P., Chebotko, I. S., Khusid, B. M., Merzhanov, A. G., and Khina, B. B., Stochastic model for a wavelike isothermal reaction in condensed heterogeneous systems. *Phys. Rev. E,* **47,** 1993 (1993).
Avakyan, P. B., Nersesyan, M. D., and Merzhanov, A. G., New Materials for electronic engineering. *Am. Ceram. Soc. Bull.,* **75,** 50 (1996).
Azatyan, T. S., Mal'tsev, V. M., Merzhanov, A. G., and Seleznev, V. A., Some principles in combustion of titanium-silicon mixtures. *Combust. Explos. Shock Waves,* **15,** 35 (1979).
Barzykin, V. V., Initiation of SHS processes. *Pure Appl. Chem.,* **64,** 909 (1992).
Boldyrev, V. V., Aleksandrov, V. V., Korchagin, M. A., Tolochko, B. P., Gusenko, S. N., Sokolov, A. S., Sheromov, M. A., and Lyakhov, N. Z., The study of dynamics of phase formation during nickel monoaluminade synthesis in the combustion regime. *Dokl. Akad. Nauk SSSR,* **259,** 722 (1981).
Booth, F., The theory of self-propagating exothermic reactions in solid systems. *Trans. Farad. Soc.,* **49,** 272 (1953).
Borovinskaya, I. P., Chemical classes of SHS processes and materials. *Pure Appl. Chem.,* **64,** 919 (1992).
Borovinskaya, I. P., Refractory compounds formation during combustion of heterogeneous condensed systems. *Proceedings of the Fourth All Union Symposium on Combustion and Explosion (Russian),* Moscow, 138 (1977).
Borovinskaya, I. P., and Loryan, V. E., Self-propagating high-temperature synthesis of titanium nitrides under high nitrogen pressures. *Sov. Powd. Metall.,* **11,** 851 (1978).
Borovinskaya, I. P., and Loryan, V. E., Self-propagating processes in the formation of solid solutions in the zirconium-nitrogen system. *Dokl. Phys. Chem.,* **231,** 1230 (1976).
Borovinskaya, I. P., and Pityulin, A. N., Combustion of hafnium in nitrogen. *Combust. Explos. Shock Waves,* **14,** 111 (1978).
Borovinskaya, I. P., Cantero, I., Estaire, L., Hernan, M. A., and Guzman, R., SHS Espana: An international experience. *Int. J. SHS,* **4,** 405 (1995).
Borovinskaya, I. P., Ivleva, T. I., Loryan, V. I., Merzhanov, A. G., and Shkadinskii, K. G., Autowave processes determined with an adsorption-diffusion mechanism of metal-gas interaction. *Int. J. SHS,* **2,** 227 (1993).
Borovinskaya, I. P., Merzhanov, A. G., Mukasyan, A. S., Rogachev, A. S., and Khusid, B. M., Macrokinetics of structure formation during infiltration combustion of titanium in nitrogen. *Dokl. Akad. Nauk SSSR,* **322,** 912 (1992a).
Borovinskaya, I. P., Ratnikov, V. I., and Vishnyakova, G. A., Some chemical aspects of powder SHS compacting. *J. Eng. Phys. Thermophys.,* **63,** 1059 (1992b).
Borovinskaya, I. P., Levashov, E. A., and Rogachev, A. S., "Physical-Chemical and Technological Base of Self-Propagating High-Temperature Synthesis, A Series of Lectures. MISiS Press, Moscow, 1991.
Borovinskaya, I. P., Merzhanov, A. G., Pityulin, A. N., and Shehtman, V. S., Self-propagating high-temperature synthesis of the tantalum nitride. *In* "Combustion Processes in Chemical Engineering and Metallurgy (Russ.)." (A. G. Merzhanov, ed.). USSR Academy of Science, Chernogolovka, Russia, 1975a, p. 113.
Borovinskaya, I. P., Vishnyakova, V. M., Maslov, V. M., and Mezhanov, A. G., On the possibility of the composite materials obtaining in combustion regime. *In* "Combustion Processes in Chemical Engineering and Metallurgy (Russ.)" (A. G. Merzhanov, ed.). USSR Academy of Science, Chernogolovka, Russia, 1975b, p. 141.

Borovinskaya, I. P., Merzhanov, A. G., Novikov, N. P., and Filonenko, A. K., Gasless combustion of mixtures of powdered transition metals with boron. *Combust. Explos. Shock Waves,* **10,** 2 (1974).

Borovinskaya, I. P., Merzhanov, A. G., Butakov, A. A., Rabin'kin, A. G., and Shehtman, V. S., Synthesis of tantalum mononotride. *Izvest. Bull.,* **9,** Certificate 264365, Patent 1311660 (1970).

Bowen, C. R., and Derby, B., Modelling of self-propagating high-temperature synthesis reactions. *In* Engineering Ceramics: Fabrication Science and Technology" (P. P. Thompson, ed.). The Institute of Metals, London, 1993, p. 29.

Brailsford, A. D., and Major, K. G., The thermal conductivity of aggregates of several phases, including porous materials. *Br. J. Appl. Phys.* **15,** 313 (1964).

Bratchikov, A. D., Merzhanov, A. G., Itin, V. I., Khachin, V. N., Dudarev, E. F., Gyunter, V. E., Maslov, V. M., and Chernov, D. B., Self-propagating high temperature synthesis of titanium nickelide. *Sov. Powd. Metall.,* **1,** 5 (1980).

Brezinsky, K., Brehm, J. A., Law, C. K., and Glassman, I., Supercritical combustion synthesis of titanium nitride. *Twenty-sixth Symposium (International) of Combustion.* The Combustion Institute, Pittsburgh, 1875 (1996).

Calcote, H. F., Felder, W., Keil, D. G., and Olson, D. B., A new flame process for synthesis of Si_3N_4 powders for advanced ceramics. *Twenty-third Symposium (International) on Combustion,* The Combustion Institute, Pittsburgh, 1739 (1990).

Cao, G., and Varma, A., A new expression for velocity of the combustion front during self-propagating high-temperature synthesis. *Combust. Sci. Tech.,* **102,** 181 (1994).

Chang, D. G., Shim, G. C., Won, C. W., and Chun, B.-S., Preparation of TiC powder by SHS (self-propagating high temperature synthesis). *J. Kor. Inst. Metals (Korean),* **29,** 626 (1991).

Chang, D. K., Won, C. W., Chun, B. S., and Shim, G. C., Purifying effects and product microstructure in the formation of TiC powder by the self-propagating high-temperature synthesis. *Metall. Mater. Trans. B,* **26B,** 176 (1995).

Chemical Engineering Progress, Process improves titanium diboride materials. **91,** 25 (1995).

Choi, Y., and Rhee, S.-W., Effect of aluminum addition on the combustion reaction of titanium and carbon to form TiC. *J. Mater. Sci.,* **28,** 6669 (1993).

Choi, Y., Mullins, M. E., Wijayatilleke, K., and Lee, J. K., Fabrication of metal matrix composites of TiC-Al through self-propagating synthesis reaction. *Metall. Trans. A,* **23A,** 2387 (1992).

Cooper, M. G., Mikic, B. B., and Yovanovich, M. M., Thermal contact conductance. *Int. J. Heat Mass Trans.,* **12,** 279 (1969).

Costantino, K., and Holt, J. B., High-pressure burning rate of silicon in nitrogen. *In* "Combustion and Plasma Synthesis" (Z. A. Munir and J. B. Holt, eds.). VCH Publishers, New York, 1990, p. 315.

Dandekar, H. W., Hlavacek, V., and Degreve, J., An explicit 3D finite-volume method for simulation of reactive flows using a hybrid moving adaptive grid. *Numer. Heat Trans., B,* **24,** 1 (1993).

Dandekar, H. W., Puszynski, J. A., and Hlavacek, V., Two-dimensional numerical study of cross-flow filtration combustion. *AIChE J.,* **36,** 1649 (1990).

Daniel, P. J., The theory of flame motion. *Proc. Roy. Soc.,* **126,** 393 (1930).

Davis, K., Brezinsky, K., and Glassman, I., Chemical equilibrium constants in the high-temperature formation of metallic nitrides. *Combust. Sci. Tech.,* **77,** 171 (1991).

DeAngelis, T. P., Advanced ceramic materials via reaction hot pressing. *Proceedings of the First US–Japanese Workshop on Combustion Synthesis,* Tokyo, Japan, 147 (1990).

DeAngelis, T. P., and Weiss, D. S., Advanced ceramics via SHS. *In* "Combustion and Plasma Synthesis of High-Temperature Materials" (Z. A. Munir and J. B. Holt, eds.). VCH Publishers, New York, 1990, p. 144.

Deevi, S. C., Diffusional reactions in the combustion synthesis of $MoSi_2$. *Mater. Sci. Eng.,* **A149,** 241 (1992).

Deevi, S. C., Structure of the combustion wave in the combustion synthesis of titanium carbides. *J. Mater. Sci.,* **26,** 2662 (1991).

Dolukhanyan, S. K., Hakobyan, H. G., and Aleksanyan, A. G., Combustion of metals in hydrogen and hydride production by SHS. *Int. J. SHS*, **1**, 530 (1992).

Dolukhanyan, S. K., Akopyan, A. G., and Merzhanov, A. G., Reaction of intermetallides based on zirconium and cobalt with hydrogen in combustion conditions. *Combust. Explos. Shock Waves*, **17**, 525 (1981).

Dolukhanyan, S. K., Nersesyan, M. D., Martirosyan, N. A., and Merzhanov, A. G., Using the SHS processes in the chemistry and technology of hydrides. *Izvestia AN SSSR, Neorganicheskie Materialy (Russ.)*, **14**, 1581 (1978).

Dolukhanyan, S. K., Nersesyan, M. D., Nalbandyan, A. B., Borovinskaya, I. P., and Merzhanov, A. G., Combustion of the transition metals in the hydrogen. *Dokl. Akad. Nauk SSSR*, **231**, 675 (1976).

Dunmead, S. D., Munir, Z. A., and Holt, J. B., Temperature profile analysis of combustion in the Zr-B system using the Boddington-Laye method. *Int. J. SHS*, **1**, 22 (1992a).

Dunmead, S. D., Munir, Z. A., and Holt, J. B., Temperature profile analysis of combustion synthesis: I. Theory and background. *J. Am. Ceram. Soc.*, **75**, 175 (1992b).

Dunmead, S. D., Munir, Z. A., and Holt, J. B., Temperature profile analysis of combustion synthesis: II. Experimental observations. *J. Am. Ceram. Soc.*, **75**, 180 (1992c).

Dunmead, S. D., Holt, B. J., and Kingman, D. D., Simultaneous combustion synthesis and densification of A1N. *In* "Combustion and Plasma Synthesis of High-Temperature Materials" (Z. A. Munir and J. B. Holt, eds.). VCH Publishers, New York, 1990a, p. 186.

Dunmead, S. D., Holt, B. J., and Kingman, D. D., Combustion synthesis in the Ti-C-Ni-Al system. *In* "Combustion and Plasma Synthesis of High-Temperature Materials" (Z. A. Munir and J. B. Holt, eds.). VCH Publishers, New York, 1990b, p. 229.

Dutta, A., Self-propagating high-temperature synthesis (SHS)-cum-superplastic forging of Fe$_3$Al-Nb alloy. *Int. J. SHS*, **4**, 309 (1995).

Eslamloo-Grami, M., and Munir, Z. A., Effect of porosity on the combustion synthesis of titanium nitride. *J. Am. Ceram. Soc.*, **73**, 1235 (1990a).

Eslamloo-Grami, M., and Munir, Z. A., Effect of nitrogen pressure and diluent content on the combustion synthesis of titanium nitride. *J. Am. Ceram. Soc.*, **73**, 2222 (1990b).

Evans, T., and James, R., A study of the transformation of diamond to graphite. *Proc. Roy. Soc. A*, **277**, 260 (1964).

Fedorov, V. M., Gordopolov, Y. A., and Merzhanov, A. G., Explosive treatment of SHS end products (of high-temperature superconductors). *J. Eng. Phys. Thermophys.*, **63**, 1166 (1992).

Feng, A., and Munir, Z. A., Field-assisted self-propagating synthesis of SiC. *J. Appl. Phys.*, **76**, 1927 (1994).

Ferreira, A., Meyers, M. A., and Thadani, N. N., Dynamic compaction of titanium aluminides by explosively generated shock waves: Microstructure and mechanical properties. *Metall. Trans. A*, **23A**, 3251 (1992).

Filatov, V. M., and Naiborodenko, Y. S., Combustion mechanism for nickel-aluminum thermites. *Combust. Explos. Shock Waves*, **28**, 47 (1992).

Filatov, V. M., Naiborodenko, Y. S., and Ivanov, I. A., Combustion of low-gas systems with redox stages. *Combust. Explos. Shock Waves*, **24**, 472 (1988).

Filonenko, A. K., Non-stationary phenomena during the combustion of heterogeneous systems with refractory products. *In* "Combustion Processes in Chemical Engineering and Metallurgy (Russ.)" (A. G. Merzhanov, ed.). USSR Academy of Science, Chernogolovka, Russia, 1975, p. 258.

Fu, Z. Y., Wang, H., Wang, W. M., Yuan, R. Z., and Munir, Z. A., Process of study on self-propagating high-temperature synthesis of ceramic-metal composites. *Int. J. SHS*, **2**, 175 (1993a).

Fu, Z. Y., Wang, W. M., Wang, H., Yuan, R. Z., and Munir, Z. A., Fabrication of cermets by SHS-QP method. *Int. J. SHS*, **2**, 307 (1993b).

Glassman, I., "Combustion." Academic Press, New York, 1977.

Glassman, I., Davis, K. A., and Brezinsky, K., A gas-phase combustion synthesis process for nonoxide ceramics. *Twenty-fourth Symposium (International) on Combustion.* The Combustion Institute, 1877 (1992).

Goel, N. S., Geroc, J. S., and Lehmann, G., A simple model for heat conduction in heterogeneous materials and irregular boundaries. *Int. Comm. Heat Mass Transfer,* **19,** 519 (1992).

Gogotsi, G. A., Savada, V. P., and Khartionov, F. Y., Strength and crack resistance of ceramics. *Problemy Prochnosti (Russian),* **12,** 7 (1984).

Golubjatnikov, K. A., Stangle, G. C., and Sprigs, R. M., The economics of advanced self-propagating high-temperature synthesis materials fabrication. *Am. Ceram. Soc. Bull.,* **72,** 96 (1993).

Gordopolov, Y. A., and Merzhanov, A. G., Shock waves in SHS research. *Proceedings of the 13th International Colloquium on Dynamics of Explosion and Reactive Systems,* Nagoya, Japan (1991).

Gotman, I., Koczak, M. J., and Shtessel, E., Fabrication of Al matrix in situ composites via self-propagating synthesis. *Mater. Sci. Eng.,* **A187,** 189 (1994).

Grachev, V. A., Ivleva, T. I., and Borovinskaya, Filtration combustion in SHS reactor. *Int. J. SHS,* **4,** 252 (1995).

Grebe, H. A., Advani, A., Thadani, N. N., and Kottke, T., Combustion synthesis and subsequent explosive densification of titanium carbide ceramics. *Metall. Trans. A,* **23A,** 2365 (1992).

Grigor'ev, Y. M., Using the electrically heated wires for studying the kinetic of high temperature interaction in metal–gas system. *In* "Combustion Processes in Chemical Engineering and Metallurgy (Russ.)" (A. G. Merzhanov, ed.). USSR Academy of Science, Chernogolovka, Russia, 1975, p. 199.

Grigor'ev, Y. M., and Merzhanov, A. G., SHS coatings. *Int. J. SHS,* **1,** 600 (1992).

Hahn, Y.-D., and Song, I.-H., Microstructural characteristics of titanium aluminides synthesized by using the wave propagation mode. *Int. J. SHS,* **4,** 293 (1995).

Hakobian, H. G., and Dolukhanyan, S. K., Combustion process in the Mo-Al-Si system and synthesis of molybdenum alumocilicide. *Int. J. SHS,* **3,** 299 (1994).

Hardt, A. P., and Holsinger, R. W., Propagation of gasless reactions in solids—II. Experimental study of exothermic intermetallic reaction rates. *Combust. Flame,* **21,** 91 (1973).

Hardt, A. P., and Phung, P. V., Propagation of gasless reactions in solids—I. Analytical study of exothermic intermetallic reaction rates. *Combust. Flame,* **21,** 77 (1973).

Hernandez-Guerrero, A., Huque, Z., and Kanury, A. M., An experimental investigation of combustive synthesis of titanium carbide. *Combust. Sci. Tech.,* **81,** 115 (1992).

Hirao, K., Miyamoto, Y., and Koizumi, M., Combustion reaction characteristics at nitridation of silicon. *Adv. Ceramics,* **2,** 780 (1987).

Hlavacek, V., Combustion synthesis: A historical perspective. *Am. Ceram. Soc. Bull.,* **70,** 240 (1991).

Hlavacek, V., and Puszynski, J. A., Chemical engineering aspects of advanced ceramic materials. *Ind. Eng. Chem. Res.,* **35,** 349 (1996).

Ho-Yi, L., Hong-Yu, Y., Shu-Xia, M., and Sheng, Y., Combustion synthesis of titanium aluminides. *Int. J. SHS,* **1,** 447 (1992).

Hoke, D. A., Meyers, M. A., Meyer, L. W., and Gray III, G. T., Reaction synthesis: Dynamic compaction of titanium diboride. *Metall. Trans. A,* **23A,** 77 (1992).

Holt, J. B., and Dunmead, S. D., Self-heating synthesis of materials. *Annu. Rev. Mater. Sci.,* **21,** 305 (1991).

Holt, J. B., and Munir, Z. A., Combustion synthesis of titanium carbide: Theory and experiment. *J. Mater. Sci.,* **21,** 251 (1986).

Holt, J. B., Wong, J., Larson, E. M., Waide, P. A., Rupp, B., and Frahm, R., A new experimental approach to study solid combustion reaction using synchrotron radiation. *Proceedings of the First US–Japanese Workshop on Combustion Synthesis,* Tokyo, Japan, 107 (1990).

Holt, J. B., Kingman, D. D., and Bianchini, G. M., Kinetics of the combustion synthesis of TiB_2. *Mater. Sci. Eng.,* **71,** 321 (1985).

Howers, V. R., Graphitization of diamond. *Proc. Roy. Soc.,* **80,** 648 (1962).

Hunter, K. R., and Moore, J. J., The effect of gravity on the combustion synthesis of ceramic and ceramic-metal composites. *J. Mater. Syn. Proc.,* **2,** 355 (1994).
Huque, Z., and Kanury, A. M., A theoretical analysis of combustive synthesis of titanium carbide and a comparison of predictions with measurements. *Combust. Sci. Tech.,* **89,** 27 (1993).
Hwang, S., Ph.D. Dissertation, University of Notre Dame (1997).
Hwang, S., Mukasyan, A. S., Rogachev, A. S., and Varma, A., Combustion wave microstructure in gas-solid system: Experiments and theory. *Combust. Sci. Tech.,* **123,** 165 (1997).
Ikeda, S., Urabe, K., Koizumi, M., and Izawa, H., In-situ formation of SiC and SiC-C blocked solids by self-combustion synthesis. *In* "Combustion and Plasma Synthesis of High-Temperature Materials" (Z. A. Munir and J. B. Holt, eds.). VCH Publishers, New York, 1990, p. 151.
International Journal of SHS, SHS Bibliography, **5,** 303 (1996).
Itin, V. I., and Naiborodenko, Y. S., "High Temperature Synthesis of Intermetallic Compounds (Russ.)." Tomsk University Publishing, Tomsk, Russia, 1989.
Itin, V. I., Khachin, V. N., Gyunter, V. E., Bratchikov, A. D., and Chernov, D. B., Production of titanium nickelide by self-propagating high-temperature synthesis. *Porosh. Metall.,* **3,** 156 (1983).
Itin, V. I., Bratchikov, A. D., and Lepinskikh, A. V., Phase transition accompanying combustion of mixtures of copper and aluminum powders. *Combust. Explos. Shock Waves,* **17,** 506 (1981a).
Itin, V. I., Bratchikov, A. D., Merzhanov, A. G., and Maslov, V. M., Principles of self-propagating high-temperature synthesis of titanium compounds with elements of the iron group. *Combust. Sci. Tech.,* **17,** 293 (1981b).
Itin, V. I., Bratchikov, A. D., and Postnikova, L. N., Use of combustion and thermal explosion for the synthesis of intermetallic compounds and their alloys. *Sov. Powd. Metall.,* **19,** 315 (1980).
Itin, V. I., Khachin, V. N., Bratchikov, A. D., Gyunter, V. E., Dudarev, E. F., Monasevich, T. V., Chernov, D. B., Timonin, G. D., and Paperskii, A. P., Structure and properties of titanium nickelide materials made by self-propagating high-temperature synthesis. *Russ. Phys. J.,* **12,** 1631 (1977).
Ivleva, T. I., and Shkadinskii, K. G., "Mathematical modeling of two-dimensional problems in filtration combustion of condensed systems." *Proceedings of the Kinetics and Mechanisms of Physicochemical Processes,* Chernogolovka, Russia, 74 (1981).
Kachelmyer, C. R., and Varma, A., Combustion synthesis of niobium aluminide matrix composites. *Mater. Res. Soc. Symp.,* **350,** 33 (1994).
Kachelmyer, C. R., Rogachev, A. S., and Varma, A., Mechanistic and processing studies in combustion synthesis of niobium aluminides. *J. Mater. Res.* **10,** 2260 (1995).
Kachelmyer, C. R., Lebrat, J.-P., Varma, A., and McGinn, P. J., Combustion synthesis of intermetallic aluminides: Processing and mechanistic studies. *In* "Heat Transfer in Fire and Combustion Systems" (B. Farouk, M. P. Pinar Menguc, R. Viskanta, C. Presser, and S. Chellaiah, eds.). ASME, New York, 1993, p. 271.
Kachin, A. R., and Yukhvid, V. I., SHS of cast composite materials and pipes in the field of centrifugal forces. *Int. J. SHS,* **1,** 168 (1992).
Kaieda, Y., Nakamura, M., Otaguchi, M., and Oguro, N., Combustion synthesis of TiAl intermetallic compounds. *Proceedings of the First US–Japanese Workshop on Combustion Synthesis,* Tokyo, Japan, 207 (1990a).
Kaieda, Y., Otaguchi, M., and Oguro, N., Combustion synthesis of intermetallic compounds. *In* "Combustion and Plasma Synthesis of High-Temperature Materials" (Z. A. Munir and J. B. Holt, eds.). VCH Publishers, New York, 1990b, p. 106.
Kanamaru, N., and Odawara, O., Combustion synthesis of phosphides of aluminum, gallium and indium. *Int. J. SHS,* **3,** 305 (1994).
Kanury, A. M., A kinetic model for metal + nonmetal reactions. *Metall. Trans. A,* **23A,** 2349 (1992).
Kasat'skiy, N. G., Filatov, V. M., and Naiborodenko, Y. S., SHS in the low exothermic and high density aluminum systems. *In* "Self-Propagating High-Temperature Synthesis" (Ya. Maksimov, ed.). Tomsk, 1991, p. 63.

Kecskes, L. J., Niiler, A., and Kottke, T., Precursor morphology effects in combustion-synthesized and dynamically consolidated titanium carbide and titanium boride. *J. Am. Ceram. Soc.*, **76**, 2961 (1993).

Kecskes, L. J., Kottke, T., and Niiler, A., Microstructural properties of combustion-synthesized and dynamically consolidated titanium boride and titanium carbide. *J. Am. Ceram. Soc.*, **73**, 1274 (1990).

Khaikin, B. I., and Khudiaev, S. I., Nonuniqueness of combustion temperature and rate when competing reactions take place. *Phys. Dokl.*, **245**, 155 (1979).

Khaikin, B. I., and Merzhanov, A. G., Theory of thermal propagation of a chemical reaction front. *Combust. Explos. Shock Waves*, **2**, 22 (1966).

Kharatyan, S. L., and Nersisyan, H. H., Combustion synthesis of silicon carbide under oxidative activation conditions. *Int. J. SHS*, **3**, 17 (1994).

Khomenko, I. O., Mukasyan, A. S., Ponomarev, V. I., Borovinskaya, I. P., and Merzhanov, A. G., Dynamics of phase-forming processes in metal-gas system during combustion. *Combust. Flame*, **92**, 201 (1993).

Khusid, B. M., Khina, B. B., Kuang, V. Z., and Bashtovaya, E. A., Numerical investigation of quenching of the state of a material in a wave of SHS during a two-step reaction. *Combust. Explos. Shock Waves*, **28**, 389 (1992).

Kirdyashkin, A. I., Maksimov, Y. M., and Merzhanov, A. G., Effects of capillary flow on combustion in a gas-free system. *Combust. Explos. Shock Waves*, **17**, 591 (1981).

Koizumi, M., and Miyamoto, Y., Recent progress on combustion synthesis of high-performance material in Japan. In "Combustion and Plasma Synthesis of High-Temperature Materials" VCH Publishers, New York, 1990, p. 54.

Korchagin, M. A., and Aleksandrov, V. V., An electron-microscope study of the interaction of titanium with carbon. *Combust. Explos. Shock Waves*, **17**, 58 (1981).

Korchagin, M. A., and Podergin, V. A., Investigation of chemical transformations in the combustion of condensed system. *Fiz. Gor. Vzriva*, **15**, 325 (1979).

Kottke, T., Kecskes, L. J., and Niiler, A., Control of TiB_2 SHS reactions by inert dilutions and mechanical constraint. *AIChE J.*, **36**, 1581 (1990).

Kottke, T., and Niiler, A., "Thermal Conductivity Effects on SHS reactions," USA Ballistic Research Laboratory. Aberdeen Proving Ground, MD (1988).

Kumar, S., Self-propagating high-temperature synthesis of ceramic materials. Ph.D. Dissertation, State University of New York, Buffalo (1988).

Kumar, S., Agrafiotis, C., Puszynski, J., and Hlavacek, V., Heat transfer characteristics in combustion synthesis of ceramics. *AIChE Symp. Ser. 263*, **84**, 50 (1988a).

Kumar, S., Puszynski, J. A., and Hlavacek, V., Combustion characteristics of solid-solid systems experiments and modeling. *Proceedings of the International Symposium on Combustion and Plasma Synthesis of High Temperature Materials*, San Francisco (1988b).

Kuroki, H., and Yamaguchi, K., Combustion synthesis of Ti/Al intermetallic compounds and dimensional changes of mixed powder compacts during sintering. *Proceedings of the First US–Japanese Workshop on Combustion Synthesis*, Tokyo, Japan, 23 (1990).

Kvanin, V. L., Gorovoi, V. A., Balikhina, N. T., Borovinskaya, I. P., and Merzhanov, A. G., Investigation of the process of the forced SHS compaction of large-scale hard-alloy articles. *Int. J. SHS*, **2**, 56 (1993).

Lakshmikantha, M. G., and Sekhar, J. A., Influence of multi-dimensional oscillation combustion fronts on thermal profiles. *J. Mater. Sci.*, **28**, 6403 (1993).

Larson, E. M., Wong, J., Holt, J. B., Waide, P. A., Nutt, G., Rupp, B., and Terminello, L. J., A time-resolved diffraction study of the Ta-C Solid combustion system. *J. Mater. Res.*, **8**, 1533 (1993).

LaSalvia, J. C., and Meyers, M. A., Microstructure, properties, and mechanisms of TiC-Mo-Ni cermet produced by SHS. *Int. J. SHS*, **4**, 43 (1995).

LaSalvia, J. C., Meyers, M. A., and Kim, D. K., Combustion synthesis/dynamic densification of TiC-Ni cermets. *J. Mater. Syn. Proc.,* **2,** 255 (1994).
Lebrat, J.-P., and Varma, A., Self-propagating high-temperature synthesis of Ni$_3$Al. *Combust. Sci. Tech.,* **88,** 211 (1992a).
Lebrat, J.-P., and Varma, A., Some further studies in combustion synthesis of the YBa$_2$Cu$_3$O$_{7-x}$ superconductor. *Combust. Sci. Tech.,* **88,** 177 (1992b).
Lebrat, J.-P., and Varma, A., Combustion synthesis of the YBa$_2$Cu$_3$O$_{7-x}$ superconductor. *Physica C,* **184,** 220 (1991).
Lebrat, J.-P., Varma, A., and McGinn, P. J., Mechanistic studies in combustion synthesis of Ni$_3$Al and Ni$_3$Al-matrix composites. *J. Mater. Res.,* **9,** 1184 (1994).
Lebrat, J.-P., Varma, A., and Miller, A. E., Combustion synthesis of Ni$_3$Al and Ni$_3$Al–matrix composites. *Metall. Trans. A,* **23A,** 69 (1992).
Lee, W.-C., and Chung, S.-L., Self-propagating high-temperature synthesis of TiC powders. *Int. J. SHS,* **1,** 211 (1992).
Lee, W.-C., Hsu, K. C., and Chung, S.-L., Combustion synthesis of Ti-Al intermetallic materials. *Int. J. SHS,* **4,** 95 (1995).
Levashov, E. A., Vijushkov, B. V., Shtanskaya, E. V., Borovinskaya, I. P., Ohyanagi, M., Hosomi, S., and Koizumi, M., Regularities of structure and phase formation of SHS diamond-containing functional gradient materials: Operational characteristics of articles based on them. *Int. J. SHS,* **3,** 287 (1994).
Levashov, E. A., Borovinskaya, I. P., Rogachev, A. S., Koizumi, M., Ohyanagi, M., and Hosomi, S., SHS: A new method for production of diamond-containing ceramics. *Int. J. SHS,* **2,** 189 (1993).
Levashov, E. A., Bogatov, Y. V., and Milovidov, A. A., Macrokinetics and mechanism of the SVS-process in systems on a titanium-carbon base. *Combust. Explos. Shock Waves,* **27,** 83 (1991).
Li, H.-P., Processing Ti-B compounds by combustion synthesis. *Int. J. SHS,* **4,** 199 (1995).
Li, J., Zhou, M., and Wang, J., The study of microstructure of porous ceramic-lined pipes made by centrifugal SHS. *Proceedings of the 3rd International Symposium on Self-Propagating High-Temperature Synthesis (Book of Abstracts),* Wuhan, China, 48 (1995).
Li, J.-T., Xia, Y.-L., and Ge, C.-C., Combustion synthesis of magnesium nitride. *Int. J. SHS,* **3,** 225 (1994).
Lin, S.-C., Richardson, J. T., Luss, D., YBa$_{2Cu}$3O$_{6+x}$ synthesis using vertical self-propagating high-temperature synthesis. *Phys. C,* **233,** 281 (1994).
Luikov, A. V., Shashkov, A. G., Vasiliev, L. L., and Fraiman, Y. E., Thermal conductivity of porous systems. *Int. J. Heat Mass Trans.,* **11,** 117 (1968).
Makhviladze, G. M., and Novozilov, B. V., Two-dimensional stability of combustion of condensed systems. *Zhurn. Prikl. Mekhaniki I Tekhn. Fiziki,* **5,** 51 (1971).
Makino, A., and Law, C. K., SHS combustion characteristics of several ceramics and intermetallic compounds. *J. Am. Ceram. Soc.,* **77,** 778 (1994).
Maksimov, Y. M., Ziatdinov, M. K., Merzhanov, A. G., Raskolenko, L. G., and Lepakova, O. K., Combustion of vanadium-iron alloys in nitrogen. *Combust. Explos. Shock Waves,* **20,** 487 (1984).
Maksimov, Y. M., Ziatdinov, M. K. Raskolenko, A. G., and Lepakova, O. K., Interaction between vanadium and nitrogen in the combustion regime. *Combust. Explos. Shock Waves,* **15,** 420 (1979).
Maksimov, E. I., Merzhanov, A. G., and Shkiro, V. M., Gasless compositions as a simple model for the combustion of nonvolatile condensed systems. *Combust. Explos. Shock Waves,* **1,** 15 (1965).
Mallard, E., and Le Chatelier, H. L., Combustion des melanges gazeux explosifs. *Ann. Mines,* **4,** 274 (1883).
Mamyan, S. S., and Vershinnikov, V. I., Specific features of combustion of SHS systems containing magnesium as a reductant. *Int. J. SHS,* **1,** 392 (1992).
Margolis, S. B., A new route of chaos in gasless combustion. *Combust. Sci. Tech.,* **88,** 223 (1992).

Margolis, S. B., An asymptotic theory of heterogeneous condensed combustion. *Combust. Sci. Tech.,* **43,** 197 (1985).
Margolis, S. B., An asymptotic theory of condensed two-phase flame propagation. *SIAM J. Appl. Math.,* **43,** 351 (1983).
Martirosyan, N. A., Dolukhanyan, S. K., and Merzhanov, A. G., Nonuniqueness of stable regimes during combustion of powder mixture of Zr and C in hydrogen. *Fizika Fizika Goreniia i Vzryva,* **5,** 39 (1983).
Martynenko, V. M., and Borovinskaya, I. P., Some characteristic features of the combustion in system niobium–carbon. *In* "Combustion Processes in Chemical Engineering and Metallurgy (Russ.)" (A. G. Merzhanov, ed.). USSR Academy of Science, Chernogolovka, Russia, 1975, p. 127.
Maslov, V. M., Voyuev, S. I., Borovinskaya, I. P., and Merzhanov, A. G., The role of inert-diluent dispersion in gasless combustion. *Combust. Explos. Shock Waves,* **26,** 441 (1990).
Maslov, V. M., Borovinskaya, I. P., and Ziatdinov, M. K., Combustion of the systems niobium–aluminium and niobium–germanium. *Combust. Explos. Shock Waves,* **15,** 41 (1979).
Maslov, V. M., Borovinskaya, I. P., and Merzhanov, A. G., Problem of the mechanism of gasless combustion. *Combust. Explos. Shock Waves,* **12,** 631 (1976).
Mason, W., and Wheeler, R. V., The uniform movement during the propagation of flame. *J. Chem. Soc.,* **CXI,** 1044 (1917).
Matkowsky, B. J., and Sivashinsky, G. I., Propagation of a pulsating reaction front in solid fuel combustion. *SIAM J. Appl. Math.,* **35,** 465 (1978).
Matsuzaki, Y., Fujioka, J., Minakata, S., and Miyamoto, Y., Fabrication of $MoSi_2$-SiC/TiAl functional gradient materials by gas-pressure combustion sintering process. *Proceedings of the 1st Int. Symposium on Functionally Gradient Materials,* Tokyo, 263 (1990).
Maupin, H. E., and Rawers, J. C., Metal-intermetallic composites formed by reaction-sintering elemental powders. *J. Mater. Sci. Lett.,* **12,** 540 (1993).
Mei, B., Wang, W., Yuan, R., and Fu, Z., Study of TiC/Ni_3Al composites prepared by combustion synthesis. *Int. J. SHS,* **3,** 79 (1994).
Mei, B., Yuan, R., and Duan, X., Investigation of Ni_3Al-matrix composites strengthened by TiC. *J. Mater. Res.,* **8,** 2830 (1993).
Mei, B., Yuan, R., and Duan, X., Self-propagating high-temperature synthesis of MoB_2. *Int. J. SHS,* **1,** 421 (1992).
Merzhanov, A. G., History of and new developments in SHS. *Ceram. Int.* **21,** 371 (1995).
Merzhanov, A. G., Combustion processes that synthesize materials. Paper presented at AMPT'93 International Conference on Advances in Materials and Processing Technology, Dublin, Ireland (1993a).
Merzhanov, A. G., Theory and practice of SHS: Worldwide state of the art and the newest results. *Int. J. SHS,* **2,** 113 (1993b).
Merzhanov, A. G., Self-propagating high-temperature synthesis and powder metallurgy: Unity of goals and competition of principles. *In* "Particulate Materials and Processes: Advances in Powder Metallurgy and Particulate Materials." Metal Powder and Industries Federation, Princeton, NJ, 1992, p. 341.
Merzhanov, A. G., Advanced SHS ceramics: Today and tomorrow morning. Invited paper presented at International Symposium devoted to the 100th Anniversary of Ceramic Society of Japan, Yokohma (1991).
Merzhanov, A. G., Self-propagating high-temperature synthesis: Twenty years of search and findings. *In* "Combustion and Plasma Synthesis of High-Temperature Materials" (Z. A. Munir and J. B. Holt, eds.). VCH Publishers, New York, 1990a, p. 1.
Merzhanov, A. G., Self-propagating high-temperature synthesis of ceramic (oxide) superconductors. *In* "Ceramic Transactions: Superconductivity and Ceramic Superconductors" (K. M. Nair and E. A. Giess, eds.). American Ceramic Society, Westerville, OH, 1990b, p. 519.

Merzhanov, A. G., Self-propagating high-temperature synthesis. *In* "Current Topics in Physical Chemistry (Russ.)" (Ya. M. Kolotyrkin, ed.). Khimiya, Moscow, 1983, p. 8.

Merzhanov, A. G., SHS-process: Combustion theory and practice. *Archivum Combustionis,* **1,** 24 (1981).

Merzhanov, A. G., New elementary models of the second kind. *Dokl. Phys. Chem.,* **223,** 430 (1977).

Merzhanov, A. G., and Borovinskaya, I. P., Self-propagating high-temperature synthesis of refractory inorganic compounds. *Dokl. Chem.,* **204,** 429 (1972).

Merzhanov, A. G., and Khaikin, B. I., Theory of combustion waves in homogeneous. *Prog. Energy Combust. Sci.,* **14,** 1 (1988).

Merzhanov, A. G., and Rogachev, A. S., Structural macrokinetics of SHS processes. *Pure Appl. Chem.,* **64,** 941 (1992).

Merzhanov, A. G., and Yukhvid, V. I., The self-propagating high temperature synthesis in the field of centrifugal forces. *Proceedings of the First US–Japanese Workshop on Combustion Synthesis,* Tokyo, Japan, 1 (1990).

Merzhanov, A. G., Borovinskaya, I. P., Khomenko, I. O., Mukasyan, A. S., Ponomarev, V. I., Rogachev, A. S., and Shkiro, V. M., Dynamics of phase formation during SHS processes. *Ann. Chimie,* **20,** 123 (1995).

Merzhanov, A. G., Borovinskaya, I. P., Ponomarev, V. I., Khomenko, I. O., Zanevskii, Y. V., Chernenko, S. P., Smykov, L. P., and Cheremukhina, G. A., Dynamic x-ray study on phase formation in the course of SHS. *Phys. Dokl.,* **328,** 11 (1993).

Merzhanov, A. G., Rogachev, A. S., Mukasyan, A. S., and Khusid, B. M., Macrokinetics of structural transformation during the gasless combustion of a titanium and carbon powder mixture. *Combust. Explos. Shock Waves,* **26,** 92 (1990a).

Merzhanov, A. G., Rogachev, A. S., Mukasyan, A. S., Khusid, B. M., Borovinskaya, I. P., and Khina, B. B., The role of gas-phase transport in combustion of tantalum–carbon system. *J. Eng. Phys. Thermophys.,* **59,** 809 (1990b).

Merzhanov, A. G., Borovinskaya, I. P., Yukhvid, V. I., and Ratnikov, V. I., New production methods of high temperature materials based on combustion. *In* "Scientific Principles of Material Science." Nauka, Moscow, 1981, p. 193.

Merzhanov, A. G., Yukhvid, V. I., and Borovinskaya, I. P., Self-propagating high-temperature synthesis of cast inorganic refractory compounds. *Dokl. Chem.,* **255,** 503 (1980).

Merzhanov, A. G., Borovinskaya, I. P., and Volodin, Y. E., Combustion of porous metallic samples in nitrogen. *Dokl. Phys. Chem.,* **206,** 833 (1972).

Merzhanov, A. G., Shkiro, V. M., and Borovinskaya, I. P., Synthesis of refractory inorganic compounds. *Izvest. Bull.* **10,** Certificate 255221. (1971). (See also French Patent 2088668, 1972; U.S. Patent 3726643, 1973; U.K. Patent 1321084, 1974; Japanese Patent 1098839, 1982.)

Meyers, M. A., Yu, L.-H., and Vecchio, K. S., Shock synthesis of silicides—II. Thermodynamics and kinetics. *Acta Metall. Mater.,* **42,** 715 (1994).

Meyers, M. A., LaSalvia, J. C., Meyer, L. W., Hoke, D., and Niiler, A., Reaction Synthesis/Dynamic Compuction of Titanium Carbide and Titanium Diboride. *J. de Physique IV (Colloque),* **1,** 123 (1991).

Michelson, V. A., *In* "Collected Works (Russ.)." Novyi Agronom, Moscow, 1930, vol. 1, p. 56.

Miyamoto, Y., Tanihata, K., Matsuzaki, Y., and Ma, X., HIP in SHS technology. *Int. J. SHS,* **1,** 147 (1992).

Miyamoto, Y., Nakanishi, H., Tanaka, I., Okamoto, T., and Yamada, O., Processing study for the functionally gradient material TiC-Ni by the gas-pressure combustion sintering. *Proceedings of the First US–Japanese Workshop on Combustion Synthesis,* Tokyo, Japan, 173 (1990b).

Miyamoto, Y., Takahura, T., Tanihata, K., Tanaka, I., Yamada, O., Saito, M., and Takahashi, H., Processing study for TiB_2-Ni FGM by gas-pressure combustion sintering. *Proceedings of the 1st International Symposium on FGMs,* Tokyo, Japan, 169 (1990b).

Miyamoto, Y., Koizumi, M., and Yamada, O., High-pressure self-combustion sintering for ceramics. *J. Am. Ceram. Soc.,* **67,** C224 (1984).

Molodetskaya, I. E., Pisarskii, V. P., and Ulanova, O. O., Effect of the parameters of the starting Zn-S mixture on the structure of ZnS synthesized in a combustion wave. *Combust. Explos. Shock Waves,* **28,** 385 (1992).

Molokov, I. V., and Mukasyan, A. S., Explosive treatment of SHS gas–solid systems. *Int. J. SHS,* **1,** 155 (1992).

Moore, J. J., Combustion synthesis of ceramic-metal composite materials in microgravity. *Proceedings of the Third International Microgravity Workshop,* Cleveland, OH, 165 (1995).

Moore, J. J., and Feng, H. J., Combustion synthesis of advanced materials. *Prog. Mater. Sci.,* **39,** 243 (1995).

Moore, J. J., Feng, H. J., Hunter, K. R., and Wirth, D. G., *Proceedings of the Second International Microgravity Combustion Workshop,* Cleveland, OH, 157 (1992).

Muchnik, S. V., Formation of nickel phosphides under combustion conditions. *Izv. Akad. Nauk SSSR, Nearg. Mater.,* **20,** 158 (1984).

Muchnik, S. V., Lomnitskaya, Y. F., Chernogorenko, V. B., and Lynchak, K. A., Interaction under combustion conditions in copper-zirconium-phosphorus mixtures. *Sov. Powd. Metall.,* **32,** 865 (1993).

Mukasyan, A. S., Structure- and phase-formation of nitrides in SHS processes. D.Sc. Dissertation, Institute of Structural Macrokinetics, Russian Academy of Science (1994).

Mukasyan, A. S., Characteristics and mechanism of silicon and boron combustion in nitrogen." Ph.D. Dissertation, Institute of Chemical Physics, USSR Academy of Science (1986).

Mukasyan, A. S., and Borovinskaya, I. P., Structure formation in SHS nitrides. *Int. J. SHS,* **1,** 55 (1992).

Mukasyan, A. S., Blinov, M. Y., Borovinskaya, I. P., and Merzhanov, A. G., Some preparation aspects of nitrogen ceramics by direct SHS methods. *Proceedings of the International Conference Engineering Ceramics,* Smolenice Castle, Czechoslovakia, 161 (1989).

Mukasyan, A. S., Khomenko, I. O., and Ponomarev, V. I., About nonuniqueness of combustion modes in the heterogeneous systems. *Combust. Sci. Tech.,* **128,** 215 (1977a).

Mukasyan, A. S., Pelekh, A. E., Varma, A., and Rogachev, A. S., The effects of gravity on combustion synthesis in heterogeneous gasless systems. *AIAA J,* **35,** 1 (1997b).

Mukasyan, A. S., Hwang, S., Rogachev, A. S., Sytchev, A. E., Merzhanov, A. G., and Varma, A., Combustion wave microstructure in heterogeneous gasless systems. *Combust. Sci. Tech.,* **115,** 335 (1996).

Mukasyan, A. S., Merzhanov, A. G., Martynenko, V. M., Borovinskaya, I. P., and Blinov, M. Y., Mechanism and principles of silicon combustion in nitrogen. *Combust. Explos. Shock Waves,* **22,** 534 (1986).

Munir, Z. A., and Anselmi-Tamburini, U., Self-propagating exothermic reactions: The synthesis of high-temperature materials by combustion. *Mater. Sci. Reports,* **3,** 277 (1989).

Munir, Z. A., and Holt, J. B., The combustion synthesis of refractory nitrides. Part 1: Theoretical analysis. *J. Mater. Sci.,* **22,** 710 (1987).

Munir, Z. A., and Lai, W., The SHS diagram for TiC. *Combust. Sci. Tech.,* **128,** 215 (1977a).

Munir, Z. A., and Sata, N., SHS diagrams: Theoretical analysis and experimental observations. *Int. J. SHS,* **1,** 355 (1992).

Naiborodenko, Y. S., and Itin, V. I., Gasless combustion of mixtures of metal powders. I Mechanism and details. *Combust. Explos. Shock Waves,* **11,** 293 (1975a).

Naiborodenko, Y. S., and Itin, V. I., Gasless combustion of mixtures of metal powders. II Effect of mixture composition of the combustion rate and the phase composition of the products. *Combust. Explos. Shock Waves,* **11,** 626 (1975b).

Naiborodenko, Y. S., Lavrenchuk, G. V., and Filatov, V. M., Self-propagating high-temperature synthesis of aluminides. *Sov. Powd. Metall.,* **21,** 909 (1982).

Naiborodenko, Y. S., Itin, V. I., and Savitskii, K. V., Reactions at phase boundaries and their effects on the sintering process—I. Exothermic effects in powder sintering. *Sov. Powd. Metall.*, **7**, 562 (1970).

Naiborodenko, Y. S., Itin, V. I., and Savitskii, K. V., Exothermic effects during sintering of a mixture of nickel and aluminium powders. *Sov. Phys. J.*, **11**, 19 (1968).

Nekrasov, E. A., Tkachenko, V. N., and Zakirov, A. E., Diffusive combustion of multicomponent gasless systems forming multi-phase peroducts. *Combust. Sci. Tech.*, **91**, 207 (1993).

Nekrasov, E. A., Smolyakov, V. K., and Maksimov, Y. M., Mathematical model for the titanium–carbon system combustion. *Combust. Explos. Shock Waves*, **17**, 513 (1981).

Nekrasov, E. A., Maksimov, Y. M., Ziatdinov, M. K., and Shteinberg, A. S., Effect of capillary spreading on combustion-wave propagation in gas-free system. *Combust. Explos. Shock Waves*, **5**, 575 (1978).

Niiler, A., Kecskes, L. J., and Kottke, T., Consolidation of combustion synthesised materials by explosive compaction," *Proceedings of the First US–Japanese Workshop on Combustion Synthesis,* Tokyo, Japan, 53 (1990).

Nishida, T., and Urabe, K., TiC ceramic prepared by SHS/HIP processing. *Int. J. SHS*, **1**, 566 (1992).

Novikov, N. P., Borovinskaya, I. P., and Merzhanov, A. G., Dependence of the product composition and combustion wave as a function of the green mixture composition in the metal–boron systems. *Fizika Gorenia I Vzriva*, 201 (1974).

Novozilov, B. V., Non-linear SHS phenomena: Experiments, theory, numerical modeling. *Pure Appl. Chem.*, **64**, 955 (1992).

Novozilov, B. V., The rate of propagation of the front of an exothermic reaction in a condensed phase. *Phys. Dokl.*, **141**, 836 (1961).

Odawara, O., SHS technology for large composite pipes. *Int. J. SHS*, **1**, 160 (1992).

Odawara, O., Long ceramic-lined pipes produced by a centrifugal-thermite process." *J. Am. Ceram. Soc.*, **73**, 629 (1990).

Odawara, O., Method for providing ceramic lining to a hollow body by thermite reaction." U.S. Patent 4363832 (1982).

Odawara, O., and Ikeuchi, J., Ceramic composite pipes produced by a centrifugal-exothermic process. *J. Am. Ceram. Soc.*, **69**, C80 (1986).

Odawara, O., Mori, K., Tanji, A., and Yoda, S., Thermite reaction in a short-time microgravity environment. *J. Mater. Syn. Process.* **1**, 203 (1993).

Ohyanagi, M., Koizumi, M., Tanihata, K., Miyamoto, Y., Yamada, O., Matsubara, I., and Yamashitaa, H., Production of superconducting NbN thin plate and wire by the self-propagating high-temperature synthesis method. *J. Mater. Sci. Lett.*, **12**, 500 (1993).

Okolovich, E. V., Merzhanov, A. G., Khaikin, B. I., and Shkadinskii, K. G., Propagation of the combustion zone in melting condensed mixtures. *Combust. Explos. Shock Waves*, **13**, 264 (1977).

Orru', R., Simoncini, B., Virdis, P. F., and Cao, G., Further studies on a centrifugal SHS process for coating preparation and structure formation in thermite reactions. *Int. J. SHS*, **4**, 137 (1995).

Osipov, E. Y., Levashov, Y. A., Chernyshev, V. N., Merzhanov, A. G., and Borovinskaya, I. P., Prospects for simultaneous use of vacuum-performed SHS processes and various hot rolling techniques for production of semifinished and finished items of ceramometallic or intermetallic composites. *Int. J. SHS*, **1**, 314 (1992).

Padyukov, K. L., and Levashov, E. A., Self-propagating high-temperature synthesis: a new method for production of diamond-containing materials. *Diamond Related Mater.*, **2**, 207 (1993).

Padyukov, K. L., Kost, A. G., Levashov, E. A., Borovinskaya, I. P., and Bogatov, Y. V., Production regularities, structure, and properties of diamond-containing and composite materials. *Int. J. SHS*, **1**, 443 (1992a).

Padyukov, K. L., Levashov, E. A., Borovinskaya, I. P., and Kost, A. G., Specific features of the behavior of synthetic diamond in a self-propagating high-temperature synthesis (SHS) combustion wave. *J. Eng. Phys. Thermophys.*, **63**, 1107 (1992b).

Padyukov, K. L., Levashov, E. A., Kost, A. G., and Borovinskaya, I. P., SHS: A new fabrication method for diamond-containing ceramics. *Indust. Diamond Rev.*, **5,** 255 (1992c).

Pampuch, R., Stobierski, L., Lis, J., and Raczka, M., Solid combustion synthesis of β SiC powders. *Mater. Res. Bull.*, **22,** 1225 (1987).

Pampuch, R., Stobierski, L., and Lis, J., Synthesis of sinterable β-SiC Powders by solid combustion method. *J. Am. Ceram. Soc.*, **72,** 1434 (1989).

Petrovskii, V. Y., Gorvits, E. I., Borovinskaya, I. P., and Martinenko, V. M., SHS silicon nitride: An attractive powder for dielectric ceramic production. *In* "Problems of Technological Combustion." Chernogolovka, 5 (1981).

Philpot, K. A., Munir, Z. A., and Holt, J. B., An investigation of the synthesis of nickel aluminides through gasless combustion. *J. Mater. Sci.*, **22,** 159 (1987).

Pityulin, A. N., Sytschev, A. E., Rogachev, A. S., and Merzhanov, A. G., One-stage production of functionally gradient materials of the metal-hard alloy type by SHS-compaction. *Proceedings of the 3rd International Symposium on Structural and Functional Gradient Materials (Book of Abstracts)*, Lausanne, Switzerland, 25 (1994).

Pityulin, A. N., Bogatov, Y. V., and Rogachev, A. S., Gradient hard alloys. *Int. J. SHS*, **1,** 111 (1992).

Pityulin, A. N., Shcherbakov, V. A., Borovinskaya, I. P., and Merzhanov, A. G., Laws and mechanism of diffusional surface burning of metals. *Combust. Explos. Shock Waves*, **15,** 432 (1979).

Podlesov, V. V., Radugin, A. V., Stolin, A. M., and Merzhanov, A. G., Technological basis of SHS extrusion. *J. Eng. Phys. Thermophys.*, **63,** 1065 (1992a).

Podlesov, V. V., Stolin, A. M., and Merzhanov, A. G., SHS extrusion of electrode materials and their application for electric-spark alloying of steel surfaces. *J. Eng. Phys. Thermophys.*, **63,** 1156 (1992b).

Prausnitz, J. M., Lichtenthaler, R. N., and de Azevedo, E. G., "Molecular Thermodynamics of Fluid-Phase Equilibria." Prentice-Hall, Englewood Cliffs, NJ, 1986.

Puszynski, J., Degreve, J., and Hlavachek, V., Modeling of exothermic solid-solid noncatalytic reactions. *Ind. Engng Chem. Res.*, **26,** 1424 (1987).

Rabin, B. H., and Wright, R. N., Microstructure and properties of iron aluminides produced from elemental powders. *Int. J. SHS*, **1,** 305 (1992).

Rabin, B. H., and Wright, R. N., Synthesis of iron aluminides from elemental powders: reaction mechanisms and densification behavior." *Metall. Trans. A*, **22A,** 277 (1991).

Rabin, B. H., Korth, G. E., and Williamson, R. L., Fabrication of TiC-Al$_2$O$_3$ composites by combustion synthesis and subsequent dynamic consolidation. *Int. J. SHS*, **1,** 336 (1992a).

Rabin, B. H., Wright, R. N., Knibloe, J. R., Raman, R. V., and Rale, S. V., Reaction processing of iron aluminides. *Mater. Sci. Eng.*, **A153,** 706 (1992b).

Rabin, B. H., Bose, A., and German, R. M., Combustion synthesis of nickel aluminides. *In* "Combustion and Plasma Synthesis of High-Temperature Materials" (Z. A. Munir and J. B. Holt, eds.). VCH Publishers, New York, 1990, p. 114.

Raman, R. V., Rele, S. V., Poland, S., LaSalvia, J., Meyers, M. A., and Niiler, A. R., The one-step synthesis of dense titanium-carbide tiles. *J. Metals*, **47,** 23 (1995).

Rawers, J. C., Wrzesinski, W. R., Roub, E. K., and Brown, R. R., TiAl-SiC composites prepared by high temperature synthesis. *Mater. Sci. Tech.*, **6,** 187 (1990).

Rice, R. W., Microstructural aspects of fabricating bodies by self-propagating synthesis. *J. Mater. Sci.*, **26,** 6533 (1991).

Rice, R. W., Richardson, G. Y., Kunetz, J. M., Schroeter, T., and McDonough, W. J., Effects of self-propagating synthesis reactant compact character on ignition, propagation and microstructure. *Adv. Ceramic Mater.*, **2,** 222 (1987).

Rice, R. W., McDonough, W. J., Richardson, G. Y., Kunetz, J. M., Schroeter, T., Hot-rolling of ceramics using self-propagating synthesis. *Ceram. Eng. Sci. Proc.*, **7,** 751 (1986).

Rogachev, A. S., D.Sc. Dissertation, Institue of Structural Macrokinetics, Russian Academy of Science (1995).

Rogachev, A. S., Shugaev, V. A., Khomenko, I. A., Varma, A., and Kachelmyer, C., On the mechanism of structure formation during combustion synthesis of titanium silicides. *Combust. Sci. Tech.,* **109,** 53 (1995).

Rogachev, A. S., Khomenko, I. O., Varma, A., Merzhanov, A. G., and Ponomarev, V. I., The mechanism of self-propagating high-temperature synthesis of nickel aluminides, Part II: Crystal structure formation in a combustion wave. *Int. J. SHS,* **3,** 239 (1994a).

Rogachev, A. S., Shugaev, V. A., Kachelmyer, C. R., and Varma, A., Mechanisms of structure formation during combustion synthesis of materials. *Chem. Eng. Sci.,* **49,** 4949 (1994b).

Rogachev, A. S., Varma, A., and Merzhanov, A. G., The mechanism of self-propagating high-temperature synthesis of nickel aluminides, Part I: Formation of the product microstructure in a combustion wave. *Int. J. SHS,* **2,** 25 (1993).

Rogachev, A. S., Shkiro, V. M., Chausskaya, I. D., and Shvetsov, M. V., Gasless combustion in titanium–carbon–nickel system. *Combust. Explos. Shock Waves,* **24,** 720 (1988).

Rogachev, A. S., Mukasyan, A. S., and Merzhanov, A. G., Structural transitions in the gasless combustion of titanium-carbon and titanium boron systems. *Dokl. Phys. Chem.,* **297,** 1240 (1987).

Rozenkranz, R., Frommeyer, G., and Smarsly, W., Microstructures and properties of high melting point intermetallic Ti$_5$Si$_3$ and TiSi$_2$ compounds. *Mater. Sci. Eng.,* **A152,** 288 (1992).

Saidi, A., Chrysanthou, A., Wood, J. V., and Kellie, J. L. F., Characteristics of the combustion synthesis of TiC and Fe-TiC composites. *J. Mater. Sci.,* **29,** 4993 (1994).

Samsonov, G. V., "Refractory Transition Metal Compounds; High Temperature Cermets. Academic Press, New York, 1964.

Samsonov, G. V., and Vinitskii, I. M., Handbook of Refractory Compounds. IFI/Plenum, New York, 1980.

Sarkisyan, A. R., Dolukhanyan, S. K., and Borovinskaya, I. P., Self-propagating high-temperature synthesis of transition metal silicides. *Sov. Powd. Metall.,* **17,** 424 (1978).

Sata, N., SHS-FGM studies on the combustion synthesis of fine composites. *Int. J. SHS,* **1,** 590 (1992).

Sata, N., Nagata, K., Yanagisawa, N., Asano, O., Sanada, N., Hirano, T., and Teraki, J., Research and development on functionally gradient materials by using a SHS process. *Proceedings of the First US–Japanese Workshop on Combustion Synthesis,* Tokyo, Japan, 139 (1990a).

Sata, N., Sanada, N., Hirano, T., and Niino, M., Fabrication of a functionally gradient material by using a self-propagating reaction process. *In* "Combustion and Plasma Synthesis of High-Temperature Materials" (Z. A. Munir and J. B. Holt, eds.). VCH Publishers, New York, 1990b, p. 195.

Semenov, N. N., Zur theorie des verbrennungsprozesses. *Zhur. Fiz. B.,* **42,** 571 (1929).

Shah, D. M., Berczik, D., Anton, D. L., and Hecht, R., Appraisal of other silicides as structural materials. *Mater. Sci. Eng.,* **A155,** 45 (1992).

Sharipova, N. S., Ermolaev, V. N., and Kahn, C. G., Study by electron microscopy, of processes occurring at boundary between Cr$_2$O$_3$ and Al. *Combust. Explos. Shock Waves,* **28,** 151 (1992).

Shcherbakov, V. A., Gryadunov, A. N., and Shteinberg, A. S., Macrokinetics of the process of SHS compaction. *J. Eng. Phys. Thermophys.,* **63,** 1111 (1992).

Shcherbakov, V. A., and Pityulin, A. N., Unique features of combustion of the system Ti-C-B. *Combust. Explos. Shock Waves,* **19,** 631 (1983).

Shingu, P. H., Ishihara, K. N., Ghonome, F., Hayakawa, T., Abe, M., and Taguchi, K., Solid state synthesis of TiAl by use of pseudo HIP. *Proceedings of the First US–Japanese Workshop on Combustion Synthesis,* Tokyo, Japan, 65 (1990).

Shirayev, A. A., Yuranov, I., and Kashireninov, O., Thermodynamics of high speed combustion processes: Some particular features. *Proceedings of the Joint meeting of the Russian and Japanese Combustion Section,* Chernogolovka, Russia, 158 (1993).

Shishkina, T. N., Podlesov, V. V., and Stolin, A. M., Microstructure and properties of extruded SHS materials. *J. Eng. Phys. Thermophys.,* **63,** 1082 (1992).

Shkadinskii, K. G., Shkadiskaya, G. V., Matkowsky, B. J., and Volpert, V. A., Combustion synthesis of a porous layer. *Combust. Sci. Tech.,* **88,** 247 (1992a).

Shkadinskii, K. G., Shkadiskaya, G. V., Matkowsky, B. J., and Volpert, V. A., Self-compaction or expansion in combustion synthesis of porous materials. *Combust. Sci. Tech.,* **88,** 271 (1992b).

Shkadinskii, K. G., Khaikin, B. I., and Merzhanov, A. G., Propagation of a pulsating exothermic reaction front in the condensed phase. *Combust. Explos. Shock Waves,* 15 (1971).

Shkiro, V. M., and Borovinskaya, I. P., Capillary flow of liquid metal during combustion of titanium mixtures with carbon. *Combust. Explos. Shock Waves,* **12,** 828 (1976).

Shkiro, V. M., and Borovinskaya, I. P., Study of the titanium and carbon mixtures combustion. In "Combustion Processes in Chemical Engineering and Metallurgy (Russ.)" (A. G. Merzhanov, ed.). USSR Academy of Science, Chernogolovka, Russia, 1975, p. 253.

Shkiro, V. M., Nersisyan, G. A., Borovinskaya, I. P., Merzhanov, A. G., and Shekhtman, V. S., Preparation of tantalum carbides by self-propagating high-temperature synthesis. *Sov. Powd. Metall.,* **18,** 227 (1979).

Shkiro, V. M., Nersisyan, G. A., and Borovinskaya, I. P., Principles of combustion of tantalum–carbon mixtures. *Combust. Explos. Shock Waves,* **14,** 455 (1978).

SHS Bibliography, *Int. J. SHS,* **5,** 309 (1996).

Shteinberg, A. S., Scherbakov, V. A., Martynov, V. V., Mukhoyan, M. Z., and Merzhanov, A. G., Self-propagating high-temperature synthesis of high-porosity materials under zero-g conditions. *Phys. Dokl.,* **36,** 385 (1991).

Shtessel, E. A., Kurylev, M. V., and Merzhanov, A. G., Features of self-propagating processes in the reaction of aluminum with iodine. *Dokl. Phys. Chem.,* **288,** 529 (1986).

Shugaev, V. A., Rogachev, A. S., and Ponomarev, V. I., A model for structure formation in SHS systems. *Int. J. SHS,* **1,** 72 (1992a).

Shugaev, V. A., Rogachev, A. S., Ponomarev, V. I., and Merzhanov, A. G., Structurization of products from the interaction of boron with niobium when subjected to rapid heating. *Dokl. Phys. Chem.,* **314,** 348 (1992b).

Skibska, M., Szulc, A., Mukasyan, A. S., and Rogachev, A. S., Microstructural peculiarities of silicon nitride formation under high nitrogen pressures. Part I: The influence of initial Si particle size distribution on Si_3N_4 SHS morphology. *Int. J. SHS,* **2,** 39 (1993a).

Skibska, M., Szulc, A., Mukasyan, A. S., Shugaev, V. A., and Shiryaev, A. A., Microstructural peculiarities of silicon nitride formation under high nitrogen pressures. Part II: The effect of nitrogen pressure on SHS Si_3N_4 morphology and phase composition. *Int. J. SHS,* **2,** 247 (1993b).

Storms, E. K., "The Refractory Carbides." Academic Press, New York, 1967.

Strutt, A. J., Vecchio, K. S., Yu, L.-H., and Meyers, M. A., Shock synthesis of nickel aluminides. *AIP Conf. Proc.,* **309,** 1259 (1994).

Subrahmanyam, J., Combustion synthesis of $MoSi_2$-Mo_5Si_3 composites. *J. Mater. Res.,* **9,** 2620 (1994).

Subrahmanyam, J., Vijaykumar, M., and Ranganath, S., Thermochemistry of self propagating high temperature synthesis of titanium diboride composites. *Metals Mater. Processes,* **1,** 105 (1989).

Tabachenko, A. N., and Kryuchkova, G. G., Self-propagating high-temperature synthesis of composite materials. The melting compounds (TiC, TiB_2)—intermetallides, their structure and properties. *J. Eng. Phys. Thermophys.,* **65,** 1026 (1993).

Taneoka, Y., Odawara, O., and Kaieda, Y., Combustion synthesis of the titanium–aluminum–boron system. *J. Am. Ceram. Soc.,* **72,** 1047 (1989).

Tanihata, K., Miyamoto, Y., Matsushita, K., Ma, X., Kawasaki, A., Watanabe, R., and Hirano, K., Fabrication of Cr_3C_2/Ni functionally gradient materials by gas-pressure combustion sintering. *Proceedings of the 2nd. Int. Symp. on FGMs,* San Francisco (1992).

Thadani, N. N., Shock-induced chemical reactions and synthesis of materials. *Prog. Mater. Sci.,* **37,** 117 (1993).

Trambukis, J., and Munir, Z. A., Effect of particle dispersion on the mechanism of combustion synthesis of titanium silicide. *J. Am. Ceram. Soc.,* **73,** 1240 (1990).
Urabe, K., Miyamoto, Y., Koizumi, M., and Ikawa, H., Microstructure of TiB$_2$ sintered by the self-combustion method. *In* "Combustion and Plasma Synthesis of High-Temperature Materials" (Z. A. Munir and J. B. Holt, eds.). VCH Publishers, New York, 1990, p. 281.
Vadchenko, S. G., Gordopolov, A. Y., and Mukasyan, A. S., The role of gas- and solid-phase conduction mechanisms in propagation of heterogeneous combustion wave. *Phys. Dokl.,* **42,** 288 (1997).
Vadchenko, S. G., Merzhanov, A. G., Mukasyan, A. S., and Sytchev, A. E., Influence of uniaxial loading on the macrokinetics of gasless systems combustion. *Phys. Dokl.,* **337,** 618 (1994).
Vadchenko, S. G., Bulaev, A. M., Gal'chenko, Y. A., and Merzhanov, A. G., Interaction mechanism in laminar bimetal nickel-titanium and nickel-aluminium systems. *Combust. Explos. Shock Waves,* **23,** 706 (1987).
Vadchenko, S. G., Grigoriev, Y. M., and Merzhanov, A. G., The kinetics of high-temperature tantalum nitriding. *Bull. Akad. Nauk SSSR, Metals,* **5,** 223 (1980).
Vadchenko, S. G., Grigoriev, Y. M., and Merzhanov, A. G., Investigation of the mechanism of the ignition and combustion of the system Ti+C, Zr+C, by an electrothermographic method. *Combust. Explos. Shock Waves,* **12,** 606 (1976).
Varma, A., and Lebrat, J.-P., Combustion synthesis of advanced materials. *Chem. Eng. Sci.,* **47,** 2179 (1992).
Vecchio, K. S., Yu, L.-H., and Meyers, M. A., Shock synthesis of silicides—I. Experimentation and microstructural evolution. *Acta Metall. Mater.,* **42,** 701 (1994).
Vecchio, K. S., LaSalvia, J. C., Meyers, M. A., and Gray, G. T., Microstructural characterization of self-propagating high-temperature synthesis: Dynamically compacted and hot-pressed titanium carbides. *Metall. Trans. A,* **23A,** 87 (1992).
Volpe, B. M., and Evstigneev, V. V., Structure formation in the SHS-system titanium–aluminium–carbon. *Combust. Explos. Shock Waves,* **28,** 173 (1992).
Walton, J. D., and Poulos, N. E., Cermets from thermite reactions. *J. Am. Ceram. Soc.,* **42,** 40 (1959).
Wang, L., Wixom, M. R., and Tompson, L. T., Structural and mechanical properties of TiB$_2$ and TiC prepared by self-propagating high-temperature synthesis/dynamic compaction. *J. Mater. Sci.,* **29,** 534 (1994a).
Wang, L. L., Munir, Z. A., and Holt, J. B., The combustion synthesis of copper aluminides. *Metall. Mater. Trans. B,* **21B,** 567 (1990).
Wang, W., Mei, B., Fu, Z., and Yuan, R., Self-propagating high-temperature synthesis and densification of intermetallic compound-matrix composites (IMCs). *Int. J. SHS,* **2,** 183 (1993).
Wang, Z., Ge, C., and Chen, L., Fabrication of TiB$_2$-Cu composites and functionally gradient materials by GPCS. *Int. J. SHS,* **3,** 85 (1994b).
Wenning, L. A., Lebrat, J.-P., and Varma, A., Some observations on unstable self-propagating high-temperature synthesis of nickel aluminides. *J. Mater. Syn. Proc.,* **2,** 125 (1994).
Wiley, J. B., and Kaner, R. B., Rapid solid-state precursor synthesis of materials. *Science,* **255,** 1093 (1992).
Williams, F. A., "Combustion Theory." Addison-Wesley, Reading, MA, 1965.
Wong, J., Larson, E. M., and Holt, J. B., Time-resolved X-ray diffraction study of solid combustion reaction. *Science,* **249,** 1406 (1990).
Work, S. J., Yu, L. H., Thadhani, N. N., Meyers, M. A., Graham, R. A., and Hammetter, W. F., Shock-induced chemical synthesis of intermetallic compounds. *In* "Combustion and Plasma Synthesis of High-Temperature Materials" (Z. A. Munir and J. B. Holt, eds.). VCH Publishers, New York, 1990, p. 133.
Xiangfeng, M., Tanihata, K., and Miyamoto, Y., Gas-pressure combustion sintering and properties of Cr$_3$C$_2$ ceramic and its composite with TiC. *J. Ceram. Soc. Japan,* **100,** 605 (1992).

Yamada, O., Fabrication of fully dense composite materials by SHS under pressurized reactive-gas atmosphere. *J. Soc. Mater. Sci. Japan*, **43,** 1059 (1994).

Yamada, O., Miyamoto, Y., and Koizumi, M., High-pressure self-combustion sintering of titanium carbide. *J. Am. Ceram. Soc.,* **70,** C206 (1987).

Yamada, O., Miyamoto, Y., and Koizumi, M., Self-propagating high-temperature synthesis of the SiC. *J. Mater. Res.,* **1,** 275 (1986).

Yamada, O., Miyamoto, Y., and Koizumi, M., High pressure self-combustion sintering of silicon carbide. *Am. Ceram. Soc. Bull.,* **64,** 319 (1985).

Yanagisawa, N., Asano, O., Sata, N., and Sanada, N., Synthesis and properties of TiB_2-TiNi composite materials by SHS process. *Proceedings of the First US–Japanese Workshop on Combustion Synthesis,* Tokyo, Japan, 157 (1990).

Yi,. H. C., and Moore, J. J., The combustion synthesis of Ni-Ti shape memory alloys. *J. Metals,* **42,** 31 (1990).

Yi,. H. C., and Moore, J. J., Combustion synthesis of TiNi intermetallic compounds. Part 1: Determination of heat of fusion of TiNi and heat capacity of liquid TiNi. *J. Mater. Sci.,* **24,** 3449 (1989).

Yi,. H. C., Varma, A., Rogachev, A. S., and McGinn, P. J., Gravity-induced microstructural nonuniformities during combustion synthesis of intermetallic-ceramic composite materials. *Ind. Eng. Chem. Res.,* **35,** 2982 (1996).

Yukhvid, V. I., Modifications of SHS processes. *Pure Appl. Chem.,* **64,** 977 (1992).

Yukhvid, V. I., Kachin, A. R., and Zakharov, G. V., Centrifugal SHS surfacing of the refractory inorganic materials. *Int. J. SHS,* **3,** 321 (1994).

Yukhvid, V. I., Borovinskaya, I. P., and Merzhanov, A. G., Influence of pressure on the laws governing the combustion of molten heterogeneous systems. *Combust. Explos. Shock Waves,* **19,** 277 (1983).

Zavitsanos, P. D., Gebhardt, J. J., and Gatti, A., The use of self-propagating high-temperature synthesis of high-density titanium diboride. *In* "Combustion and Plasma Synthesis of High-Temperature Materials" (Z. A. Munir and J. B. Holt, eds.). VCH Publishers, New York, 1990, p. 170.

Zeldovich, Y. B., Theory for the limit of the quiet flame propagation. *J. Exper. Theor. Phys.,* **11,** 159 (1941).

Zeldovich, Y. B., and Frank-Kamenetskii, D. A., The theory of thermal propagation of flames. *Zh. Fiz. Khim.,* **12,** 100 (1938).

Zeldovich, Y. B., Barenblatt, G. I., Librovich, V. B., and Makhviladze, G. M., "Mathematic Theory of Combustion and Explosions" (D. H. McNeil, trans.). Consultant Bureau, New York, 1985.

Zenin, A. A., Merzhanov, A. G., and Nersisyan, G. A., Thermal wave structure in SHS processes (by the example of boride synthesis). *Combust. Explos. Shock Waves,* **17,** 63 (1981).

Zenin, A. A., Merzhanov, A. G., and Nersisyan, G. A., Structure of the heat wave in some self-propagating high-temperature processes. *Dokl. Phys. Chem.,* **250,** 83 (1980).

Zhang, X., Yin, W., and Guo, J., Exploration of combustion synthesis of TiNi intermetallic compound. *Int. J. SHS,* **4,** 301 (1995).

COMPUTATIONAL FLUID DYNAMICS APPLIED TO CHEMICAL REACTION ENGINEERING

J. A. M. Kuipers and W. P. M. van Swaaij

Department of Chemical Engineering
Twente University of Technology
7500 AE Enschede, The Netherlands

I. Introduction	227
II. Traditional Approaches Followed within Chemical Reaction Engineering	228
A. Role and Types of Modeling in Chemical Engineering	232
B. Relation with Experimental Work	233
C. CFD in Chemical Engineering Education	234
III. Computational Fluid Dynamics	234
A. Definition and Theoretical Framework	236
B. Numerical Techniques	244
C. Existing Software Packages	251
IV. Application to Chemical Reaction Engineering	253
A. Single-Phase Systems	254
B. Multiphase Systems	265
C. State of the Art of CFD in Chemical Reaction Engineering	280
V. Experimental Validation	282
VI. Selected Applications of CFD Work Conducted at Twente University	287
A. Two-Fluid Simulation of Gas Fluidized Beds	287
B. Discrete Particle Simulation of Gas Fluidized Beds	291
C. Circulating Fluidized Beds	296
D. Bubble Columns	298
E. Modeling of a Laminar Entrained Flow Reactor	311
VII. Conclusion	313
Nomenclature	316
References	319

I. Introduction

In the last few decades computational fluid dynamics has become a very powerful and versatile tool for the analysis and solution of problems that are of considerable interest to the chemical engineer, despite the fact that CFD has not yet reached its full potential. The chemical engineering discipline has developed many valuable semiempirical strategies to solve problems of practical interest. Prior to expanding on the nature of CFD and its current possibilities and limitations, a brief outline on these strategies will be presented.

II. Traditional Approaches Followed within Chemical Reaction Engineering

In the process technology raw materials are converted into desired products via chemical and physical processes. These products are often intermediates that are subsequently converted in other production processes. This general scheme (see Fig. 1) from raw materials like minerals, crude oil, natural gas, agricultural products, etc., via preprocessing and purification is very common. A process engineer developing, designing, or optimizing a process has to deal with many disciplines varying from chemistry or biochemistry to economy. His or her ultimate aim is to produce a valuable product in a safe way at acceptable cost and burden to the environment. Also time is a scarce quantity because he or she has to take the changing market and competitors into account. The traditional chemical engineering sciences like transport phenomena and chemical reaction engineering play an essential role in achieving the aforementioned aims. In these sciences elements from physics (like transport theories, fluid dynamics, and thermodynamics), chemistry (kinetics, catalysis), and mathematics have been integrated to form dedicated tools to tackle extremely complicated problems that come up in studying and developing processes.

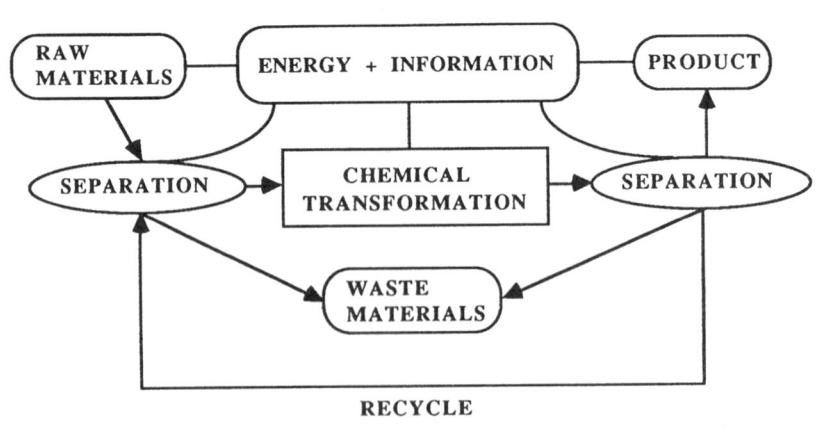

FIG. 1. Schematic representation of a production scheme typically encountered in process technology.

A new emerging tool is a combination of fluid dynamics and numerical mathematics backed up by the immense growth of computer power: computational fluid dynamics (CFD). It is affecting the chemical engineering sciences and the art of the chemical engineer in a profound way. We also think that a special branch of CFD will originate from the confrontation with chemical engineering science. This is because of special demands in the applications, the application-oriented attitude of the chemical engineer, and the way the knowledge has been organized traditionally.

Originally, technical scientific knowledge was arranged around the individual process steps: distillation, absorption, extraction, crystallization, heat exchange, mass exchange, etc.; the so-called "unit operations" (± 1920). For a short period of time also the notion of unit processes was used for chemical process steps such as chlorination, oxydation, hydrogenation, etc. However, because of lack of coherence this never became a big success. Specifically after the second world war it became increasingly clear that the different unit operations could be strongly unified by combining the principle of microbalance formulation with laws of linear transport of mass, momentum, and energy while adding elements of technical fluid dynamics and radiative heat transfer. This culminated in the textbook *Transport Phenomena* by Bird, Stewart, and Lightfoot (1960), which is still a standard for chemical engineers. Chemical reactions often form a considerable complication in process description and modeling. Integration of chemical reaction kinetics and thermodynamics in transport phenomena resulted in the discipline of chemical reaction engineering. The first textbooks on chemical reaction engineering appeared in the early 1960s (Levenspiel, 1962). Of course, these sciences have grown since then and the widespread use of computing facilities changed their nature allowing, amongst others, the chemical engineer to use large databases, advanced design and optimization methods, and complex and extended models that were previously difficult to handle. The increased use of the Maxwell–Stefan transport model instead of the traditional Fickian model represents an example of this development.

As mentioned earlier, in the past many powerful tools with a strong empirical base had been devised by chemical engineers to (approximately) solve extremely complicated problems encountered in chemical reaction engineering such as the design of process equipment (including chemical reactors) involving nonideal flow (see Fig. 2; Levenspiel, 1962). Very often the two idealized flow patterns, that is, plug flow and mixed flow, do not occur in reality but nevertheless these concepts have proven valuable when dealing with the design of chemical reactors because the conversion obtained in these two extreme cases provides the boundaries for the conversion in a chemical reactor in which nonideal flow prevails. Levenspiel (1962) stated that "If we know precisely what is happening within the vessel, thus if we have a complete velocity distribution map for the fluid, then we are able to predict the behavior of a vessel as a reactor. Though fine in principle, the attendant

230   J. A. M. KUIPERS AND W. P. M. VAN SWAAIJ

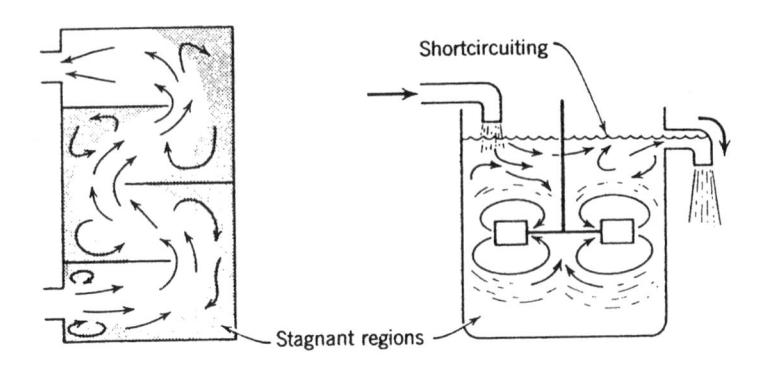

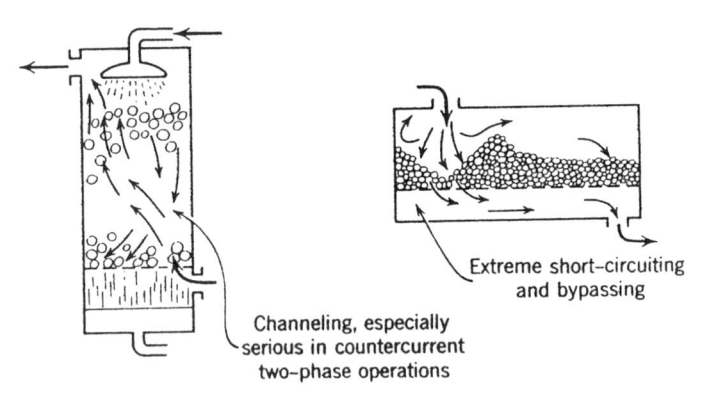

FIG. 2. Some typical examples of nonideal flow patterns that can occur in process equipment. (From Levenspiel, O., "Chemical Reaction Engineering." John Wiley & Sons, New York, 1962. Reprinted by permission of John Wiley & Sons, Inc.)

complexities make it impractical to use this approach." This situation has clearly changed because nowadays the "complete velocity distribution map" can in principle be obtained by computing the velocity distribution in the system of interest.

A well-known traditional approach adopted in chemical engineering to circumvent the intrinsic difficulties in obtaining the "complete velocity distribution map" is the characterization of nonideal flow patterns by means of residence time distribution (RTD) experiments where typically the response of a piece of process equipment is measured due to a disturbance of the inlet concentration of a tracer. From the measured response of the system (i.e., the concentration of the tracer measured in the outlet stream of the relevant piece of process equipment) the differential residence time distribution $E(t)$ can be obtained where $E(t)dt$ represents the fraction $dF(t)$ of the volume elements with a residence time between t and $t + dt$:

$$dF(t) = E(t)dt. \quad (1)$$

The fraction of the fluid elements with a residence time of less than t is given by the cumulative residence distribution function $F(t)$ given by:

$$F(t) = \int_0^t E(t)dt. \quad (2)$$

For a continuous reactor with a nonideal flow pattern, characterized by the differential residence time distribution $E(t)$, the following expression holds for the conversion $\xi_{nonideal}$, which is attained in case complete segregation of all fluid elements passing through the reactor can be assumed:

$$\xi_{nonideal} = \int_{t_{min}}^{t_{max}} \xi_{batch}(t) E(t) dt, \quad (3)$$

where $\xi_{batch}(t)$ represents the conversion attained in a batch reactor with batch time t. Of course, in reality a certain degree of mixing between the fluid elements passing through the reactor occurs with simultaneous chemical transformation and therefore information on the macromixing patterns is generally not sufficient to enable accurate prediction of the extent of chemical conversion in a reactor possessing a nonideal flow pattern. This mixing phenomenon necessitates the use of a micromixing model to account for the finite rate with which segregated fluid elements eventually achieve mixing at the molecular level (i.e., segregation decay) to permit the occurrence of a chemical transformation.

Due to the advent of CFD the aforementioned approach can still be followed but now the $E(t)$ and $F(t)$ functions can in principle be computed from the computed velocity distribution. Alternatively, the species conservation equations can be solved simultaneously with the fluid flow equations and thereby the extent of chemical conversion can also be obtained directly without invoking the concept of residence time distributions.

The strong demand for data reduction also applies to results of CFD. Moreover chemical engineers have a sound distrust of, based on a lot of experience, calculated results that have not been validated with experimental data. If we compare the traditional chemical engineering approach to the pure computational fluid dynamics-based procedure several advantages and disadvantages become apparent and at least for a long period of time both approaches have to (and probably will) merge to obtain an optimal result. However, more and more the traditional empirical and phenomenological models are being replaced by more fundamental descriptions, which are based on the full microbalances for mass, momentum, and (thermal) energy. In the next section a brief discussion is presented of the role and types of modeling encountered in chemical engineering.

A. ROLE AND TYPES OF MODELING IN CHEMICAL ENGINEERING

The widespread availability of fast computing facilities and the rapid advance of powerful numerical techniques and software offer the possibility of numerical simulation of many processes of interest in chemical engineering science and the process industry. Problems often encountered in process operations can be described by balance equations for conservation of mass, momentum, and (thermal) energy in combination with transport equations for chemical conversion or phase transition. Because the equations are in most cases coupled and nonlinear, the problem is complex. Moreover, transport rates are dependent on local fluid properties, which are often themselves a function of process variables. Therefore, only in the simplest cases can analytical solutions be found and numerical solution procedures are welcomed. CFD has been applied successfully in analyzing complex single-phase laminar flow problems. Here the accuracy of the simulations is so high that validation experiments are often considered unnecessary. This is certainly not the case for single-phase turbulent flow problems in complex geometries, although considerable progress has been made here in recent years in turbulence modeling.

Several computer codes are now commercially available (see Section III,C) that are applicable for laminar and turbulent single-phase fluid flow problems. They show variable success in different applications. For two-phase flow the situation is still more complicated and detailed microbalance modeling is still in its infancy. Before we discuss these types of models in more detail, we first consider their advantages and disadvantages in comparison with other type of models. Specifically a comparison between detailed microbalance models (i.e., fluid flow equation-based microbalance models) and global system models (i.e., models that are not based on fluid flow equations) is made; Table I gives a short overview. A similar table has been presented by Villermaux (1996). The arguments given in Table I are global ones, because the boundaries between the different models are not clear-cut. Moreover, a wide range of model types is possible that shows an important overlap in properties.

The most powerful property of the detailed microbalance models, especially in combination with visualization techniques, is the *a priori* prediction of (observable) macroscale phenomena. This can be particularly helpful in reducing the required experimental effort. Important problems are the amount of detailed information required for the microscale transport equations and the large programming and computational efforts required to solve specific problems. Nevertheless, these types of models, by generating insight in the micro- and

TABLE I
Comparison of Detailed Microbalance Models with Global System Models

Detailed Microbalance Models Advantage	Global System Models Advantage
More exact solution available	Simple models and simple solutions facilitate understanding
Phenomena follow from calculations *a priori*	Can be adapted to the detail of information required
Formal balance equations can be written straightforwardly (for single phase systems)	Limited calculation capacity often sufficient
Processes can be visualized	After adjustment of parameters accurate macroscale behavior prediction
Disadvantage	Disadvantage
Detailed knowledge required about the elementary process	Lot of *a priori* knowledge together with imaginative power required
Massive data production, thus data reduction required	Experimental validation and adjustment of parameters necessary
Large calculation capacity required	Meaning of parameters sometimes unclear due to lumping
Difficult to generalize. Each specific problem requires additional (computational) effort	
Macroscopic behavior not always accurately predicted	

mesoscale mechanisms, can play an important role in the preparation of experimental programs and in innovations to improve existing processes.

B. Relation with Experimental Work

Despite its great potential, in the near future CFD will not completely replace experimental work or standard approaches currently used by the chemical engineering community. In this connection it is even not sure that CFD is guaranteed to succeed or even be an approach that will lead to improved results in comparison with standard approaches. For single-phase turbulent flows and especially for multiphase flows, it is imperative that the results of CFD analysis somehow be compared with experimental data in order to assess the validity of the physical models and the computational algorithms. In this connection we should mention that only computational results that possess invariance with respect to spatial and temporal discretization should be confronted with experimental data. A CFD model usually gives very detailed information on the temporal and spatial variation of many key quantities (i.e., velocity components, phase volume fractions, temperatures, species concentrations, turbulence parameters), which leads to increasing demands for experimental methods. In this way CFD is fruitful because it can lead to the development of new techniques to measure quantities that previously have not been considered in detail.

C. CFD in Chemical Engineering Education

Due to the rapid advances in CFD and the potential it provides to analyze, on a fundamental basis, systems of considerable interest to the chemical engineer it can be anticipated that the importance of CFD as a "workhorse" for the chemical engineering community will rapidly increase in the near future. This development implies that the chemical engineer working with CFD will need a good knowledge in a large number of disciplines, including physics, chemistry, thermodynamics, materials science, fluid dynamics, chemical reaction engineering, and numerical and experimental methods. Thus broader university education is important and the ability to participate in interdisciplinary research teams is considered very important to meet the demands of an integrated approach combining experimental and theoretical methods with numerical simulation techniques. In this connection it is considered of crucial importance for (introductory) CFD courses to be implemented in the curricula of technical universities at (preferably) the undergraduate level. As far as the authors know most technical universities do not yet offer such courses.

III. Computational Fluid Dynamics

First we give a brief introduction to the historical development of the science of fluid mechanics, and subsequently the development and present areas of application of CFD are highlighted.

The equations that form the theoretical foundation for the whole science of fluid mechanics were derived more than one century ago by Navier (1827) and Poisson (1831) on the basis of molecular hypotheses. Later the same equations were derived by de Saint Venant (1843) and Stokes (1845) without using such hypotheses. These equations are commonly referred to as the Navier–Stokes equations. Despite the fact that these equations have been known of for more than a century, no general analytical solution of the Navier–Stokes equations is known. This state of the art is due to the complex mathematical (i.e., nonlinearity) nature of these equations.

Toward the end of the nineteenth century the science of fluid mechanics began to develop in two branches, namely theoretical hydrodynamics and hydraulics. The first branch evolved from Euler's equations of motion for a frictionless, nonviscous fluid, whereas the development of the second branch was driven by the rapid progress in technology were engineers, faced with the solution of practical problems for which the "classical" science of theoretical hydrodynamics had no answers, developed their own highly empirical science of hydraulics. At the beginning of this century Prandtl showed how the unification of the aforementioned divergent branches of fluid mechanics could be obtained. He showed both experimentally and theoretically that the neglect of the viscous forces, which are indeed very small compared to the remaining forces for the two most important fluids encountered in practice (water and air), leads to incorrect results in thin fluid regions near solid walls. In this region, the *boundary layer,* viscous forces are very important and therefore should be taken into account.

During the first half of this century a spectacular development in the boundary layer theory took place that was driven mainly by the needs of the aerodynamics community. Most of the initial developments involved approximate analytical solution or transformation and subsequent numerical integration of the relevant fluid flow equations (Schlichting, 1975). Until 1960 CFD was virtually absent in the aerodynamics discipline but gradually the aforementioned solution procedures were replaced by CFD-based approaches where the full conservation equations were solved. However, until 1970 the storage capacity and the speed of digital computers were not sufficient to enable efficient calculation of full three-dimensional flow fields around airplanes. This situation has by now definitely changed since a number of computer programs for the calculation of three-dimensional flow fields around airplanes have become industry standards, resulting in their use as a tool in the design process. It can be anticipated that for the (future) design of hypersonic aircraft, where the ground test facilities—wind tunnels—do not exist to cover all relevant flow regimes, CFD will be the principal workhorse for the actual design. An interesting review of CFD applications in aeronautics has been presented by Jameson (1988), whereas the book of Anderson (1995) gives an excellent introduction to the basics of CFD and its applications in, amongst others, aeronautical engineering.

Although the initial development of the CFD discipline was driven by the aerodynamics community, nowadays CFD is truly interdisciplinary since its cuts across all disciplines where the analysis of fluid flow and associated phenomena is of importance. For example, CFD has found application in the automobile industries to study both the internal flow in combustion engines (Griffin *et al.,* 1978) and the external flow (Shaw, 1988, Matsunaga *et al.,* 1992). Also in civil engineering CFD has found application in the study of problems involving flow dynamics of rivers, lakes, and estuaries and external flow around buildings. In environmental engineering CFD has been used to analyze the complex flow patterns that exist in various types of furnaces (Bai and Fuchs, 1992). Here the CFD approach has proven useful to optimize furnace performance (i.e., improved thermal efficiency and reduction of emission of pollutants). CFD has also been applied to calculate air currents throughout buildings in order to arrive at improved designs of (natural) ventilation systems (Alamdari *et al.*, 1991).

In industrial manufacturing applications a myriad of applications exist of which the modeling of chemical vapor deposition reactors in the semiconductor industries (Steijsiger *et al.*, 1992) and the modeling of the casting process of liquid metals (Mampaey and Xu, 1992) can be mentioned as examples. For further details on CFD applications in various industrial manufacturing processes the reader is referred to previous reviews (Colenbrander, 1991; Trambouze, 1993; Johansen and Kolbeinsen, 1996). Colenbrander has prepared a review on CFD applications in the petrochemical industries with specific emphasis on CFD applications and related experimental work carried out in the Shell Group laboratories. The application of CFD to chemical reaction engineering has been reviewed by Trambouze (1993). During the last two decades CFD has also become a powerful tool for analyzing and designing metallurgical processes. Johansen and Kolbeinsen (1996) have recently prepared a review on this subject, where CFD applications at SINTEF Materials and Technology were highlighted. Finally, Harris *et al.* (1995) have recently presented a review on the application of CFD in chemical reaction engineering (CRE) with emphasis on single-phase flow applications. Before the application of CFD to CRE is discussed, a brief outline of CFD and its theoretical framework are first presented.

A. Definition and Theoretical Framework

Computational fluid dynamics involves the analysis of fluid flow and related phenomena such as heat and/or mass transfer, mixing, and chemical reaction using numerical solution methods. Usually the domain of interest is divided into a large number of control volumes (or computational cells or elements) which have a relatively small size in comparison with the macroscopic volume of the domain of interest. For each control volume a discrete representation of the relevant conservation equations is made after which an iterative solution procedure is invoked to obtain the solution of the nonlinear equations. Due to the advent of high-speed digital computers and the availability of powerful numerical algorithms the CFD approach has become feasible. CFD can be seen as a hybrid branch of mechanics and mathematics. CFD is based on the conservation laws for mass, momentum, and (thermal) energy, which can be expressed as follows:

1. Mass is conserved.
2. Newton's second law: $\overline{F} = m \cdot \overline{a}$.
3. Energy is conserved.

Subsequently, the theoretical foundation is briefly explained where the authors have chosen to make a distinction between single-phase systems and multiphase systems.

1. Single-Phase Systems

For single-phase systems involving laminar flows the conservation equations are firmly established. The mass and momentum conservation equations are respectively given by (Bird *et al.*, (1960):

$$\frac{\partial}{\partial t}\rho + (\nabla \cdot \rho \bar{u}) = 0, \tag{4}$$

$$\frac{\partial}{\partial t}\rho \bar{u} + (\nabla \cdot \rho \bar{u}\bar{u}) = -\nabla p - (\nabla \cdot \tau) + \rho \bar{g}, \tag{5}$$

where ρ, $\bar{u}$, p, and τ, respectively, represent the fluid density, fluid velocity, pressure, and viscous stress tensor. For nonisothermal systems the mass and momentum conservation equations, Eq. (4) and (5), have to be supplemented with an energy equation for which either the total (i.e., the sum of internal energy and mechanical energy) or thermal energy equation can be used. Since in many chemical engineering applications (especially those dealing with chemical reactors) the mechanical energy changes are relatively small compared to the changes in thermal energy, the thermal energy equation is commonly used. For systems involving chemical transformations, conservation equations for all species i involved have to be added. The thermal energy equation and the conservation equation for species i are respectively given by:

$$\frac{\partial}{\partial t}(\rho e) + (\nabla \cdot \rho e \bar{u}) = -(\nabla \cdot \bar{q}) - p(\nabla \cdot \bar{u}) + (-\tau:\nabla \bar{u}) + S_h \tag{6}$$

and

$$\frac{\partial}{\partial t}(\rho \omega_i) + (\nabla \cdot \rho \bar{u} \omega_i) = -(\nabla \cdot \bar{J}_i) + S_i. \tag{7}$$

In Eqs. (6) and (7) e represents the internal energy per unit mas, $\bar{q}$ the heat flux vector due to molecular transport, S_h the volumetric heat production rate, ω_i the mass fraction of species i, $\bar{J}_i$ the mass flux vector of species i due to molecular transport, and S_i the net production rate of species i per unit volume. In many chemical engineering applications the viscous dissipation term $(-\tau:\nabla \bar{u})$ appearing in Eq. (6) can safely be neglected. For closure of the above set of equations, an equation of state for the density ρ and constitutive equations for the viscous stress tensor τ, the heat flux vector $\bar{q}$, and the mass flux vector $\bar{J}_i$ are required. In the absence of detailed knowledge on the true rheology of the fluid, Newtonian behavior is often assumed. Thus, for τ the following expression is used:

$$\tau = -\left\{(\lambda - \frac{2}{3}\mu)(\nabla \cdot \bar{u})I + \mu((\nabla \bar{u}) + (\nabla \bar{u})^T)\right\}, \tag{8}$$

where λ and μ, respectively, represent the bulk viscosity and the shear viscosity. In dense gases and liquids the bulk viscosity λ can probably be neglected (Bird *et al.*, 1960). For the heat flux vector $\bar{q}$ and the mass flux vector $\bar{J}_i$, respectively, Fourier's law and Fick's law are often used:

$$\bar{q} = -k\nabla T, \qquad (9)$$

$$\bar{J}_i = -D_i \rho \nabla \omega_i, \qquad (10)$$

where k represents the thermal conductivity, T the temperature, and D_i the Fickian diffusion coefficient of species i. Equation (9) for the description of the conductive heat flux is only valid for isotropic media; for anisotropic media the heat flux vector $\bar{q}$ should be written as the dot product of the conductivity tensor and the temperature gradient. Fick's law for the description of diffusional transport is, strictly speaking, only valid for the description of systems where (very) low concentrations of the relevant species prevail. For more general expressions the interested reader is referred to Bird *et al.* (1960).

The conservation equations, Eqs. (4)–(7) are also valid for the description of turbulent flows, but within the context of CFD a very high resolution in space and time would be required to capture all the details of the turbulent flow field. Although some exciting results have been obtained in recent years (see Section IV,A,2) in the so-called direct numerical simulation (DNS) in most industrial applications, which involve turbulent flows at (very) high Reynolds numbers, this approach cannot be followed. For the description of such turbulent flows, with their rapid temporal and spatial changes of pressure and velocity, usually the concept of Reynolds decomposition (Tennekes and Lumley, 1977; Warsi, 1993) is invoked where the instantaneous value of each variable X appearing in the balance equations is represented as the sum of its time-averaged $\overline{X}$ and a fluctuating component X':

$$X = \overline{X} + X'. \qquad (11)$$

Substitution of the expressions for the instantaneous X values in the conservation equations, Eq. (4)–(7), followed by some kind of suitable time-averaging procedure leads to the time-averaged conservation equations for turbulent flow. These equations in fact show a very close resemblance to the original equations: The instantaneous pressure and velocity are replaced by the corresponding time-averaged quantities, whereas the viscous stress tensor τ is replaced by the sum of τ (now appearing with time-averaged velocity) and the so-called Reynolds stress tensor $\tau^{(t)}$. The physical origin of the Reynolds stresses can be related to the interaction of the fluctuating fluid motion with the time-averaged fluid motion. Similarly, the heat flux and mass flux appearing in the time-averaged conservation equations for turbulent flow [corresponding, respectively, to Eqs. (6) and (7)] consist of the sum of the molecular flux expressions for $\bar{q}$ and $\bar{J}_i$ (now appearing with, respectively, time-averaged temperature and mass fraction of species i) and the turbulent or eddy contributions. For the treatment of the source terms appearing in Eq. (7), which very often depend in a nonlinear manner on the mass fractions of the species involved in the chemical transformation, special strategies are required depending on ratio of the characteristic time for chemical transformation and the characteristic time for mixing (Pope, 1994; Fox, 1996). For nonisothermal chemically reactive systems, the proper averaging of the heat source term S_h, which depends on the temperature via an Arrhenius type expression (Levenspiel, 1962), should also be taken into account. The principal difficulty in modeling turbulent flow lies in the specification of the Reynolds stresses (and turbulent energy and species i mass flux), which should be specified in terms of the time-averaged variables. This so-called closure problem constitutes a formidable problem for the fluid dynamicist and considerable effort has been made to develop the corresponding closure laws. For the description of turbulent flows a semiempirical turbulence model is often invoked to calculate the Reynolds stresses. A turbulence model is defined as a set of equations (algebraic or differential) that determines the turbulent (momentum) transport terms in the flow equations governing the time-averaged or mean flow. It should be stressed that the turbulence model does not simulate the details of the fluctuating fluid motion, merely its effect on the time-averaged fluid motion. The nature and the degree of refinement contained in the turbulence model will determine the range of flows for which its has true predictive power. Among the turbulence models most often applied one can mention (Rodi, 1980):

- Constant eddy viscosity model
- Prandtl's mixing length model
- Prandtl–Kolmogorov model
- k-ϵ model
- Algebraic stress model (ASM)
- Reynolds stress model (RSM)

All of these models require some form of empirical input information, which implies that they are not general applicable to *any* type of turbulent flow problem. However, in general it can be stated that the most complex models such as the ASM and RSM models offer the greatest predictive power. Many of the older turbulence models are based on Boussinesq's (1877) eddy-viscosity concept, which assumes that, in analogy with the viscous stresses in laminar flows, the Reynolds stresses are proportional to the gradients of the time-averaged velocity components:

$$\tau_{ij}^{(t)} = -\mu^{(t)}\left[\frac{\partial U_i}{\partial x_j} + \frac{\partial U_j}{\partial x_i}\right] + \delta_{ij}\frac{2}{3}\rho k, \qquad (12)$$

where $\mu^{(t)}$ represents the turbulent viscosity, which, in contrast with the molecular viscosity μ, is not a fluid property but a property of the turbulent flow. In Eq. (12) U_i and U_j represent time-averaged velocity components. Within the framework of the models that use the eddy-viscosity concept, the task of the turbulence model is the description, by means of algebraic or differential equations, of the turbulent viscosity $\mu^{(t)}$.

The constant eddy-viscosity model and Prandtl's mixing length model belong to the class of zero-equation models since no transport equations are involved for the turbulence quantities. In fact, the constant eddy-viscosity model cannot be regarded as a true turbulence model since the appropriate value of $\mu^{(t)}$ is usually fitted from experimental data. Experience has shown that mixing length models are not suitable when convective or diffusive transport processes of turbulence are important. The Prandtl–Kolmogorov model belongs to the class of single-equation models because one conservation equation for the turbulent kinetic energy k is solved. It overcomes the aforementioned problem of the mixing length model because a transport equation for the characteristic velocity scale of turbulence is solved for. The weak point in this model is the specification of the characteristic length scale L, which is required in the Kolmogorov–Prandtl expression for the turbulent viscosity:

$$\mu^{(t)} = C_\mu \rho \sqrt{k} L, \qquad (13)$$

where C_μ represents an empirical constant. Especially for complex flows it is difficult to specify the length scale (distribution) and therefore two-equation models such as the k-ϵ model have become more popular because here an additional transport equation is invoked to obtain the length scale (distribution) L. In the k-ϵ model two transport equations are solved for the turbulent kinetic energy k and the viscous dissipation rate ϵ. The expression for the turbulent viscosity is given by:

$$\mu^{(t)} = C_\mu \rho \frac{k^2}{\epsilon}. \qquad (14)$$

Computational experience has revealed that the two-equation models, employing transport equations for the velocity and length scales of the fluctuating motion, often offer the best compromise between width of application and computational economy. There are, however, certain types of flows where the k-ϵ model fails, such as complex swirling flows, and in such situations more advanced turbulence models (ASM or RSM) are required that do not involve the eddy-viscosity concept (Launder, 1991). According to the ASM and the RSM the six components of the Reynolds stress tensor are obtained from a complete set of algebraic equations and a complete set of transport equations. These models are conceptually superior with respect to the older turbulence models such as the k-ϵ model but computationally they are also (much) more involved.

Due to the advances in computer technology and numerical solution procedures two powerful simulation types of turbulent flows have recently received particular attention, namely, direct numerical simulation (DNS) and large eddy simulation (LES). As stated earlier turbulent flows are also governed by the Navier–Stokes equations and in principle the solution of these equations with a sufficiently high temporal and spatial resolution should provide all the details of the turbulent flow without the necessity of turbulence modeling. Due to the fact that a turbulent motion contains elements with a linear dimension which is typically $O(10^{-3})$ smaller than the linear dimension of the macroscopic flow domain, a DNS simulation in three dimensions would require roughly $O(10^9)$ grid points, which is still far beyond the current storage capacity of present-day computers. The estimate given here is only a very rough one and corresponds to relatively high Reynolds numbers. It should be mentioned here that the required number of grid points strongly depends on the nature of the turbulent flow (homogeneous isotropic turbulent flow or homogeneous shear flow) and also on the Reynolds number since the ratio of the dimension of the macroscopic system and the dimension of the smallest eddies present in the turbulent flow depends on the Reynolds number: The smaller the Reynolds number the smaller this ratio. For turbulent channel flow the ratio of the channel width 2δ to the scale of the smallest eddies λ is proportional to $(Re_\delta)^{0.9}$ where the Reynolds number Re_δ is defined as:

$$Re_\delta = \frac{U_c \delta}{\nu}, \tag{15}$$

where U_c is the time-averaged velocity in the centre of the channel. On basis of the DNS performed by Kim *et al.* (1987) at $Re_\delta = 3300$ using $2 \cdot 10^6$ grid points, Reynolds (1991) has estimated the number of grid points N_{xyz} required for a DNS computation with comparable resolution to that of Kim *et al.* (1987) as:

$$N_{xyz} = 2 \cdot 10^6 \left[\frac{Re_\delta}{3300} \right]^{2.7}. \tag{16}$$

In a LES the spatial resolution of the computational mesh is (deliberately) chosen in such a manner that only the large-scale turbulent motion (eddies) is resolved. The consequence of this approach is the need to use subgrid models, which in fact model the turbulent stresses on a scale smaller than the computational grid. Due to the fact that the small-scale turbulence is isotropic, the specification of subgrid models is (far) less difficult in comparison with the aforementioned closure models for the Reynolds stresses. The advantage of LES in comparison with DNS is its possibility to study (with a given number of grid points or control volumes) turbulent flows at (significant) higher Reynolds numbers.

2. Multiphase Systems

For each continuous phase k present in a multiphase system consisting of N phases, in principle the set of conservation equations formulated in the previous section can be applied. If one or more of the N phases consists of solid particles, the Newtonian conservation laws for linear and angular momentum should be used instead. The resulting formulation of a multiphase system will be termed the local instant formulation. Through the specification of the proper initial and boundary conditions and appropriate constitutive laws for the viscous stress tensor, the hydrodynamics of a multiphase system can in principle be obtained from the solution of the governing equations.

However, for most systems of practical interest, the analysis of multiphase systems on basis of the local instant formulation is intractable, even for existing and near-future supercomputers, and consequently some kind of simplification must be made. From a computational point of view this state of the art bears some resemblance to the problems encountered in DNS of turbulent flows.

The aforementioned simplification can be achieved through a continuum mathematical description of the multiphase system. There is extensive literature on the derivation of continuum equations for multiphase systems; the interested reader is referred to Ishii (1975). The derivation of the continuum equations is usually based on spatial averaging techniques where the point hydrodynamic variables are replaced by local averaged variables. The resulting multifluid formulation can be solved by appropriate numerical methods, which in fact generalize the well-known single-phase solution procedure of Patankar (1980). For multiphase isothermal systems involving laminar flow the conservation equations for mass and momentum are respectively given by:

$$\frac{\partial}{\partial t}\rho_k + (\nabla \cdot \rho_k \bar{u}_k) = R_k, \qquad (17)$$

$$\frac{\partial}{\partial t}(\rho_k \bar{u}_k) + (\nabla \cdot \rho_k \bar{u}_k \bar{u}_k) = -\epsilon_k \nabla p - (\nabla \cdot \epsilon_k \tau_k) + \sum_{l=1}^{N} \overline{M}_{kl} + \bar{S}_k + \rho_k \bar{g}, \qquad (18)$$

where ρ_k, $\bar{u}_k$, ϵ_k, and τ_k represent, respectively, the macroscopic density, velocity, volume fraction, and viscous stress tensor of the kth phase, p the pressure, R_k a source term describing mass exchange between phase k and the other $N-1$ phases, $\overline{M}_{kl}$ the interphase momentum exchange term between phase k and phase l, and $\bar{S}_k$ a momentum source term of phase k due to phase changes and external forces other than gravity. For nonisothermal multiphase systems the transport equations (17) and (18) have to be supplemented with N thermal energy equations (one for each phase), whereas for multiphase systems involving chemical conversion of M species, in the most general case NM species conservation equations have to be added. The thermal energy equation and the conservation equation for species i present in phase k are respectively given by:

$$\frac{\partial}{\partial t}(\rho_k e_k) + (\nabla \cdot \rho_k e_k \bar{u}_k) = -p\left[\frac{\partial \epsilon_k}{\partial t} + (\nabla \cdot \epsilon_k \bar{u}_k)\right] - (\nabla \cdot \bar{q}_k)$$
$$+ \sum_{l=1}^{N} E_{kl} + S_{h,k}, \quad (19)$$

$$\frac{\partial}{\partial t}(\rho_k \omega_{k,i}) + (\nabla \cdot \rho_k \omega_{k,i} \bar{u}_k) = -(\nabla \cdot \bar{J}_{k,i}) + S_{k,i}, \quad (20)$$

where e_k represents the internal energy per unit mass of phase k, $\bar{q}_k$ the heat flux vector due to molecular transport in phase k, E_{kl} the interphase energy transfer term between phase k and phase l, $S_{h,k}$ the volumetric heat production rate in phase k, $\omega_{k,i}$ the mass fraction of species i in phase k, $J_{k,i}$ the mass flux vector of species i in phase k due to molecular transport, and $S_{k,i}$ the net production rate of species i in phase k per unit volume. For closure of the conservation equations expressions similar to Eqs. (9) and (10) are used but it should be kept in mind that the transport coefficients (conductivities and diffusivities) now represent effective transport coefficients that depend, amongst others, on the volume fractions of the N phases. It should be mentioned here that in many situations the M species are not present in all phases and in such circumstances (far) fewer conservation equations have to be formulated (and solved). Analogous to the situation for single-phase flows, the conservation equations (17)–(20) are also valid for the description of turbulent multiphase flows, but problems similar to those encountered in single-phase turbulent flows have to be circumvented. In this connection the strategy parallels the development presented in the previous section. However, due to the complexity of multiphase flows, the uncertainties introduced through the modeling of, for instance, the Reynolds stresses significantly increase. Additional difficulties arise due to the fact that closure equations for interphase transport of mass, momentum, and heat have also to be specified.

In multiphase systems involving one or more dispersed phases an alternative to the aforementioned *complete* continuum representation is possible by adopting a Lagrangian description for these phases. The advantages of this mixed Eulerian–Lagrangian approach are its greater generality and flexibility with respect to the incorporation of microscopic transport phenomena, whereas its relatively high (compared to completely Eulerian approaches) computational load constitutes its most important disadvantage. However, also from a computational point of view a mixed Eulerian–Lagrangian approach can offer certain advantages (see Section IV,B,1). If a Lagrangian description is adopted to represent the dispersed phase, for each individual particle (or bubble or droplet) an equation of motion is solved:

$$m_i \frac{d}{dt} \bar{v}_i = \sum \bar{F}_i, \quad (21)$$

where m_i, $\bar{v}_i$ represent, respectively, the mass and velocity of the ith particle and $\Sigma \bar{F}_i$ the sum of the forces acting on the ith particle. Forces due to gravity, drag, virtual mass, vorticity in the continuous phases, and electrical forces can be included in this term. The particle position vector is calculated from:

$$\frac{d}{dt}\bar{x}_i = \bar{v}_i. \tag{22}$$

The solution of differential equations (21) and (22) can be obtained with standard numerical integration techniques.

Depending on the volume fraction of the dispersed phase, one-way coupling or two-way coupling between the dispersed phase and the continuous phase prevails. In systems involving (turbulent) multiphase flow at very small volume fraction of the dispersed phase, say, smaller than 10^{-6}, one-way coupling may be assumed. At such low volume fractions the effect of the particles on the turbulence structure in the continuous phase is negligible while particle–particle interactions (i.e., collisions) do not play a role. For systems with higher volume fractions (10^{-6} to 10^{-3}) the turbulence structure of the continuous phase is influenced by the dispersed phase while particle–particle interaction can still be neglected and two-way coupling between the phases has to be accounted for. With respect to the effect of the dispersed phase on the turbulence structure it can be mentioned that the ratio of the particle response time τ_p and the Kolmogorov time scale τ_K determines whether the particles will enhance the production rate of turbulence energy ($\tau_p/\tau_K > 100$) or increase the dissipation rate of turbulence energy ($\tau_p/\tau_K < 100$). For still higher volume fractions of the dispersed phase particle–particle interaction (i.e., collisions) becomes important and four-way coupling has to be accounted for (see Fig. 3; from Elgobashi, 1991). In this case an integrated modeling approach is required combining features of molecular dynamics (MD) to deal effectively with the huge number of particle–particle and/or particle–wall collisions, and CFD to obtain the velocity distribution in the continuous phase (see Section VI).

B. Numerical Techniques

We will not attempt here to give a detailed explanation of the numerical aspects (fundamentals of discretization, error estimates, and error control) of CFD since a number of excellent texts are available in the literature that deal in depth with this matter (Fletcher, 1988a,b; Hirsch, 1988, 1990). First some general aspects of the numerical techniques used for solving fluid flow problems are discussed and, subsequently, a distinction is made between single-phase flows and

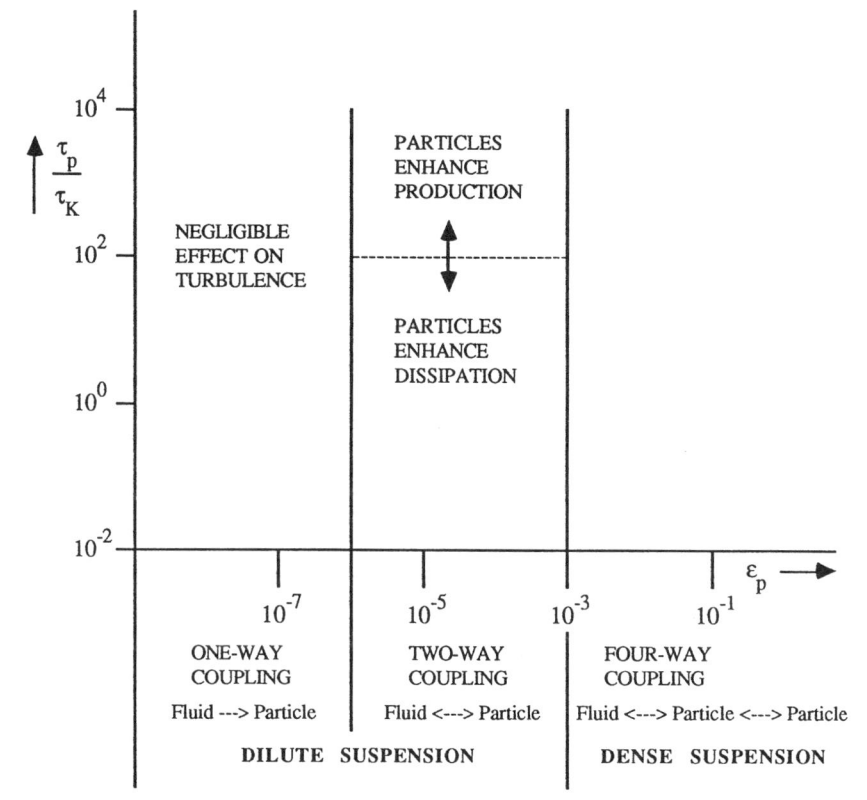

FIG. 3. Map of flow regimes in particle-laden flows. (From Levenspiel, O., "Chemical Reaction Engineering." John Wiley & Sons, New York, 1962. Reprinted by permission of John Wiley & Sons, Inc.)

multiphase flows. In general the solution of a fluid flow problem involves the following steps:

1. grid generation with the aid of:

- Algebraic methods (Wang and Hoffman, 1986; Marcum and Hoffman, 1988; Hoffman, 1992)
- Conformal mapping (Thompson et al., 1982, 1985; Thompson, 1982)
- Systems of (elliptic) partial differential equations (Thompson et al., 1982, 1985; Thompson, 1982)

Numerical grid generation can be seen as a highly specialized branch of CFD. The grids generated by either of the above-mentioned methods can be divided into structured grids and unstructured grids (see Fig. 4). In a structured grid the connectivity is constant throughout the interior computational domain, whereas unstructured grids do not necessarily have a constant connectivity. In general, un-

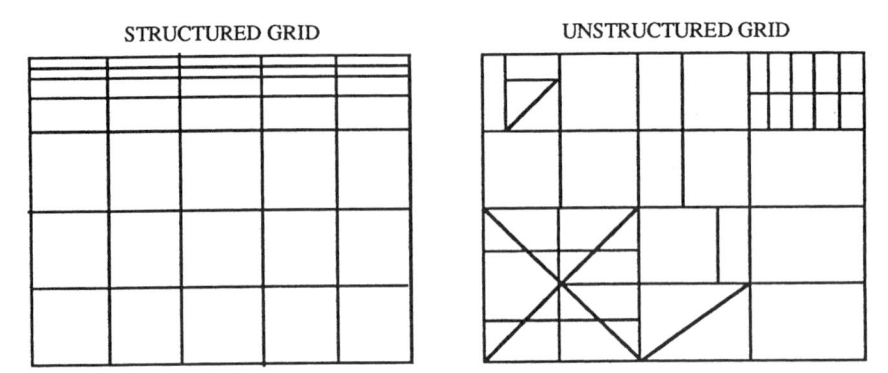

FIG. 4. Structured and unstructured grids.

structured grids offer better possibilities to represent systems with complex geometric features. Furthermore, staggered and nonstaggered (collocated) computational grids can be distinguished (see Fig. 5).

2. Discrete representation of the transport equations using:

- Finite difference methods (FDM) (Roache, 1972)
- Finite volume methods (FVM) (Patankar, 1980)

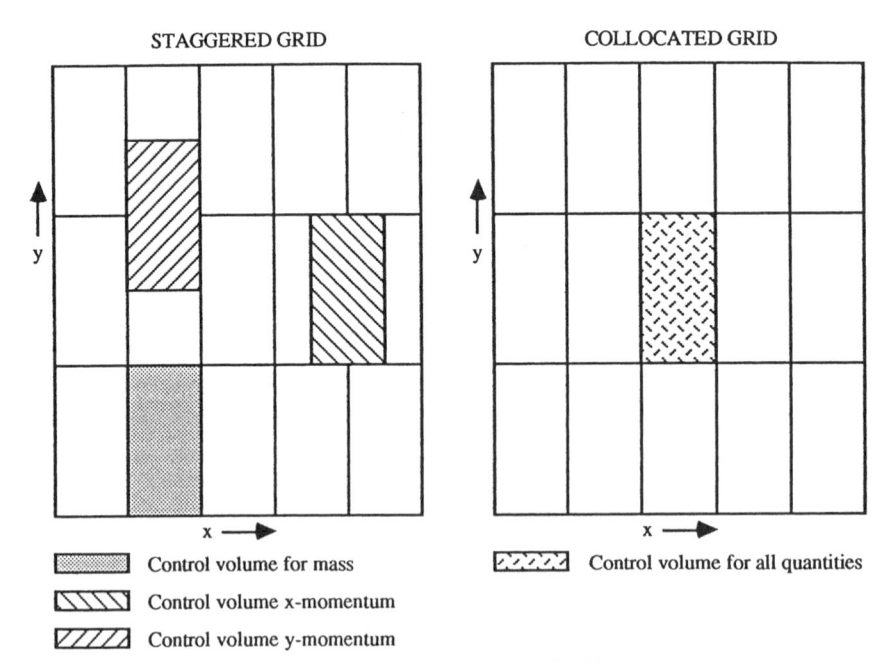

FIG. 5. Staggered and collocated grids.

- Finite element methods (FEM) (Baker, 1985; Zienkiewicz and Taylor, 1989a,b).
- Boundary element methods (BEM) (Becker, 1992)

The solution of the resulting nonlinear equations is usually achieved via an iterative algorithm. Once a converged solution has been obtained it is essential to assess the invariance of the computational results with respect to the temporal and/or spatial discretisation. This aspect is unfortunately often not addressed in computational studies due to, amongst others, computer time constraints.

3. Postprocessing:

- Contour and vector plots
- Data reduction
- Animations

Postprocessing constitutes a very important final step, especially for the chemical engineer because in many chemical processes besides fluid phenomena many other aspects have to be considered (i.e., fouling, catalyst deactivation).

1. Single-Phase Systems

It should be mentioned here that the discrete representation of the transport equations can be based on the so-called primary variables such as pressure and velocity components but also on the stream function vorticity formulation (Roache, 1972). The SIMPLE algorithm (Patankar, 1980) is probably the most widely employed algorithm for the solution of fluid flow problems and forms the basis of many commercial FVM-based codes (PHOENICS, FLUENT, and FLOW3D). The SIMPLE algorithm is based on a sequential solution method of the nonlinear equations, which is advantageous from a modeling point of view since additional transport equations can be added with relative ease. This solution method can cause problems in case a (very) strong coupling between chemistry and heat transfer processes prevails. Finite element-based CFD codes are mostly based on the method of weighted residuals (Baker, 1985; Zienkiewicz and Taylor, 1989a,b) and use a simultaneous solution procedure for the transport equations where typically the multidimensional Newton–Raphson method is used.

A totally different approach to solving the Navier–Stokes equations is made in alternating direction implicit (ADI) methods (Briley and McDonald, 1975) and approximate factorization implicit (AFI) methods (Beam and Warming, 1977, 1978). These methods apply an approximate spatial factorization technique to avoid the inversion of huge banded matrices and are computationally very efficient.

Another class of powerful numerical techniques that should be mentioned here are spectral and pseudospectral methods, which are often used in DNS because of their superior accuracy (relative to finite difference methods) in case the same resolution is used. At the same time these techniques are competitive in computational efficiency. For a detailed description of many spectral algorithms and exhaustive discussion of the theoretical aspects of these numerical methods, the interested reader is referred to Canuto *et al.* (1988).

An important class of fluid flow problems involves free surface flows where only the flow phenomena in a single phase are of importance despite the fact that two or more phases are present in the system. Such situations are encountered, for example, in gas–liquid two-phase flows where, due to the large density difference, from a hydrodynamic point of view the flow can be treated as a liquid flow in a vacuum. The analysis of such free surface flows is rather complex due to the fact that in addition to the fluid flow problem the position of the interface and enforcement of appropriate boundary conditions have to be dealt with.

A number of computational techniques have been proposed in the past such as the marker and cell (MAC) method and the simplified marker and cell (SMAC) method developed, respectively, by Welch *et al.* (1965) and Amsden and Harlow (1970). In these methods a fixed computational mesh is employed through which the fluid moves; in addition a large number of massless marker particles are used to track the interface dynamically (see Fig. 6). The movement of these marker particles is performed on basis of the local fluid flow field where area (2D) or volume (3D) weighting is applied with respect to the velocity components, which are available only at the Eulerian nodes of the computational mesh. Especially for 3D simulations the storage requirements for the marker particle based methods can become problematic and therefore more efficient (from the point of view of storage requirements) free surface computational algorithms have been devised. An interesting development in this connection is the (single-material) volume of fluid

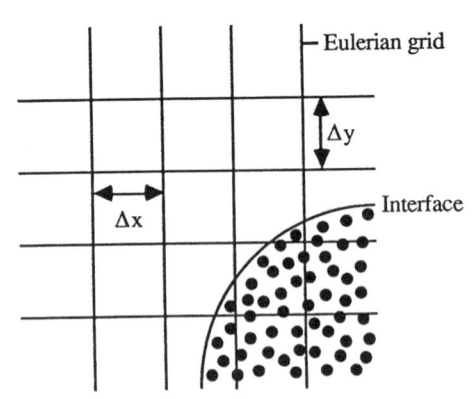

FIG. 6. Marker particles in an Eulerian grid. The marker particles are used to track the interface dynamically.

(VOF) method (Hirt and Nichols, 1981; Nichols et al., 1980) where in addition to the Navier–Stokes equations a scalar transport equation (the so-called F equation) for the fractional amount of fluid F is solved for:

$$\frac{\partial F}{\partial t} + (\nabla \cdot \bar{u}F) = 0, \qquad (23)$$

where $\bar{u}$ represents the local fluid velocity. The solution of the F equation yields the F values at all computational cells of the domain, and with the aid of an interface reconstruction algorithm the position and orientation of the interface can be obtained. With the VOF method it is possible to incorporate, for example, surface tension via the condition for the normal stress at the interface (Hirt and Nichols, 1981) or via the continuum surface force (CSF) model (Kothe et al., 1991).

Recently Rudman (1997) has compared the performance of various volume-tracking methods for interfacial flow calculations (including the VOF method of Hirt and Nichols, 1981) and concluded that the method originally proposed by Youngs (1982) provides the best results (i.e., conservation of a sharp and well-defined interface) on various "artificial" problems including solid body rotation and shearing flow. In addition this method yielded the best results for the numerical simulation of Rayleigh–Taylor instability. Nakayama and Mori (1996) combined a marker particle approach and a finite element method (FEM) based solution methodology to model time-dependent free surface flows in complex geometries. Udaykumar et al. (1996) presented a mixed Eulerian–Lagrangian method for fluid flows with complex and moving boundaries. They also used marker particles to represent and track the interface in the Eulerian grid. A special feature of their computational approach is the possibility to embed solidification or melting processes, with the associated dynamic evolution of the phase boundaries, in the flow simulation.

2. Multiphase Systems

For multiphase systems a rough distinction can be made between systems with separated flows and those with dispersed flows. This classification is not only important from a physical point of view but also from a computational perspective since for each class different computational approaches are required. For multiphase systems involving multiphase flow both Eulerian, mixed Eulerian–Lagrangian, and two-material free surface methods can be used. An excellent review on models and numerical methods for multiphase flow has been presented by Stewart and Wendroff (1984). A similar review with emphasis on dilute gas-particle flows has been presented by Crowe (1982).

a. Eulerian Methods. The development of Eulerian computational methods for multiphase flows was pioneered by Harlow and Amsden (1974, 1975) at the Los

Alamos scientific laboratory. Their implicit continuous Eulerian (ICE) method formed the basis for many (Rivard and Torrey, 1977, 1979; Cook et al., 1981) later developments.

b. Eulerian-Lagrangian Methods. For dispersed multiphase flow roughly speaking three different situations and corresponding computational strategies can be distinguished (also see Fig. 7):

1. Dilute flows where on the average less than one particle is present in a computational cell
2. Dense flows where a relatively high number of particles are present in a computational cell
3. Dilute or dense flows where a large number of computational cells is contained in a single particle

We should mention here that the dispersed phase could also consists of drops or bubbles, which could, in principle, be deformable.

The situation for case 1 arises when suspensions are relatively dilute and the particles are small. Depending on the exact value of the volume fraction of the dispersed phase, one-way coupling or two-way coupling prevails (Pan and Banerjee, 1996a). With one-way coupling particles are being moved in response to the fluid motion without feeding back effects to the continuous phase, whereas in two-way coupling feedback effects are taken into account. As discussed in detail by Pan and Banerjee (1996a) care must be taken to implement two-way coupling correctly. Examples of one-way coupled calculations are those of Squires and Eaton (1990) and Pedionotti *et al.* (1992, 1993) who both used DNS to study, respectively, particle segregation in homogeneous isotropic turbulent flow and particle segregation in a turbulent channel flow. For two-way coupling, different computational strategies are used depending on whether there is only one particle in a computational cell or many.

In case 2 we deal with dense flows and in our computational strategy four-way coupling has to be accounted for since there is not only mutual interaction between the suspended particles and the continuous phase but also particle–particle interactions (i.e., collisions).

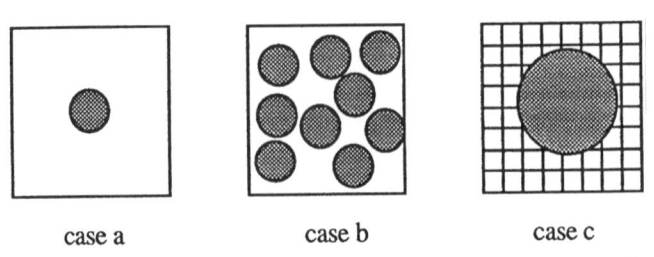

FIG. 7. Different situations that can be distinguished in modeling dispersed multiphase flow.

In case 3 the relative size of the particles (with respect to the computational cells) is large enough that they contain many hundreds or even thousands of computational cells. It should be noted that the geometry of the particles is not exactly represented by the computational mesh and special, approximate techniques (i.e., body force methods) have to be used to satisfy the appropriate boundary conditions for the continuous phase at the particle surface (see Pan and Banerjee, 1996b). Despite this approximate method, the empirically known dependence of the drag coefficient versus Reynolds number for an isolated sphere could be correctly reproduced using the body force method. Although these computations are at present limited to a relatively low number of particles they clearly have their utility because they can provide detailed information on fluid–particle interaction phenomena (i.e., wake interactions) in turbulent flows.

c. Two-Material Free Surface Methods. The MAC or SMAC and VOF techniques can also be applied to study two-material or two-phase free surface flows. For details the interested reader is respectively referred to Welch *et al.* (1965) or Amsden and Harlow (1970), Hirt and Nichols (1981), and Nichols *et al.* (1980).

C. Existing Software Packages

Because of the mathematical and numerical difficulties involved in the development of a CFD package this activity has been undertaken mainly by specialists where the user is kept away from the details of the solution procedures. Nowadays there exist many user-friendly CFD packages which allow the engineer to set up and solve complex fluid flow problems with relative ease. First we would like to make a few comments on the selection of commercially available CFD packages and the aspects which should be kept in mind in this connection:

1. *User requirements.* Does the user merely want to use a CFD package in its existing form to study fluid flow and related phenomena or does the user intend to modify and further develop the physical models incorporated in this package? For the second class of users it would be very beneficial to have access to the source code of the CFD package. In this last respect it is worrying that the claimed future trend is toward fewer and fewer CFD codes being available in source code.
2. *Geometry.* Do we have to deal with relative simple geometries or very complex ones? In equipment encountered in many industrial processes we have to deal with complex geometries and in such situations body fitted coordinates (BFCs) may be used to fit the contours of the volume in question. To provide for greater flexibility with respect to distribution of grid points in regions of the domain (possibly adaptive) where steep gradients are

expected, nonstructured grids might be advantageous. Both CFD codes based on finite volume and finite element formulations are suitable for analysis of fluid flow in complex geometries.

3. *Dimensions.* The number of relevant dimensions for the class of fluid flow problems to be studied should be identified. Although in many cases it will be necessary to solve the full three-dimensional transport equations it is often wise to first perform preliminary two-dimensional calculations in order to gain some computational experience for the system of interest.

4. *Preprocessing.* Especially for analysis of fluid flow phenomena in complex geometries, the availability of efficient and easy-to-use preprocessors for both problem definition and grid generation is of crucial importance. Fortunately, most commercially available CFD packages meet this requirement.

5. *Numerical method.* The choice of the proper numerical solution procedure should be left to specialists. Even when a particular CFD package has been chosen, the user can usually choose different solution strategies for the linearized equations (point or line relaxation techniques versus whole field solution techniques) and associated values of the relaxation parameters. The proper choice of relaxation factors to obtain converged solutions (at all) within reasonable CPU constraints is a matter of experience where cooperation between the engineer and the specialist is required. This is especially true for new classes of fluid flow problems where previous experience is nonexistent.

6. *Turbulence models.* Do we have to consider turbulent flow (in complex) geometries or not? Especially in systems with complex geometries where turbulent flows prevail, simple turbulence models may fail and in such circumstances advanced turbulence models (ASM or RSM) are required.

7. *Multiphase capabilities.* For CFD analysis of multiphase flows using commercially available CFD packages it is very important that the CFD package provide for an open programming environment where the user can implement his or her own physical submodels.

8. *Postprocessing.* Usually the direct result of a simulation consists of large arrays of numbers that have to be converted somehow into more appealing information. Fortunately, most CFD packages incorporate various postprocessing facilities to convert the basic data into beautiful color plots which show the spatial distribution of key variables. Despite these postprocessing facilities it is very important that some kind of further data reduction can be achieved which enables the engineer to present the computational results in an efficient and compact manner. For multiphase systems, where the flow is often nonstationary, it is furthermore of importance that the simulation results can be visualized dynamically, that is, with computer-generated movies.

The existing software packages can be divided into two broad categories, namely, commercially available general-purpose codes such as PHOENICS,

FLUENT, FLOW3D, ASTEC, and FIDAP, and codes that have been developed at universities and industrial laboratories for more specific applications:

1. FLUFIX, for modeling fluidized bed combusters (Chang et al., 1989), at the Argonne National Laboratory (USA)
2. KIVA I and II, for analysis of internal combustion engines (Amsden, 1985, 1989), at the Los Alamos National Laboratory (USA)
3. MELODIF and ESTET-ASTRID, for various applications in the field of energy production, at Electricité de France (Simonin, 1996)

The development of the modeling capability of commercially available codes such as PHOENICS, FLUENT, and FLOW3D has proceeded rapidly and it is to be expected that this development will continue in the near future. It is interesting to note that some vendors of general-purpose CFD packages have recently started to offer specialized versions of their codes for certain applications: POLYFLOW for CFD analysis of polymer materials processing applications (i.e., extruders), NEKTON for CFD analysis of flow phenomena in thin-film coating processes, and RAMPANT for CFD analysis of complex (external) flows. This section concludes with the specification of some of the current modeling capabilities embedded in these codes:

1. Steady-state or transient two-dimensional (2D) and three-dimensional (3D) flows in standard geometries involving Cartesian, cylindrical, or spherical coordinates and complex geometries involving BFCs with adaptive grids
2. Laminar flows involving Newtonian and non-Newtonian fluids
3. Turbulent flows with simple closure models (eddy viscosity, mixing length, k-ϵ) or complex closure models (ASM, RSM, RNG) for the Reynolds stresses
4. Compressible and incompressible flows
5. Flows involving mixing and/or chemical reaction
6. Flows in porous media
7. Multiphase flows using Eulerian multifluid approaches
8. Multiphase flows involving dispersed phases (particles, droplets or bubbles) using mixed Eulerian–Lagrangian approaches both with one-way and two-way coupling
9. Flows involving radiative heat transfer
10. Complex boundary conditions

IV. Application to Chemical Reaction Engineering

As stated in the introductory section, applications of CFD may be divided into broad categories, namely, those involving single-phase systems and those involving multiphase systems. The motivation for this distinction is due to (1) large differences in degree of complexity of physical description and (2) large differences in numerical solution strategies.

A. SINGLE-PHASE SYSTEMS

Within single-phase systems a further distinction can be made between systems involving (1) laminar flows, (2) turbulent flows, (3) flows with complex rheology, and (4) fast chemical reactions. Of course certain systems exist that fall into several of these classes and therefore also possess their main characteristics. These types of systems are discussed in more detail.

1. Laminar Flows

In many practical applications the chemical engineer has to deal with flows in complex geometries of which flows in curved pipes or channels and bends and flows around bodies with various shapes (i.e., cylinders and spheres) positioned inside ducts are well-known examples. As far as laminar flow systems are concerned, accurate flow field prediction, even in systems with great geometrical complexity, can be achieved nowadays with the aid of CFD and probably the accuracy of the simulations exceeds that of experiments. For systems with great geometrical complexity a number of computational approaches are possible such as finite difference or finite volume methods that invoke curvilinear coordinates (Thompson *et al.*, 1982) or finite element methods (Baker, 1985; Zienkiewicz and Taylor, 1989a,b). In the excellent books written by Baker (1985) and Zienkiewicz and Taylor (1989a,b) and the references cited therein numerous applications of (laminar) flows involving complex geometries (ducts with complex shape and/or immersed bodies, stirred vessels, etc.) can be found. A few illustrative examples of laminar flow around bodies present inside ducts and laminar flow in stirred vessels are highlighted.

a. Laminar Flow in Ducts. An example of the application of curvilinear coordinates can be found in the paper by Wang and Andrews (1995) who presented simulation results for laminar flow of a Newtonian liquid in a helical duct. They used a helical coordinate system to describe the geometry of the flow domain and used a modified MAC method (Welch *et al.*, 1965) to solve the two-dimensional incompressible Navier–Stokes equations. Specific features of the flow in this geometry are a relatively high frictional flow resistance and the occurrence of a so-called secondary flow pattern. In Fig. 8a the helical coordinate system is shown, whereas in Fig. 8b the effect of the pressure gradient in the circumferential direction ($\partial \phi / \partial \theta$) on the structure of the secondary flow is shown in terms of the streamline pattern. Note that with increasing circumferential pressure gradient the

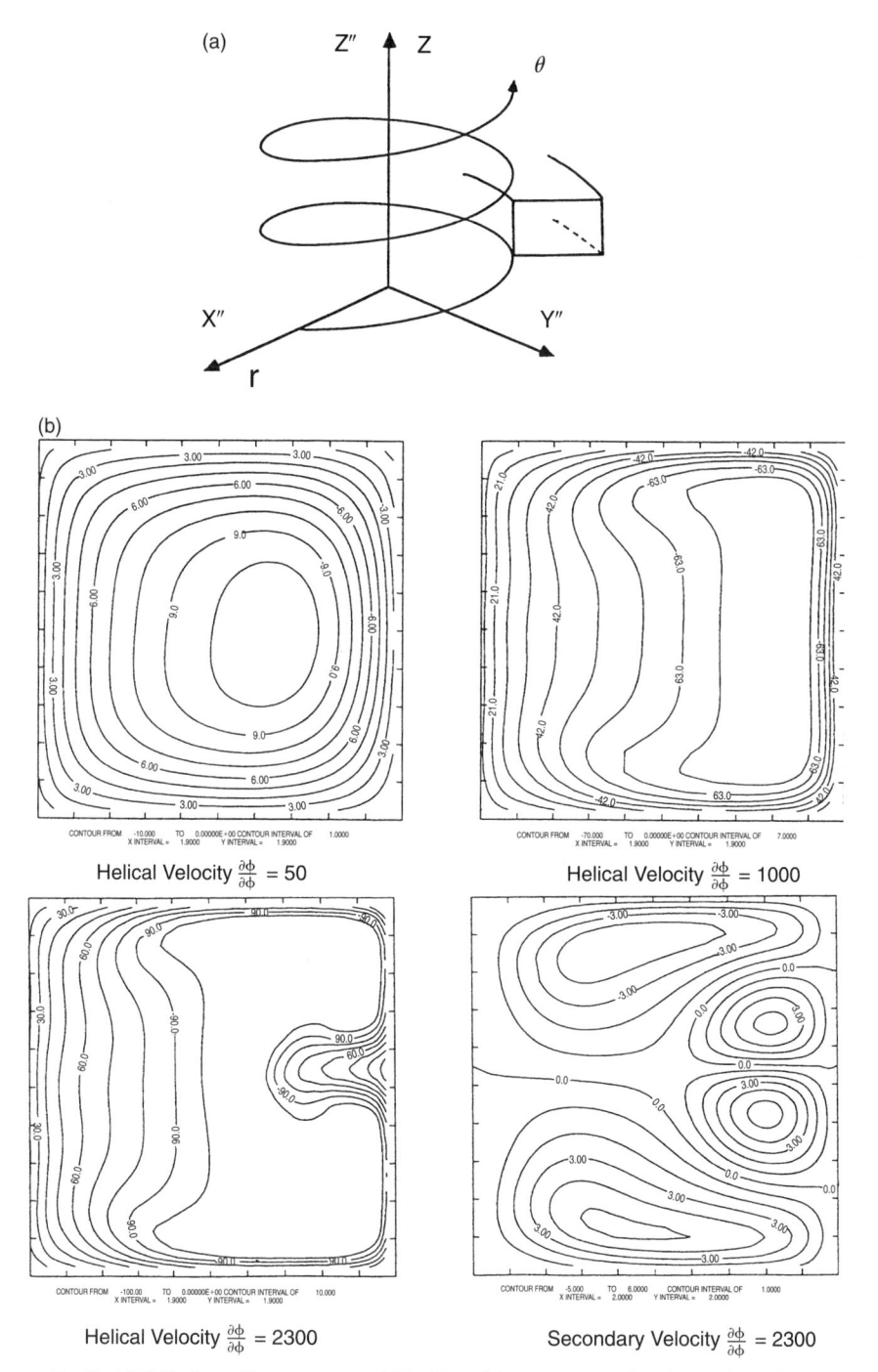

FIG. 8. (a) Helical coordinate system and (b) effect of the pressure gradient in the circumferential direction ($\partial\phi/\partial\theta$) on the secondary flow structure. [From Wang, J.W., and Andrews, J.R.G., Numerical simulation of flow in helical ducts. *AIChE J.* **41**(5), 1071 (1995). Reproduced with permission of the American Institute of Chemical Engineers. Copyright © 1995 AIChE. All rights reserved.]

secondary flow changes significantly. These calculations are highly relevant to tubular reactors where the "tube" has been wrapped in the form of a coil due to space limitations. The occurrence of secondary flow can also be used to promote transport of mass and/or heat to the tube wall and in this case it is important that a reliable prediction of the associated enhancement factors can be made.

Further examples of recent applications of the finite element method can be found in Targett *et al.* (1995) who studied flow through curved rectangular channels of large aspect ratio using the FEM-based FIDAP code. They also compared their computational results with experimental data obtained from visualization experiments and found good agreement between theory and experiment.

b. Laminar Flow around Bluff Bodies Inserted in Channels. In the literature numerous computational studies can be found that deal with the flow around bluff bodies. Anagnostopoulos and Iliadis (1996) used a FEM to study the vortex shedding behind a cylinder positioned in a channel, whereas Xu and Michaelides (1996) performed a numerical study of the flow over an ellipsoidal object inside a cylindrical tube. The extension of such calculations to systems that involve arrays of cylinders or "tube bundles" as encountered in many heat transfer applications is in principle possible but they are very CPU intensive. Figure 9 shows as an illustration of the Karman vortex street formed behind a cylinder (diameter d) positioned in a channel (width $2h$) (Anagnostopoulos and Iliadis, 1996) at a Reynolds number of 106. From the pattern of the solid lines (equivorticity lines) shown in this figure the vortical structures behind the cylinder can be clearly recognized. In addition, the increasing influence of the channel walls on the vortical structures with increasing d/h ratio can be inferred from Fig. 9.

c. Laminar Flow in Stirred Vessels. The blending or mixing of highly viscous (miscible) liquids is encountered in a variety of industrial operations and has been studied computationally by, among others, Abid *et al.* (1992). They studied the effect of agitator geometry on the flow patterns generated in a stirred vessel containing a very viscous liquid and performed both 2D and 3D numerical simulations. Especially in the case of very viscous fluids, which are often not transparent and therefore not accessible to advanced experimental techniques such as laser Doppler anemometry (LDA), a CFD approach can be very beneficial to generate useful data, provided that some kind of experimental validation of the computational strategy has been undertaken. This is especially important for these types of flows where (considerable) simplifications of the agitator geometry are often made in the computations.

d. Concluding Remarks. The simulation of processes involving 3D laminar fluid flow and associated phenomena such as mass and heat transport and chemical

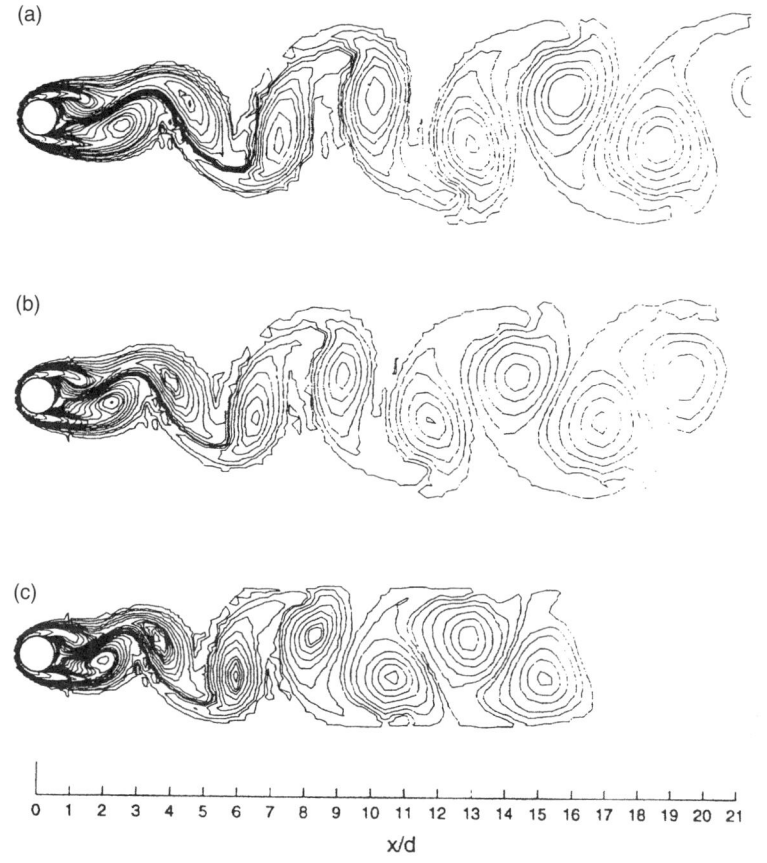

FIG. 9. Karman vortex street formed behind a cylinder (diameter d) positioned in a channel (width $2h$) at a Reynolds number of 106. [From Anagnostopoulos, P., and Iliadis, G. Numerical study of the blockage effect on viscous flow past a circular cylinder. *Int. J. Num. Methods Fluids* **22**, 1061 (1996). Copyright John Wiley & Sons Limited. Reproduced with permission.]

transformation can nowadays be performed with great accuracy. The main difficulties that can emerge here are due to (1) the use of insufficiently refined computational grids to resolve all relevant details of the flow, (2) the accurate treatment of moving boundaries as encountered in agitated vessels, and (3) the use of inappropriate numerical methods which cause severe numerical diffusion.

2. *Turbulent Flows*

For systems involving turbulent flows accurate modeling is much more complicated in comparison with those involving laminar flow and here it is considered very important to interpret and use the computational results with great care, especially when one deals with turbulent flows in complex geometries. Whenever possible a (limited) comparison with experimental data should be made to assess the correctness of the computations. This state of the art is due to the fact that our understanding of and our capability to predict turbulent fluid flow phenomena is unfortunately still limited as pointed out by Banerjee (1991) in a very lucid manner. Other reviews on the modeling and numerical simulation of turbulent flows can be found in Rogallo and Moin (1984), Launder (1991), and Reynolds (1991).

In studies that involve the CFD analysis of turbulent fluid flow, the k-ϵ model is most frequently used because it offers the best compromise between width of application and computational economy (Launder, 1991). Despite its widespread popularity the k-ϵ model, if used to generate an isotropic turbulent viscosity, is inappropriate for simulation of turbulent swirling flows as encountered in process equipment such as cyclones and hydrocyclones (Hargreaves and Silvester, 1990) and more advanced turbulence models such as the ASM or the RSM should be considered. Because these models are computationally much more demanding and involve an increased number of empirical parameters compared to the k-ϵ model, other strategies have been worked out (Boysan *et al.*, 1982; Hargreaves and Silvester, 1990) to avoid the isotropic nature of the classical k-ϵ model.

Due to advances in computer technology and numerical solution techniques both DNS and LES have become feasible. Although DNS is still limited to relatively low Reynolds numbers, it is a very powerful tool for understanding turbulent flow structure and a tool to generate databases of turbulent flows which can be used, for example, to validate conventional turbulence models such as k-ϵ or Reynolds stress models. A good example in this respect is the study conducted by Hrenya *et al.* (1995) who tested 10 turbulence models against DNS results and experimental data. Hrenya *et al.* found that the model proposed by Myong and Kasagi (1990) showed the best overall performance in predicting fully developed, turbulent pipe flow. Eggels (1994) performed both DNS and LES simulations of turbulent pipe flow and also made a detailed comparison with experimental data obtained with a digital particle image velocimetry (DPIV) technique developed by Westerweel (1993).

The extension of this approach to higher Reynolds numbers and more complex geometries is very important and deserves further attention in the future but depends critically on the future advances in computer hardware and numerical solution methods. As far as LES or partially resolved simulations are concerned, similar limitations exist although these type of simulations can be carried to (much) higher Reynolds numbers. DNS can also be used to obtain insight in turbulence producing mechanisms as shown in the study of Lyons *et al.* (1989).

A flow configuration of particular interest to the chemical engineer is the (baffled) stirred vessel and very significant efforts have been documented in the literature (Ranade and Joshi, 1990a,b), Kresta and Wood, 1991) to compute the flow patterns inside the vessel using CFD models. Here models with a varying degree of sophistication both from a physical and numerical point of view have been used. In most studies the k-ϵ model was used with standard values for the turbulence parameters to compute the turbulent flow patterns despite the fact that these parameter values were obtained for (much) simpler geometries. Another principal difficulty in modeling turbulent flows in stirred vessels is the accurate description of the impeller geometry, whereas the resolution of the vortical structures behind the baffles is also problematic due to the coarse grids which are commonly employed in the simulations. A possible approach to describe the impeller geometry in more detail is the application of so-called sliding meshes where two meshes are applied: one mesh for the central impeller region, which rotates with the speed of the impeller, and another one for the exterior region, which is fixed in space. An alternative approach is the use of a single moving deforming mesh as reported by Perng and Murthy (1992). Here the grid is attached to the impeller and moves with it and as a consequence mesh deformation occurs.

CFD has also been applied to analyze the flow patterns in a special countercurrent solvent extraction column (Angelov *et al.*, 1990). They used a single-phase flow representation and a k-ϵ turbulence model to compute the flow patterns in a periodic structure of the column. Validation of the computational results was achieved by applying LDA to obtain experimental data on the velocity profiles. CFD is a very useful tool here because the optimization of the performance of the extraction column from a geometrical point of view can be achieved with relative ease in comparison with a pure empirical strategy.

In the last decade very significant progress has been made in modeling turbulent fluid flow. There remain, however, very significant problems of which, in addition to the problems mentioned in the previous section, we would like to mention the following problem areas: (1) availability of accurate turbulence models which can be used with confidence in complex geometries while at the same time the computational cost should be acceptable and (2) availability of DNS-generated databases to validate semiempirical turbulence models.

3. Complex Rheology

In several industrial processes the engineer has to deal with non-Newtonian fluids of which food processing and production processes involving glues, colors, polymer solutions, and pure polymers are well-known examples.

The simulation of non-Newtonian fluid flow is significantly more complex in comparison with the simulation of Newtonian fluid flow due to the possible occurrence of sharp stress gradients which necessitates the use of (local) mesh refinement techniques. Also the coupling between momentum and constitutive equations makes the problem extremely stiff and often time-dependent calculations have to be performed due to memory effects and also due to the possible occurrence of bifurcations. These requirements explain the existence of specialized (often FEM-based) CFD packages for non-Newtonian flow such as POLYFLOW.

Both FVM and FEM based packages have been applied successfully to study non-Newtonian fluid flow.

The progress and challenges in computational rheology have been recently reviewed by Keunings (1990). In literature viscoelastic liquids have received particular attention (Crochet, 1987; Boger, 1987; Keunings, 1990; Yoo and Na, 1991; Van Kemenade and Deville, 1994; Mompean and Deville, 1996) due to the fact that many industrially relevant non-Newtonian materials such as polymer solutions and polymers exhibit viscoelastic behavior. In addition the strong interplay between modeling and mathematical and numerical considerations makes this particular field fascinating but also very complex, which requires good cooperation among rheologists, mathematicians, and CFD experts.

Boger (1987) has reviewed viscoelastic flows through contractions and has pointed out that three-dimensional and time-dependence characteristics have to be taken into account for flows at high Weissenberg number We defined by:

$$We = \frac{\lambda V}{L}, \qquad (24)$$

where λ is the relaxation time of the fluid and L/V the characteristic inertial time of the flow based on a reference length L and reference velocity V.

Examples of CFD applications involving non-Newtonian flow can be found, for example, in papers by Keunings and Crochet (1984), Van Kemenade and Deville (1994), and Mompean and Deville (1996). Van Kemenade and Deville used a spectral FEM and experienced severe numerical problems at high values of the Weissenberg number. In a later study Mompean and Deville (1996) could surmount these numerical difficulties by using a semi-implicit finite volume method.

Similar to the role that DNS and discrete particle models (see Section IV,B,3) might play in the development of improved turbulence models, which can be used in engineering applications, and closure laws for gas–solid continuum models, Brownian dynamics (BD) should be mentioned as a powerful tool to develop closure models for non-Newtonian fluids (Brady and Bossis, 1988).

The CFD-analysis of non-Newtonian flow has made progress in the last decade but the extension of the calculations to other types of non-Newtonian fluid flow and more complex geometries is highly desirable from the perspective of the chemical engineer.

4. Mixing and Chemically Reactive Flows

By the very nature of the profession, the chemical engineer has to deal very frequently with chemically reactive flows in various types of single-phase and multiphase reactors. Before the advent of CFD he or she typically had to use highly idealized and approximate solution strategies supplemented with empirical information to obtain solutions for practical problems. A very well-known approach in this connection is the combined use of macromixing and micromixing models to predict the performance (i.e., conversion and/or selectivity) of a chemical reactor. Here, with the aid of stimulus–response experimental techniques, information on the residence time distribution (RTD) is obtained that is subsequently used to devise a macromixing model whereas a postulated micromixing model is invoked to account for the finite rate with which segregated fluid parcels eventually achieve mixing at the molecular level (i.e., segregation decay) to permit the occurrence of a chemical transformation. During the last 30 years a host of micromixing models has been proposed in the literature of which the coalescence dispersion (CD) model (Curl, 1963; Spielman and Levenspiel, 1965), the interaction by exchange with the mean model (IEM) (Villermaux and Devillon, 1972), and the general micromixing model (GMM) (Villermaux and Falk, 1994) should be mentioned as typical examples. For further references in this area, the interested reader is referred to the reviews by Ottino (1994), Villermaux and Falk (1996), and Fox (1996). It should also be mentioned here that very important contributions toward a better understanding and description of chemically reactive flows in general and related CFD work in particular have been made by the mechanical engineering community with combustion research as an important driver. For a review from this perspective we refer to Correa and Shyy (1987).

Until 10 to 15 years ago the combined approach of macromixing and micromixing models was very widely used in the field of CRE but gradually CFD-based strategies have replaced the first mentioned strategy. In this respect it should be noted that this change also introduced big conceptual differences because the traditional CRE approach is usually formulated in the "age" space of fluid parcels whereas in CFD approaches a Eulerian framework is often adopted. Subsequently a brief overview of CFD-based approaches for reacting flows is presented and the current limitations are also indicated.

a. Direct Numerical Simulation. Direct numerical simulation (DNS) has also been applied to chemically reactive flows (Givi and McMurtry, 1988; Leonard and Hill, 1988; McMurtry and Givi, 1989) but due to current limitations in computer capacity it is very difficult to apply DNS at high Reynolds and Schmidt numbers. Since such situations are frequently encountered in chemical process engineering where in addition the geometry is often (very) complex, DNS is not considered a practical tool for the simulation of industrial reactors. However, it is a very interesting research tool to study chemistry–turbulence interactions and to generate databases which can be used for the verification of mixing closure laws used in process engineering models.

b. Large Eddy Simulation. Although LES can be applied at (much) higher Reynolds numbers and has the potential to predict accurately time-dependent macromixing patterns in process equipment with complex geometries (i.e., stirred tanks), its principal weakness lies in the fact that turbulence–chemistry interaction at the subgrid scale (SGS) (i.e., the scale smaller than the smallest resolved scale) has to be modeled. These SGS models themselves are computationally demanding, which hampers the combined use with LES, which is also computationally intensive.

c. Moment Methods. Turbulent reacting flows are frequently encountered in the chemical process industry and very often (i.e., in most commercial CFD codes) the theoretical description of these flows is based on the Reynolds averaged Navier–Stokes (RANS) equations (Brodkey and Lawelle, 1985; Launder, 1991). The consequence of this approach is the necessity to specify closure laws for the Reynolds stresses and fluxes of scalar quantities (i.e., mass or molar fluxes) and most importantly the average reaction rate. This last term is the most difficult term due to its nonlinearity, and its treatment depends on the magnitude of the Damköhler number Da (Fox, 1996). For chemically reactive flows a distinction should be made between three important classes, which can be classified in terms of the Damköhler number Da $= t_m/t_r$ where t_m and t_r, respectively, represent a characteristic time scale for turbulent micromixing and a characteristic time scale for chemical transformation:

- *Da* $\ll$ *1:* flows where the time scale for chemical conversion t_r is relatively large compared to the time scale for turbulent micromixing t_m
- *Da* $\gg$ *1:* flows where the time scale for chemical conversion t_r is relatively small compared to the time scale for turbulent micromixing t_m
- *Da* $\approx$ *1:* flows where the time scale for chemical conversion t_r is comparable in magnitude to the time scale for turbulent micromixing t_m.

For the first two cases Da $\ll$ 1 (slow reactions) and Da $\gg$ 1 (very fast reactions) adequate closure models are available in many commercial CFD codes. For the third case, where the time scale for chemical conversion approximately equals the time scale for turbulent micromixing, moment methods are inappropriate and other methods should be used. In this situation the reactor performance may be significantly affected by mixing efficiency. Here the engineer is faced with the difficult problem of predicting the overall conversion and/or selectivity of the chemical process. In the last three decades this problem has received considerable attention in three scientific areas, namely, chemical reaction engineering, fluid mechanics and combustion, and various approaches have been followed.

d. Micromixing Models. Micromixing models offer the advantage that the chemical reaction rate expression is treated in an exact manner but they suffer from the fact that highly idealized descriptions of the fluid dynamics (backed up by empirical information) are used. This latter point certainly constitutes a major weakness of micromixing models, however, by incorporating micromixing models in RANS models this disadvantage can be partly overcome. For complex inhomogeneous flows this approach again introduces significant difficulties and recourse should be made to methods that treat the chemistry and turbulence and their mutual interaction on a more fundamental basis.

e. Probability Density Function Methods. Probability density function (PDF) methods combine the strong points of exact treatment of the chemical reaction term and second-order turbulence closure (Fox, 1996). PDF methods make use of a probability density distribution function f which can take the form of a *presumed* function or a transport equation which has to be *solved* simultaneously with the fluid flow equations. Once the function f is known, the expected value $<c>$ of an arbitrary function c (including, for example, the chemical reaction rate) can be computed as follows:

$$<c> = \int cf d\overline{x}, \quad (25)$$

where the integration has to be performed over all possible values of all elements constituting the vector $\overline{x}$. This vector usually consists of the three velocity components and all concentrations (or mass fractions, etc.) of the species involved in the chemical transformation, where all quantities are evaluated at a fixed point. Full PDF methods have proven very successful in the description of complex turbulent reacting flow but, unfortunately, due to limitations in computer hardware, at present full PDF methods are limited to the analysis of turbulent reacting flows involving three spatial variables and for five to six chemical species with arbitrary chemical kinetics. This limitation is due to the fact that full PDF methods typically make use of a so-called "chemical look-up table" which contains information of precomputed composition changes due to chemical conversion. For further details on full PDF methods the interested reader is referred to Pope (1994) and the excellent review paper by Fox (1996). These classes of chemically reactive flows are briefly discussed together with some illustrative applications.

f. Slow Chemical Reactions Da $\ll$ 1. De Saegher et al. (1996) developed a comprehensive 3D reactor model for thermal cracking of hydrocarbons in internally finned tube reactors where both tubes with longitudinal and helicoidal fins were studied. Their model combines detailed descriptions of the 3D turbulent fluid flow in tubes with a complex internal geometry, endothermic multispecies chemical conversion, and heat transfer. Despite the fact that thermal cracking processes of hydrocarbons typically involve high temperatures, they could show that the ratio of t_r and t_m was sufficiently small to justify the neglect of turbulence–chemistry interactions. Although the gas temperatures did not differ much between tubes with longitudinal fins and tubes with helicoidal fins, the nonuniformity of the process gas over the tube cross section was much more pronounced for tubes with longitudinal fins. Traditional reactor models, which assume plug flow, are inadequate to predict the performance of cracking furnaces employing such new internal configurations of the tubes. Moreover these simple reactor models are not able to predict the circumferential nonuniform coke deposition in the tubes.

g. Very Fast Chemical Reactions Da $\gg$ 1. Very fast reactions are encountered, for example, in the domain of acid–base chemistry. Pipino and Fox (1994), for example, studied acid–base (i.e., HCl–NaOH) neutralization in a laboratory tubular reactor and used two different PDF approaches to describe quantitatively the effect of turbulent mixing on the chemical conversion. Their first description was based on the Lagrangian joint PDF of velocity and composition, whereas the second description was based on the Eulerian composition PDF. Pipino and Fox used the CFD package FLUENT to obtain the mean velocity field and the turbulence quantities and proposed a new model for molecular mixing that explicitly accounts for relaxation of the concentration spectrum. They used a spectroscopic method to determine the absorbance of the reacting mixture, containing phenol red as pH indicator, along the reactor axis. In their paper they reported good agreement between simulation results and experimental data, provided that their new molecular mixing model was invoked in the simulations. The study of Pipino and Fox clearly demonstrates that PDF simulations are a powerful tool for simulating mixing-sensitive chemical transformations in a turbulent flow field. The extension to (significantly) more complex reaction schemes and geometries is considered highly desirable.

h. Fast Chemical Reactions Da $\approx$ 1. A very difficult and challenging (both from a theoretical and a computational point of view) problem arises when the time scale for the chemical reaction is comparable to the time scale for (turbulent) mixing of the fluid. A well-known example of this class is encountered in the diazocoupling of 1-naphtol (A) and diazotized sulfanilic acid (B) in an alkaline-buffered solution:

$$A + B \xrightarrow{k_1} R$$

$$R + B \xrightarrow{k_2} S$$

In this case $k_1 \gg k_2 \approx 1/t_m$ and consequently full PDF methods are required to describe accurately the turbulence–chemistry interactions. Tsai and Fox (1994) have reported full PDF simulations for the above-mentioned chemical system and have compared their simulation results with experimental data reported by Li and Toor (1986).

i. Concluding Remarks. With respect to chemically reactive flows it can be stated that significant progress has been made toward a fundamental CFD-based description of these systems. Particularly promising are the so-called full PDF models (Pope, 1994; Fox, 1996). However, the industrial demands are in the area of (much) more complex reaction schemes involving a large number of species and reactions, very viscous fluids, non-Newtonian flow, and last but not least multiphase flows. In these areas much (CFD) work remains to be done. Keep in mind, however, that in many processes encountered in chemical engineering detailed information on the intrinsic chemical kinetics is lacking and therefore a good balance should be maintained between (very) detailed modeling of turbulence and turbulence–chemistry interactions and available information on the rate of the chemical transformations. In addition the availability of nonintrusive experimental techniques to obtain whole field concentration distributions is critical for the future development of this important class of CRE problems (see Section V).

B. Multiphase Systems

In many processes encountered in industrial practice, multiphase flows are encountered and in general it can be stated that, due to the inherent complexity of such flows, general applicable models and related CFD codes are nonexistent. The reason for this relatively unsatisfactory state of the art is due to the following causes:

- Many types of multiphase flow exist (i.e., gas–liquid, gas–solid, liquid–liquid, gas–liquid–solid) where within one type of flow several possible flow regimes exist. In Fig. 10 (Ishii, 1975) a classification is given for two-phase flow.
- The detailed physical laws and correct mathematical representation of phenomena taking place in the vicinity of interfaces (coalescence, breakup, accumulation of impurities) are still largely undeveloped.

Very often multiphase flow systems show inherent oscillatory behavior that necessitates the use of transient solution algorithms. Examples of such flows are encountered in bubbling gas-fluidized beds, circulating gas-fluidized beds, and bubble columns where, respectively, bubbles, clusters, or strands and bubble plumes are present that continuously change the flow pattern.

Class	Typical regimes	Geometry	Configuration	Examples
Separated flows	Film flow	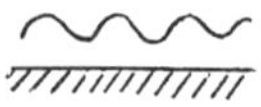	liquid film in gas / gas film in liquid	film cooling / film boiling
	Annular flow	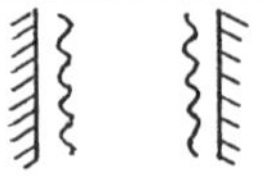	liquid core and gas film / gas core and liquid film	film boiling / condensers
	Jet flow	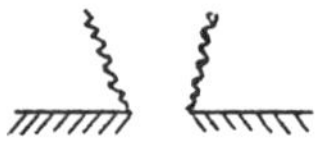	liquid jet in gas / gas jet in liquid	atomization / jet condenser
Mixed or transitional flows	Slug (plug) flow	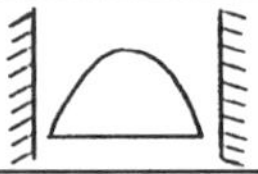	gas pocket in liquid	sodium boiling in forced convection (liquid metal fast breeder reactor)
	Bubbly annular flow	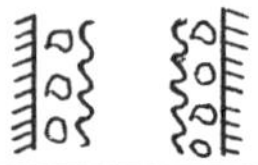	gas bubbles in liquid film with gas core	evaporators with wall nucleation
	Droplet annular flow		gas core with droplets and liquid film	steam generator
	Bubbly droplet annular flow		gas core with droplets and liquid film with gas bubbles	boiling nuclear reactor channel
Dispersed flows	Bubbly flow		gas bubbles in liquid	bubble columns
	Droplet flow	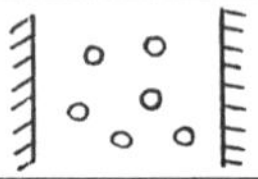	liquid droplets in gas	spray towers
	Particulate flow		solid particles in gas / solid particles in liquid	gas fluidized beds / liquid fluidized beds

FIG. 10. Flow regime classification for two-phase flow (Adapted from Ishii (1975)).

Prior to the discussion of the progress in the CFD analysis of multiphase flows, first some general requirements for modeling multiphase systems are mentioned:

- Regime characterization and flow regime transition
- Spatial distribution of the phases
- Sensitivity of system behavior for physicochemical parameters
- Prediction of effect of internals

1. Gas–Liquid (G-L) Systems

Gas–liquid systems are encountered very frequently in a variety of industrial applications. For example, the production of crude oil and natural gas involves the transportation of a gas and a liquid phase in pipes. Although very significant efforts have been made to arrive at a fundamental description and subsequent CFD modeling of these type of flows, unfortunately the progress is still very limited and the engineer, faced with the solution of practical problems, very often has to resort to semiempirical methods. This state of the art is mainly due to the fact that numerous flow regimes, with their specific hydrodynamic characteristics, can prevail.

Gas–liquid systems of particular interest to the chemical engineer are encountered in bubble columns, spray columns, air lift, falling film, and stirred tank reactors. Usually the form of these reactors corresponds to that of vessels or columns. From the perspective of the chemical engineer, who is concerned with the conversion and selectivity of chemical transformations, it is of utmost importance that an intensive contact between a gas and a liquid be achieved and therefore very often one phase is continuous whereas the other is disperse. Therefore, the interfacial area and the size of the disperse phase elements constitute very important aspects of CFD modeling of these types of systems.

As mentioned earlier many two-phase gas–liquid flows are highly dynamic and this property requires the use of dynamic simulation methods. A well-known example of such a type of flow is encountered in bubble columns where recirculating flow structures are present which are not stationary but which continuously change their size and location in the column. Due to their frequent application in chemical reaction engineering and their relatively simple geometry, CFD analyses of bubble columns have received significant attention and are discussed in more detail here.

a. Bubble Columns. Both two-fluid Eulerian models (Gasche *et al.*, 1990; Torvik and Svendsen, 1990; Svendsen *et al.*, 1992; Hjertager and Morud, 1993; Sokolichin and Eigenberger, 1994) and mixed Eulerian–Lagrangian models (Trapp and Mortensen, 1993; Lapin and Lübbert, 1994; Devanathan *et al.*, 1995; Delnoij *et al.*, 1997a) have been used to study gas–liquid two-phase flow. The general advantages of mixed Eulerian–Lagrangian models in comparison with completely Eulerian models are as follows: (1) Phenomena such as bubble breakup and coalescence can be implemented in a more direct way, and (2) the computational smearing or numerical diffusion can be significantly reduced (see, for example, Lapin and Lübbert, 1994). However, by applying higher order numerical schemes for the evaluation of the convective fluxes instead of the often used first-order UPWIND numerical scheme, the computational smearing of the

Eulerian approach can be reduced significantly (Sokolichin et al., 1997). Especially for dispersed flows with a high volume fraction of the dispersed phase, the increased computational requirements of mixed Eulerian–Lagrangian approaches should be mentioned as a disadvantage.

Figure 11 shows as an example some typical computational results obtained by Lapin and Lübbert with a mixed Eulerian–Lagrangian approach. This figure shows the combined results for the liquid phase velocity pattern (left) and the bubble positions (right) in a wafer column (diameter, 1.0 m; height, 1.5 m) where the bubbles are generated uniformly over its entire bottom. The simulation was started at $T = 0$ (s) with a bubble-free liquid. Initially the bubble clusters were found to rise in the liquid in the shape of a plane front but after approximately 2 s the front becomes unstable and attains a wavy shape. After 5 s a considerable instability has developed near the wall, which eventually leads to macroscopic flow instability (i.e., very irregular flow structure with large density differences). However, about 70 s after startup fairly regular flow patterns are established where several circulation cells can be recognized which were found to change their size and location continuously during the rest of the simulation. The flow in the bubble column is thus inherently instationary and a true steady-state flow pattern does not exist. Lapin and Lübbert stress in their paper the use of a sufficiently refined computational mesh in order to avoid excessive computational smearing of the resolved flow structures. With respect to the model presented by Lapin and Lübbert it should be mentioned here that their approach is essentially a pseudo-Lagrangian one due to the fact that they track bubble clusters instead of individual bubbles. Moreover, their model does not account for bubble–bubble interactions and momentum exchange between the bubbles and the liquid. The induced flow in the liquid phase is thus in fact assumed to be due to buoyancy effects only, which is probably a reasonable assumption in this type of flow. A more refined model in this respect has been developed by Delnoij et al. (1997a) and is discussed in more detail in Section VI.

Kumar et al. (1995) used the CFDLIB code developed at Los Alamos Scientific Laboratory to simulate the gas–liquid flow in bubble columns. Their model, which is based on the Eulerian approach, could successfully predict the experimentally observed von Karman vortices (Chen et al., 1989) in a 2D bubble column with large aspect ratio (i.e., ratio of column height and column diameter).

Besides the modeling of bubble columns with either Eulerian or mixed Eulerian–Lagrangian models the dynamics of individual bubbles has also been studied extensively in literature using various CFD approaches (Ryskin and Leal, 1984a,b,c; Tomiyama et al., 1993; and Hoffman and van den Bogaard, 1995). Ryskin and Leal solved the Navier–Stokes equations for the liquid flowing around a deformable bubble where the orthogonal curvilinear coordinate system was constructed in accordance with the shape which the bubble attained during its rise through the liquid, whereas Tomiyama et al. and Hoffman et al., respectively, used the VOF method (Hirt and Nichols, 1981) and the finite element based

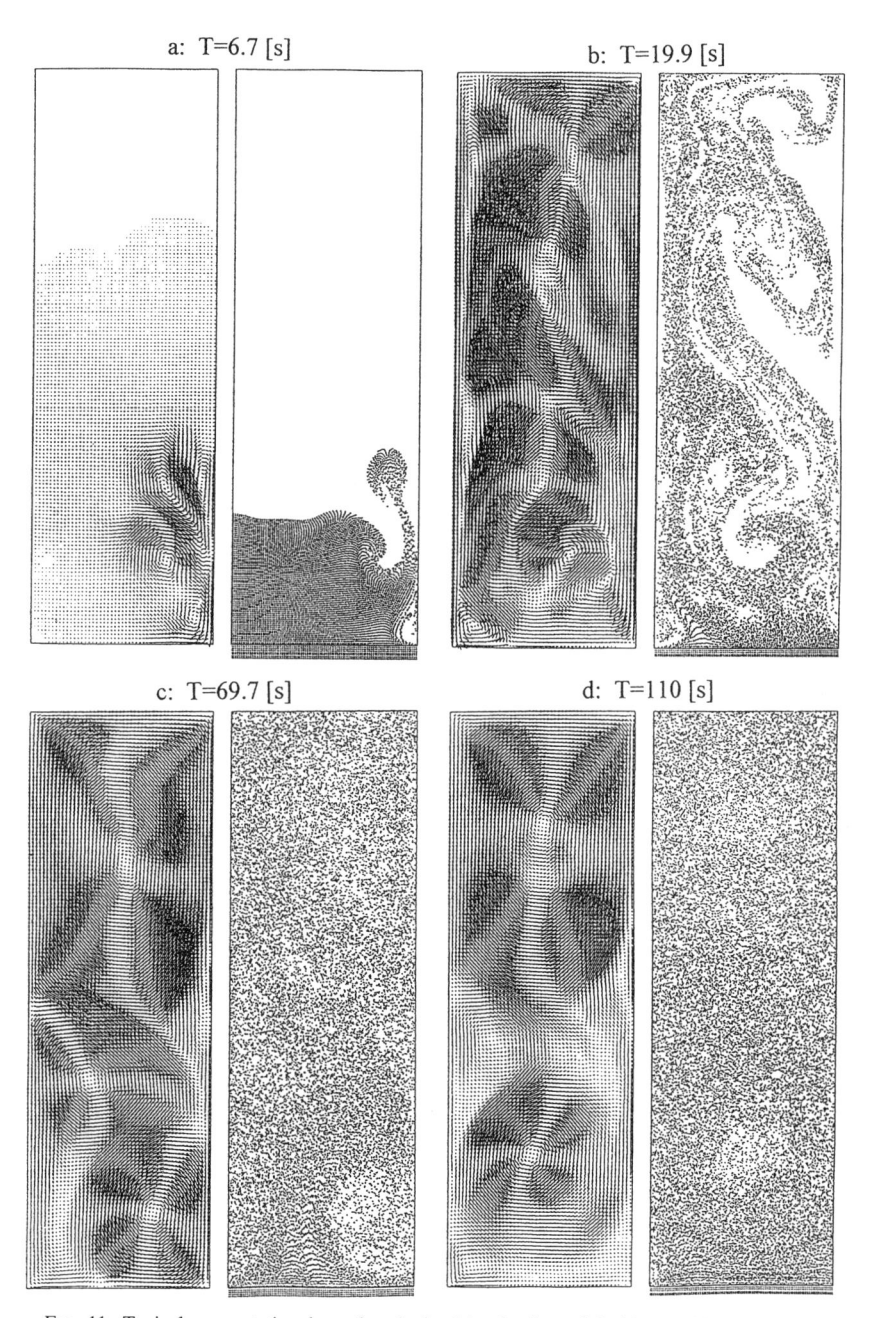

Fig. 11. Typical computational results obtained by Lapin and Lübbert (1994) with a mixed Eulerian–Lagrangian approach. Liquid phase velocity pattern (left) and the bubble positions (right) in a wafer column (diameter, 1.0 m; height, 1.5 m) where the bubbles are generated uniformly over its entire bottom. (Reprinted from *Chemical Engineering Science,* Volume 49, Lapin, A. and Lübbert, A., Numerical simulations of the dynamics of two-phase gas-liquid flows in bubble columns, p. 3661, copyright 1994 with permission from Elsevier Science.)

SEPRAN package (Cuvelier *et al.*, 1986). Recently Delnoij *et al.* (1997b) applied the VOF method to study the dynamics of single gas bubbles rising in a quiescent Newtonian liquid. They were able to demonstrate that the predicted bubble shape and the induced flow patterns in the liquid phase could be predicted very well as a function of the key physical properties of the liquid phase such as density, viscosity, and surface tension (see Section VI.D). The results reported by Tomiyama *et al.* are discussed in more detail below.

Tomiyama *et al.* (1993) applied the VOF method to analyze the motion of a single gas bubble rising in a liquid. They were able to show that the shape and terminal velocities of the gas bubbles could be predicted satisfactorily well over a wide range of Eotvös and Morton numbers. Figure 12a shows some typical results reported by Tomiyama *et al.* (1993) on the effect of the Morton number M on the shape and dynamics of a single bubble rising in a Newtonian liquid. For the purpose of reference in Fig. 12b the graphical correlation presented by Grace (1973) and Grace *et al.* (1976) is included, which takes into account the effect of fluid properties and equivalent bubble diameter d_b on the bubble shape and the terminal velocity V_t using three dimensionless quantities: the Reynolds number Re, the Morton number M, and the Eotvös number Eo defined by:

$$\text{Re} = \frac{\rho V_t d_b}{\mu}, \tag{26}$$

$$\text{M} = \frac{g \mu^4 \Delta \rho}{\rho^2 \sigma^3}, \tag{27}$$

$$\text{Eo} = \frac{g \Delta \rho d_b^2}{\sigma}, \tag{28}$$

where ρ, $\Delta \rho$, μ, and σ represent, respectively, the liquid density, the density difference between the liquid and gas, the dynamic shear viscosity of the liquid, and the surface tension of the liquid. From inspection of Figs. 12a and b it can be concluded that the predicted effect of the Morton number at least qualitatively agrees, with those obtained from the graphical correlation presented by Grace (1973) and Grace *et al.* (1976).

Recently Lin *et al.* (1996) applied the VOF method to study the time-dependent behavior of bubbly flows and compared their computational results with experimental data obtained with a particle image velocimetry (PIV) technique. In their study the VOF technique was applied to track several bubbles emanating from a small number of orifices. Lin *et al.* reported satisfactory agreement between theory and experiment.

b. Gas–Liquid Stirred Tank Reactors. CFD has also been applied to study the flow phenomena in gas–liquid stirred tank reactors. In many studies one-way cou-

FIG. 12. Typical results reported by Tomiyama et al. (1993) on the effect of the Morton number M (at Eötvös number Eo = 10) on the shape and dynamics of a single bubble rising in (a) a Newtonian liquid, and (b) graphical correlation due to Grace (1973) and Grace et al. (1976). [Part (a) reprinted from *Nuclear Engineering and Design*, Volume 141, Tomiyama, A., Zun, I., Sou, A., and Sakaguchi, T., Numerical analysis of bubble motion with the VOF method, pp. 69–82, Copyright 1993, with permission from Elsevier Science. Part (b) reprinted from Grace, R., Clift, R., and Weber, M.E., "Bubbles, Drops, and Particles." Academic Press, Orlando, 1976. Reprinted by permission of Academic Press.]

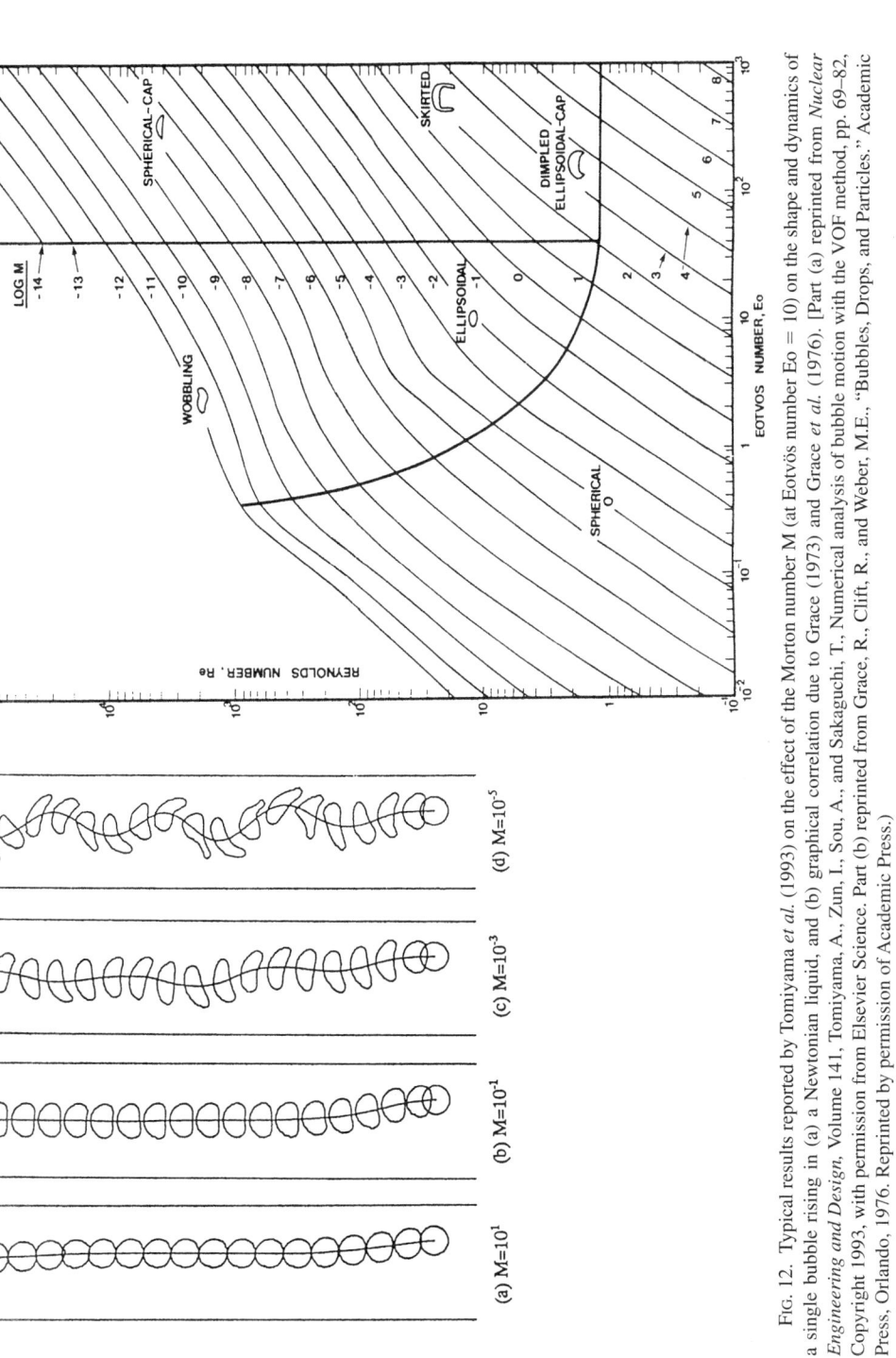

pling (Bakker and van den Akker, 1994) was assumed in the simulations although, in view of the reported values of the gas phase volume fraction (5–10%), it would have been more appropriate to account for two-way coupling. This is to some extent due to the possibilities offered in (early versions) of commercial CFD packages. Another problem is the turbulence modeling, where due to the absence of more detailed knowledge often the standard single-phase k-ϵ model is used. Finally difficulties in accurately representing the impeller geometry are also encountered in this type of flow (also see Sections IV,A,1 and 2).

c. Other Gas–Liquid Reactors. In literature a variety of other CFD applications involving gas–liquid reactors have been reported such as hydrocyclones (Hargreaves and Silvester, 1990; Hsieh and Rajamani, 1991) and loop reactors (Sokolichin and Eigenberger, 1994).

d. Concluding Remarks. The studies reported in this section clearly show that for gas–liquid two-phase flow very promising results have been obtained in recent years. The results obtained with the mixed Eulerian–Lagrangian method and the VOF method in particular are very promising despite the fact that due to CPU constraints the analysis is limited respectively to moderate (typically 10^5) and a relatively small (typically 10) number of bubbles.

Despite this significant progress, complicated problems still remain such as the prediction of the rate of bubble formation at gas distributors and the coalescence and breakup of bubbles in the gas–liquid dispersion. More general it is still impossible to predict flow regime transition in various types of gas–liquid two-phase flow. Also the coupling to chemical transformation remains extremely difficult due to the fact that often an extremely fine spatial and temporal resolution has to be obtained especially in cases where chemically enhanced mass transfer takes place (Versteeg *et al.,* 1989, 1990; Frank *et al.,* 1995a,b). Nevertheless the coupling of mass transfer models described by Versteeg *et al.* and Frank *et al.* with mixed Eulerian–Lagrangian models as developed by Delnoij *et al.* (1997a) offers a potential to arrive at fundamental reactor models for bubble columns in the foreseeable future.

However, with respect to the mixed Eulerian–Lagrangian approach it should be mentioned that disagreement still exists on rather fundamental issues such as the correct description of the forces acting on a single bubble and the effect of other bubbles on these forces. In this area much fundamental work remains to be done. Moreover the extension of the mixed Eulerian–Lagrangian approach to the churn-turbulent or heterogeneous bubbly regime is of crucial importance due to the fact that this flow regime is frequently encountered in industrial applications involving bubble columns. In this respect a modeling approach is required that can simultaneously deal with "large" bubbles and "small" bubbles. In this respect a model that combines the features of the VOF method and the mixed Eulerian–Lagrangian method deserves attention. Last but not least the incorporation of turbulence effects is still problematic and should receive more attention in the future.

2. Liquid–Liquid (L-L) Systems

Liquid–liquid systems are encountered in many practical applications involving physical separations of which extraction processes performed in both sieve-tray and packed columns are well-known examples. In principle, all three methods discussed in Section III,B,2 can be used to model liquid–liquid two-phase flow problems. The added complexity in this case is the possible deformation of the interface and the occurrence of flow inside the droplet.

Of particular interest in this connection is the work of Ohta *et al.* (1995) who used a two-material VOF method to study the formation of a single droplet at an orifice in a pulsed sieve-plate column. They also compared the theoretically computed droplet sizes with those obtained from experiments using a high-speed video camera and reported good agreement between theory and experiment. In Fig. 13 the observed (a) and computed (b) droplet formation reported by Ohta *et al.* (1995) is shown. Note that the agreement between theory and experiment is reasonable although subtle differences between the computed and observed droplet shape can be clearly recognized. One of the great advantages of the two-material VOF method is its possibility to provide detailed information on the fluid flow both inside and outside the droplet. The VOF method is a very powerful numerical tool for analyzing complex free surface flows and its possible ability to study systems involving for example droplet interaction should be further explored. In this respect the required computational time may become problematic.

3. Fluid–Solid (G-S and L-S) Systems

Fluid–solid systems, especially in situations where the fluid is a gas, are very frequently encountered in various important industrial processes such as packed-bed reactors, moving-bed reactors, fluidized-bed reactors, and entrained reactors.

a. Packed-Bed Reactors. Although flows in packed-bed reactors can also be classified as a chemically reactive single-phase flow in a complex geometry, the authors think that it is more appropriate to treat this type of flow as a special type of two-phase flow. The analysis of heat transfer processes and coupled chemical conversion in packed-bed reactors has traditionally received considerable attention from chemical engineers. In most of these studies rather simple representations of the prevailing flow patterns (i.e., plug flow) were used despite the fact that flow inhomogeneities are known to exist due to (local) porosity disturbances.

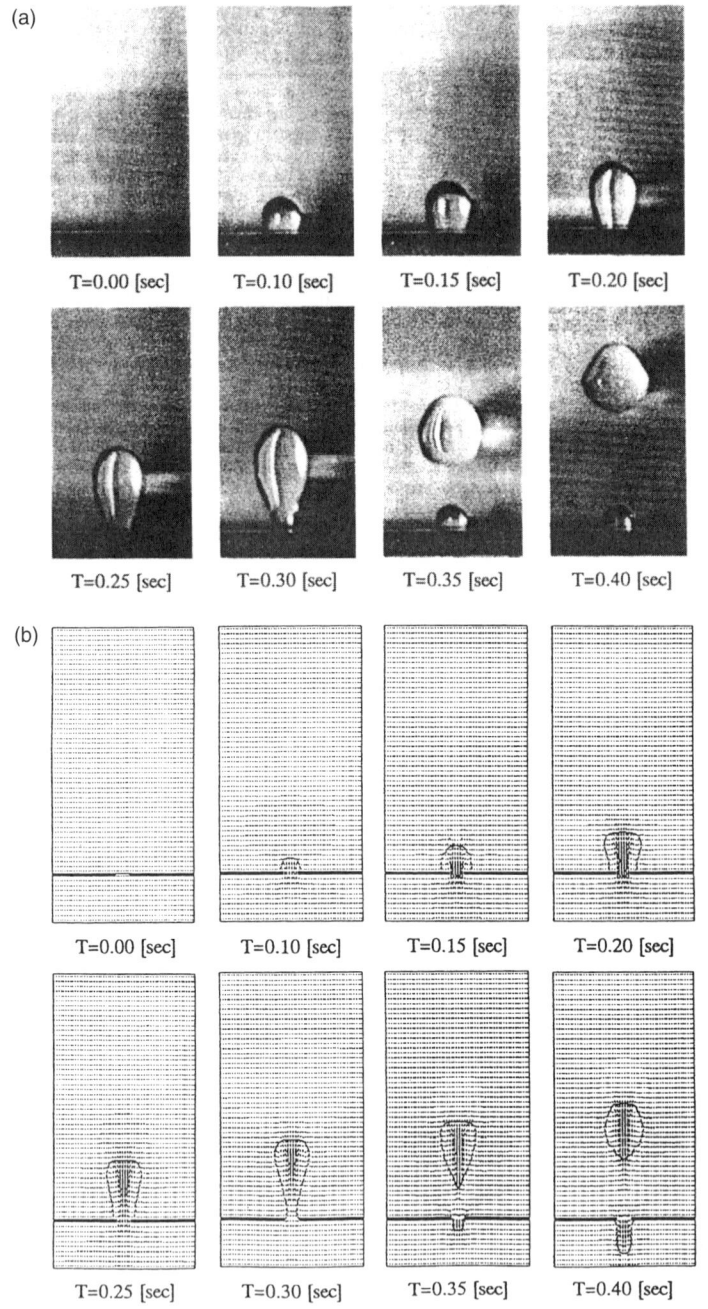

FIG. 13. (a) Observed and (b) computed formation of a droplet at an orifice in a pulsed sieve-plate extraction column. (Reprinted from *Chemical Engineering Science,* Volume 50, Ohta M., *et al.*, Numerical analysis of a single drop formation process under pressure pulse condition, pp. 2923–2931, copyright 1995, with permission from Elsevier Science.)

A very important form of such disturbances is caused by the presence of the wall of the tube containing the packed bed. Vortmeyer and Schuster (1983) have used a variational approach to evaluate the steady two-dimensional velocity profiles for isothermal incompressible flow in rectangular and circular packed beds. They used the continuity equation, Brinkman's equation (1947), and a semiempirical expression for the radial porosity profile in the packed bed to compute these profiles. They were able to show that significant preferential wall flow occurs when the ratio of the channel diameter to the particle diameter becomes sufficiently small. Although their study was done for an idealized situation it has laid the foundation for more detailed studies. Here CFD has definitely contributed to the improvements of theoretical prediction of reactor performance.

Another possibility for modeling packed-bed reactors involves the use of a so-called unit cell approach where a suitable periodic structure in the packing is identified and subsequently used to define the boundaries of the computational domain. Due to the geometrical complexity the fluid flow (and other relevant equations have to be formulated and solved in curvilinear coordinates. In fact this approach has been followed for example, by Guj and De Matteis (1986) who used a MAC-like scheme (Welch *et al.*, 1965) to solve the Navier–Stokes equations. For random packings the unit cell approach becomes much more difficult due to the fact that a suitable periodic structure is difficult to define.

b. Fluidized Bed Reactors. Another type of chemical reactor where the CFD approach has proven fruitful is the gas fluidized-bed reactor. These reactors find a widespread application in the petroleum, chemical, metallurgical, and energy industries (Kunii and Levenspiel, 1991) and significant research efforts have been made in both academic and industrial research laboratories to develop detailed micobalance models of gas-fluidized beds.

A very lucid review on fluidized-bed modeling and analysis of existing hydrodynamic models in terms of Occam's razor (a philosophical maxim based on the following principle: "It is futile to do with more what can be done with fewer") has been given by Clift (1993). Jackson (1994) has recently prepared a review on the state of the art of our understanding of the mechanics of fluidized beds and emphasizes that despite the fact that the goal of deriving exactly correct equations on the basis of the local instant formulation may be unattainable in the future, for many practical purposes it is acceptable to use a set of equations that is good enough (i.e., with acceptable engineering accuracy) to describe the behavior of a range of situations relevant to the chemical engineer.

Pritchett *et al.* (1978) were the first to report numerical solutions of the nonlinear equations of change for fluidized suspensions. With their computer code, for the first time, bubbles issuing from a jet with continuous gas through-flow, could be calculated theoretically. Figure 14 shows as an illustration the computed mo-

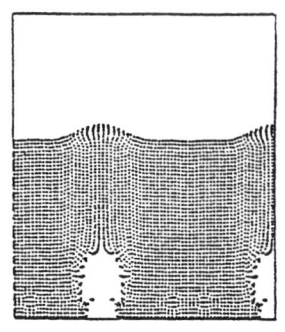

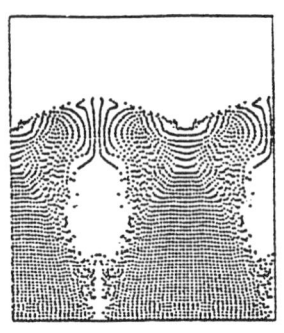

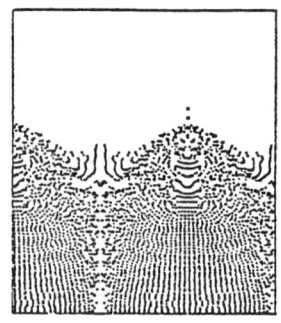

FIG. 14. Computed motion of a collection of pseudo-Lagrangian particles due to the evolution and propagation of gas bubbles in a section of a hypothetical two-dimensional gas fluidized bed [From Pritchett, J.W., Blake, T.R., and Garg, S.K. A numerical model of gas fluidized beds. *AIChE Symp. Ser.* 176, **74**, 134 (1978). Reproduced with permission of the American Institute of Chemical Engineers. Copyright © 1978 AIChE. All rights reserved.]

tion of a collection of pseduo-Lagrangian particles due to the evolution and propagation of bubbles in a section of a hypothetical two-dimensional gas fluidized bed. Schneyer *et al.* (1981) extended this hydrodynamic model to a complete reactor model for a fluidized-bed coal gasifier. Their model incorporates the kinetics of the heterogeneous and homogeneous chemistry of combustion and gasification and probably represents the most ambitious and comprehensive fluidization modeling effort to date. A more or less parallel development has been made by the Jaycor group (Henline *et al.*, 1981; Scharff *et al.*, 1982) who also made an attempt to model a fluidized bed coal gasifier from first principles.

In the literature a relatively large number of publications exist which deal with the CFD analysis of gas-fluidized beds using two-fluid models (Gidaspow and Ettehadieh, 1983; Ettehadieh *et al.*, 1984; Syamlal and Gidaspow, 1985; Gidaspow, 1986; Bouillard *et al.*, 1989; Kuipers 1990; Kuipers *et al.*, 1991, 1992a,b,c, 1993; Nieuwland *et al.*, 1996a). These studies have clearly revealed that many important key properties of bubbling fluidized beds, that is, gas bubble behavior, spatial voidage distributions, and wall-to-bed heat transfer processes can be predicted satisfactorily from two-fluid simulations (see Section VI,A). However, the agreement between theory and experiments is not perfect and has been attributed among other factors to the simple solids rheology (i.e., Newtonian behaviour) incorporated in the two-fluid models. Further theoretical developments have led to the application of the kinetic theory of granular flow (KTGF) to gas-fluidized beds (Ding and Gidaspow, 1990; Manger, 1996) of which its ability to provide *a priori* information on the viscosities of the particulate phase can be mentioned as an important advantage over the constant-solids-viscosity two-fluid models. It has however not yet been demonstrated that the KTGF based model predicts more realistic gas bubble behavior in comparison with the constant-solids-viscosity two-fluid models.

CFD has also been applied to predict the flow patterns in the circulating fluidized bed (CFB) riser (Tsuo and Gidaspow, 1990). Despite their widespread industrial application in, for example, petrochemical and energy production processes, our understanding of the hydrodynamics of these systems is, unfortunately, still very limited. For reliable design of large-scale CFB risers the designer requires accurate information regarding the spatial solids distribution, which is generally inhomogeneous in both axial and radial direction. The inhomogeneous solids distribution determines the gas–solid phase flow patterns to a large extent and thereby significantly influences riser reactor performance. Understanding and *a priori* prediction of this and related hydrodynamic aspects of riser flow are of crucial importance for design purposes. In literature lateral solids segregation has been attributed to interaction between gas phase eddies and dispersed particles and additionally to direct collisional interaction between particles. In most studies the direct particle–particle collisions have been described using KTGF, which is basically an extension of the Chapman–Enskog theory of dense gases. The KTGF was first applied by Sinclair and Jackson (1989) to predict successfully lateral segregation in CFB risers. Pita and Sundaresan (1991, 1993) further developed this model and found strong parametric sensitivity of their model predictions with respect to the value of the restitution coefficient e for particle–particle collisions. In cases where realistic e values were used, no lateral segregation could be predicted. In none of these models was turbulence accounted for, and very limited comparison with experimental data was reported in these studies.

Theologos and Markatos (1992) used the PHOENICS program to model the flow and heat transfer in fluidized catalytic cracking (FCC) riser-type reactors. They did not account for collisional particle–particle and particle–wall interactions and therefore it seems unlikely that this type of simulation will produce the correct flow structure in the riser reactor. Nevertheless it is one of the first attempts to integrate multiphase hydrodynamics and heat transfer.

A more recent and very promising development in the modeling of (dense) gas-particle flows, as encountered in bubbling fluidized beds and circulating fluidized beds, is the discrete particle approach (Tsuji *et al.*, 1993; Kawaguchi *et al.*, 1995; Tanaka *et al.*, 1996; Hoomans *et al.*, 1996) where the motion of individual spherical particles is directly calculated from the forces acting on them, accounting for the particle–particle and particle–wall interactions and drag between the particles and the interstitial gas phase. Tsuji *et al.* used the distinct element model (DEM), originally developed by Cundall and Strack (1979), to represent the interaction forces resulting from particle–particle and particle–wall encounters. The model developed by Tsuji *et al.* in fact represents a so-called "soft sphere" model because the particles are thought to undergo deformation during their contact where the interaction forces in both the normal and tangential direction are calculated from simple mechanical representations involving a spring, a dashpot, and a slider. Usually this model type requires the spring constant, the damping coefficient of the dashpot, and the friction coefficient of the slider as input parameters.

The model developed by Hoomans *et al.* (1997) represents a so-called "hard sphere" model because in their model quasi-rigid spheres are implied which undergo a quasi-instantaneous collision where point contact prevails. Their model requires specification of two phenomenological parameters, namely, the coefficient of restitution and the friction coefficient. Some illustrative results that have been obtained from this type of modeling of gas fluidized beds are presented in Section VI,B. Hoomans *et al.* (1998) have compared the predictions obtained from hard sphere models with those of soft sphere models and found that the macroscopic system behavior predicted by both types of models is very similar provided that the spring constant used in the DEM model is given a sufficiently large value.

In the discrete particle models discussed so far the hydrodynamic interaction between particles was neglected. For gas–solid systems this assumption will be valid but for liquid–solid systems this type of interaction will become important and has to be accounted for. Stokesian dynamics (Brady and Bossis, 1988) constitutes a very powerful tool and offers significant possibilities for understanding of suspension dynamics on a more fundamental basis. For example, Ichiki and Hayakawa (1995) presented a Stokesian dynamics based model in which hydrodynamically interacting granular particles were considered. They neglected the inertial terms in the Navier–Stokes equations describing the interstitial fluid flow and therefore their model is only valid for relatively small particle Reynolds numbers (i.e., $Re_p < 1$). These computations are very CPU intensive and at present only simulations with a relatively low number of particles are possible. Despite these (serious) limitations this type of work is very important because it can lead to improved constitutive equations for fluid–particle interactions, which can be used subsequently in both discrete particle models which do not treat the flow of the continuous phase at the scale of a single particle.

Koelman and Hoogerbrugge (1993) have developed a particle-based method that combines features from molecular dynamics (MD) and lattice-gas automata (LGA) to simulate the dynamics of hard sphere suspensions. A similar approach has been followed by Ge and Li (1996) who used a pseudo-particle approach to study the hydrodynamics of gas–solid two-phase flow. In both studies, instead of the Navier–Stokes equations, fictitious gas particles were used to represent and model the flow behavior of the interstial fluid while collisional particle–particle interactions were also accounted for. The power of these approaches is given by the fact that both particle–particle interactions (i.e., collisions) and hydrodynamic interactions in the particle assembly are taken into account. Moreover, these modeling approaches do not require the specification of closure laws for the interphase momentum transfer between the particles and the interstitial fluid. Although these types of models cannot yet be applied to macroscopic systems of interest to the chemical engineer they can provide detailed information which can subsequently be used in (continuum) models which are suited for simulation of macroscopic systems. In this context improved rheological models and boundary condition descriptions can be mentioned as examples.

c. Concluding Remarks. On the basis of the aforementioned results it is clear that significant progress has been made in modeling gas–particle flows and to a lesser extent liquid–particle flows. There remain however *very* significant challenges of which we would like to mention the following:

- Detailed modeling of the (mutual) interaction between turbulent flowing fluids and suspended particles
- Accurate description of particle–particle and particle–wall interactions including effects due to surface roughness and deviations from spherical particle shape
- Development of improved closure laws for the rheological description of fluidized suspensions on basis of discrete particle models
- Prediction of regime transition in gas–particle flows
- Incorporation of chemical conversion models in multiphase flow simulation

4. Gas–Liquid–Solid Systems

In many commercial processes efficient transfer of soluble substances or heat between a gas and a (reactive) liquid is achieved by contacting the phases in the presence of a packing where the function of the packing is the creation of sufficient contact area between the gas and liquid phase. In addition the packing can also act as a catalyst to enhance desired chemical transformations. Although the flow of the gas–liquid mixture through the packing can be classified as a two-phase flow in a complex geometry the authors think that it is more appropriate to treat this type of flow as a special type of three-phase flow. Such three-phase flows are encountered for example in trickle-bed reactors. In addition to these reactors other three-phase reactors find frequent application in commercial processes such as slurry reactors and three-phase fluidized-bed reactors. Contrary to the situation for trickle-bed reactors, in this case the solid phase (usually a catalyst) is suspended in the (continuous) liquid phase and can move freely in the reactor. The state of the art of CFD analysis of these three-phase reactors is briefly discussed.

a. Trickle-Bed Reactors. Unfortunately, very little progress has been made up to now on the CFD analysis of trickle-bed reactors. The reason for this state of the art is due to (1) the great geometrical complexity of the packing (often a catalyst), (2) the fact that we have to deal with a complex free surface flow with a very complex interfacial structure (i.e., partially wetted packing) where in certain flow regimes the liquid film interacts with the gas phase, and (3) the prevailing flow regime in a trickle-bed reactor can depend on the flow history experienced by the packing. Due to the aforementioned difficulties it will be very difficult to obtain significant progress in CFD modeling of trickle-bed reactors in the near future, and as a first step toward a more fundamental understanding of these complex systems approaches as followed by Melli and Scriven (1991) will probably be more fruitful.

b. Slurry Reactors. As far as the computation of macroscopic flow patterns is concerned, the CFD analysis of slurry reactors shows some progress but unfortunately very little work has been done yet in this area, and in the few existing studies no comparison with experimental data is reported. For example, very recently Hamill *et al.* (1995) applied the CFDS-FLOW3D package to study the three-phase flow of solids, liquid, and a gas in a mixing vessel agitated by five impellers. They used the standard k-ϵ turbulence model and an explicit model for the modeling of the fluid flow around the impellers. Although their computational results appear reasonable, no comparison with experimental data was made in this case. Torvik and Svendsen (1990) used a multifluid formulation to model slurry reactors.

Despite the fact that there is some progress in modeling the macroscopic flow structure of slurry reactors, a number of microscopic phenomena are very difficult to capture in macroscopic flow simulation models such as the possible accumulation of solid particles near the gas–liquid interface, which significantly affects the mass transfer characteristics of the slurry system (Beenackers and van Swaaij, 1993).

c. Three-Phase Fluidized Beds. Similar to the situation for slurry reactors very little progress is seen in this area, which seems understandable in view of the remaining difficulties in modeling gas–liquid and liquid–solid two-phase flow.

C. STATE OF THE ART OF CFD IN CHEMICAL REACTION ENGINEERING

As is evident from the examples mentioned in the preceding sections it has hopefully become apparent that in various areas of single-phase and multiphase flows significant progress has been made in CFD-based system descriptions. In

TABLE II
PRESENT SITUATION AND PROSPECTS FOR CFD IN CRE INVOLVING SINGLE-PHASE FLOW

Flow Category	Progress	Future Trends and Areas of Interest
Laminar flows	Significant	More complex geometries + interaction with mixing and chemical transformation processes
Turbulent flows	Reasonable	Development of accurate and reliable turbulence models for enginering applications + use of LES and DNS for generation of databases for turbulence quantities
Non-Newtonian flows	Reasonable	More complex materials and more complex geometries + improved numerical solution procedures + extension to multiphase flow
Chemically reactive flows	Reasonable	Application to more complex reaction schemes and more complex geometries + extension to multiphase flow

Tables II and III, respectively, the progress and the future trends and remaining difficulties in systems involving single-phase flows and multiphase flows are briefly summarized.

TABLE III
PRESENT SITUATION AND PROSPECTS FOR CFD IN CRE INVOLVING MULTIPHASE FLOW

Flow Category	Progress	Future Trends or Problem Areas
Gas–liquid	Reasonable	Turbulence modeling + modeling of interfacial transport phenomena + prediction of flow regime transition + interaction of hydrodynamics with chemical transformation processes
Bubble columns	Reasonable	Modeling of churn-turbulent flow regime
Loop reactors	Reasonable	Modeling of churn-turbulent flow regime
Stirred tanks	Reasonable	Improved geometrical representation of impeller and baffles
Hydrocyclones	Reasonable	Improved geometrical representation of system
Liquid–liquid	Little	Turbulence modeling + modeling of interfacial transport phenomena + interaction of hydrodynamics with chemical transformation processes
Fluid–solid	Reasonable	Turbulence modeling + refined models for particle–particle and particle–wall interaction + prediction of flow regime transition + interaction of hydrodynamics with chemical transformation processes
Packed beds	Some reasonable	Unit cell approach
Fluidized beds		Extension to industrial fluidized beds with complex geometries
Gas–liquid–solid		Turbulence modeling + prediction of flow regime transition + interaction of hydrodynamics with chemical transformation processes
Trickle bed reactors	No	Modeling on basis of unit cell approach + development of correspondence rules for macroscopic system behavior
Slurry reactors	Little	Modeling of the effect of the solids phase on interfacial transport phenomena
Three-phase fluidized beds	Little	Modeling of the effect of the solids phase on interfacial transport phenomena + development of refined models for particle–particle and particle–wall interaction

V. Experimental Validation

Experimental validation of CFD results is considered a prerequisite to paving the road for widespread acceptance of CFD in the chemical engineering community, especially in connection with multiphase flow applications. The authors do not intend to give here a complete review on available measuring techniques for single-phase and multiphase flows; only the more advanced techniques are briefly discussed. For an overview of the latest advances realized in noninvasive measurement of multiphase systems the interested reader is referred to Chaouki *et al.* (1997). The available experimental techniques can be classified according to the following aspects:

- Type of quantity measured
- Local or whole field measuring method
- Instantaneous or time-averaged quantities measured
- Intrusive or nonintrusive method

For the measurement of pressure, temperature, phase concentration, composition, and velocities in single-phase and multiphase systems a variety of experimental methods are available ranging from simple probe techniques to sophisticated whole field measuring methods. Thermal anemometry, electrical sensing techniques, light scattering and optical methods, electromagnetic wave techniques, and ultrasonic techniques have all been used to study complex fluid flows. For an overview of these techniques the interested reader is referred to Cheremisinoff (1986a).

Especially for multiphase systems flow visualization (Wen-Jei Yang, 1989; Merzkirch, 1987) can provide valuable initial information on the prevailing flow patterns and should at least always be considered as a first step. Of course, in applications that involve extreme conditions such as high temperature and/or pressure it is very difficult if not impossible to apply flow visualization and other techniques should be considered. Here the use of cold flow models which permit visual observation might be considered as an alternative as an important first step to obtain (qualitative) information on the flow regime and associated flow pattern. Of course, multiphase flows exist such as dense gas–solid flows that do not permit visual observation and in such cases the application of idealized flow geometries should be considered. A well-known example in this respect is the application of so-called 2D gas fluidized beds to study gas bubble behavior (Rowe, 1971).

Due to the fact that the CFD approach usually leads to very detailed information on the temporal and spatial distribution of key variables it has also led to the development of very advanced experimental techniques to obtain the relevant experimental data. Some of these techniques, which reflect the recent progress in experimental fluid dynamics are discussed in more detail. Point-measuring techniques such as hot-wire anemometry (HWA) and laser Doppler anemometry (LDA) cannot give information on the instantaneous spatial structure of the flow. Especially in situations where one is interested in the dynamics of coherent structures, as encountered in, for example, turbulent single phase flows and dispersed gas–liquid two-phase flow, point-measuring techniques are inadequate and more advanced techniques are required.

a. Laser Doppler Anemometry. With LDA the velocities of flowing (small) particles can be determined from the frequency information contained in light scattered by the particles as they pass through a fringe or interference pattern. In single-phase flows these particles, which have to be sufficiently small to prevent separation effects, are deliberately added to the fluid (seeding of the flow). Since the intensity of the scattered light also contains information on the particle size LDA has also found application in atomized and dispersed flows to obtain particle size information (Farmer, 1972, 1974).

b. Particle Image Velocimetry. A very powerful class of velocity measuring techniques, termed pulsed light velocimetry, has become available in experimental fluid dynamics. In these techniques typically the motion of small, marked regions of a fluid is measured by observing the locations of the images of the markers at two or more times. The velocity is obtained from the ratio of the observed displacement $\Delta \bar{x}$ and the time interval Δt separating the subsequent observations of the marker images:

$$\bar{u} = \frac{\Delta \bar{x}}{\Delta t}. \tag{29}$$

The marker fluid parcels can consist of small particles, bubbles, or droplets but can also be generated *in situ* by activating molecules constituting the fluid with laser beams, causing them to fluoresce (Gharib *et al.*, 1985).

A very powerful method that belongs to the class of PLV techniques to obtain quantitative information on the instantaneous structure of the flow is particle image velocimetry (PIV) and its digital counterpart (DPIV). In these techniques a suitably chosen number of neutrally buoyant tracer particles are suspended in the fluid (usually a liquid) and subsequently a section of the flow is illuminated with the aid of a laser whereby the tracer particles become visible. By making high-resolution video recordings of the illuminated section of the flow and subsequent processing of the digitized images using advanced statistical techniques (Westerweel, 1993) the instantaneous flow field of the fluid in the test section can be obtained. Due to this ability DPIV can provide detailed information on the dynamics of coherent flow structures, which is typically also the type of information that can be obtained from (advanced) CFD approaches. A critical aspect of PIV is the so-called seeding of the fluid with marker particles. For further details the interested reader is referred to the review paper of Adrian (1991) and Westerweel (1993) and the references cited therein.

PIV has also been applied to gas–liquid systems (Reese and Fan, 1994; Lin *et al.*, 1996) to study the flow structure in bubble columns. A specific complication here is caused by the presence of the gas bubbles. On the basis of a prior knowledge of the size distribution of the tracer particles and the gas bubbles it is possible to discriminate bubbles from particles and thus phase-specific postprocessing of the images can be undertaken whereby both the flow pattern of the bubbles and the liquid in principle can be obtained. Particle image velocimetry has also been applied (Chen and Fan, 1992) to study the flow structure in 3D gas–liquid–solid fluidized beds.

c. Nuclear Magnetic Resonance Imaging. Although LDA and HWA are powerful methods for studying spatial and temporal variations of fluid velocities under a wide range of conditions, unfortunately these methods cannot be applied in optically opaque fluids (dense slurries, pastes, etc.) and flows which are inaccessible due to geometric constraints (i.e., porous media). In these situations magnetic resonance imaging (MRI) offers a convenient means to obtain detailed information on the flow (see, for example, Derbyshire *et al.*, 1994).

MRI techniques can be broadly divided into magnitude and phase-based methods which respectively employ the magnitude and orientation of the local nuclear magnetization vector M as tags which can be monitored in space and time. After creation and evolution of the tags the local magnetization state can be imaged. By creating subsequent images the displacement of fluid parcels over distances ranging from micrometers to centimeters occurring on a time scale varying from milliseconds to seconds can be registered. Especially for multiphase flows this type of measuring method possesses great potential. An example of the application of MRI to determine concentration and velocity profiles in solid suspensions in rotating geometries can be found in the paper of Corbett *et al.* (1995).

d. Laser-Induced Fluorescence. Another technique that is particularly powerful in studies involving (turbulent) mixing processes, possibly in the presence of fast chemical transformations, involves fluorescent tracers together with laser-induced fluorescence (LIF) techniques (Gaskey *et al.*, 1990; André *et al.*, 1992). Through the application of laser sheet techniques it is possible to make measurements in a very thin cross section of the mixing zone, which is of crucial importance to validate closure models for turbulent reactive flow involving fast chemical reactions. A very recent development in this connection is the work by Dahm *et al.* (1995) where a very high temporal and spatial resolution was achieved. Recently Lemoine *et al.* 1996 used LIF and LDA to measure simultaneously species concentration and velocity in turbulent flows.

e. Particle Tracking Methods. In particle tracking methods typically the motion of a particle is tracked in time using a set of detectors with a suitable geometrical arrangement where the behavior of the particle is (considered) representative for the phase under consideration. In literature several particle tracking techniques have been successfully employed to study multiphase flow systems such as bubble columns [Devanathan *et al.*, 1990, using computer-aided radioactive particle tracking (CARPT)], fluidized beds [Larachi *et al.*, 1995, 1996 using radioactive particle tracking (RPT)], and rotary mixers [Bridgwater *et al.*, 1993, using positron emission particle tracking (PEPT)].

Devanathan *et al.* (1990) employed their CARPT technique to obtain information on the 3D path of a particle which from a hydrodynamic point of view can be considered as representative for the liquid phase in a bubble column. This technique thus offers the possibility to obtain important quantitative information on a key quantity of the Lagrangian representation of the flow. CARPT is essentially a particle tracking technique, in which the position of an essentially neutrally buoyant tiny particle, suspended in a two-phase gas–liquid flow, is registered as a function of time with a set of detectors with a proper geometrical arrangement around the bubble column. In principle, this technique can also be applied to other multiphase reactors such as gas fluidized or liquid fluidized beds and three-phase fluidized beds. Figure 15 shows the time-averaged streamlines and velocity vectors of the liquid phase obtained with the CARPT technique (Denavathan *et al.*, 1990) in a 12-in.-diameter bubble column operated at a superficial gas velocity of 0.105 m/s. From this figure the existence of a single circulation cell can be clearly inferred, which is contrary to the hypothesis of the existence of multiple circulation cells with cell height equal to the column diameter (Joshi and Sharma, 1979).

Larachi *et al.* (1996) applied their radioactive particle tracking technique to determine the time-averaged solids velocity in a three-phase fluidized bed. Techniques which are similar to CARPT, such as positron emission particle tracking and radioactive particle tracking, have been applied successfully to systems involving particulate solids.

f. Tomographic Techniques. Tomography refers to the cross-sectional imaging of an object from data collected by illuminating this object from many directions. This class of techniques is particularly powerful to obtain information on the spatial distribution of phases in a multiphase system in a nonintrusive manner. Thus tomographic techniques usually provide information on the flow morphology only and not on the motion of the flowing phases. X-ray tomography, γ-ray tomography, and electrical capacitance tomography constitute some of the better known tomography methods. Toye *et al.* (1995), for example, applied X-ray tomography to study liquid flow distribution in a trickle-bed reactor. Such measurements are considered very important in connection with the future development

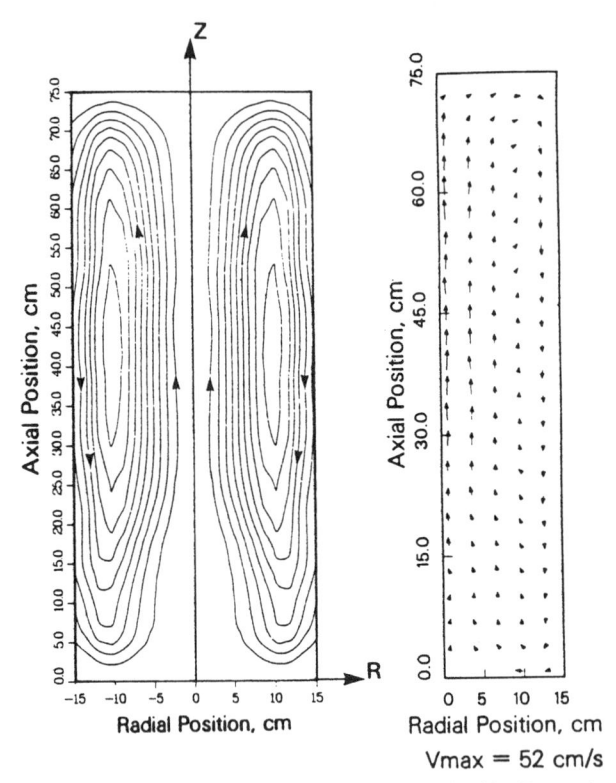

FIG. 15. Time-averaged streamlines and velocity vectors obtained by Devanathan et al. (1990) using the CARPT technique. (From *Chemical Engineering Science*, Volume 45, Devanathan, N., Moslemian, D., and Dudukovicz, M.P., Flow mapping in bubble columns using CARPT, pp. 2285–2291, copyright 1990, with permission from Elsevier Science.)

of fundamental hydrodynamic models for trickle-bed reactors. Martin et al. (1992) applied γ-ray tomography to determine the spatial distribution of cracking catalyst in a riser tube (0.19 m diameter) of a cold-flow circulating fluidized bed and could assess the existence of the so-called core-annulus structure of the flow with a relative dense region near the wall of the riser tube.

g. *γ-Ray Densitometry.* A powerful method used to obtain information on the spatial solids distribution is γ-ray densitometry, which has been successfully applied in bubbling fluidized beds (Gidaspow et al., 1983) and circulating fluidized beds.

h. *Miscellaneous Methods.* For the measurement of key parameters, such as the solids volume fraction and solids velocity, in systems involving (dense) gas–solid two-phase flow a number of (specific) experimental techniques are available (Cheremisinoff, 1986b). Unfortunately, a number of powerful techniques such as

LDA and PIV, which are applicable in single-phase systems, cannot be applied in dense gas-solid systems due to the fact that they are not optically transparent. Due to this fact a large number of specific techniques have been developed for these industrially important flows. Very often probe techniques have been used to measure locally the particle concentration and/or the particle velocity (see Nieuwland *et al.*, 1996d, for a brief review). Due to the fact that probe techniques are usually intrusive, they may cause significant disturbances of the local flow behavior, which is especially disturbing in connection with validation of CFD models. Lim *et al.* (1995) have recently reviewed the hydrodynamics of gas–solid fluidization with particular emphasis on experimental findings and phenomenological modeling of bubbling and circulating fluidized beds. For gas–liquid systems a relatively large number of experimental techniques are available to obtain information on the gas holdup distribution. Also specific methods such as the "chemical method" should be mentioned here to determine the intefacial area by measuring the absorption rate of a gas in a suitably chosen absorption regime.

i. Concluding Remarks. Significant progress has been made in recent years with respect to the development of experimental techniques for both single-phase and multiphase flow applications. In Table IV a brief overview and classification of available experimental methods is presented. Clearly CFD has partly generated the driving force that has led to the development of some of these advanced experimental techniques and will continue to do so in the future. For multiphase flow there still exists a strong demand for nonintrusive experimental techniques which permit measurement of both time-averaged and fluctuating components of key quantities such as the local velocity and concentration of the phases present in the mixture.

VI. Selected Applications of CFD Work Conducted at Twente University

In this section some selected results are reported on CFD work that has been carried out within the context of several Ph.D. programs completed at Twente University. Most of this work has been done on hydrodynamic modeling of multiphase systems.

A. Two-Fluid Simulation of Gas Fluidized Beds

Our previous studies have clearly shown that two-fluid models are able to capture much of the complex system behavior featured by gas fluidized beds. For example, Kuipers *et al.* (1991) and Nieuwland *et al.* (1996a) studied bubble forma-

TABLE IV
OVERVIEW AND CLASSIFICATION OF EXPERIMENTAL METHODS

Measuring Technique	Quantity Measured	Point/Whole Field Technique	Instantaneous/ Time-Averaged	Intrusive/ Nonintrusive	Type of Flow Application
HWA	Velocity	Point	Instantaneous	Intrusive	Single phase G + L
LDA	Velocity	Point	Instantaneous	Nonintrusive	Single phase G + L Multiphase GL + dilute GS
LDA	Particle size	Point	Instantaneous	Nonintrusive	Single phase G + L multiphase dilute GS
PDA	Velocity	Point	Instantaneous	Nonintrusive	Single phase G + L Multiphase GL + dilute GS
PDA	Particle size	Point	Instantaneous	Nonintrusive	Single phase G + L Multiphase dilute GS
PIV	Velocity	Whole field	Instantaneous	Nonintrusive	Single phase G + L Multiphase GL + GLS
MRI	Velocity	Whole field	Instantaneous	Nonintrusive	Single phase G + L
LIF	Concentration	Point	Instantaneous	Nonintrusive	Single phase G + L
CARPT	Velocity	Whole field	Time-averaged	Nonintrusive	Multiphase GL
RPT	Velocity	Whole field	Time-averaged	Nonintrusive	Multiphase GS + GLS

tion at a single orifice in gas fluidized beds and have reported good agreement between experimentally observed bubble sizes and those obtained from two-fluid simulations. They have also shown that the two-fluid model, compared to existing approximate bubble formation models, in general, produced superior results. Figure 16 shows, as an example, the theoretically calculated growth, rise, and eruption of a single air bubble injected into a 2D air fluidized bed operated at minimum fluidization conditions, whereas Fig. 17 shows the experimentally observed and theoretically calculated bubble growth at a single orifice in a 2D air fluidized bed. From Fig. 17 it can be seen that the agreement between theory and experiment is satisfactory especially when one realizes that none of the parameter values required in the numerical simulation was derived from experimental data. Both theory and experiment indicate that especially during the initial stage of bubble formation a very sig-

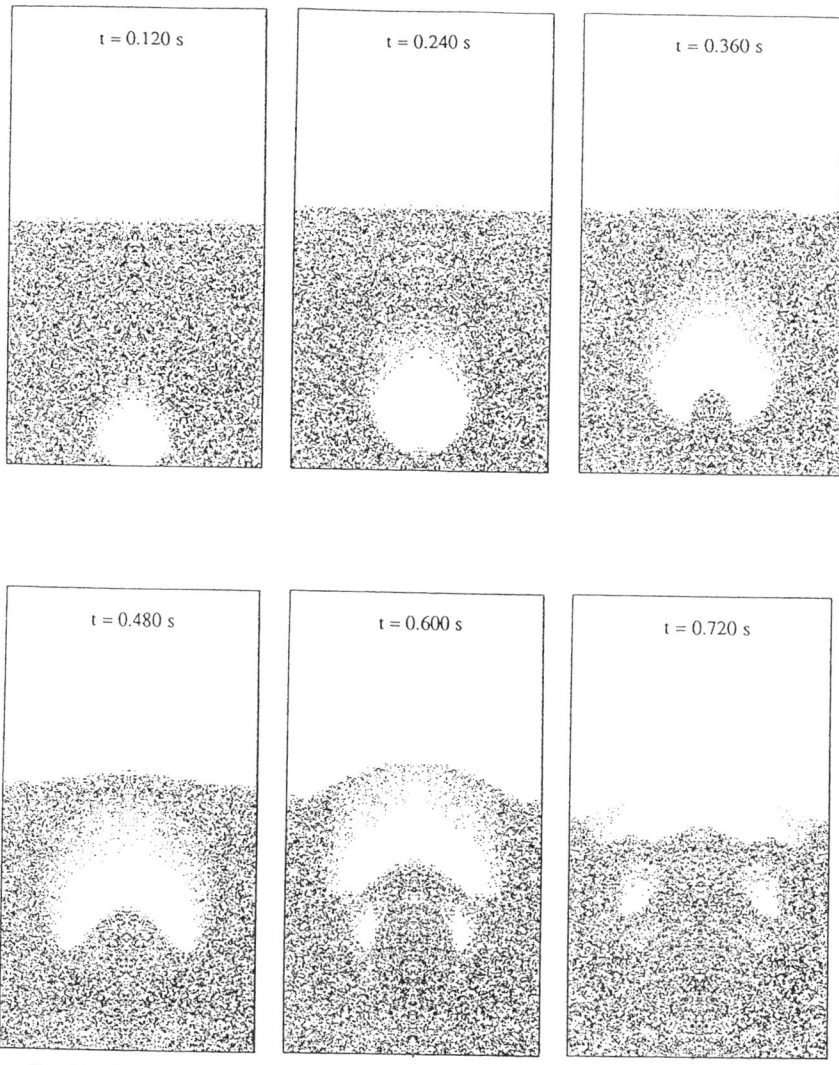

FIG. 16. Computer-generated solidity distribution showing the formation, rise, and eruption of a single bubble in a 2D gas fluidized bed. Physical properties of the particles: diameter, 500 μm; density, 2660 kg/m^3. Bed dimensions: width, 0.58 m; height, 1.0 m.

nificant gas leakage from the bubble to the emulsion phase occurs. A quantitative description of this phenomenon is of considerable importance to gain insight in grid conversion phenomena in fluidized-bed reactors.

Nieuwland *et al.* (1995) studied bubble formation at a single orifice in a 2D gas fluidized bed operated at elevated pressure and found very good agreement between theory (two-fluid simulations) and experiment. Similar results were ob-

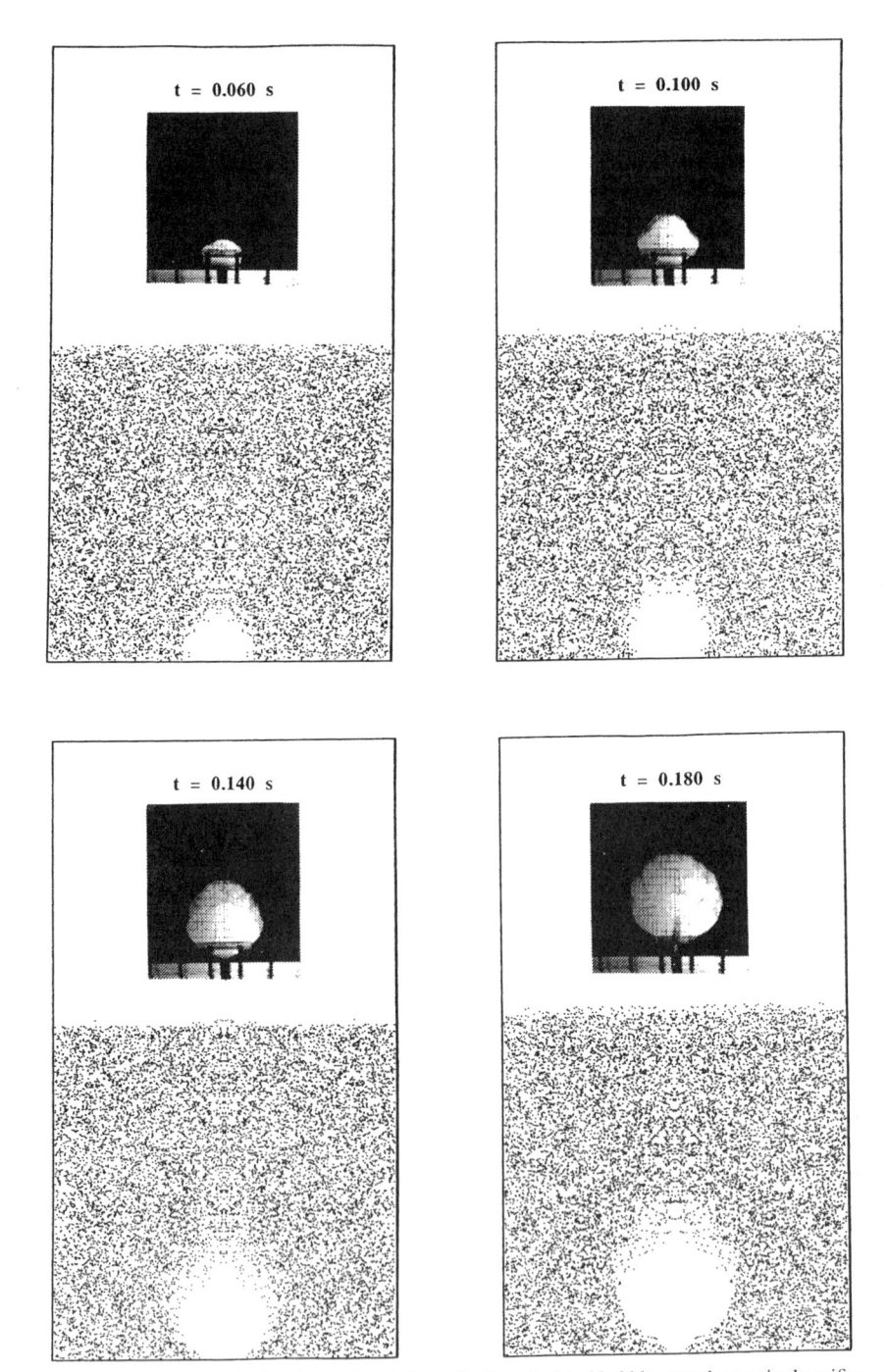

FIG. 17. Photographically observed and theoretically calculated bubble growth at a single orifice in a 2D gas fluidized bed. Physical properties of the particles: diameter, 500 μm; density, 2660 kg/m^3. Bed dimensions: width, 0.58 m; height, 1.0 m. Injection velocity through orifice 10.0 m/s.

tained by Huttenhuis *et al.* (1996) who studied the effect of gas phase density on bubble formation by injecting He, air, or SF_6 into an incipiently air-fluidized bed. Huttenhuis *et al.* (1996) also reported results of 3D simulations and found a clear effect of the front and back wall of their pseudo 2D bed. Their results stress the importance of 3D hydrodynamic modeling.

Two-fluid simulations have also been performed to predict void profiles (Kuipers *et al.*, 1992b) and local wall-to-bed heat transfer coefficients in gas fluidized beds (Kuipers *et al.*, 1992c). In Fig. 18 a comparison is shown between experimental (a) and theoretical (b) time-averaged porosity distributions obtained for a 2D air fluidized bed with a central jet (air injection velocity through the orifice: 10.0 m/s which corresponds to $40u_{mf}$). The experimental porosity distributions were obtained with the aid of a nonintrusive light transmission technique where the principles of liquid–solid fluidization and vibrofluidization were employed to perform the necessary calibration. The principal differences between theory and experiment can be attributed to the simplified solids rheology assumed in the hydrodynamic model and to asymmetries present in the experiment.

Figure 19 shows, as an example, the evolution and propagation of bubbles in a 2D gas-fluidized bed with a heated wall. The bubbles originate from an orifice near the heated right wall (air injection velocity through the orifice's 5.25 m/s, which corresponds to $21u_{mf}$). The instantaneous axial profile of the wall-to-bed heat transfer coefficient is included in Fig. 19. From this figure the role of the developing bubble wake and the associated bed material refreshment along the heated wall, and its consequences for the local instantaneous heat transfer coefficient, can be clearly seen. In this study it became clear that CFD based models can be used as a tool (i.e., a learning model) to gain insight into complex system behavior.

B. Discrete Particle Simulation of Gas Fluidized Beds

Hoomans *et al.* (1996) developed a discrete particle model of gas fluidized beds where the 2D motion of individual, spherical particles was directly calculated from the forces acting on them, accounting for particle–particle and particle–wall interaction and interaction with the interstitial gas phase. Their collision model is based on the conservation laws for linear and angular momentum and requires, apart from geometrical parameters, a restitution coefficient and a friction coefficient. Techniques which are well known within the field of molecular dynamics were used to process a sequence of collisions. In fact, the model developed by Hoomans *et al.* can be seen as a combination of a so-called granular dynamics approach for the colliding particles and a CFD approach for the gas phase percolating through the particles. Discrete particle models offer certain distinct advantages over two-fluid models since they require no specific assumptions concerning the solids rheology. Furthermore, the incorporation of a particle size distribution can be accommodated with relative ease.

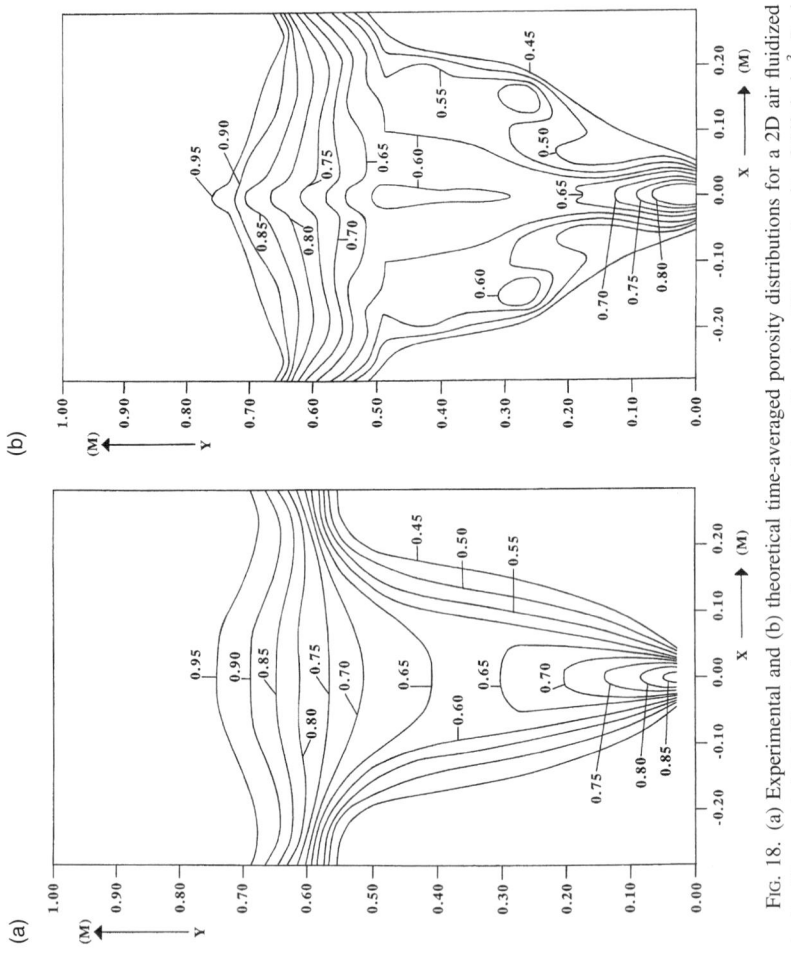

FIG. 18. (a) Experimental and (b) theoretical time-averaged porosity distributions for a 2D air fluidized bed with a central jet. Physical properties of the particles: diameter, 500 μm; density, 2660 kg/m^3. Bed dimensions: width, 0.57 m; height, 1.0 m. Injection velocity through central orifice: 10.0 m/s.

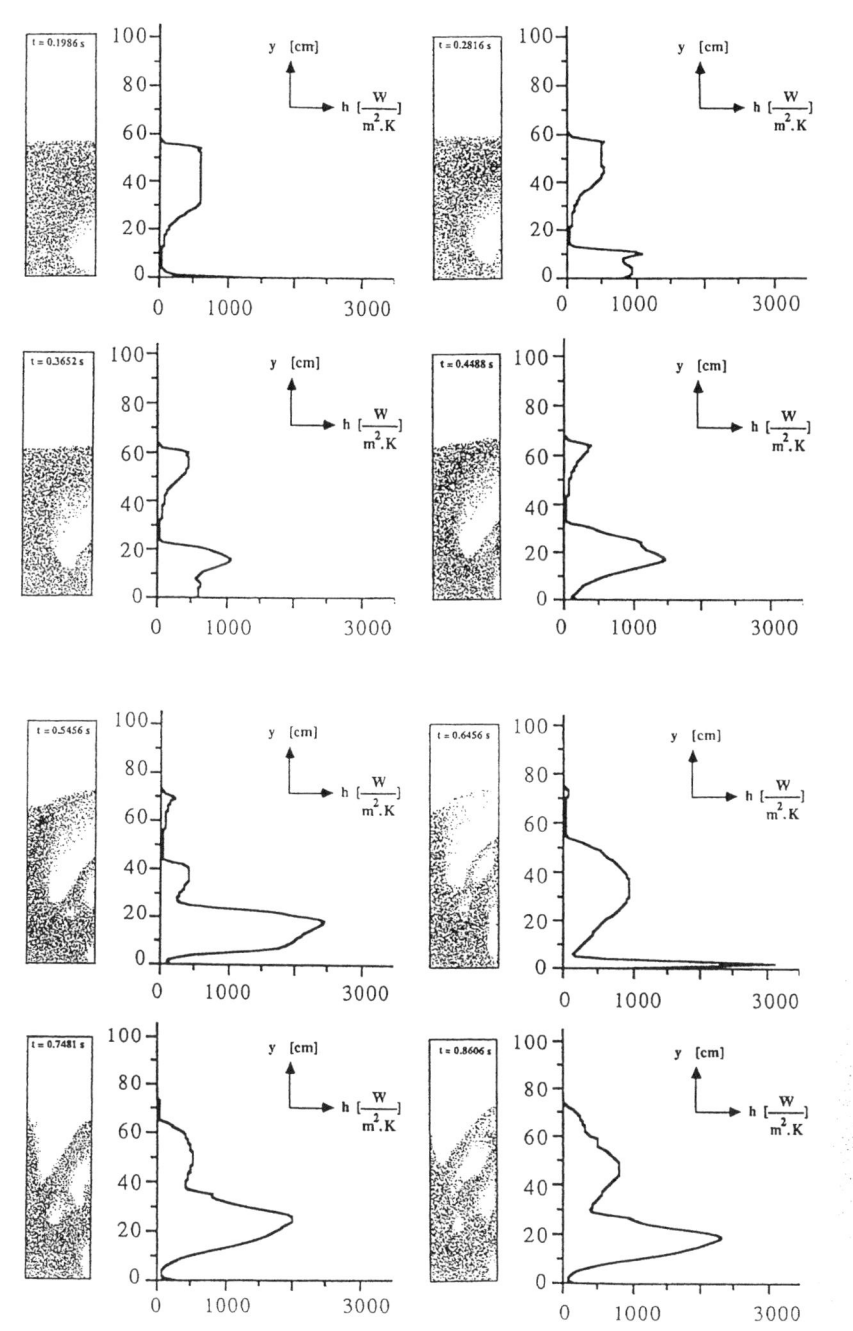

FIG. 19. Evolution and propagation of bubbles in a 2D air fluidized bed with a heated wall. The bubbles originate from an orifice near the heated right wall. Physical properties of the particles: diameter, 500 μm; density, 2660 kg/m^3. Bed dimensions: width, 0.285 m; height, 1.0 m. Injection velocity through orifice 5.25 m/s.

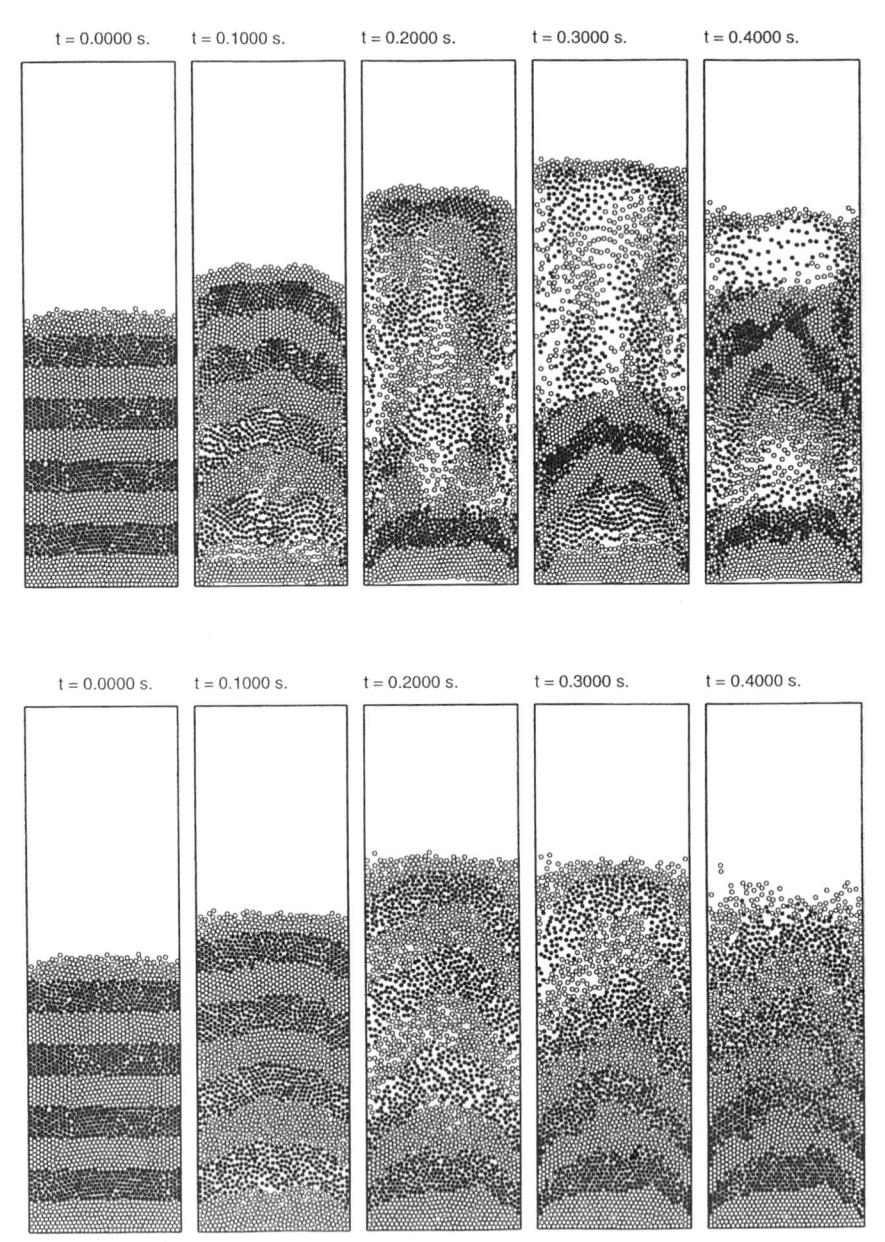

FIG. 20. Snapshots of particle configurations for the simulation of slug formation with homogeneous inflow conditions in a 2D gas fluidized bed. Top: Nonideal particles ($e = e_w = 0.9$ and $\mu = \mu_w = 0.3$); bottom: ideal particles ($e = e_w = 1.0$ and $\mu = \mu_w = 0.0$).

t = 0.5000 s. t = 0.6000 s. t = 0.8000 s. t = 1.0000 s.

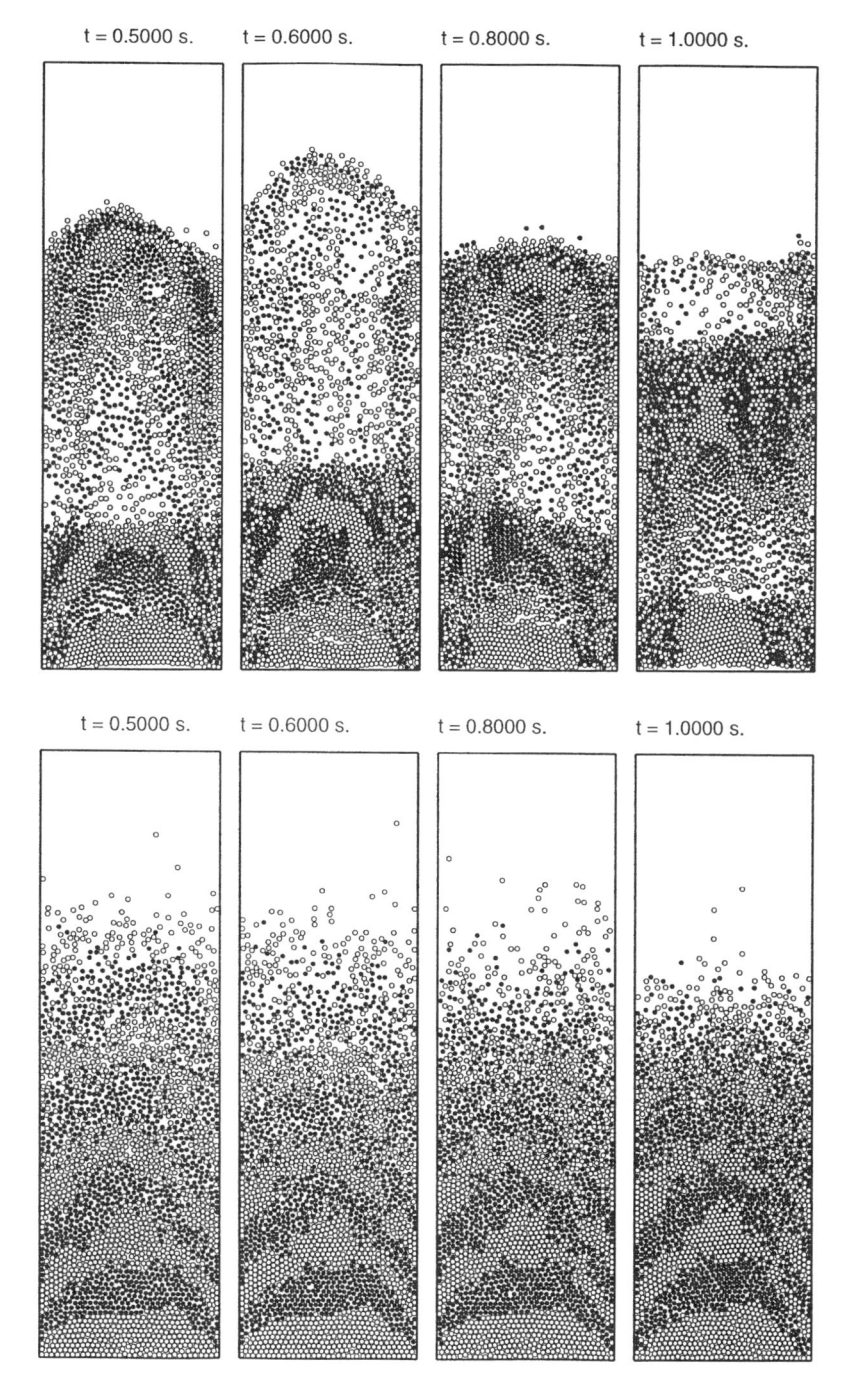

t = 0.5000 s. t = 0.6000 s. t = 0.8000 s. t = 1.0000 s.

FIG. 20. Continued.

Figure 20 shows a sequence of particle configurations obtained from a discrete particle simulation of a two-dimensional gas fluidized bed where homogeneous inflow conditions were set at the bottom of the bed. The number of particles was 2400 whereas the diameter and the density of the particles were, respectively, 4 mm and 2700 kg/m^3. The particle configurations shown in Fig. 20 (top) represent the predicted bed behavior in case realistic values were used for the restitution and friction coefficient ($e = 0.9$ and $\mu = 0.3$), whereas the particle configurations shown in Fig. 20 (bottom) refer to ideal particles ($e = 1$ and $\mu = 0$). From this figure it can clearly be seen that the global system behavior, that is, the occurrence of bubbles and slugs and the associated mixing rate, differs significantly between the two cases. These results clearly indicate that dissipative processes on a microscale have a decisive effect on the global system dynamics.

C. CIRCULATING FLUIDIZED BEDS

Nieuwland *et al.* (1996b,c) applied the kinetic theory of granular flow (KTGF) to study the hydrodynamics in a CFB riser tube. Their hydrodynamic model is based on the concept of two fully interpenetrating continua and consists of the two-fluid mass and momentum conservation equations describing the mean motion of the gas–solid dispersion and the granular temperature equation describing the fluctuating motion in the solid phase (Nieuwland, 1995). They used a modified Prandtl mixing length model to account for turbulent momentum transport in the gas phase. Nieuwland *et al.* assumed ideal (i.e., elastic) particle–particle collisions and nonideal particle–wall collisions because preliminary computations in which inelastic mutual particle collisions were assumed failed to produce the experimentally observed lateral solids segregation (Nieuwland *et al.*, 1996b). This phenomenon is most likely due to the fact that the interaction between the fluctuating motions in the gas phase and solids phase is not accounted for in the model used by Nieuwland *et al.*

Figure 21a shows a comparison between computed and measured radial profiles of solids concentration for three superficial gas velocities [$U = 7.5$ m/s, $U = 10.0$ m/s, and $U = 15.0$ m/s) at a constant solids mass flux G_s of 300 kg/(m^2 · s)]. The corresponding radial profiles of the axial solids velocity are shown in Figure 21b. The radial profiles of solids volume fraction and axial solids velocity were obtained with the aid of an optical probe developed by Nieuwland *et al.* (1996d) and similar in concept to the probe developed by Hartge *et al.* (1986). From Fig. 21a it can be seen that the radial solids segregation significantly increases with decreasing superficial gas velocity, which is in accordance with expectations. In general, their hydrodynamic model seems to underpredict the experimentally observed lateral solids segregation especially at the lowest superficial gas velocity of 7.5 m/s. However, it should be mentioned that in this case the average solids mass flux obtained from the experimental data showed a considerable positive (48%) deviation form the imposed solids mass flux of 300 kg/(m^2 · s). In

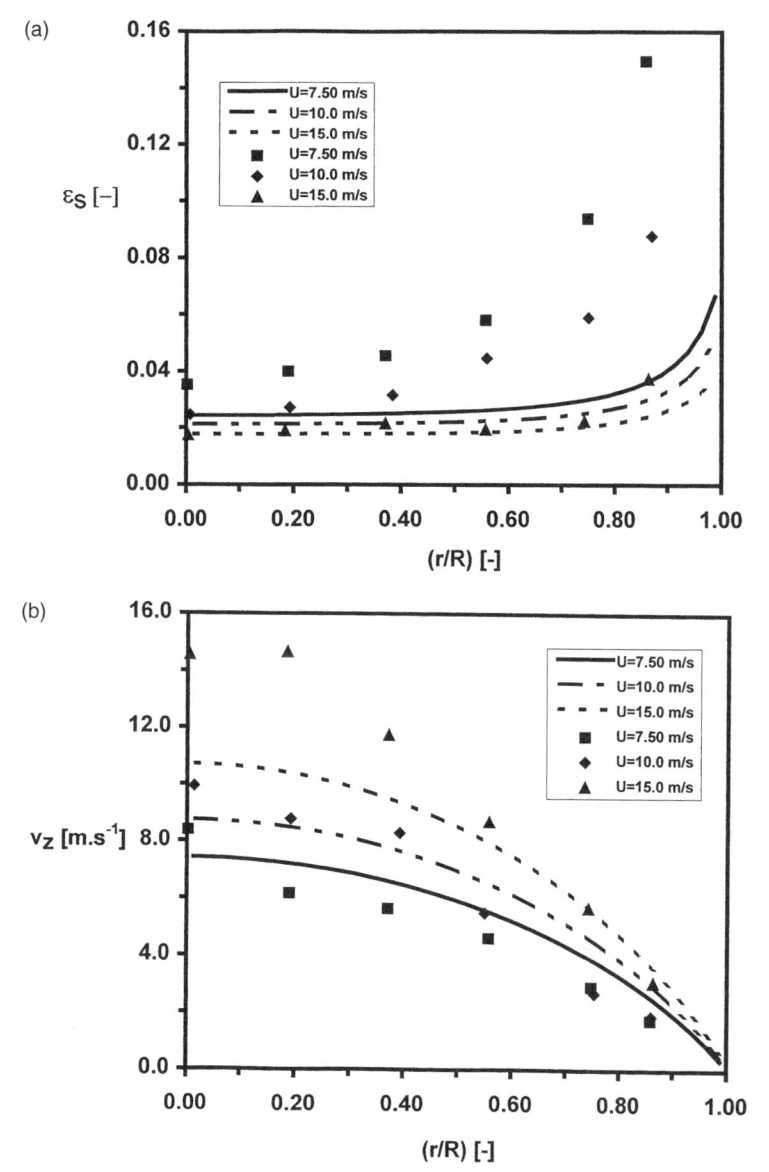

FIG. 21. Computed and measured radial profiles of (a) solids concentration and (b) axial solids velocity in a CFB riser for three superficial gas velocities (U = 7.5 m/s, U = 10.0 m/s, and U = 15.0 m/s) at a constant mass flux G_s of 300 kg/(m^2 · s). Riser diameter D = 0.0536 m, physical properties of the particles: diameter, 126 μm; density, 2540 kg/m^3.

general, the agreement between theory and experiment is quite reasonable especially when it is borne in mind that their model contains no adjustable parameters. From Figure 21b it can be seen that the model predicts, in accordance with the experimental data, no solids downflow near the tube wall.

In addition Nieuwland *et al.* (1996b,c) compared their computational results with the experimental data reported by Bader *et al.* (1988). They performed experiments in a cold-flow CFB unit (riser tube: $D = 0.304$ m, $L = 10.0$ m) with FCC ($d_p = 76$ μm, $\rho_s = 1714$ kg/m^3) as bed material and air as fluidizing agent. Preliminary calculations without using the KTGF failed to produce the experimentally observed lateral solids segregation, which demonstrates the importance of accounting for the collisional interaction between particles. Figure 22a shows a comparison between computed and measured radial profiles of solids concentration for a superficial gas velocity U of 3.7 m/s and a solids mass flux G_s of 98 kg/(m$^2 \cdot$ s). The corresponding radial profiles of the axial solids velocity are shown in Fig. 22b. Note that in this case the model predicts, in accordance with experiments, solids downflow near the tube wall.

In both figures the results of two additional computations are shown. In one of these calculations the constitutive equations derived by Ding and Gidaspow (1990) were used, which neglect the kinetic contributions in, respectively, the expressions for the solid phase shear viscosity and the solid phase pseudoconductivity. For the intermediate diameter riser ($D = 0.304$ m) used by Bader *et al.* (1988) the total (i.e., for both gas and solid phase) momentum transport in the radial direction is dominated by the turbulent contribution in the gas phase, which explains the small effect of neglecting the aforementioned kinetic contributions in the solid phase. In the other additional calculation shown in Figs 22a and b the turbulent contribution in the expression for the gas phase shear viscosity was neglected. In this case the solids downflow near the tube wall is significantly overpredicted, which demonstrates the importance of accounting for macroscopic turbulent transport in the gas phase.

The results presented so far in this section correspond to the regime of fully developed riser flow. Kuipers and van Swaaij (1996) applied the KTGF-based model developed by Nieuwland *et al.* (1996b,c) to study the effect of riser inlet configuration on the (developing) flow in CFB riser tubes and found that the differences in computed radial profiles of hydrodynamic key variables (i.e., gas and solids phase mass fluxes) rapidly disappear with increasing elevation in the riser tube.

D. BUBBLE COLUMNS

Bubble columns find frequent application in the process industries due to their relatively simple construction and advantageous properties such as excellent heat transfer characteristics to immersed surfaces. Despite their frequent use in a variety of industrial processes, many important fluid dynamical aspects of the prevailing gas–liquid two-phase flow in bubble columns are unfortunately poorly un-

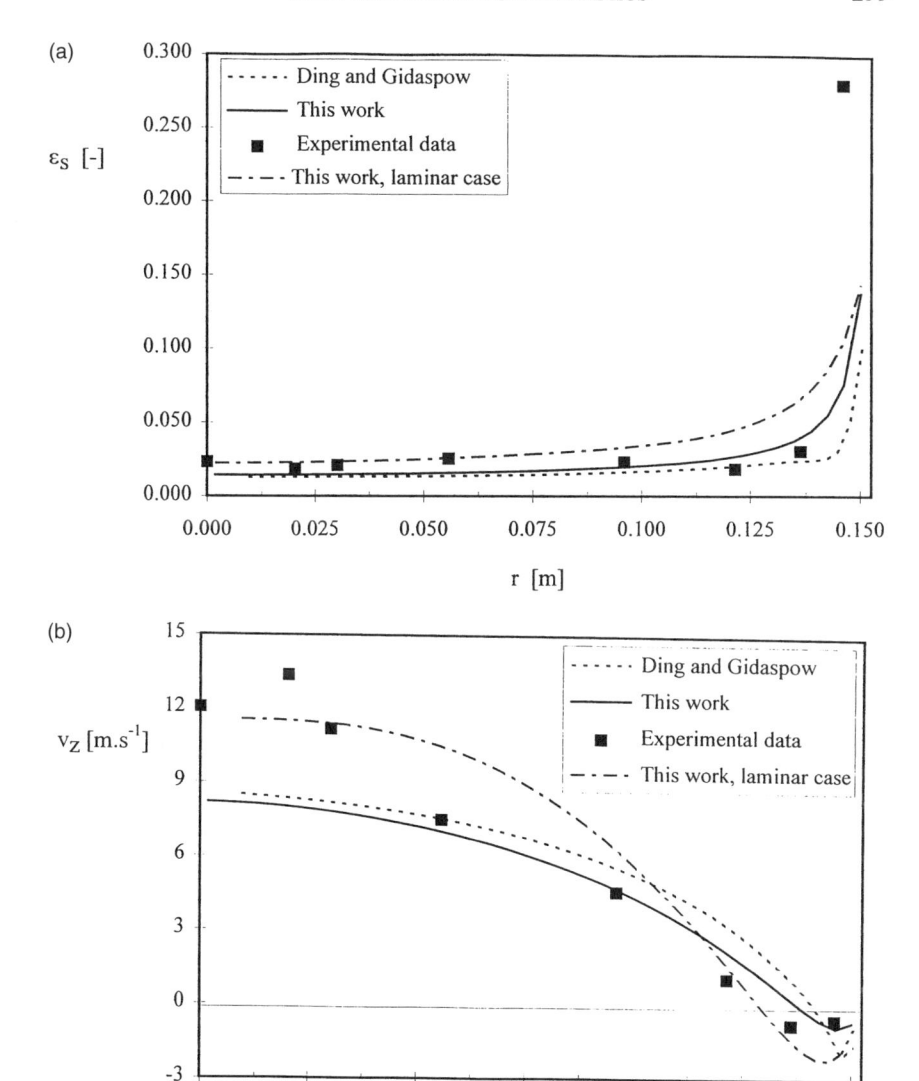

FIG. 22. Computed and measured (Bader et al., 1988) radial profiles of (a) solids concentration and (b) axial solids velocity in a CFB riser for a superficial gas velocity U of 3.7 m/s and a mass flux G_s of 98 kg/(m$^2 \cdot$ s). Riser diameter $D = 0.304$ m, physical properties of the particles: diameter, 76 μm; density, 1714 kg/m^3.

derstood. This relative unsatisfactory state of the art has led to an increased interest in recent years in development of advanced experimental tools (Devanathan *et al.*, 1990; Lin *et al.*, 1996) and detailed modeling of bubble columns on the basis of the hydrodynamic equations of change (see Sect. IV, B,1).

Depending on the magnitude of the superficial gas velocity, three flow regimes (Fan, 1989) can be distinguished, namely, (1) dispersed bubble regime, (2) vortical-spiral flow regime, and (3) turbulent flow regime. In the dispersed bubble or homogeneous regime, relatively small nearly spherical gas bubbles are present with a more or less uniform size, whereas in the vortical-spiral and turbulent flow regimes besides the small gas bubbles relatively (very) large gas bubbles are seen with a size-dependent characteristic shape. In the next two subsections two different models will be discussed which apply to a situation where only many small (spherical) gas bubbles are present and a situation where a few large (nonspherical) gas bubbles are present.

1. Discrete Bubble Model

Delnoij *et al.* (1997a) developed a detailed hydrodynamic model for dispersed gas–liquid two–phase flow based on a mixed Eulerian–Lagrangian approach. Their model describes the time-dependent motion of small, spherical gas bubbles in a bubble column operating in the homogeneous regime where all relevant forces acting on the bubble (drag, virtual mass, lift, and gravity forces) were accounted for. Direct bubble–bubble interactions were accounted for via an interaction model, which resembles the collision approach followed by Hoomans *et al.* (1996) to model gas fluidized beds. Delnoij *et al.* (1997a) simulated two experiments reported by Becker *et al.* (1994), one for a large superficial gas velocity and another one for a small superficial gas velocity. The geometry used by Becker *et al.* comprises a pseudo two-dimensional bubble column equipped with a gas distributor section containing five, individually controllable porous plates. Becker *et al.* fed air to the column through only one of the five porous plates during their experiments.

a. Results for a Large Superficial Gas Velocity. Figure 23 shows a sequence of plots showing both the instantaneous configuration of the bubbles and the associated flow field in the liquid phase. In accordance with visual observations reported by Becker *et al.* the model correctly predicts a powerful liquid circulation, which pushes the bubble swarm firmly toward the left wall of the column. This strong liquid circulation is induced by the large number of bubbles that rise through the column.

b. Results for a Small Superficial Gas Velocity. At lower superficial gas velocities Becker *et al.* (1994) observed a remarkable transition in the liquid phase flow pattern in their bubble column. Contrary to the experiment with a large superficial gas velocity, in this case the bubble swarm was found to move upward in a meandering manner. Several liquid circulation cells were reported that changed their location and size continuously. The flow was observed to be highly dynamic with a period of oscillation of the plume of approximately 41 s. Figure 24 shows a sequence of plots showing both the instantaneous configuration of the bubbles and the associated flow field in the liquid phase. The meandering behavior and main liquid phase flow characteristics reported by Becker *et al.* could clearly be reproduced by the model developed by Delnoij *et al.* However, the period of oscillation calculated by their model was approximately 30 s and was felt to be due to the two-dimensional nature of the model.

c. Effect of Column Aspect Ratio on Flow Structure. Delnoij *et al.* also investigated the effect of the bubble column aspect ratio on the prevailing flow structure and compared their computational results with experimental data reported by Chen *et al.* (1989). A transition in the gas–liquid flow pattern was predicted in case the aspect ratio of the column changed from two to four. For an aspect ratio of two the Gulfstream type of liquid circulation was predicted, whereas for an aspect ratio of four a highly dynamic liquid flow pattern with multiple vortices was computed. From computer animations it could be seen that these vortices were generated at the free surface. Furthermore, these vortices were found to be positioned staggered with respect to each other in the column. In part these computational results were supported by Chen's findings. The only major difference with the experimental observations of Chen *et al.* was the fact that the aforementioned transition already occurred at an aspect ratio of one. Again this discrepancy is most likely due to the two-dimensional nature of the model developed by Delnoij *et al.* Finally as an illustration in Figure 25 a few plots of instantaneous bubble configurations are shown for bubble columns with aspect ratios ranging from 4.8 to 11.0. From these figures it can be clearly seen that the flow structure is significantly affected by the column aspect ratio. In Figure 25 for one case ($L/D = 7.7$) the time-averaged liquid velocity is also shown. The existence of the characteristic large-scale circulation with liquid upflow in the center of the column and down-flow near the walls can be recognized clearly.

2. *Volume of Fluid (VOF) Model*

Delnoij *et al.* (1997b) developed a computer code based on the volume of fluid (VOF) method (Hirt and Nichols, 1981; Nichols *et al.*, 1980) to describe the dynamics of single gas bubbles rising in a quiescent Newtonian liquid. They were able to show that the predicted bubble shape and associated flow patterns induced in the liquid phase could be predicted very well with the VOF method over a wide range of Eotvös (Eo) and Morton (M) numbers (see Sec. IV,B,1 for the definitions of Eo and M). In Fig. 26 the formation and rise of single gas bubbles emanating from a central orifice is shown for various values of the Eotvös and Morton num-

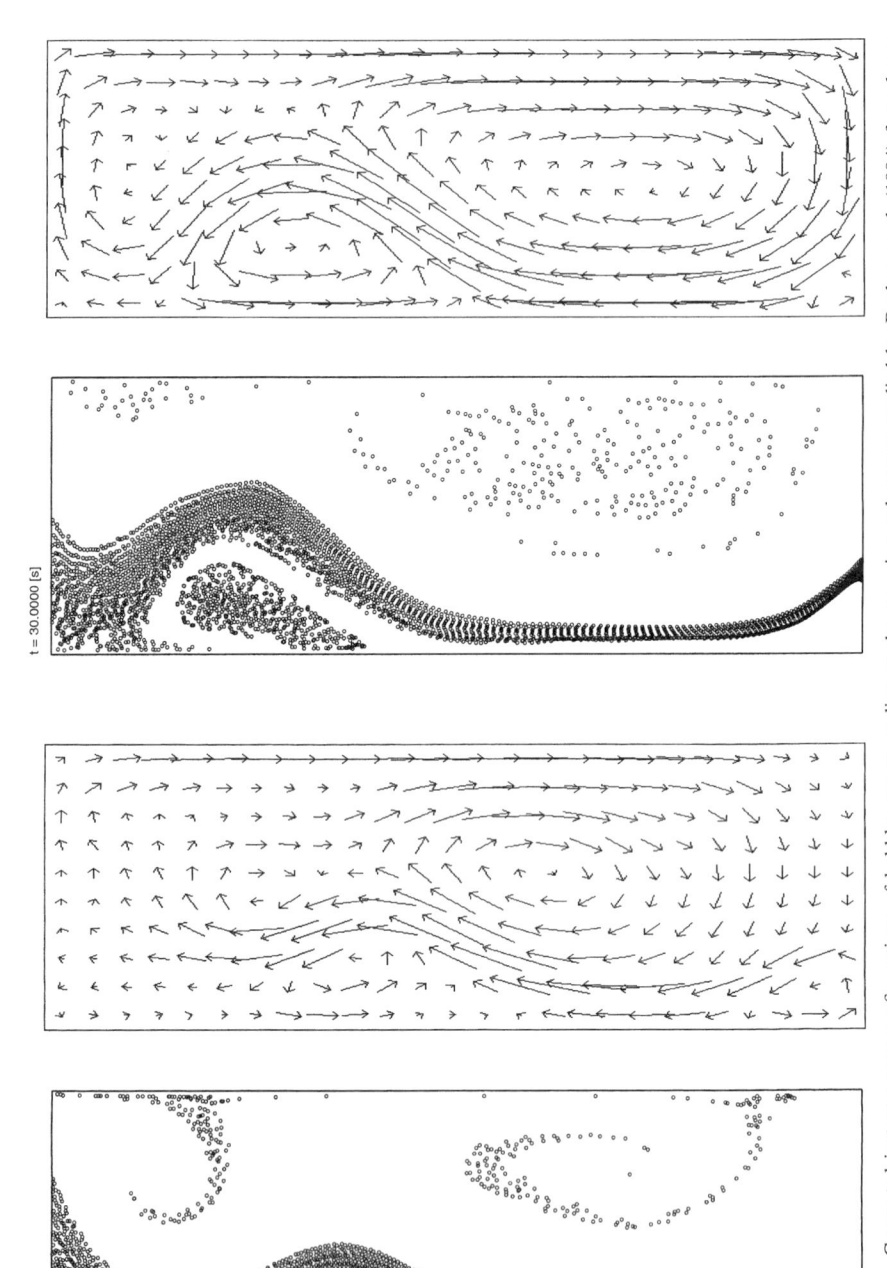

FIG. 23. Computed instantaneous configurations of bubbles corresponding to the experimental system studied by Becker *et al.* (1994) for a large superficial gas velocity.

FIG. 23. Continued

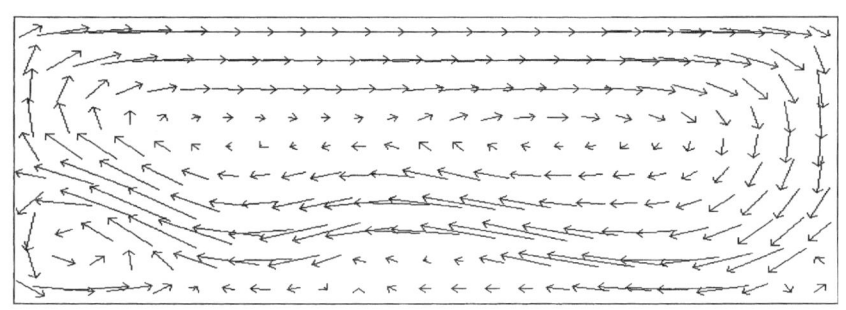

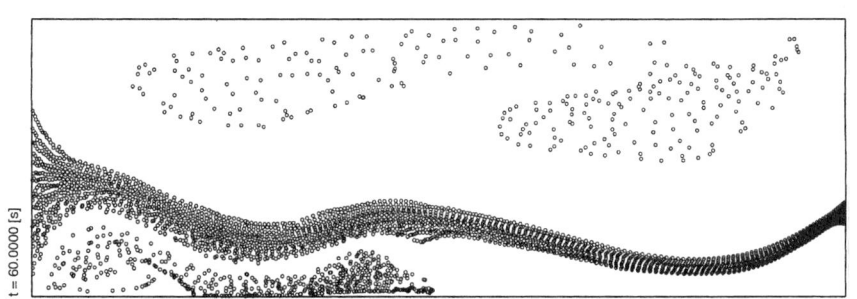

FIG. 24. Computed instantaneous configurations of bubbles corresponding to the experimental system studied by Becker *et al.* (1994) for a small superficial gas velocity.

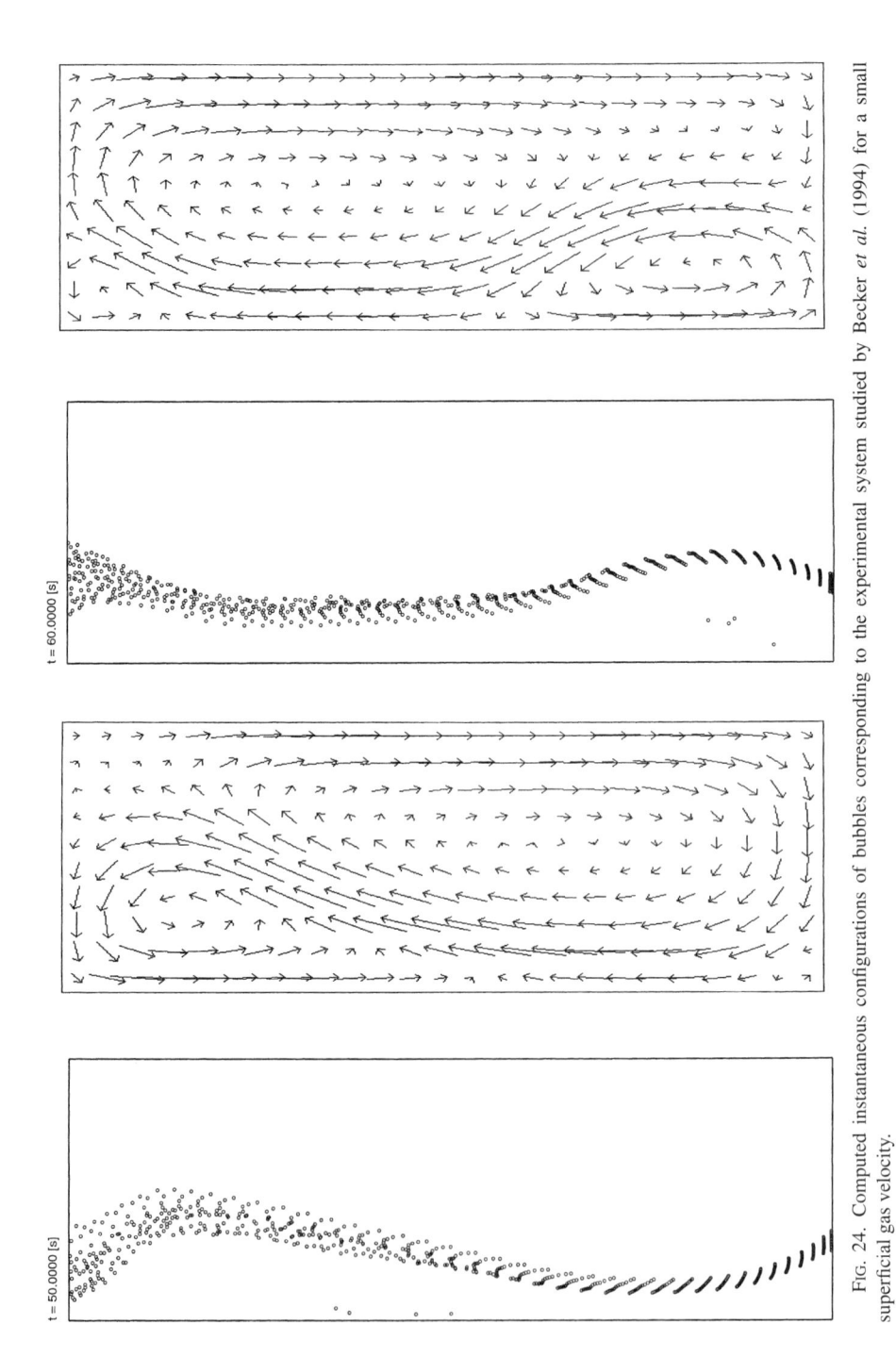

Continued

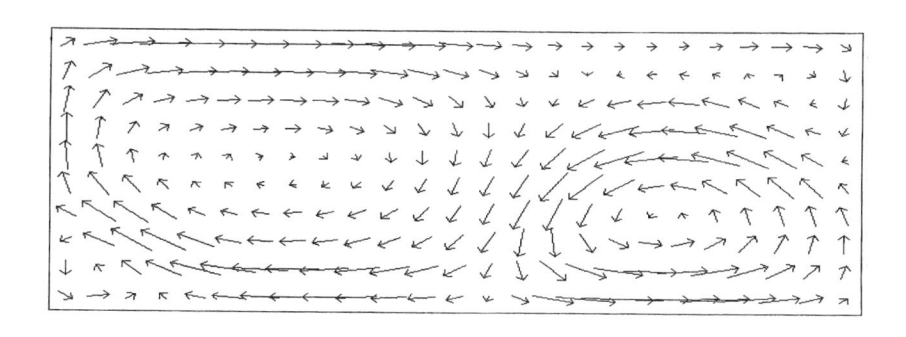

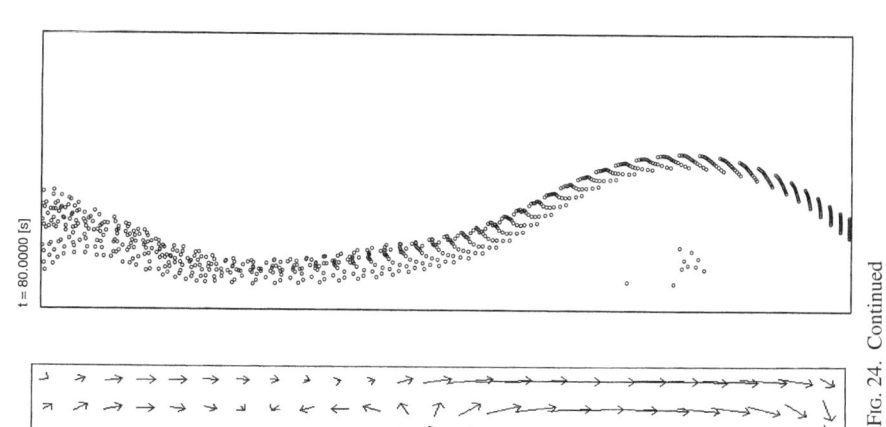

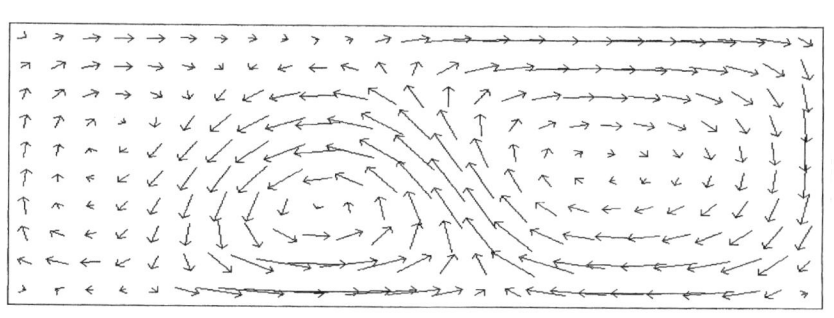

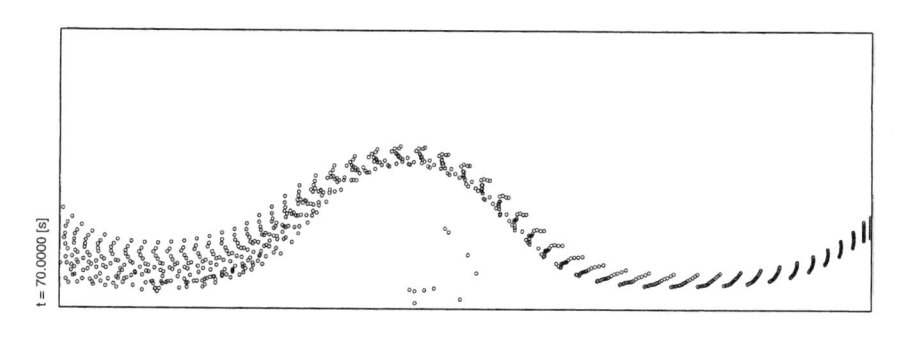

FIG. 24. Continued

305

306 J. A. M. KUIPERS AND W. P. M. VAN SWAAIJ

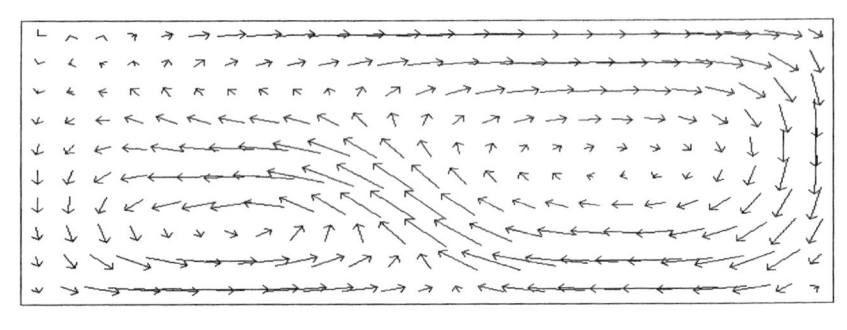

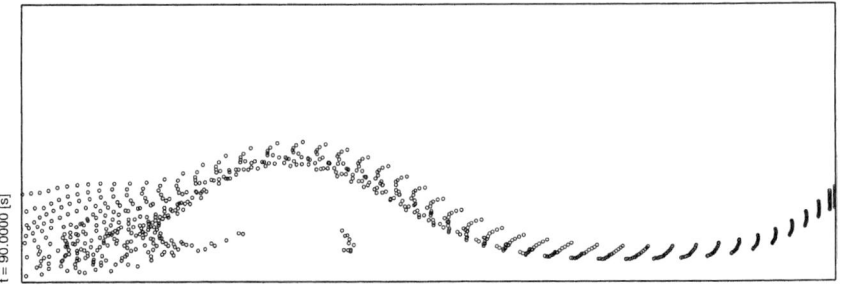

FIG. 24. Continued

COMPUTATIONAL FLUID DYNAMICS 307

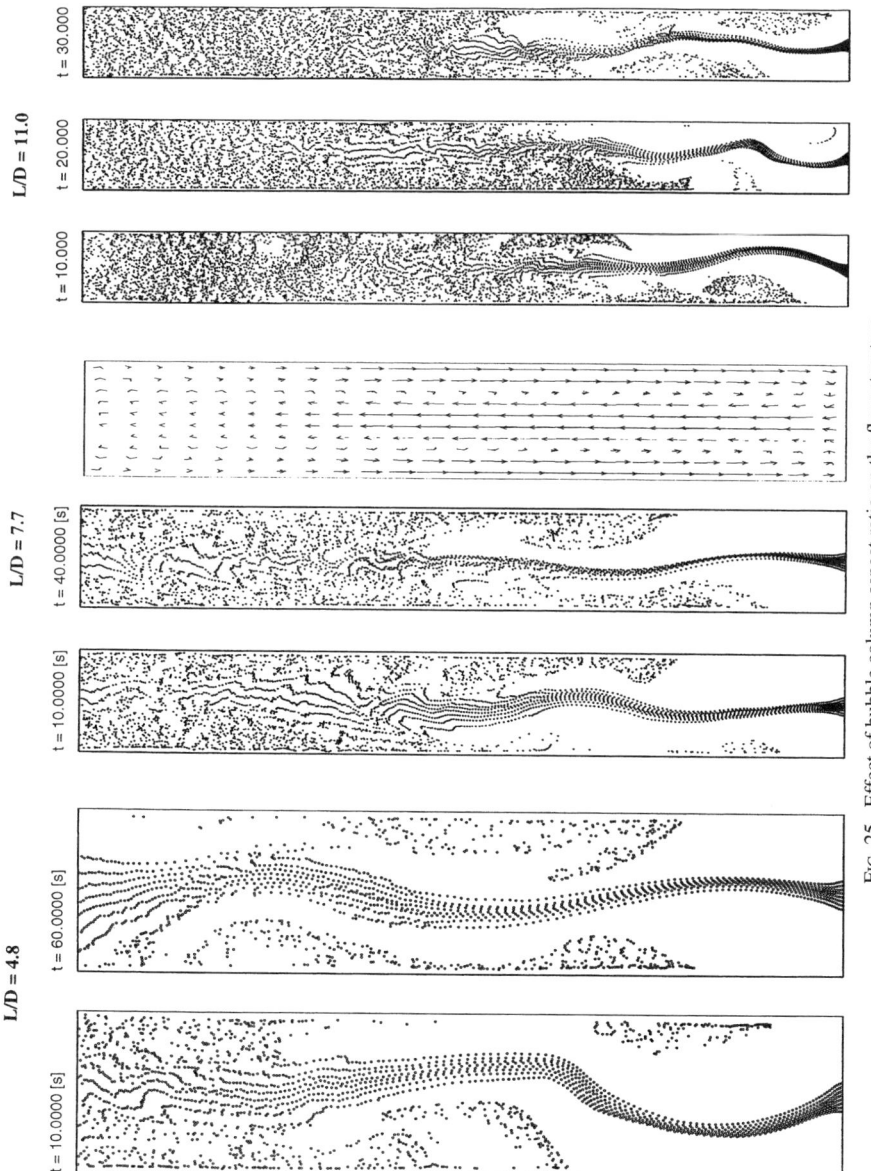

FIG. 25. Effect of bubble column aspect ratio on the flow structure.

ber. For the purpose of reference in this figure the diagram presented by Grace (1973) and Grace *et al.* (1976) is included, which shows the effect of fluid properties and equivalent bubble diameter on the bubble characteristics (i.e., shape and terminal velocity). The relevant parameter values corresponding to the simulation results presented in Fig. 26 are listed in Table V. From Fig. 26 it can be seen that the bubble corresponding to case *d* (the wobbling regime) exhibits an oscillatory motion during its rise through the liquid. From computer animations it could be clearly seen that this phenomenon is caused by vortices that are shed in an alternating mode at the left and right rear part of the bubble, similar to vortices that are shed behind a (stationary) circular cylinder subjected to cross-flow (Schlichting, 1975). As evident from inspection of Fig. 26 the computed bubble shapes show very close resemblance with those expected on the basis of the aforementioned diagram. It should be mentioned here that the diagram presented by Grace (1973) and Grace *et al.* (1976) is valid for three-dimensional gas bubbles rising in an unbounded Newtonian liquid, whereas the model is based on a two-dimensional approach where the bubbles rise in a system with finite lateral dimensions. These two differences were felt to be responsible for the fact that the bubble rise velocities inferred from this diagram systematically exceeded the computed bubble rise velocity found by Delnoij *et al.* Therefore they also studied the effect of the column width on the magnitude of bubble rise velocity and found an increasing bubble rise velocity with increasing column width at fixed (equivalent) bubble diameter. This is expected behavior because in a system with relatively small lateral dimension there is less space for the downward-moving liquid induced by the rising gas bubble.

In addition Delnoij *et al.* used the VOF method to compute the coalescence of two coaxial gas bubbles of identical size generated at the same orifice. Figure 27a shows the temporal evolution of the positions of the two gas bubbles. From the sequence of bubble positions it can be seen that the trailing bubble moves faster than the leading bubble and eventually at $t = 0.42$ s coalescence of the two gas bubbles commences. From computer animations it could clearly be seen that just after completion of the coalescence process (at $t = 0.45$ s) a "splashing" liquid

TABLE V
VALUES OF THE EÖTVÖS (Eo) AND MORTON (M) NUMBERS USED IN THE SIMULATIONS DEPICTED IN FIG. 6.11

Case	Regime	Eo	M	D^a (m)	L^a (m)
a	Spherical	1	$1.4 \cdot 10^{-4}$	0.01	0.02
b	Ellipsoidal	10	0.137	0.05	0.10
c	Spherical cap	100	$2.5 \cdot 10^{-11}$	0.10	0.20
d	Wobbling	2	$2.5 \cdot 10^{-11}$	0.02	0.06

[a] D and L, respectively, represent the lateral and vertical dimensions of the computational domain.

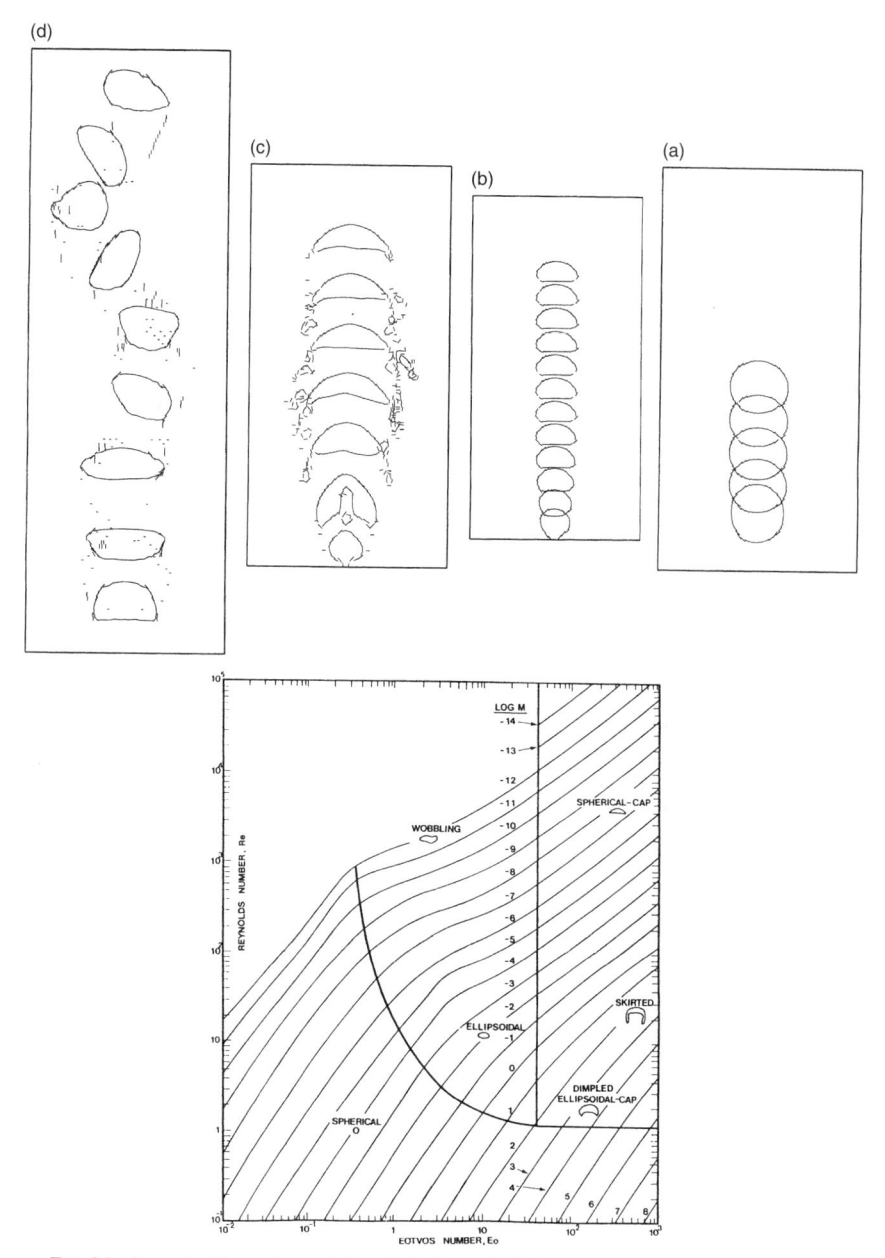

FIG. 26. Computed formation and rise of single gas bubbles emanating from a central orifice for (top) various values of the Eotvös (Eo) and Morton (M) number and (bottom) graphical correlation due to Grace (1973) and Grace et al. (1976).

310   J. A. M. KUIPERS AND W. P. M. VAN SWAAIJ

(a)

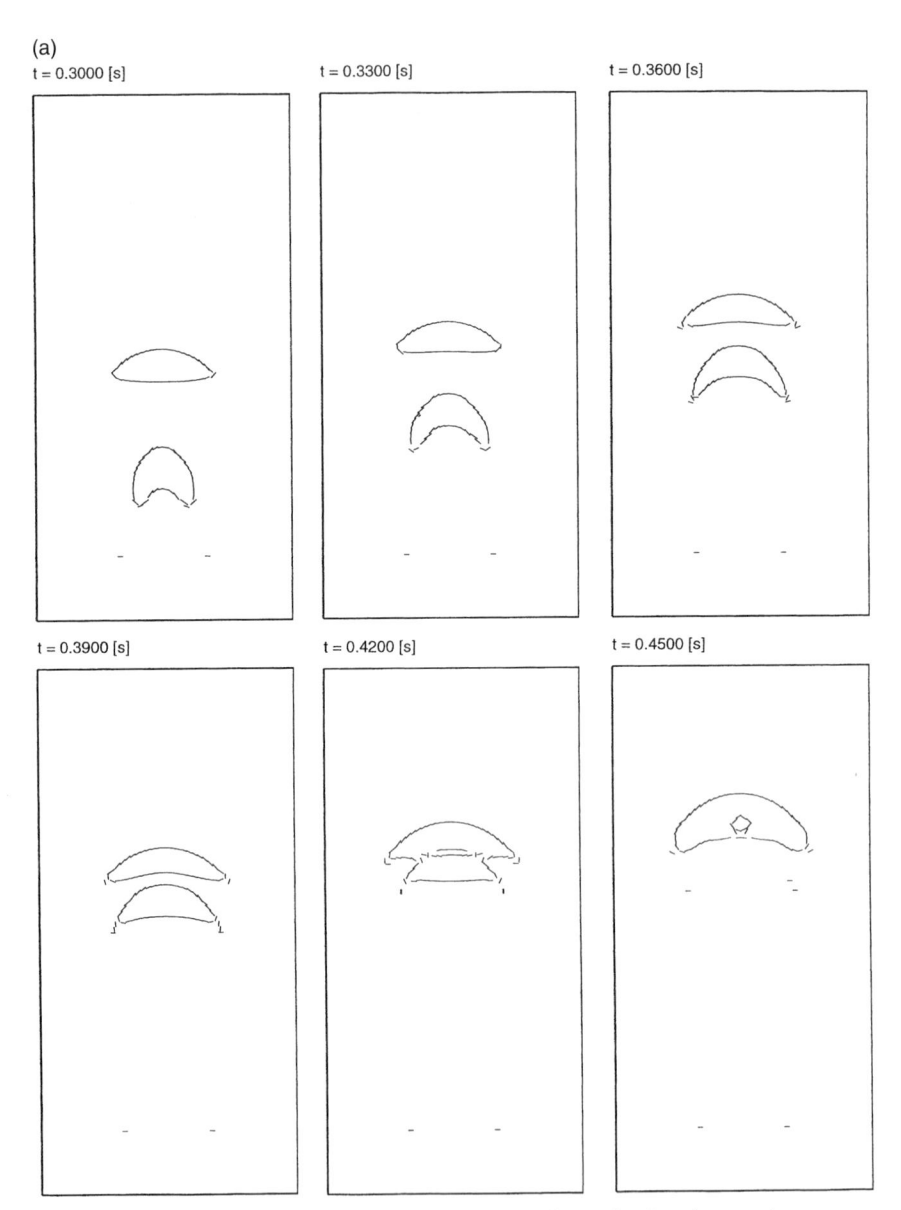

FIG. 27. (a) Experimentally observed and (b) theoretically predicted coalescence between two coaxial bubbles generated at an orifice (Eo = 16, M = $2.0.10^{-4}$, lateral (D) and vertical (L) dimensions of the computational domain: $D = 0.05$ m, $L = 0.10$ m).

(b)

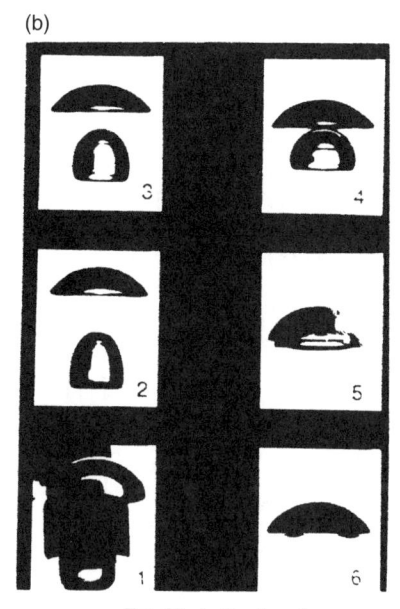

FIG. 27. b, Continued

pocket forms at the bubble base. Due to the fact that the trailing bubble rises in the wake region of the leading bubble, a difference in bubble shape develops. In Figure 27b the results of the corresponding experiment, reported by Brereton and Korotney (1991), are shown. From a comparison of the theoretical and experimental results one can conclude that a reasonable agreement is obtained, especially in view of the complexity of the coalescence process. Moreover, one should keep in mind that no adjustable parameters were used in the VOF model. Finally Fig. 28 shows the positions of the two coaxial gas bubbles at $t = 0.33$ s together with the distribution of the relative velocity field of the liquid phase. In this figure the instantaneous velocity of the leading gas bubble was subtracted from the instantaneous liquid phase velocity distribution. From Fig. 28 it can be seen that at $t = 0.33$ s the trailing bubble rises in the wake region of the leading bubble with upward (instead of downward) flowing liquid near its nose. As a consequence of this phenomenon, the trailing bubble attains a higher rise velocity than the leading bubble with bubble coalescence as the final result.

E. Modeling of a Laminar Entrained Flow Reactor

Within the context of a study aimed at the determination of the pyrolysis kinetics of polymers using a laminar entrained flow reactor (LEFR), Westerhout *et al.* (1996) developed a comprehensive model that couples a single-particle cont = 0.3300 [s]

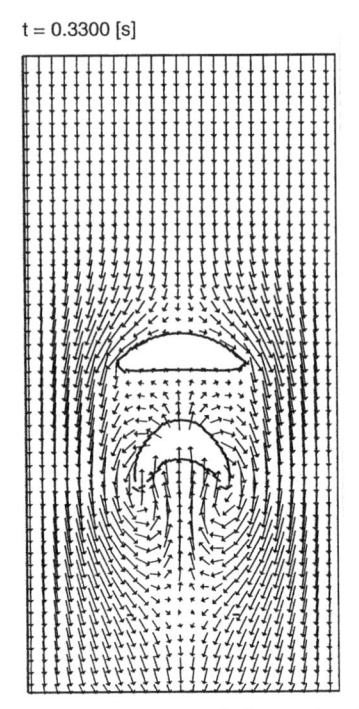

FIG. 28. Positions of two coaxial gas bubbles generated subsequently at the same orifice together with the distribution of the relative (i.e., with respect to the velocity of the leading gas bubble) liquid phase velocity.

version model with a CFD model describing the prevailing velocity and temperature distributions inside the reactor. The main characteristics of the reactor model developed by Westerhout *et al.* are as follows:

- The conversion of the polymer particles obeys first-order kinetics.
- The finite rate of heat penetration in the spherical polymer particles is accounted for.
- Due to the very low particle loading one-way coupling between the solid phase (i.e., the polymer particles) and the gas phase (nitrogen) prevails.
- The residence time of the polymer particles in the reactor and the temperature history experienced by these particles during their journey through the reactor is governed by the compressible nonisothermal Navier–Stokes equations.
- The gas phase flowing through the reactor is transparent with respect to thermal radiation.

In Fig. 29 a schematic drawing of the LEFR used by Westerhout *et al.* is shown. The polymer particles are fed to the reactor through a central cooled pipe ("cold finger") with the aid of a small nitrogen stream and subsequently they are heated by the annular nitrogen stream supplied through the flow distributor. Due to the

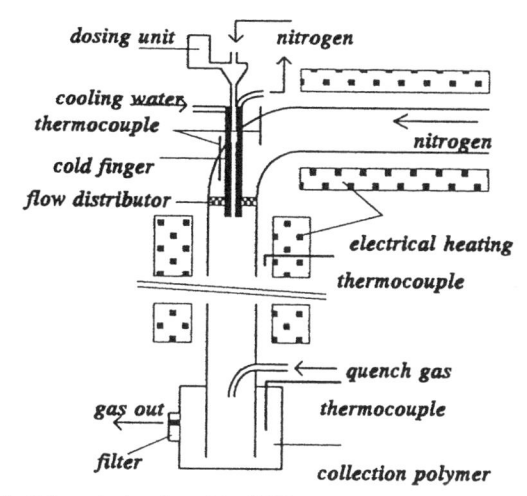

FIG. 29. Schematic drawing of the LEFR used by Westerhout et al. (1996).

velocity and temperature differences that exist between the central nitrogen stream (containing the polymer particles) and the annular nitrogen stream, the polymer particles experience a nonuniform (axial) velocity and temperature history which possibly influences their chemical conversion.

In Fig. 30a the radial profiles of the axial component of the gas velocity are shown at various axial locations, whereas the corresponding temperature profiles are shown in Fig. 30b. From these figures it can be seen that significant radial and axial velocity and temperature differences exist in the LEFR. The consequences of these phenomena for the chemical conversion of the polymer particles are depicted in Fig. 31. In this figure the computed conversion of the polymer particles is shown as a function of the axial coordinate for a case where the aforementioned velocity and temperature gradients are neglected and another case where they are taken into account. As evident from inspection of Fig. 31, significant differences exist between these two cases. By using the information on the velocity and temperature profiles prevailing in the LEFR it is possible to correct for these nonuniformities. This application demonstrates that CFD can also be used as a tool to interpret experimental data correctly.

VII. Conclusion

a. Future Role of CFD in Chemical Reaction Engineering. From the review presented in this paper it should have become apparent that CFD offers great potential for the chemical engineer and that this rapidly emerging new hybrid science of mathematics and mechanics at present already has a profound impact on chemical reaction engineering. It is expected that the role of CFD in the future design of

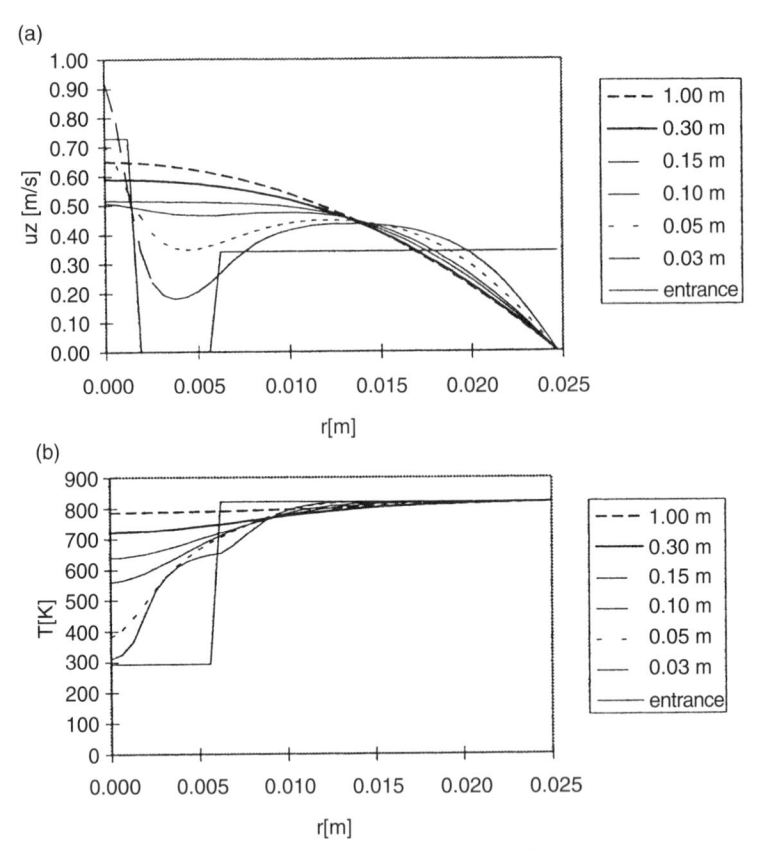

FIG. 30. (a) Computed radial profiles of the axial component of the gas velocity at various axial locations and (b) corresponding temperature profiles for LEFR studied by Westerhout et al. (1996).

chemical reactors will increase substantially and that CFD will reduce the experimental effort required to develop industrial (multiphase) reactors. Due to the anticipated future role of CFD in chemical reaction engineering the authors believe that the incorporation of CFD courses in chemical engineering curricula is of crucial importance. These courses should provide the chemical engineer at least with up-to-date knowledge on the foundations, possibilities, and limitations of CFD in the context of chemical reaction engineering.

b. Directions for Future Research. The variety and degree of complexity of systems encountered in industrial practice demands an integrated modeling approach where models with an increasing degree of sophistication should be used to feed models that invoke submodels with a strong empirical base. As far as turbulent flows are concerned, in this respect both LES and DNS offer the possibility to gener-

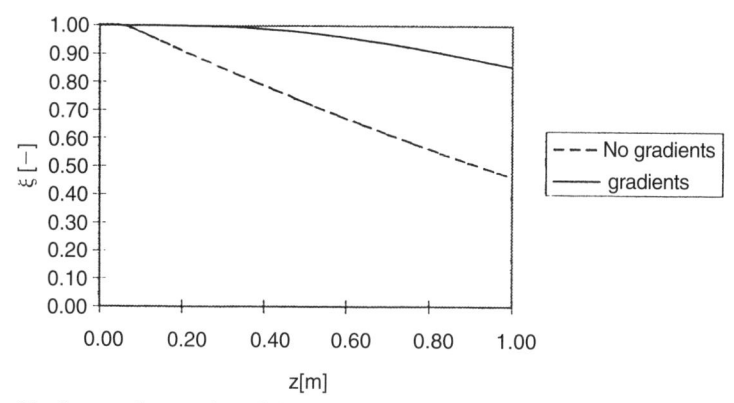

FIG. 31. Computed conversion of the polymer particle as a function of the axial coordinate for cases with and without accounting for existing velocity and temperature gradients in the LEFR.

ate databases to test and develop phenomenological turbulence models, whereas for dispersed multiphase flows mixed Eulerian–Lagrangian models offer the possibility to develop closure laws that can be used in a multifluid framework suited for the simulation of macroscopic systems of interest. In this connection Stokesian dynamics should also be mentioned because it offers great potential for simulation of concentrated suspensions in which hydrodynamic interaction has to be accounted for.

Non-Newtonian flows need much more attention since in many industrial applications the chemical engineer has to deal with materials that exhibit a complex rheological behavior that cannot (even approximately) be described with Newtonian closure models.

As is evident from inspection of Table III turbulence modeling of multiphase flow systems requires major attention in the near future. Also the development of closure laws for phenomena taking place in the vicinity of interfaces such as coalescence, breakup, and accumulation of impurities should be considered in more detail. Once these requirements have been met, in principle, it would be possible to predict a.o. flow regime transition and the spatial distribution of the phases with confidence, which is of utmost importance to the chemical engineer dealing with the design of (novel) multiphase reactors.

c. Experimental Validation. As mentioned earlier, at the present state of theoretical development *careful* experimental validation of *all* CFD predictions involving turbulent (multiphase) flows in complex geometries is of crucial importance. Due to this necessity there exists, especially for multiphase flow systems, a strong demand for the further development of nonintrusive experimental techniques that permit measurement of both the time-average and the fluctuating component of key quantities such as phase volume fractions, phase velocities, and species concentration. The requirement for the measurement of fluctuating components of key quantities is due to the fact that multiphase flows very often exhibit oscillatory behavior.

d. Commercial CFD Packages. At present a large number of commercial CFD packages are available, each with their own specific areas of application. In the near future the role of these packages in both academic and industrial research applications will most likely expand. At the same time it appears that the most significant efforts of the vendors of commercial CFD packages are directed toward the development of more advanced preprocessing and postprocessing facilities including a.o. grid generation and visualization of the computational data and not toward the development of new and more detailed physical models, an activity which is apparently left to the users of these packages. As far as this last activity is concerned, it is of crucial importance that the CFD package provide for an open programming environment where the user has full access to all relevant parts of the code so he or she can successfully implement and validate submodels.

Nomenclature

$\bar{a}$	Acceleration, m/s^2
C_μ	Empirical constant
D	diameter, m
Da	Damköhler number
D_i	Fickian diffusion coefficient of species i, m^2/s
e	Restitution coefficient
	Internal energy per unit mass, J/kg
e_k	Internal energy per unit mass for phase, k, J/kg
E	Differential residence time distribution function, s^{-1}
E_{kl}	Energy transfer rate between phase k and l per unit volume, W/m^3
Eo	Eotvös number
F	Cumulative residence time distribution function
	Fractional amount of fluid
$\bar{F}_i$	Force acting on ith particle, kg · m/s^2
$\bar{g}$	Gravitational force per unit mass, m/s^2
G_s	Solids mass flux, kg/(m^2 · s)
h	Wall-to-bed heat transfer coefficient, W/(m^2 · K)
I	Identity tensor
$\bar{J}_i$	Mass flux vector of species i due to molecular transport, kg/(m^2 · s)
$\bar{J}_{k,i}$	Mass flux vector of species i in phase k due to molecular transport, kg/(m^2 · s)
k	Turbulent kinetic energy, m^2/s^2
	Thermal conductivity, W/(m · K)

L	Length, m
m_i	Mass of ith particle, kg
M	Morton number
$\overline{M}_{kl}$	Momentum source term due to interaction between phase k and phase l, kg/(m^2 · s^2)
N_{xyz}	Required number of grid points in DNS simulation
p	Pressure, Pa
$\overline{q}$	Heat flux vector due to molecular transport, W/m^2
$\overline{q}_k$	Heat flux vector due to molecular transport in phase k, W/m^2
r	r coordinate, m
R	Radius, m
	Mass source term, kg/(m^3 · s)
Re	Reynolds number
Re$_\delta$	Reynolds number for channel flow
S_{h}	Heat production rate per unit volume, W/m^3
$S_{h,k}$	Heat production rate in phase k per unit volume, W/m^3
$\overline{S}_k$	Momentum source term due to phase changes and external forces other than gravity in phase k, kg/(m^2 · s^2)
S_i	Net production rate of species i per unit volume, kg/(m^3 · s)
$S_{k,i}$	Net production rate of species i in phase k per unit volume, kg/(m^3 · s)
t	Time, s
t_{m}	Characteristic time scale for turbulent mixing, s
t_{r}	Characteristic time scale for chemical transformation, s
T	Time, s
	Temperature, K
$\overline{u}$	Velocity, m/s
U	Superficial gas velocity, m/s
	Time-average velocity component, m/s
U_{c}	Centreline velocity in channel flow, m/s
$\overline{v}_i$	Velocity of ith particle, m/s
V	Bubble rise velocity, m/s
	Reference velocity, m/s
We	Weissenberg number
x	x coordinate, m
y	y coordinate, m
z	z coordinate, m

Greek Letters

δ	Channel half-width, m
ϵ	Volume fraction
	Viscous dissipation rate, m^2/s^3

γ	Bulk viscosity, kg/(m · s)
	Relaxation time in viscoelastic flow, s
	Linear dimension of smallest eddies in turbulent flow, m
μ	Shear viscosity, kg/(m · s)
	Dynamic friction coefficient
ν	Kinematic viscosity, m²/s
ω_i	Mass fraction of species i
$\omega_{k,i}$	Mass fraction of species i in phase k
ρ	Density, kg/m³
σ	Surface tension, Pa · m
τ	Stress tensor, Pa
τ_p	Particle response time, s
τ_K	Kolmogorov time scale, s
ξ	Chemical conversion

Subscripts

b	Bed
	Bubble
c	Centerline
h	Heat
i	Particle index number
	Species index
k	kth phase in multiphase system
K	Kolmogorov
m	Turbulent mixing
p	Particle
r	Chemical transformation
s	Solids
t	Terminal
w	Wall
z	Axial direction

Superscripts

t	Turbulent
$\underline{T}$	Transpose
	Vector quantity
′	Fluctuating component
-	Time average

Operator

∇ Gradient
$\nabla\cdot$ Divergence
$<>$ Average

References

Abid, M., Xuereb, C., and Bertrand, J., Hydrodynamics in vessels stirred with anchors and gate agitators: Necessity of 3-D modeling. *Trans. I. Chem. E.* **70**(Part A), 377 (1992).

Adrian, R. J., Particle-imaging techniques for experimental fluid mechanics. *Annu. Rev. Fluid Mech.* **23**, 261, (1991).

Alamdari, F., Edwards, S. C., and Hammond, S. P., Microclimate performance of an open atrium office building: A case study in thermo-fluid modeling. *In* "Computational Fluid Dynamics for the Environmental and Building Services Engineer—Tool or Toy?," The Institution of Mechanical Engineers, London, 1991, p. 81.

Amsden, A. A., "KIVA II: A Computer Program for Chemically Reactive Flows with Sprays," U.S. Department of Commerce NTIS Report LA-11560-MS, May 1989.

Amsden, A. A., "KIVA: A Computer Program for Two- and Three-Dimensional Fluid Flows with Chemical Reactions and Fuel Sprays," U.S. Department of Energy Report LA-10245-MS/UC-32 and UC-34, Feb. 1985.

Amsden, A. A., and Harlow, F. H., "The SMAC Method: A Numerical Technique for Calculating Incompressible Fluid Flows," Los Alamos Scientific Laboratory Report LA-4370, 1970.

Anagnostopoulos, P., and Iliadis, G., Numerical study of the blockage effect on viscous flow past a circular cylinder. *Int. J. Num. Methods Fluids* **22**, 1061 (1996).

Anderson, J. D., "Computational Fluid Dynamics: The Basics with Applications," McGraw-Hill, New York, 1995.

André, C., David, R., André, J., and Villermaux, J., A new fluorescence method for measuring cross-fluctuations of two non-reactive components in a mixing device. *Chem. Eng. Technol.* **15**, 182 (1992).

Angelov, G., Journe, E., and Gourdon, C., Simulation of the flow patterns in a disc and doughnut column. *Chem. Eng. J.* **45**, 87 (1990).

Bader, R., Findlay, J., and Knowlton, T., Gas/solids flow patterns in a 30.5 cm diameter circulating bed. *In* "Circulating Fluidized Bed Technology II" (P. Basu, and J. F. Large, eds.). Pergamon Press, New York, 1988, p. 123.

Bai, X. S., and Fuchs, L., Numerical model for turbulent diffusion flames with applications. *In* "Computational Fluid Dynamics" (C. Hirsch, J. Periaux and W. Kordulla, eds.). Elsevier, Amsterdam, 1992, vol. 1, p. 169.

Baker, A. J., "Finite Element Computational Fluid Mechanics." Hemisphere Publishing Corporation, New York, 1985.

Bakker, A., and van den Akker, H. E. A., A computational model for the gas-liquid flow in stirred reactors. *Trans. I. Chem. E.* **72**(Part A), 594 (1994).

Banerjee, S., Turbulence structures, sixth P. V. Danckwerts memorial lecture. *Chem. Eng. Sci.*, **47**(8), 1793 (1991).

Beam, R. M., and Warming, R. F., An implicit factored scheme for the compressible Navier-Stokes equations. *AIAA J.* **16**(4), 393 (1978).

Beam, R. M., and Warming, R. F., An implicit finite difference algorithm for hyperbolic systems in conservation-law form. *J. Comp. Phys.* **22**(1), 87 (1977).

Becker, A. A., "The Boundary Element Method in Engineering: A Complete Course." McGraw-Hill, London, 1992.

Becker, S., Sokolichin, A., and Eigenberger, G., Gas-liquid flow in bubble columns and loop reactors: Part II: Comparison of detailed experiments and flow simulation. *Chem. Eng. Sci.,* **49,** 5747 (1994).

Beenackers, A. A. C. M., and van Swaaij, W. P. M., Review article number 42. Mass transfer in gasliquid slurry reactors. *Chem. Eng. Sci.,* **48**(18), 3109 (1993).

Bird, R. B., Stewart, W. E., and Lightfoot, E. N., "Transport Phenomena." John Wiley & Sons, New York, 1960.

Boger, D. V., Viscoelastic flows through contractions. *Annu. Rev. Fluid Mech.* **19,** 157 (1987).

Bouillard, J. X., Lyczkowski, R. W., and Gidaspow, D., Porosity distributions in a fluidized bed with an immersed obstacle. *AIChE. J.* **35**(6), 908 (1989).

Boussinesq, J., Essai sur la Théorie des Eaux Courantes. Mém. Prés. Acad. Sci. XXIII, 46, Paris, 1877.

Boysan, F., Ayers, W. H., and Swithenbank, J., A fundamental mathematical modeling approach to cyclone design. *Trans. I. Chem. E.* **60,** 222 (1982).

Brady, J. F., and Bossis, G., Stokesian dynamics. *Annu. Rev. Fluid Mech.* **20,** 111 (1988).

Brereton, G., and Korotney, D., Co-axial and oblique coalescence of two rising bubbles. In "Dynamics of Bubbles and Vortices Near a Free Surface," *ASME,* AMD **119,** 1 (1991).

Bridgwater, J., Broadbent, C. J., and Parker, D. J., Study of the influence of blade speed on the performance of a powder mixer using positron emmission particle tracking. *Trans. I. Chem. E.* **71**(Part A), 675 (1993).

Briley, W. R., and McDonald, H., Solution of the three-dimensional compressible Navier-Stokes equations by an implicit technique. *Proc. 4th Int. Conf. Num. Methods in Fluid Dynamics, Lecture Notes in Physics,* Springer-Verlag, Berlin, 1975, vol. 35, p. 105.

Brinkman, H. C., A calculation of the viscous force exerted by a flowing fluid on a dense swarm of particles. *Appl. Sci. Res.* **A1,** 27, 81 (1947).

Brodkey, R. S., and Lawelle, J., Reactor selectivity based on first order closures of the turbulent concentration equations. *AIChE. J.* **31,** 111 (1985).

Canuto, C., Hussaini, M. Y., Quarteroni, A., and Zang, T. A., "Spectral Methods in Fluid Dynamics." Springer-Verlag, Berlin, 1988.

Chang, S. L., Lyczkowski, R. W., and Berry, G. F., Spectral dynamics of computer simulated twodimensional fewtube fluidized beds. *AIChE. Symp. Ser. 2b9,* 2b7 (1989).

Chaouki, J., Larachi, F., and Dudukovic, M. P. (eds.), "Non-Invasive Monitoring of Multiphase Flows." Elsevier, Amsterdam, 1997.

Chen, J. J. J., Jamialahmadi, M., and Li, S. M., Effect of liquid depth on circulation in bubble columns: A visual study. *Chem. Eng. Res. Des.* **67,** 203 (1989).

Chen, R. C., and Fan, L. S., Particle image velocimetry for characterizing the flow structures in threedimensional gas-liquid-solid fluidized beds. *Chem. Eng. Sci.* **47,** 3615 (1992).

Cheremisinoff, N. P., "Instrumentation for Complex Fluid Flows." Technomic Publishing Company, Lancaster, PA, 1986a.

Cheremisinoff, N. P., Review of experimental methods for studying the hydrodynamics of gas-solid fluidized beds, *I&EC Process Des. Devel.,* **25,** 329 (1986b).

Clift, R., An Occamist review of fluidized bed modelling. *AIChE. Symp. Ser. 296* **89,** 1 (1993).

Colenbrander, G. W., CFD research for the petrochemical industry. *Appl. Sci. Res.* **48,** 211 (1991).

Cook, T. L., Demuth, R. B., and Harlow, F. H., PIC calculations of multiphase flow. *J. Comp. Phys.* **41,** 51 (1981).

Corbett, A. M., Phillips, R. J., Kauten, R. J. and McCarthy, K. L., Magnetic resonance imaging of concentration and velocity profiles of pure fluids and solid suspensions in rotating geometries. *J. Rheol.* **39**(5), 907 (1995).

Correa, S. M., and Shyy, W., Computational models and methods for continuous gaseous turbulent combustion. *Prog. Energy Combust. Sci.* **13**, 249 (1987).
Crochet, M. J., Numerical simulation of flow processes. *Chem. Eng. Sci* **42**(5), 979 (1987).
Crowe, C. T., Review—Numerical models for dilute gas-particle flows. *J. Fluids Eng.* **104**, 297 (1982).
Cundall, P. D., and Strack, O. D. L., A discrete numerical model for granular assemblies. *Geotechnique* **29**, 47 (1979).
Curl, R. L., Dispersed phase mixing: I. Theory and effects in simple reactors. *AIChE. J.* **9**(2), 175 (1963).
Cuvelier, C., Segal, A., and van Steenhoven, A. A., "Finite Element Methods and Navier-Stokes Equations." D. Reidel Publishing Company, Dordrecht, 1986.
Dahm, W., Southerland, L., and Su, L., Fully-resolved four dimensional spatio-temporal micromeasurements of the fine scale structure and dynamics of mixing in turbulent flows. *AIChE Meet.*, Miami Beach, FL, paper 262c (1995).
Delnoij, E., Lammers, F. A., Kuipers, J. A. M., and Van Swaaij, W. P. M., Dynamic simulation of dispersed gas-liquid two-phase flow using a discrete bubble model. *Chem. Eng. Sci.* **52**, 1429 (1997a).
Delnoij, E., Kuipers, J. A. M., and van Swaaij, W. P. M., Stimulation of bubble dynamics in gas-liquid two-phase flow using the volume of fluid (VOF) method. Submitted for publication (1997b).
Derbyshire, J. A., Gibbs, S. J., Carpenter, T. A., and Hall, L. D., Rapid three-dimensional velocimetry by nuclear magnetic resonance imaging. *AIChE. J.* **40**(8), 1404 (1994).
De Saegher, J. J., Detemmerman, T., and Froment, G. F., Three dimensional simulation of high severity internally finned cracking coils for olefin production. *Rev. Inst. Francais Du Pétrole* **51**(2), 245 (1996).
de Saint Venant, B., Note a Joindre un Mémoire sur la Dynamique des Fluides, *Comptes Rendus* **17**, 1240 (1843).
Devanathan, N., Dudukovicz, M. P., Lapin, A., and Lübbert, A., Chaotic flow in bubble column reactors. *Chem. Eng. Sci.* **50**, 2661 (1995).
Devanthan, N., Moslemian, D., and Dudukovicz, M. P., Flow mapping in bubble columns using CARPT. *Chem. Eng. Sci.* **45**, 2285 (1990).
Ding, J., and Gidaspow, D., A bubble fluidisation model using kinetic theory of granular flow. *AIChE. J.* **36**(4), 523 (1990).
Eggels, J. G. M., Direct and large eddy simulation of turbulent flow in a cylindrical pipe geometry, Ph.D. Thesis, Delft University (1994).
Elgobashi, S., Particle-laden turbulent flows: Direct numerical simulation and closure models. *Appl. Sci. Res.* **48**, 301 (1991).
Ettehadieh, B., Gidaspow, D., and Lyczkowski, R. W., Hydrodynamics of fluidization in a semicircular bed with a jet. *AIChE. J.* **30**(4), 529 (1984).
Fan, L. S., Bubble dynamics in liquid-solid suspensions. *AIChE. Symp. Ser. 305* **91**, 1 (1995).
Fan, L. S., "Gas-Liquid-Solid Fluidization Engineering." Butterworth Publishers, New York, 1989.
Farmer, W. M., *Appl. Opt.* **13**, 610 (1974).
Farmer, W. M., *Appl. Opt.* **11**, 2603 (1972).
Fletcher, C. A., "Computational Techniques for Fluid Dynamics, Vol. I, Fundamental and General Techniques. Springer Verlag, Berlin, 1988a.
Fletcher, C. A., "Computational Techniques for Fluid Dynamics, Vol. II, Specific Techniques for Different Flow Categories." Springer Verlag, Berlin, 1988b.
Fox, R. O., Computational methods for turbulent reacting flows in the chemical process industry. *Rev. Inst. Francais Du Pétrole* **51**(2), 215 (1996).
Frank, M. J. W., Kuipers, J. A. M., Versteeg, G. F., and van Swaaij, W. P. M., Modelling of simultane-

ous mass and heat transfer with chemical reactions using the Maxwell-Stefan theory—I. Model development and isothermal study. *Chem. Eng. Sci.* **50**(10), 1645 (1995a).

Frank, M. J. W., Kuipers, J. A. M., Krishna, R., and van Swaaij, W. P. M., Modelling of simultaneous mass and heat transfer with chemical reactions using the Maxwell-Stefan theory—II. Nonisothermal study. *Chem. Eng. Sci.* **50**(10), 1661 (1995b).

Gasche, H. E., Edinger, C., Kömpel, H., and Hofmann, A fluid-dynamically based model of bubble column reactors. *Chem. Eng. Technol.* **13**, 341 (1990).

Gaskey, S., Vacus, P., David, R., André, J. C., and Villermaux, J., A method for the study of turbulent mixing using fluorescence spectroscopy. *Exp. Fluids,* **9**, 137 (1990).

Ge, W., and Li, J., Pseudo-particle approach to hydrodynamics of gas-solid two-phase flow, *Proc. CFB-5,* Beijing, paper DT8 (1996).

Gharib, M., Hernan, M. A., Yavrouian, A. H., and Sarohia, V., Flow velocity measurement by image processing of optically activated tracers. AIAA paper no. 85-0172 (1985).

Gidaspow, D., Hydrodynamics of fluidization and heat transfer: super computer modelling. *Appl. Mech. Rev.* **39**(1), 1 (1986).

Gidaspow, D., and Ettehadieh, B., Fluidization in a two-dimensional bed with a jet, Part 2: Hydrodynamic modelling. *Ind. Eng. Chem. Fundam.* **22**, 193 (1983).

Gidaspow, D., Lin, C., and Seo, Y. C., Fluidization in a two-dimensional bed with a jet, Part 1: Experimental porosity distributions. *Ind. Eng. Chem. Fundam.* **22**, 187 (1983).

Givi, P., and McMurtry, P. A., Nonpremixed reaction in homogeneous turbulence: Direct numerical simulation. *AIChE. J.* **34**, 1039 (1988).

Grace, J. R., Shapes and velocities of bubbles rising in infinite liquids. *Trans. Inst. Chem. Eng.* **51**, 116 (1973).

Grace, J. R., Wairegi, T., and Nguygen, T. H., Shapes and velocities of single drops and bubbles moving freely through immiscible liquids. *Trans. Inst. Chem. Eng.* **54**, 167 (1976).

Grier, M. R., and Fox, R. O., A particle-conserving composition joint probability density function code. Report 259, Kansas State University, 1993.

Griffin, M. E., Diwaker, R., Anderson, J. D., and Jones, E., Computational fluid dynamics applied to flows in an internal combustion engine. *AIAA 16th Aerospace Sciences Meet.,* paper 78-57, January 1978.

Guj, G., and De Matteis, G., Fluid-particles interaction in particulate fluidized beds: Numerical and experimental analysis. *Physico-Chemical Hydrodynamics* **7**(2/3), 145 (1986).

Hamill, I. S., Hawkins, I. R., Jones, I. P., Lo, S. M., Splawski, B. A., and Fontenot, K., The application of CFDS-FLOW3D to single and multiphase flows in mixing vessels. *AIChE Symp. Ser. 305* **91**, 150 (1995).

Hargreaves, J. H., and Silvester, R. S., Computational fluid dynamics applied to the analysis of deoiling hydrocyclone performance. *Trans. I. Chem. E.* **68**, 365 (1990).

Harlow, F. H., and Amsden, A. A., Numerical calculation of multiphase fluid flow. *J. Comp. Phys.* **17**, 19 (1975).

Harlow, F. H., and Amsden, A. A., KACHINA: An eulerian computer program for multifield fluid flows. Los Alamos Scientific Laboratory Report LA-5680, 1974.

Harris, C. K., Roekaerts, D., and Rosendal, F. J. J., Computational fluid dynamics for chemical reaction engineering. *Chem. Eng. Sci.* **51**, 1569 (1995).

Hartge, E. U., Li, Y., and Werther, J., Analysis of the local structure of two-phase flow in a fast fluidized bed. In "CFB Technology I" (P. Basu, ed.). Pergamon Press, 1986, p. 153.

Henline, W. D., Klein, H. H., Scharff, M. F., and Srinivas, B., "Final report on computer modelling of the U-Gas Reactor," DOE, DE-ACO2-77ET13406, Jaycor, 1981.

Hirsch, C., "Numerical Computation of Internal and External Flows, Vol. II: Computational Methods for Inviscid and Viscous Flows." Wiley, New York, 1990.

Hirsch, C., "Numerical Computation of Internal and External Flows, Vol. I: Fundamentals of Numerical Discretisation." Wiley, New York, 1988.

Hirt, C. W., and Nichols, B. D., Volume of fluid (VOF) method for the dynamics of free boundaries. *J. Comp. Phys.* **39,** 201 (1981).

Hjertager, B. H., and Morud, K., Computational fluid dynamics simulation of bioreactors. *In* "Bioreactor Performance" (U. Mortensen and H. J. Noorman, eds.). Ideon, Lund, 1993, p. 47.

Hoffman, J. D., "Numerical Methods for Engineers and Scientists." McGraw-Hill, New York, 1992.

Hoffmann, A. C., and van den Bogaard, H. A., A numerical investigation of bubbles rising at intermediate Reynolds and large Weber numbers. *Ind. Eng. Chem. Res.* **34,** 366 (1995).

Hoomans, B. P. B., Kuipers, J. A. M., and van Swaaij, W. P. M., Discrete particle simulation of a two-dimensional gas-fluidised bed: Comparison between a soft sphere and a hard sphere approach. Submitted for publication (1998).

Hoomans, B. P. B., Kuipers, J. A. M., Briels, W. J., and van Swaaij, W. P. M., Discrete particle simulation of bubble and slug formation in a two-dimensional gas-fluidised bed: A hard sphere approach. *Chem. Eng. Sci.* **51**(1), 99 (1996).

Hrenya, C. M., Bolio, E. J., Chakrabarti, D., and Sinclair, J. L., Comparison of low Reynolds k-ϵ turbulence models in predicting fully developed pipe flow. *Chem. Eng. Sci.* **50**(12), 1923 (1995).

Hsieh, K. T., and Rajamani, R. K., Mathematical model of the hydrocyclone based on physics of fluid flow. *AIChE. J.* **37**(5), 735 (1991).

Huttenhuis, P. J. G., Kuipers, J. A. M., and W. P. M. van Swaaij, The effect of gas phase density on bubble formation at a single orifice in a two-dimensional gas-fluidized bed. *Chem. Eng. Sci.* **51**(24), 5273 (1996).

Ichiki, K., and Hayakawa, H., Dynamical simulation of fluidized beds: Hydrodynamically interacting granular particles. *Phys. Rev. E* **52**(1), 658 (1995).

Ishii, M., "Thermo-Fluid Dynamic Theory of Two-Phase Flow." Eyrolles, Paris, 1975.

Jackson, R., Progress toward a mechanics of dense suspensions of solid particles. *AIChE. Symp. Ser. 301* **90,** 1 (1994).

Jameson, A., Computational aerodynamics for aircraft design. *Science* **245,** 361 (1988).

Johansen, S. T., and Kolbeinsen, L., "Applications of Computational Fluid Dynamics in Optimisation and Design of Metallurgical Processes," SINTEF Materials Technology, N-7034 Trondheim-NTH, Norway, 1996.

Joshi, J. B. and Sharma, M. M., A circulation cell model for bubble columns. *Trans. Inst. Chem. Engrs.* **57,** 244 (1979).

Kawaguchi, T., Yamamoto, Y., Tanaka, T., and Tsuji, Y., Numerical simulation of a single rising bubble in a two-dimensional fluidized bed. *Proc. 2nd Int. Conf. Multiphase Flow,* Kyoto/Japan, FB2-17-FB2-22 (1995).

Keunings, R., Progress and challenges in computational rheology. *Rheol. Acta* **29,** 556 (1990).

Keunings, R., and Crochet, M. J., Numerical simulation of the flow of a viscoelastic fluid through an abrupt contraction. *J. Non-Newtonian Fluid Mech.* **14,** 279 (1984).

Kim, J., Moin, P., and Moser, R. D., Turbulence Statistics in fully-developed channel flow at low Reynolds number. *J. Fluid Mech.* **177,** 133 (1987).

Koelman, J. M. V. A., and Hoogerbrugge, P. J., Dynamic simulations of hard-sphere suspensions under steady shear. *EuroPhys. Lett* **21**(3), 363 (1993).

Kothe, D. B., Mjolsness, R. C., and Torrey, M. D., "RIPPLE: A Computer Program for Incompressible Flows with Free Surfaces," Los Alamos Scientific Laboratory Report LA-12007-MS, 1991.

Kresta, S. M., and Wood, P. E., Prediction of the three-dimensional turbulent flow in stirred tanks. *AIChE. J.* **37**(3), 448 (1991).

Kuipers, J. A. M., A two-fluid micro balance model of fluidized beds. Ph.D. Thesis, University of Twente (1990).

Kuipers, J. A. M., and van Swaaij, W. P. M., Developing flow in CFB-risers: A computational study of the effect of riser inlet configuration on flow structure development. *Workshop II on Modelling and Control of Fluidized Bed Systems, CFB-5,* Beijing (1996).

Kuipers, J. A. M., van Duin, K. J., van Beckum, F. P. H., and van Swaaij, W. P. M., Computer simulation of the hydrodynamics of a two-dimensional gas-fluidized bed. *Comp. Chem. Eng.* **17**(8), 839 (1993).

Kuipers, J. A. M., van Duin, K. J., van Beckum, F. P. H., and van Swaaij, W. P. M., A numerical model of gas-fluidized beds. *Chem. Eng. Sci.* **47**(8), 1913 (1992a).

Kuipers, J. A. M., Tammes, H., Prins, W., and van Swaaij, W. P. M., Experimental and theoretical porosity profiles in a two-dimensional gas-fluidized bed with a central jet. *Powder Technol.* **71**, 87 (1992b).

Kuipers, J. A. M., Prins, W., and van Swaaij, W. P. M., Numerical calculation of wall-to-bed heat transfer coefficients in gas-fluidized beds. *AIChE. J.* **38**(7), 1079 (1992c).

Kuipers, J. A. M., Prins, W., and van Swaaij, W. P. M., Theoretical and experimental bubble formation at a single orifice in a two-dimensional gas-fluidized bed. *Chem. Eng. Sci.* **46**(11), 2881 (1991).

Kumar, S., VanderHeyden, W. B., Devanathan, N., Padial, N. T., Dudukovicz, M. P., and Kashiwa, B. A., Numerical simulation and experimental verification of the gas-liquid flow in bubble columns. *AIChE Symp. Ser 305* **91**, 11 (1995).

Kunii, D., and Levenspiel, O., "Fluidization Engineering," 2nd ed. Butterworth-Heinemann, Boston, 1991.

Lapin, A., and Lübbert, A., Numerical simulations of the dynamics of two-phase gas-liquid flows in bubble columns. *Chem. Eng. Sci.* **49**, 3661 (1994).

Larachi, F., Cassanello, M., Chaouki, J., and Guy, C., Flow structure of the solids in a 3-D gas-liquid-solid fluidized bed. *AIChE. J.* **42**(9), 2439 (1996).

Larachi, F., Chaouki, J., and Kennedy, G., 3-D mapping of solids flow fields in multiphase reactors with RPT. *AIChE. J.* **41**(2), 439 (1995).

Launder, B. E., Current capabilities for modelling turbulence in industrial flows. *Appl. Sci. Res.* **48**, 247 (1991).

Lemoine, F., Wolff, M., and Lebouche, M., Simultaneous concentration and velocity measurement using combined laser-induced fluorescence and laser Doppler velocimetry: Application to turbulent transport. *Exp Fluids* **20**, 319 (1996).

Leonard, A. D., and Hill, J. C., Direct numerical simulation of turbulent flows with chemical reaction. *J. Sci. Comp.* **3**, 25 (1988).

Levenspiel, O., "Chemical Reaction Engineering." John Wiley & Sons, New York, 1962.

Li, K. T., and Toor, H. L., Turbulent reactive mixing with a series-parallel reaction: Effect of mixing on yield. *AIChE. J.* **32**, 1312 (1986).

Lim, K. S., Zhu, J. X., and Grace, J. R., Hydrodynamics of gas-solid fluidization. *Int. J. Multiphase Flow* **21**, 141 (1995).

Lin, T. J., Reese, J., Hong, T., and Fan, L. S., Quantitative analysis and computation of two-dimensional bubble columns. *AIChE. J.* **42**(2), 301 (1996).

Lyons, S. L., Hanratty, T. J., and McLaughlin, J. B., Turbulence-producing eddies in the viscous wall region. *AIChE. J.* **35**(12), 1962 (1989).

Mampaey, F., and Xu, Z. A., An experimental and simulation study of mould filling combined with heat transfer. *In* "Computational Fluid Dynamics '92, (C. Hirsch, J. Periaux and W. Kordulla, eds.), Elsevier, Amsterdam, 1992, Vol. 1, 421.

Manger, E., Modelling and simulation of gas-solids flow in curvilinear coordinates. Ph.D. Thesis, Telemark College, 1996.

Marcum, D. L., and Hoffman, J. D., Calculation of three-dimensional inviscid flow-fields in propulsive nozzles with centerbodies. *Am. Inst. Aeronautics Astronautics. Propulsion Power* **4**(2), 172 (1988).

Martin, M. P., Turlier, P., and Bertrand, J. R., Gas and solid behavior in cracking circulating fluidized beds. *Powder Technol.* **70**, 249 (1992).

Matsunaga, K., Mijata, H., Aoki, K., and Zhu, M., Finite-difference simulation of 3D vortical flows past road vehicles. *Vehicle Hydrodynamics,* SAE Special Publication 908, 65 (1992).

McMurtry, P. A., and Givi, P., Direct numerical simulations of mixing and reaction in a nonpremixed homogeneous turbulent flow. *Combust. Flame* **77**, 171 (1989).
Melli, T. R., and Scriven, L. E., Theory of two-phase cocurrent downflow in networks of passages. *Ind. Eng. Chem. Res.* **30**, 951 (1991).
Merzkirch, W., "Flow Visualization." Academic Press, Orlando, 1987.
Mompean, G., and Deville, M. O., Recent developments in three-dimensional unsteady flows of non-Newtonian fluids. *Rev. Inst. Francais Du Pétrole* **51**(2), 261 (1996).
Myong, H. K., and Kasagi, N., A new approach to the improvement of k-ε models for wall-bounded shear flows. *JSME Int. J. (Series II)* **33**, 63 (1990).
Nakayama, T., and Mori, M., An Eulerian finite element method for time-dependent free surface problems in hydrodynamics. *Int. J. Num. Methods Fluids* **22**, 175 (1996).
Navier, M., Mémoire sur les Lois du Mouvement des Fluides. *Mem. de l'Acad. de Sci.* **6**, 389 (1827).
Nichols, B. D., Hirt, C. W., and Hotchkiss, R. S., "SOLA-VOF: A Solution Algorithm for Transient Fluid Flow with Multiple Free Boundaries," Los Alamos Scientific Laboratory Report LA-8355, 1980.
Nieuwland, J. J., Hydrodynamic modelling of gas-solid two-phase flows. Ph.D. Thesis, University of Twente, 1995.
Nieuwland, J. J., Veenendaal, M. L., Kuipers, J. A. M., and van Swaaij, W. P. M., Bubble formation at a single orifice in a two-dimensional gas-fluidized bed. *Chem. Eng. Sci.* **51**(17), 4087 (1996a).
Nieuwland, J. J., Sing Annaland van, M., Kuipers, J. A. M., and van Swaaij, W. P. M., Hydrodynamic modelling of gas-particle flows in riser reactors. *AIChE. J.* **42**(6), 1569 (1996b).
Nieuwland, J. J., Kuipers, J. A. M., and van Swaaij, W. P. M., Theoretical and experimental study of CFB riser hydrodynamics. *Proc. CFB-5,* Beijing, paper MSS3 (1996c).
Nieuwland, J. J., Meijer, R., Kuipers, J. A. M., and van Swaaij, W. P. M., Measurement of solids concentration and axial solids velocity in gas-solid two-phase flows. *Powder Technol.* **87**, 127 (1996d).
Nieuwland, J. J., Kuipers, J. A. M., and van Swaaij, W. P. M., Bubble formation in a two-dimensional gas-fluidized bed at elevated pressures, *Proc. Fluidization VIII,* Tours (1995).
Ohta, M., Yamamoto, M., and Suzuki, M., Numerical analysis of a single drop formation process under pressure pulse condition. *Chem. Eng. Sci.* **50**(18), 2923 (1995).
Ottino, J. M., Mixing and chemical reactions—A tutorial. *Chem. Eng. Sci.* **49**(24A), 4005 (1994).
Pain, Y., and Banerjee, S., Numerical simulation of particle interactions with wall turbulence. *Phys. Fluids* **8**(10), 2733 (1996a).
Pan, Y., and Banerjee, S., Numerical investigation of the effects of large particles on wall-turbulence. Submitted for publication (1996b).
Patankar, S. V., "Numerical Heat Transfer and Fluid Flow." McGraw-Hill, New York, 1980.
Pedinotti, S., Mariotti, G., and Banerjee, S., Effect of Reynolds number on particle behavior near walls in two-phase turbulent flows. *Proc. 5th. Int. Symp. Flow Mod. Turb. Meas.,* Paris, Presses Pont et Chausees, 425 (1993).
Pedinotti, S., Mariotti, G., and Bamerjee, S., Direct numerical simulation of particle behavior in the wall region of turbulent flows in horizontal channels. *Int. J. Mult. Flow* **18**, 927 (1992).
Perng, C. Y., and Murthy, J. Y., A moving-deforming-mesh technique for simulation of flow in mixing tanks. *AIChE. Symp. Ser. 293* **89**, 37 (1992).
Pipino, M., and Fox, R., Reactive mixing in a tubular jet reactor: A comparison of PDF simulations with experimental data. *Chem. Eng. Sci.* **49**(24B), 5229 (1994).
Pita, J. A., and Sundaresan, S., Developing flow of a gas-particle mixture in a vertical riser. *AIChE. J.* **39**(4), 541 (1993).
Pita, J. A., and Sundaresan, S., Gas solid flow in vertical tubes. *AIChE. J.* **37**(7), 1009 (1991).
Poisson, S. D., Mémoire sur les Equations Générales de l'Equilibre et du Mouvement des Corps Solides Elastiques et des Fluides. *J. de L'Ecole Polytechn.* **13**, 139 (1831).

Pope, S. B., Lagrangian PDF methods for turbulent flows. *Annu. Rev. Fluid Mech.* **26,** 23 (1994).
Pritchett, J. W., Blake, T. R., and Garg, S. K., A numerical model of gas fluidized beds. *AIChE. Symp. Ser. 176,* **74,** 134 (1978).
Ranade, V. V., and Joshi, J. B., Flow generated by a disc turbine: Part I. Experimental. *Trans. I. Chem. E.* **68**(Part A), 19 (1990a).
Ranade, V. V., and Joshi, J. B., Flow generated by a disc turbine: Part II. Mathematical modelling and comparison with experimental data. *Trans. I. Chem. E.* **68**(Part A), 34 (1990b).
Reese, J., and Fan, L. S., Transient flow structure in the entrance region of a bubble column using particle image velocimetry. *Chem. Eng. Sci.* **49**(24B), 5623 (1994).
Reynolds, W. C., The potential and limitations of direct and large eddy simulations. *In* "Whither Turbulence? Turbulence at the Crossroads" (J. L. Lumley, ed.). Springer Verlag, Berlin, 1991, p. 313.
Rivard, W. C., and Torrey, M. D., "THREED: An Extension of the K-FIX Code for Three-Dimensions, "Los Alamos Scientific Laboratory Report AL-NUREG-6623, 1979.
Rivard, W. C., and Torrey, M. D., "K-FIX: A Computer Program for Transient, Two-DImensional, Two-Fluid Flow," Los Alamos Scientific Laboratory Report LA-NUREG-6623, 1977.
Roache, P. J., "Computational Fluid Mechanics," Hermosa, Albuquerque, NM, 1972.
Rodi, W., "Turbulence Models and Their Application in Hydraulics—A State of the Art Review," International Association for Hydraulic Research Section on Fundm. of Division II: Exp. and Math. Fluid Dynamics.
Rogallo, R. S., and Moin, P., Numerical simulation of turbulent flows. *Annu. Rev. Fluid Mech.* **16,** 99 (1984).
Rowe, P. N., *In* "Fluidization" (J. F. Davison and D. Harrison, eds.). Academic Press, London, 1971.
Rudman, M., Volume-tracking methods for interfacial flow calculations. *Int. J. Num. Methods Fluids* **24,** 671 (1997).
Ryskin, G., and Leal, L. G., Numerical solution of free-boundary problems in fluid mechanics. Part I. The finite difference technique. *J. Fluid Mech.* **148,** 1 (1984a).
Ryskin, G., and Leal, L. G., Numerical solution of free-boundary problems in fluid mechanics. Part II, Buoyancy-driven motion of a gas bubble through a quiescent liquid. *J. Fluid Mech.* **148,** 19 (1984b).
Ryskin, G., and Leal, L. G., Numerical solution of free-boundary problems in fluid mechanics. Part III, Bubble deformation in an axisymmetric straining flow. *J. Fluid Mech.* **148,** 37 (1984c).
Scharff, M. F., Chan, R. K-C., Chiou, M. J., Dietrich, D. E., Dion, D. D., Klein, H. H., Laird, D. N., Levine, H. R., Meister, C. A., and Srinivas, B., "Computer Modelling of Mixing and Agglomeration in Coal Conversion Reactors," Vols. I & II, DOE/ET/10329-1211, Jaycor, 1982.
Schlichting, H., "Boundary Layer Theory." McGraw-Hill, New York, 1975.
Schneyer, G. P., Peterson, E. W., Chen, P. J., Brownell, D. H., and Blake, T. R., "Computer Modelling of Coal Gasification Reactors. "Final Report for June 1975-1980 DOE/ET/10247, Systems, Science and Software, 1981.
Shaw, C. T., "Predicting Vehicle Aerodynamics Using Computational Fluid Dynamics—A User's Perspective." Research in Automotive Aerodynamics, SAE Special Publication 747, Feb. 1988, p. 119.
Simonin, O., Modelling turbulent reactive dispersed two-phase flows in industrial equipments. *Proc. Third World Conf. Applied Computational Fluid Dynamics,* May 19–23, Freiburg, Germany, Workshop E, 17.9, 1996.
Sinclair, J., and Jackson, R., Gas-particle flow in a vertical pipe with particle-particle interactions. *AIChE. J.* **35**(5), 1473 (1989).
Sokolichin, A., and Eigenberger, G., Gas-liquid flow in bubble columns and loop reactors: Part I: Detailed modelling and numerical simulation. *Chem. Eng. Sci.* **49,** 5735 (1994).
Sokolichin, A., Eigenberger, G., Lapin, A., and Lübbert, A., Dynamic simulation of gas-liquid two-phase flows—Euler/Euler versus Euler/Lagrange. *Chem. Eng. Sci.* **52,** 611 (1997).

Spielman, L. A., and Levenspiel, O., A Monte-Carlo treatment for reacting and coalescing dispersed phase systems. *Chem. Eng. Sci.* **20**, 247 (1965).
Squires, K. D., and Eaton, J. K., Particle response and turbulence modification in isotropic turbulence. *Phys. Fluids* **A2**, 1191 (1990).
Steijsiger, C., Lankhorst, C. M., and Roman, Y. R., Influence of gas phase reactions of the deposition rate of silicon carbide from the precursors methyltrichlorosilane and hydrogen. *In* "Numerical Methods in Engineering '92" (C. Hirsch and O. C. Zienkiewicz, eds.). Elsevier, Amsterdam, 1992, p. 857.
Stewart, H. B., and Wendroff, B., Two-phase flow: Models and methods. *J. Comp. Phys.* **56**, 363 (1984).
Stokes, G. G., On the theories of internal friction of fluids in motion. *Trans. Cambr. Phil. Soc.* **8**, 287 (1845).
Svendsen, H. F., Jakobsen, H. A., and Torvik, R., Local flow structure in internal loop and bubble column reactors. *Chem. Eng. Sci.* **47**(13/14), 3297 (1992).
Syamlal, M., and Gidaspow, D., Hydrodynamics of fluidization: Prediction of wall-to-bed heat transfer coefficients. *AIChE. J.* **31**(1), 127 (1985).
Tanaka, T., Yonemura, S., Kiribayashi, K., and Tsuji, Y., Cluster formation and particle-induced instability in gas-solid two-phase flows predicted by the DSMC method. *JSME, Ser. B* **39**(2), 239 (1996).
Targett, M. J., Retallick, W. B., and Churchill, S. W., Flow through curved rectangular channels of large aspect ratio. *AIChE. J.* **41**(5), 1061 (1995).
Tennekes, H., and Lumley, J. L., "A First Course in Turbulence." The MIT Press, Cambridge, MA, 1977.
Theologos, K. N., and Markatos, N. C., Modelling of flow and heat transfer in fluidized catalytic cracking riser-type reactors. *Trans. I. Chem. E.* **70**(Part A), 239 (1992).
Thompson, J. F., ed., "Numerical Grid Generation." North-Holland, New York, 1982.
Thompson, J. F., Warsi, Z. U. A., and Mastin, C. W., "Numerical Grid Generation," Foundations and Applications." North-Holland, New York, 1985.
Thompson, J. F., Warsi, Z. U. A., and Mastin, C. W., Boundary-fitted coordinate systems for numerical solution of partial differential equations—A review. *J. Comp. Phys.* **47**, 1 (1982).
Tomiyama, A., Zun, I., Sou, A., and Sakaguchi, T., Numerical analysis of bubble motion with the VOF method. *Nucl. Eng. Des.* **141**, 69 (1993).
Torvik, R., and Svendsen, H. F., Modelling of slurry reactors, a fundamental approach. *Chem. Eng. Sci* **45**, 2325 (1990).
Toye, D., Marchot, P., Crine, M., and L'Homme, G., Analasys of liquid flow distribution in trickling flow reactor using computer assisted x-ray tomography. *Trans. I. Chem. E.* **73**(Part A), 258 (1995).
Trambouze, P., Computational fluid dynamics applied to chemical reaction engineering. *Rev. Inst. Francais du Petrole* **48**(6), 595 (1993).
Trapp, J. A., and Mortensen, G. A., A discrete particle model for bubble slug two-phase flow. *J. Comp. Phys.* **107**, 367 (1993).
Tsai, K., and Fox, R. O., PDF simulations of a turbulent series-parallel reaction in an axisymmetric reactor. *Chem. Eng. Sci.* **49**(24B), 5141 (1994).
Tsuji, Y., Kawaguchi, T., and Tanaka, T., Discrete particle simulation of two-dimensional fluidized bed. *Powder Technol* **77**, 79 (1993).
Tsuo, Y. P., and Gidaspow, D., Computation of flow patterns in circulating fluidized beds. *AIChE. J.* **36**(6), 885 (1990).
Udaykumar, H. S., Shyy, W., and Rao, M. M., ELAFINT: A mixed Eulerian-Lagrangian method for fluid flows with complex and moving boundaries. *Int. J. Num. Methods Fluids* **22**, 691 (1996).
Van Kemenade, V., and Deville, M. O., Spectral elements for viscoelastic flows with change of type. *J. Rheol.* **38**(2), 291 (1994).

Versteeg, G. F., Kuipers, J. A. M., van Beckum, F. P. H., and van Swaaij, W. P. M., Mass transfer with complex reversible chemical reactioms—II. Parallel reversible chemical reactions. *Chem. Eng. Sci.* **45**(1), 183 (1990).

Versteeg, G. F., Kuipers, J. A. M., van Beckum, F. P. H., and van Swaaij, W. P. M., Mass transfer with complex reversible chemical reactions—I. Single reversible chemical reaction. *Chem. Eng. Sci.* **44**(10), 2295 (1989).

Villermaux, J., Personal Communication, 1996.

Villermaux, J., and Devillon, J. C., Représentation de la Coalescence et de la Redispersion des Domaines de Ségrégation dans un Fluide par un Modele d'Interaction Phénomenologique. *Proc. 2nd Int. Symp. Chem. React Engin.*, Amsterdam, B1-B13 (1972).

Villermaux, J., and Falk, L., Recent advances in modelling micromixing and chemical reaction. *Rev. Inst. Francais Du Pétrole* **51**(2), 205 (1996).

Villermaux, J., and Falk, L., A generalized mixing model for initial contacting of reactive fluids. *Chem. Eng. Sci.* **49**(24B), 5127 (1994).

Vortmeyer, D., and Schuster, J., Evaluation of steady flow profiles in rectangular and circular packed beds by a variational method. *Chem. Eng. Sci.* 1691 (1983).

Wang, J. W., and Andrews, J. R. G., Numerical simulation of flow in helical ducts. *AIChE. J.* **41**(5), 1071 (1995).

Wang, B. N., and Hoffman, J. D., Algebraic grid generation for annular nozzle flowfield prediction. In "Proceedings of the First International Conference on Numerical Grid Generation in Computational Fluid Dynamics," Pineridge Press, Swansea, United Kingdom, 399 (1986).

Warsi, Z. U. A., "Fluid Dynamics, Theoretical and Computational Approaches." CRC Press, Boca Raton, FL (1993).

Welch, J. E., Harlow, F. H., Shannon, J. P., and Daly, B. J., "The MAC Method: A Computing Technique for Solving Viscous Incompressible Transient Fluid Flow Problems Involving Free Surfaces." Los Alamos Scientific Laboratory Report LA-3425, 1965.

Wen-Jei Yang, "Handbook of Flow Visualization." Hemisphere Publishing Corporation, New York, 1989.

Westerhout, R. W. J., Kuipers, J. A. M. and van Swaaij, W. P. M., Development, Modelling and Evaluation of a (Laminar) Entrained Flow Reactor for the Determination of the Pyrolysis Kinetics of Polymers, *Chem. Engng. Sci.,* **51**(10), 2221, (1996).

Westerweel, J., Digital particle image velocimetry—theory and application. Ph.D. Thesis, Delft University, 1993.

Xu, Q., and Michaelides, E. E., A numerical study of the flow over ellipsoidal objects inside a cylindrical tube. *Int. J. Num. Methods Fluids* **22**, 1075 (1996).

Yoo, J. Y., and Na, Y., A numerical study of the planar contraction flow of a viscoelastic fluid using the SIMPLER algorithm. *J. Non-Newtonian Fluid Mech.* **39**, 89 (1991).

Youngs, D. L., Time-dependent multi-material flow with large fluid distortion. *In* "Numerical Methods for Fluid Dynamics" (K. W. Morton and M. J. Baines, eds.). Academic Press, New York, 1982, p. 273.

Zienkiewicz, O. C., and Taylor, R. L., "The Finite Element Method, Basic Formulation and Linear Problems," 4th ed. McGraw Hill, London, 1989a, Vol. I.

Zienkiewicz, O. C., and Taylor, R. L., "The Finite Element Method, Solid and Fluid Mechanics, Dynamics and Non-Linearity," 4th ed., McGraw Hill, London, 1989b, Vol. II.

USING RELATIVE RISK ANALYSIS TO SET PRIORITIES FOR POLLUTION PREVENTION AT A PETROLEUM REFINERY

Ronald E. Schmitt, Howard Klee, and Debora M. Sparks

Environmental Technical Services
Amoco Corporation
Warrenville, Illinois 60555

Mahesh K. Podar

Office of Water
U.S. Environmental Protection Agency
Washington, DC 20460

I.	Introduction	330
II.	Summary	331
	A. Developing a Detailed Refinery Release Inventory	331
	B. Identifying Options for Reducing Releases	332
	C. Evaluating and Ranking the Options	333
	D. Discussion of Findings	334
III.	Background	335
IV.	Assembling the Inventory of Refinery Releases	337
	A. Distribution of Releases within the Refinery	338
	B. Distribution of Releases Leaving the Refinery	341
	C. Characterization of Releases	345
	D. Identification of Sources	348
V.	Assessing the Risks Posed by Refinery Releases	351
	A. Air	352
	B. Surface Water	353
	C. Drinking Water	354
	D. Groundwater	354
VI.	Measuring Public Perception of the Refinery	354
VII.	Developing and Evaluating Options for Reducing Releases	355
	A. Option Characteristics	357
	B. Ranking the Options	360
VIII.	Ranking Methods and Results	369
	A. Options for Reducing Total Releases	369
	B. Analytical Hierarchy Process	376

        C. Regulatory Requirements and Project Options 378
        D. Summary of Ranking Results 382
    IX. Implementation of Selected Options: Obstacles and
        Incentives 384
        A. Background 384
        B. Obstacles 385
        C. General Observations on Five Highly Ranked Options 388
        D. Follow-up to the Original Project 397
        References 398

I. Introduction

In 1989, the Amoco Corporation (Amoco) and the U.S. Environmental Protection Agency (EPA) began a voluntary, joint project to identify opportunities for preventing pollution at the Amoco refinery at Yorktown, Virginia. The project produced one of the most detailed release inventories ever assembled for a major petrochemical facility and identified several previously unrecognized opportunities for significant pollution prevention. More importantly, perhaps, the study raised questions about the cost effectiveness of current government policies for regulating pollution and the barriers they may present to implementing cost-effective prevention strategies. It also suggested potential improvements in technical methods for inventorying refinery emissions and facility management and design.

The study provides both the practicing chemical engineer and the student engineer considering a career in industry or government with important insights into the technical and policy challenges posed by pollution prevention. In particular, it is food for thought for engineers who increasingly find themselves involved in policy or regulatory development and implementation.

This paper also serves as a resource guide for similar future efforts. It identifies tools that engineers may need to accomplish the following:

- Determine sampling and monitoring requirements to fully inventory environmental releases to all media.
- Select sampling methodologies and protocols.
- Include science and technology in a consensus-building process that utilizes the inputs of the entire stakeholder community.
- Rank and weight criteria allowing such diverse elements as costs, tons of releases, environmental impacts, public perceptions, and political factors to be quantitatively assessed.

II. Summary

In late 1989, Amoco and the EPA began a voluntary, joint project to identify opportunities for preventing pollution at an Amoco refinery located along the York River at Yorktown, Virginia. The two-year, $2.3 million study (70% Amoco and 30% EPA funding) involved more than 200 participants from Amoco, government agencies, educational institutions, environmental organizations, and the public.

The study focused on four major tasks:

1. Developing a detailed refinery release inventory that identified sources and quantities of releases.
2. Identifying options for preventing releases and minimizing health and environmental risks.
3. Developing a system for evaluating and ranking the options in light of cost, risk, regulatory requirements, and other factors.
4. Evaluating the incentives and obstacles to implementing the pollution prevention options.

A. Developing a Detailed Refinery Release Inventory

To develop the detailed release inventory, a study team reviewed results from the refinery's monthly mass balance calculation (inputs minus outputs), studied records maintained for compliance with state and federal laws, and undertook a massive sampling program. Mass balance calculations indicated that the refinery had releases of 8400 tons per year. In contrast, compliance records—which cover only those emissions regulated by state or federal law—documented smaller quantities. In addition, emissions of substances covered by the federal Toxic Release Inventory (TRI) were reported to be 371 tons in 1989—or 23 times less than the emissions indicated by the mass balance calculation.

To obtain a more accurate inventory of the refinery's releases, the study team undertook a massive sampling program. Almost 1000 samples of air, groundwater, surface water, soils, and solid waste were collected and analyzed; the database generated by this effort represents one of the most detailed emissions inventories ever assembled for a major petrochemical facility.

Based on results from the sampling program, the study team estimated that the refinery generated 27,500 tons/year of waste materials. Airborne emissions accounted for almost half of the total: 48% (13,200 tons). Solid waste accounted for 29% (8100 tons), surface water pollutants 14% (3700 tons), and biosolids from wastewater treatment 9% (2400 tons).

Roughly half of all wastes generated—44% (12,100 tons)—were reused, recycled, or recovered on the refinery site; 56% (15,400 tons) left the refinery. Of the materials leaving the refinery, more than half—51% (7900 tons)—were nonmethane airborne hydrocarbons; 37% (5700 tons) were airborne criteria pollutants such as SO_2, NO_x, CO, and particulates; 11% were materials disposed of on land (1700 tons); and 0.3% (50 tons) were waterborne substances.

The sampling further revealed the following:

- The refinery's existing water treatment plant was very effective in removing waterborne contaminants, with overall removal efficiencies greater than 99% for most organics and inorganics. Except for methyl tertiary-butyl ether (MTBE), most contaminants were not detectable in the treated effluent discharged into the York River.
- The treated effluent discharge contained 50 tons/year of suspended solids and other material. This is 10% of the amount permitted under the refinery's National Pollutant Discharge Elimination System (NPDES) permit (required under the Clean Water Act).
- At the plant boundary, concentrations of benzene—a chemical of concern due to the risk it poses to human and ecological health—were similar to those measured in rural environments; at a residence near the refinery, benzene concentrations were similar to those measured in remote environments.
- Groundwater contamination was significantly less than that documented at other refineries. In part, this finding can be explained by a combination of the original refinery construction methods (atypical of most older refineries), lack of petroleum spills, and the passive action of the refinery's underground sewer system, in which groundwater collects and flows to the wastewater treatment plant.
- The plant released 900 tons of substances covered by the TRI—or 2.4 times more than the plant reported in 1989. Blowdown stacks, whose contribution to releases was previously unknown, accounted for 430 tons of TRI releases; barge loading accounted for 165 tons, although they are not required to be reported under TRI regulations.

B. IDENTIFYING OPTIONS FOR REDUCING RELEASES

Having quantified the refinery's releases and identified major sources, the study team began to identify options for reducing releases and minimizing the environmental and human health risks they might pose. In March 1991 the project sponsored a three-day workshop to assist in option identification and development. It involved more than 120 representatives from EPA, Amoco, the Commonwealth of Virginia, academic, environmental and consulting organizations, and the public.

In developing pollution prevention options, workshop participants considered strategies for source reduction (preventing the creation of emissions), recycling, reuse, treatment, and—in the least attractive case—disposal. Overall, 50 distinct options for preventing emissions were identified. Twelve of these options were selected for more detailed analysis. Those chosen were felt to (1) be feasible with current technology, (2) offer significant potential for emissions reductions, (3) have manageable (or no adverse) impact on worker safety, (4) be amenable to more quantitative analysis in the time available, and (5) address concerns in different environmental media.

C. EVALUATING AND RANKING THE OPTIONS

A variety of factors and methods were used to evaluate and rank the twelve options. For example, the study team considered the reduction in relative risk to human health achieved by different options. Generally, an option's effectiveness in reducing health risks was evaluated by calculating its effect on exposure to benzene emissions. The study team selected benzene emissions as an indicator because benzene can be found in all waste media (air, water, groundwater, and surface water) and poses a known threat to human health.

Thirteen other factors were also considered in evaluating each option: (1) capital costs, (2) operating and maintenance costs, (3) liability cost rating, (4) timeliness, (5) transferability, (6) revenues, (7) equivalent annual costs, (8) pollution prevention mode, (9) net release reduction, (10) recovery costs, (11) cost effectiveness, (12) resource utilization, and (13) effects of secondary emissions. Finally, data from public opinion polling in the community near the refinery were considered in evaluating potential public support or opposition to different pollution prevention strategies.

1. Ranking Methodologies

A variety of methodologies were employed to rank the twelve options. They included single criteria rankings, multiple criteria rankings and the analytical hierarchy process (AHP) using ranking and weighting criteria. An AHP analysis was performed to compare how the options ranked from both an industry and a regulator's viewpoint. Despite assigning different weights to the AHP ranking criteria, both groups identified the same first choice option for pollution prevention: reducing emissions during barge loading. Implementing this option, which would produce a 55% reduction in benzene exposure, received a score that was two times greater than the next best option.

Overall, five options received generally high rankings: (1) reducing emissions during barge loading, (2) installing secondary seals on storage tanks, (3) upgrading blowdown stacks, (4) reducing soil intrusion into the drainage system, and (5) instituting a periodic leak detection and repair program.

2. Evaluating Costs and Benefits

After evaluating the costs and benefits of various options, the study team concluded that implementing options that received high rankings could achieve lower cost pollution prevention than options that are mandated by current law. Existing mandates, for example, will require the refinery to install specific technologies to implement eight of the identified pollution prevention options; this will reduce releases by 7300 tons per year at an annual cost of $17.5 million—or $2400/ton or $8.88/gallon. In contrast, implementing the five high-ranked options (four of which are required by current or anticipated rules) would reduce releases by 6700 tons per year at an annual cost of $2.2 million—or $328/ton or $1.21/gallon, thereby achieving more than 90% of the pollution reduction at 14% of the cost of all the mandated options.

When reducing benzene emissions was considered to be the primary goal, the study team calculated that six high-ranked options together reduced benzene exposure by 90% at an annual cost of $4.5 million. This was only 20% of the annual cost of all the options mandated by regulation.

3. Evaluating Obstacles and Incentives

The study team examined current obstacles to and incentives for implementing five of the highly ranked options. At least three major obstacles were identified: (1) limited resources, (2) poor economic return, and (3) regulatory disincentives. The study team concluded that these obstacles create a formidable constraint on implementing pollution prevention strategies that go beyond current regulatory requirements.

D. Discussion of Findings

The effort to identify options for pollution prevention at Amoco's Yorktown refinery provided important insights into both technical and policy issues.

In the technical arena, for example, the project highlights how little may be known about the true quantities and sources of emissions at major petrochemical facilities. It underscores the limitations of using existing measures, such as mass balance calculations and regulatory compliance records, to calculate releases from a major industrial facility. It also emphasizes the need for continued innovation in developing technologies for pollution prevention and new methods for evaluating their effectiveness in reducing health and ecological risks.

In the policy arena, project results suggest that current regulatory requirements may ignore major emissions sources and actually hinder—not promote—the implementation of cost-effective pollution prevention strategies. It suggests that current government efforts to develop alternatives to "command and control" regulatory systems, which mandate specific pollution control technologies, should be encouraged. Establishing programs that incorporate incentives or market-based systems that allow industries and facilities flexibility to achieve environmental performance goals will make the best use of limited resources, encourage innovation, and foster technology development.

Finally, the project highlights the progress that can occur in identifying creative, cost-effective options for pollution prevention when government, industry, and the public establish partnerships rather than operate as adversaries. While each sector of society may have a different perspective on the challenge of preventing pollution, techniques for developing and evaluating potential solutions can often help the parties identify unexpected areas of agreement and establish a common ground for moving forward.

III. Background

Amoco's Yorktown refinery, which began operations in 1956, has a processing capacity of 53,000 barrels/day. Currently, the refinery manufactures gasoline, heating oil, LPG, sulfur, and coke. The refinery is located on the York River in Yorktown, Virginia, near the Chesapeake Bay. The refinery was selected as the study site because it uses "typical" refinery processing technology, is located in an environmentally sensitive area, and is small enough to allow for a relatively complete inventory of emissions during the two-year time frame allotted for the study. The refinery's proximity to Washington, D.C., also allowed easy access by federal agency officials involved in the project.

During the course of this project, more than 100 EPA and Commonwealth of Virginia regulatory personnel visited the refinery for a first-hand view of operations and practices. Figure 1 provides a schematic flowchart of the refinery and illustrates potential releases to all environmental media (air, land, surface water, and groundwater).

The study was designed to assess comprehensively the refinery's releases to all environmental media (i.e., air, water, land, etc.), and then develop and evaluate options for reducing these releases.

A study team identified five specific tasks required to complete the project:

1. Inventory refinery releases to the environment to define their chemical type, quantity, source, and medium of release.

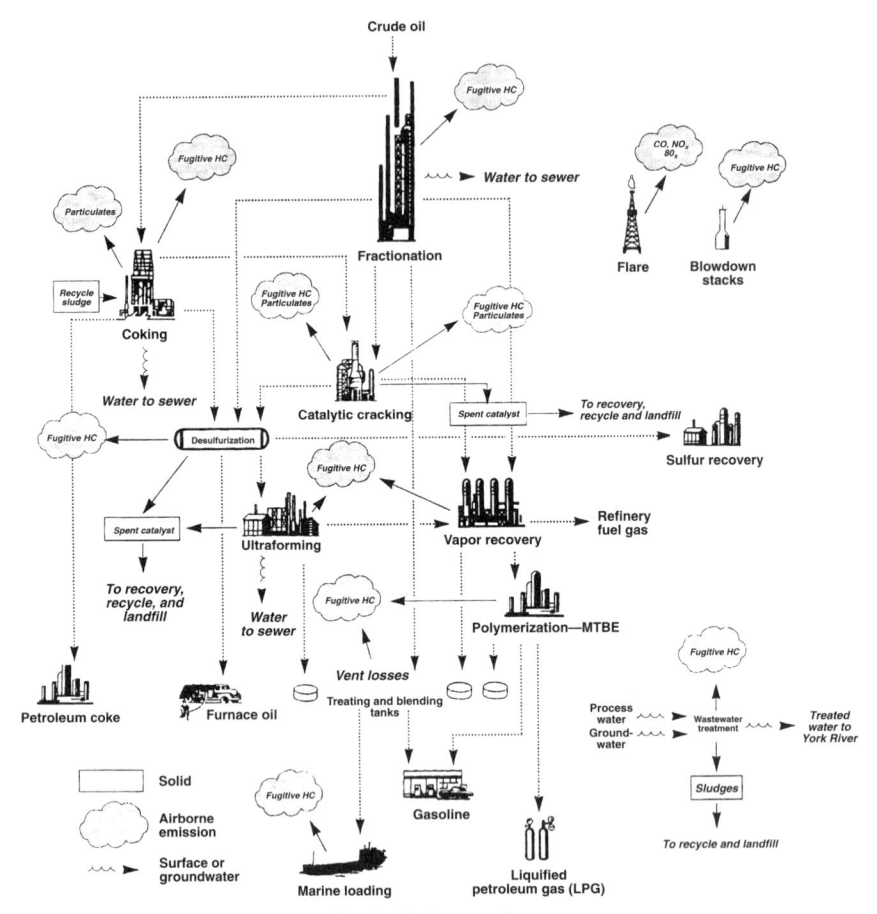

FIG. 1. Yorktown refinery.

2. Develop options to reduce selected releases.
3. Rank and prioritize the options using a variety of criteria and perspectives.
4. Identify and evaluate factors that impede or encourage the implementation of pollution prevention strategies.
5. Enhance participants' knowledge of refinery and regulatory systems.

When the project began, pollution prevention was a concept predicated on reducing or eliminating releases of materials into the environment rather than managing the releases later. The project adopted this general concept but considered all opportunities—source reduction, recycling, treatment, and environmentally sound disposal—as potential methods for pollution management. Since then, Congress, in the Pollution Prevention Act of 1990, and other organizations have put greater emphasis on source reduction as the primary, if not exclusive, means to accomplish pollution prevention.

A necessary requirement of the project was to identify evaluation criteria and develop a system for ranking opportunities for pollution prevention at the Yorktown refinery. It was important that the system recognize such factors as release reduction potential, technical feasibility, cost, environmental impact, human health risk, and risk reduction potential. Due to the inherent uncertainties in risk assessments, the project focused on relative changes in risk compared to current levels, rather than establishing absolute risk levels. Because of difficulties in quantifying changes in ecological impact from airborne emissions, changes in relative risk were based primarily on human health effects indicated by changes in exposure to benzene.

This project focused on normal refinery operations; it did not directly consider options for minimizing emergency events—although this a high priority of Amoco's facility managers—because (1) prevention and control of such events involves significantly different skills, technical resources, and analyses than controlling releases from day-to-day operations (AIChE, 1985); (2) the number, type, and frequency of incidents at Yorktown had been very low; and (3) adequate meteorological and emissions data from emergency releases were not available.

IV. Assembling the Inventory of Refinery Releases

Pollution prevention cannot be adequately implemented or monitored for effectiveness unless facility operators and regulators know what is being released from the facility and its origin. Therefore one of the study team's first tasks was to assemble a detailed inventory of releases from the refinery. At the start of the project, information on all of the refinery's release sources was not available. This was understandable, considering that complex industrial sources such as the refinery contain hundreds, sometimes thousands, of potential release points. It is technically difficult and impractical to monitor and measure each of these points.

Instead, government regulatory systems—such as those established by the Clean Water Act or Resource Conservation and Recovery Act (RCRA)—require refineries and other facilities to monitor and measure releases from a few specific points, such as the end of a discharge pipe, or in specific media, such as groundwater. As a result, monitoring resources are typically allocated to meet permit requirements rather than to measure releases at the point of generation.

To bridge the gaps in existing data, the study team designed and carried out a multimedia sample collection and analysis effort. This sampling program differed from many existing regulatory programs, which dictate the kinds of compounds or mixtures to be reported. Instead, each medium was sampled for selected chemicals such as benzene, toluene, ethylbenzene, and xylene (BTEX), as well as for particular chemical species expected to be present in specific media, such as metals and polynuclear aromatics. One goal was to identify specific chemicals present in all media, both within the refinery and entering the environment beyond the refinery property line.

About 1000 separate samples were collected during the program. They provided the first major database showing all releases from a single facility into all environmental media at one point in time. Figure 2 shows the sample distribution by media, excluding duplicates and field banks required for quality assurance/quality control purposes. The probable accuracy of most measurements is +100 tons/year.

A. DISTRIBUTION OF RELEASES WITHIN THE REFINERY

Sampling data indicated that the refinery generated an estimated 27,500 tons/year of materials that reached all four environmental media—air, surface water, groundwater, and land. Figure 3 summarizes the generation of pollutants prior to any internal recycling, transfer, or disposal.

Airborne emissions accounted for 48% of the total; solid waste 29%; and surface water 14%. In addition, biosolids from wastewater treatment accounted for 9% of solid wastes. However, the quantity and distribution of the pollutants among the media can shift, depending on natural conditions and the operation of pollution management systems designed for materials recovery and regulatory compliance.

Figure 4 shows how the releases just described were managed in the refinery. About 44% of the material generated did not leave the facility; it was handled onsite through treatment, recovery, and recycling.

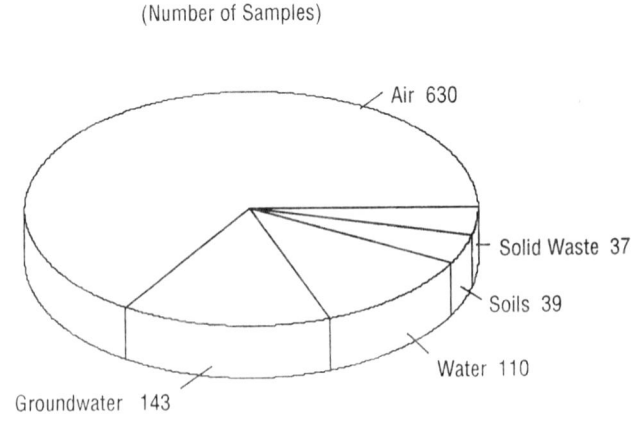

(Number of Samples)

Air 630
Solid Waste 37
Soils 39
Water 110
Groundwater 143

FIG. 2. Pollution prevention sampling program.

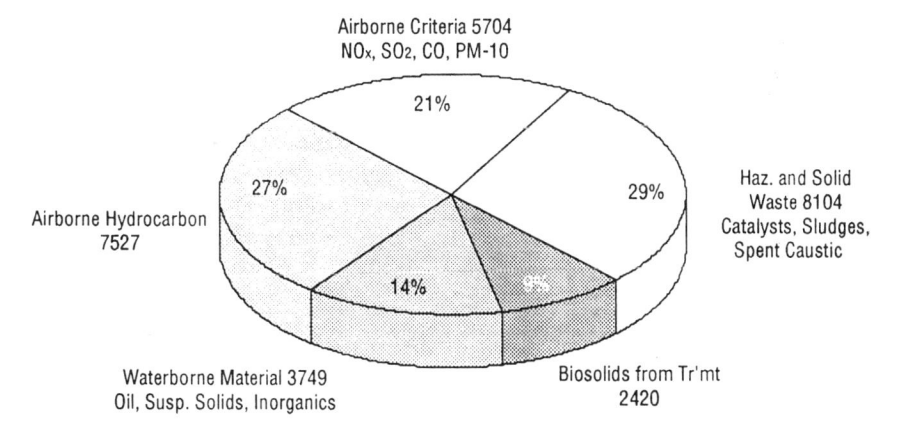

(Prior to Recycling, Transfer and Disposal)

Airborne Criteria 5704
NOx, SO2, CO, PM-10

Airborne Hydrocarbon
7527

Haz. and Solid
Waste 8104
Catalysts, Sludges,
Spent Caustic

Waterborne Material 3749
Oil, Susp. Solids, Inorganics

Biosolids from Tr'mt
2420

Units are ton/year
Total Generation = 27,504 tons/year

FIG. 3. Pollutant generation within the Yorktown refinery.

1. Water

Wastewater flows to the oil/water separator, where an average of 2700 tons/year of oil (about 50 barrels per day) was recovered and recycled. A small amount of groundwater was also recovered by the drainage system and mixed with process wastewater. The activated sludge wastewater treatment system generated 2400 tons/year of biosolids, which were ultimately recycled on-site with other solid and hazardous wastes. Treated effluent discharged to the York River contained about 46 tons/year of suspended solids and other material. This discharge is normally about 10% of the amount permitted under the refinery's NPDES permit; under infrequent upset conditions, however, the discharge may approach permit limits.

2. Solid and Hazardous Wastes

Spent caustic (3800 tons/year) is sent off-site for recovery of remaining caustic value and naphthenic acids. Most catalysts are recycled for recovery of additional activity or metals. Spent cracking catalyst (600 tons/year) is sent to Amoco's Whiting, Indiana, refinery for use as equilibrium catalyst. Spent ultraforming catalyst is returned to metals reclaimers to recover platinum for reuse in new catalyst. Spent desulfurization catalyst and polymer catalyst are nonhazardous and are buried in an on-site landfill. Sludges from the oil/water separator are a listed hazardous waste under RCRA regulations. They are combined with other solid wastes, such

340 RONALD E. SCHMITT ET AL.

Pollutant Generation
at Refinery: 27,504

Pollutant Releases from
Refinery: 15,380

Air
Airborne criteria: 5704
NO$_x$, SO$_2$, CO, PM-10
Airborne hydrocarbons:
7527

13,231

Airborne criteria:
5704

Airborne hydrocarbons:
7905

21%
27%
37%
51%

Hydrocarbon evaporation
from water: 378

Settling basin

Stormwater to York River

Water treatment

Treated effluent to York River: 46

0.3%

Water
Stormwater
Waterborne material: 3749
Oil, susp. solids, inorganics
Groundwater

14%
9%

Recovered oil:
2690

Biosolids
from
treatment
2420
(wet tons)

Sludges and oils to
on-site recycle:
4398

On-site
landfill:
713

Land disposal:
1725

Off site

11%

Solid Waste
Hazardous and
solid waste:
8104
Catalysts, sludges,
spent caustic

29%

Caustic to
off-site recycle: 3788

Catalysts to
off-site recycle: 613

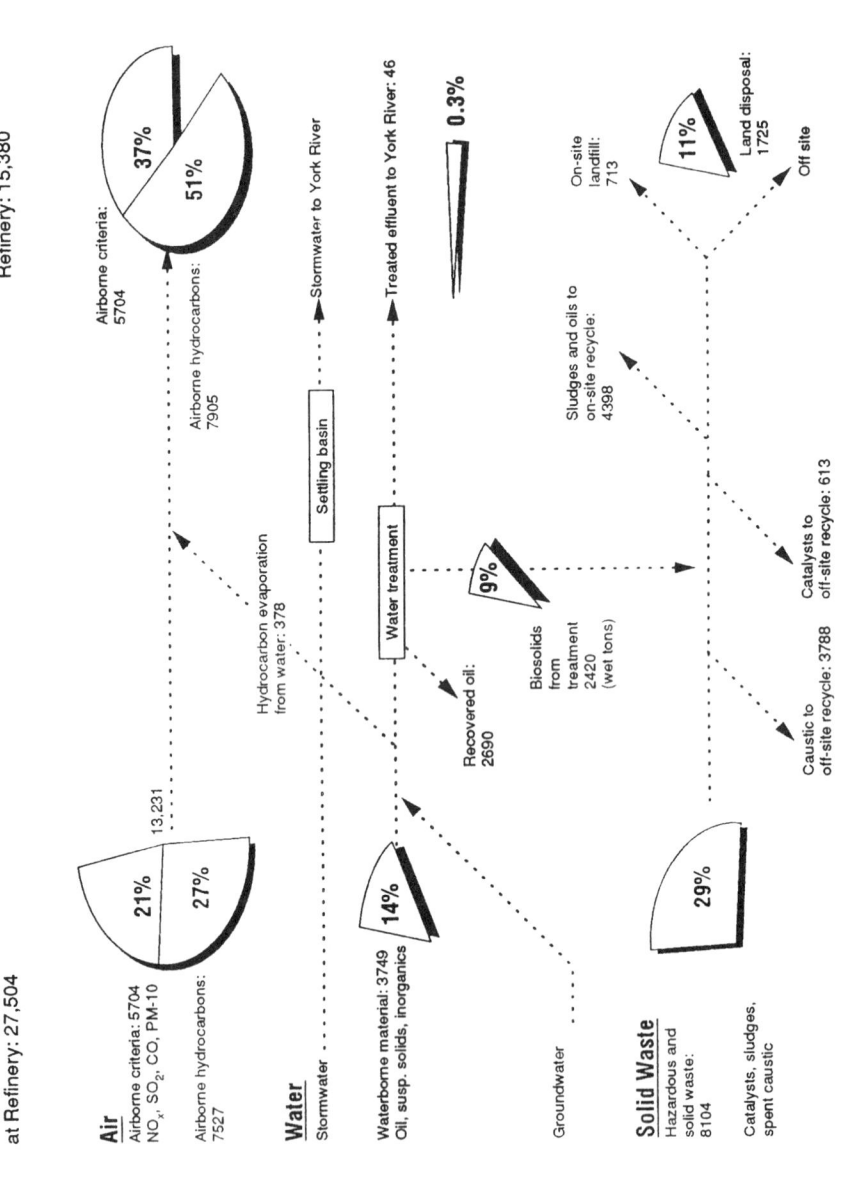

FIG. 4. Pollutant transfers/recycle/treatment within the Yorktown refinery.

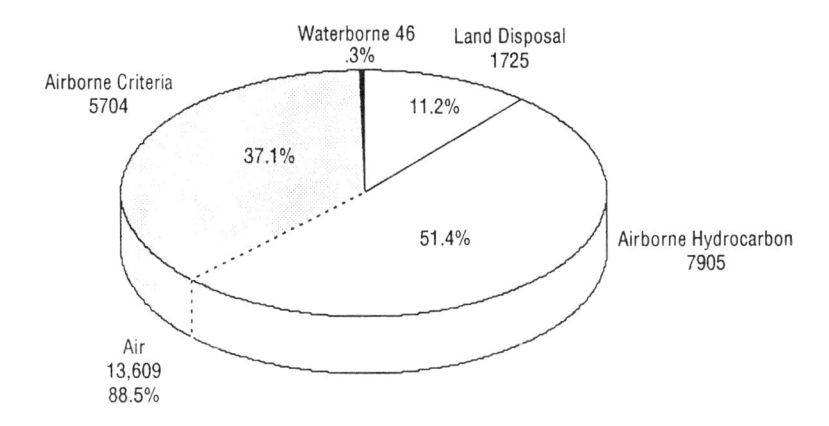

USING RELATIVE RISK ANALYSIS TO SET PRIORITIES   341

```
                    Waterborne 46    Land Disposal
                         .3%            1725
  Airborne Criteria
       5704                            11.2%
                      37.1%

                                51.4%        Airborne Hydrocarbon
                                                    7905

       Air
     13,609
      88.5%
```

Total Releases = 15380 tons/year

Units are ton/year

FIG. 5. Total releases entering the environment from Yorktown refinery.

as biosolids from the wastewater treatment plant, and recycled to the refinery's coker (4400 tons/year). In the coker, hydrocarbons are converted into salable products while water is recovered for treatment. Most remaining solid waste is construction debris and contaminated solids, which are landfilled on- and off-site. A total of 1700 tons/year of solid waste was landfilled in 1990.

B. DISTRIBUTION OF RELEASES LEAVING THE REFINERY

Figure 5 shows total releases to all environmental media leaving the refinery. Nearly 89% (13,600 tons) of the releases to the environment were airborne. The high percentage of airborne emissions leaving the facility focused both the sampling program and subsequent identification of pollution prevention options on this medium. As noted later, groundwater contamination under the facility is small, and was not moving off-property. Four hundred tons of hydrocarbons evaporate each year from refinery drainage and wastewater treatment systems, mixing with other airborne hydrocarbons.

1. Airborne Emissions

Figure 6 shows the division of airborne emissions between criteria pollutants—SO_2, NO_x, CO, and particulates (PM_{10})—and volatile organic compounds (VOCs). Criteria pollutants result primarily from combustion or other

SO₂
3802

28%

NOx
785

6%

4%

58%

PM-10
560

4%

CO
557

Hydrocarbons
7905

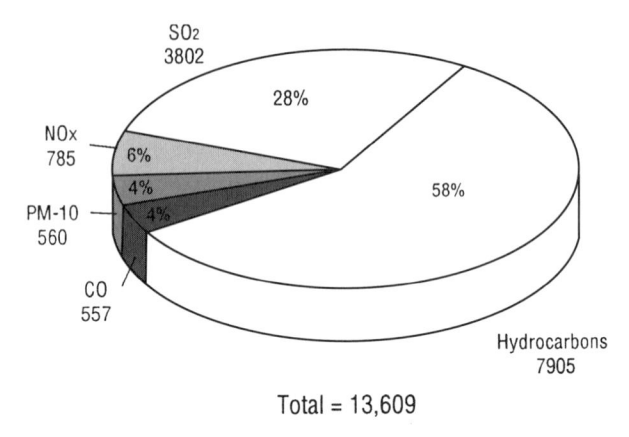

Total = 13,609

Units are ton/year

Note: The adjacent Virginia Power Plant releases about 36,466 tons / year

FIG. 6. Total air emissions for Yorktown refinery.

stacks. Nearly 60% of air emissions were VOCs. The sampling program focused on VOCs and their sources because much less was known about these sources.

Within the refinery fence line, maximum observed and/or calculated airborne concentrations of chemicals reported in the annual TRI were below OSHA action and permissible exposure levels for 8-hour time-weighted average exposures. Impacts on air quality by the refinery were calculated using air dispersion modeling techniques and the emissions inventory developed during this study.

Table I compares calculated concentrations of benzene, toluene, and ethylbenzene at several locations near the refinery with reported values for typical urban, rural, and remote settings from past EPA studies (Shah and Heyerdahl, 1988). For benzene, refinery impacts at the fenceline were similar to those observed in a rural environment. At the nearby residence, benzene concentrations were similar to those observed in a remote pristine setting. Ethylbenzene impacts were similar to benzene. Toluene impacts were somewhat higher, falling between typical rural and urban air quality. No comparable data were available for xylene. Automobiles and an adjacent power plant contribute some of these chemicals to the air. Biogenic (natural) sources also contribute. In the entire middle Atlantic region, natural sources provide about 40% of airborne hydrocarbons, with a higher percentage in more rural areas like Yorktown (Placet and Streets, 1989).

2. Surface Water

The existing water treatment plant is very effective in removing contaminants from process waters prior to discharge, with overall removal efficiencies greater

TABLE I
COMPARISON OF ANNUAL AVERAGE PREDICTED IMPACTS TO TYPICAL MEASURED CONCENTRATIONS FOR DIFFERENT TYPES OF ENVIRONMENTS FOR BTEX CHEMICALS

Chemical	Maximum Predicted Concentration at the Fence Line[a] ($\mu g/m^3$)	Maximum Predicted Concentration at the Closest Residence ($\mu g/m^3$)	Typical Remote Concentration[b] ($\mu g/m^3$)	Typical Rural Concentration[b] ($\mu g/m^3$)	Typical Urban Concentration[b] ($\mu g/m^3$)
Benzene	1.3	0.6	0.51	1.5	5.7
Toluene	5.4	2.4	0.19	1.3	7.7–12.0
Ethylbenzene	1.6	0.7	0.06	0.7	2.7

[a]Maximum on land.
[b]Shah and Heyerdahl, 1988.

than 99% for most organics and inorganics. Except for MTBE, most contaminants were not detectable in the treated effluent (Amoco/EPA, 1991a).

3. Groundwater

Subsurface contamination, detected during the sampling period, was significantly less than that observed at other petroleum refining facilities (*Los Angles Times,* 1988). Contamination appeared limited to shallow soils and/or groundwater. These unusually low levels of contamination are the result of (1) natural soil conditions, (2) no significant spills, (3) original refinery construction practice which utilized above-grade (rather than below-grade) welded piping (rather than threaded fitting construction), and (4) the underground process sewer system acting as a continuous groundwater recovery system. This sewer system passively collects about 35,000 gallons per day of groundwater and routes it to the wastewater treatment plant. This was an unexpected finding of the study (Amoco/EPA, 1991b).

4. Solid Waste Management

The refinery generated more than 10,500 tons of solid waste and spent caustic in 1990. More than 80% of the solid waste was recycled or treated either on- or off-site and does not enter the environment. Remaining materials are disposed of in approved landfill sites. Most solid wastes result from activities associated with the refinery's process water collection and treatment system. Nearly 1000 tons/year of soils enter the drainage system where they become oil-coated sludge.

5. Cross-Media Transport

Several pollution control technologies at the refinery promote *cross-media transport,* the transfer of pollutants from one medium to another. Wastewater, for example, may contain hydrocarbons that volatilize into the air; at the refinery, the wastewater treatment plant converted these waterborne hydrocarbons into 2400 tons/year of sludge, which were recycled to the coker. Cross-media transport from air to water is not significant for hydrocarbons or chemical that are only slightly soluble in water (Allen *et al.,* 1989). Studies performed by the National Center for Intermedia Transport at the University of California, Los Angeles, for instance, showed that most hydrocarbons released into the air do not transfer rapidly into other media. Therefore, ignoring intermedia transfer when examining air quality impacts is a reasonable analytical approach. Water-soluble compounds, such as methanol and MTBE, can transfer from air into water and soil media under certain conditions (Cohen *et al.,* 1991).

C. CHARACTERIZATION OF RELEASES

1. Diversity of Emissions

Crude oil contains thousands of individual hydrocarbon species. Emissions from oil refining operations reflect this diversity. Despite an extended sampling and analysis program, all species emitted could not be identified. Selected samples were analyzed for 150 organic compounds. The analysis identified about 90% of the compounds present. The remaining 10% were probably structural isomers of typical volatile organic hydrocarbons found in petroleum product mixtures. A number of small quantity emissions of unidentified hydrocarbons were found in most of the airborne samples. These compounds would be expected to exhibit similar physical and toxicological properties to compounds identified.

2. Comparison with TRI Emissions

The refinery's 1989 TRI report showed 370 tons of reportable chemicals released from all sources to all media. Based on measurements and modeling conducted for this project, releases of TRI chemicals were 900 tons, about 2.4 times higher than reported. This difference reflects (1) the identification of blowdown stacks as a significant source (430 tons) whose contribution was previously unknown and unrecognized, (2) the addition of marine loading losses (160 tons) which are not reportable for TRI (Oge, 1988), and (3) lower emissions from the oil/water separator (−90 tons). Emissions from the inactive landfarm, a coker pond, and sewer vents were also identified as new sources. Figure 7 illustrates these changes.

Table II reconciles the reported TRI values and measurements made for 12 reported chemicals. When reported values and the project's measurements are com-

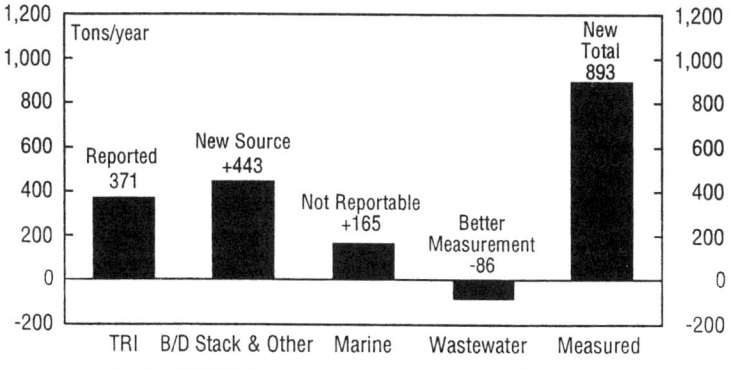

FIG. 7. 1989 TRI inventory compared to measured emissions.

TABLE II
RECONCILIATION OF 1989 TOXIC RELEASE INVENTORY REPORT AND POLLUTION PREVENTION INVENTORY YORKTOWN REFINERY (UNITS OF TONS/YEAR)

Chemical	1989 TRI Report	B/D Stack Additions	Coker Additions	Barge Loading Additions	Wastewater Subtractions	Total
Benzene	41.0	32.4	1.8	15.0	−0.1	90.1
Toluene	91.5	56.6	3.1	52.8	−23.2	180.7
Ethylbenzene	24.5	45.7	1.0	14.0	−7.3	77.9
Xylenes	107.0	121.6	6.3	69.2	−33.2	270.9
Cyclohexane	3.8	26.5			0.3	30.6
Naphthalene	2.8			5.1		7.9
Trimethylbenzene	45.5	42.2	1.1		−22.2	66.6
Ethylene	8.5	40.6				49.1
Propylene	30.5	64.1				94.6
Butadiene	0.036	0.18				0.2
Methanol	3.3					3.3
MTBE	12.1			8.8		20.9
Total	371	430	13	165	−86	893

[a] TRI as % of total VOCs = 11%.
[b] VOCs as % of crude run = 0.3%.
[c] TRI column excludes 0.175 tons chlorine.
[d] TRI reports only 1,2,4-trimethylbenzene. Other columns report 1,2,4- as well as 1,3,5- and 1,2,3-isomers.
[e] Wastewater subtractions include the net effect of reduced emissions from the API separator and increased emissions from sewer vents.
[f] The totals are rounded.

pared on the same basis (excluding marine loading losses and blowdown stack emissions), measured values (300 tons/year) are about 20% lower than reported values (370 tons/year).

3. Emissions Excluded from TRI

The sampling program helped identify and quantify emissions excluded from TRI reporting requirements. Some of these are excluded because they are below the threshold amounts that trigger reporting. Some are excluded because certain operations, such as barge loading, are not considered reportable under some circumstances by EPA (Oge, 1988). Some chemicals are excluded because they are not listed, although they have substantially similar physical and toxicological properties to chemicals that are listed. The isomers of trimethylbenzene illustrate this last point: 1,2,4-trimethylbenzene is a reportable chemical; the 1,3,5- and 1,2,3-trimethylbenzenes are not. All three occur in crude oil, gasoline, and refinery emissions and have similar physical and toxicological properties. Because of the wide diversity of emissions coming from crude oil, the refinery's TRI covered only about 11% of the total hydrocarbon emissions leaving the facility. The unreported emissions were primarily VOCs associated with pertroleum products and processing.

4. Comparison with EPA's AP-42 Emission Factors

Emission factors established by the EPA are frequently used to estimate total airborne emissions from different types of refinery equipment. These are called AP-42 emission factors. Most AP-42 factors do not provide information about the composition of emissions. The project's measurement program allowed for direct comparison between several measured or inferred emission rates and emissions calculated using these factors.

For example, actual measurements were taken that supported and validated the base assumptions of the AP-42 emission factors for quantitative assessment of fugitive and tank vent emissions. Measured emissions from the coker pond were about 40% greater than estimated using AP-42 factors. Measured emissions from the oil/water separator were 2100% lower than estimated with AP-42 factors. A combination of reasons can probably explain this discrepancy: (1) the limited database for this emission factor (no measurements), (2) improved refinery operating practices since the original data were collected in 1959, and (3) improved measurement techniques during the last 30 years (API, 1990, 1991). The EPA Office of Air Quality Planning and Standards (OAQPS) has given this emission factor a D rating on a scale of A (good) to E (poor, no data) (EPA, 1985a, 1988). Overall measured emissions from fugitive, tank vent, coker pond, oil/water separator emissions were about 60% of the amounts estimated.

D. IDENTIFICATION OF SOURCES

1. Air Emission Sources

To identify air emission sources, the study team utilized a variety of measurement techniques. Table III summarizes the different techniques used to define the airborne emissions. Emissions from sewer vents, water ponds, the inactive landfarm, and oil/water separator were measured directly (Amoco/EPA, 1991c). In general, mass balance techniques are not sufficiently accurate for most inventory calculations (NRC, 1990). However, since their flow rates were small, easily measurable, and reasonably constant over time, mass balances and inlet/outlet water analyses were used to determine emissions from the cooling tower and wastewater tanks.

Ambient monitoring both upwind and downwind of the refinery was used to infer information about emissions from such fugitive sources as leaks from process valves, flanges, pump seals, and tank vents. Emissions from barge and truck/rail loading operations were calculated using standard AP-42 emission factors (EPA, 1988) and actual refinery loadings for 1990. Emissions from blowdown stacks were calculated using the AP-42 emission factor, which gives a total rate based on refinery throughput. Composition measurements made at Yorktown were used to define chemicals in the total flow.

Figure 8 identifies and quantifies specific air emission sources. The chart reveals a number of useful facts about airborne emissions from the facility. First,

TABLE III
TECHNIQUE USED TO DETERMINE AIRBORNE EMISSIONS

Source Type	Basis of Emission Estimate	Speciated Emissions
API separator	Direct measurement	Direct measurement
Barge loading	AP-42	Proportional to product compositions loaded in 1990
Blowdown stacks	AP-42	AP-42 and direct measurements of composition
Coker pond	Direct measurement	Direct measurement
Cooling tower	Water, sampling, mass balance	Direct measurement
Inactive landfarm	Direct measurement	Direct measurement
Loading rack	AP-42	Proportional to product compositions loaded in 1990
Process fugitive	AP-42 default components	Proportional to compositions measured for similar in-stream equipment at another refinery
Sewer vents	Direct measurement	Direct measurement
Tanks	AP-42	Proportional to product compositions loaded in 1990

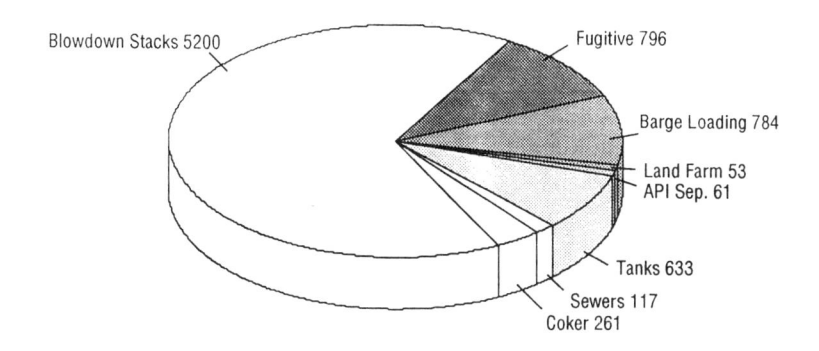

Blowdown Stacks 5200
Fugitive 796
Barge Loading 784
Land Farm 53
API Sep. 61
Tanks 633
Sewers 117
Coker 261

Total = 7905

Units are ton/year

FIG. 8. Yorktown refinery VOC air emission sources.

the three process blowdown stacks were identified as the largest source of airborne hydrocarbon emissions. At the beginning of this study, these were thought to be minor sources. Barge loading losses represented a second major source of emissions. Fugitive losses from process equipment and from tank vents were the third and fourth largest sources, respectively. The coker cooling pond was also found to be a significant emissions source. The size and significance of the coker cooling pond in relation to other sources was unknown prior to the study since few of the chemicals in pond emissions are required to be reported in the annual TRI. The refinery sewer system (sewer vents and oil/water separator) is a relatively small contributor to emissions. A single measurement is the basis for the landfarm emission estimate; it should be considered with caution.

2. Surface Water Sources

Identification of surface water sources was complicated by an old underground drainage system that was not designed for access and sample collection. Nevertheless, samples from major arteries helped identify the sources of primary pollutants of concern, such as oil and grease, biological oxygen demand (BOD), ammonia, total suspended solids, sulfides, metals, BTEX, and phenol. The crude unit desalter and crude tank water draws were found to have the highest pollutant loadings.

3. Subsurface Sources

A combination of groundwater well samples and computer modeling helped identify potential sources to and sinks for the subsurface aquifer. The most signif-

350   RONALD E. SCHMITT ET AL.

FIG. 9. Major sources of refinery solids.

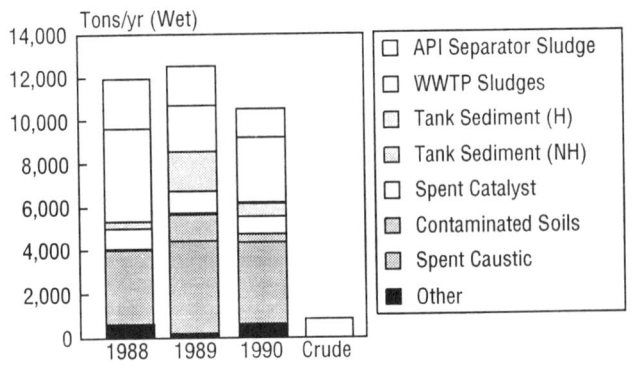

Note: About 1,000 tons of sediment in crude oil is recovered as solid waste.

icant sources of water reaching the subsurface where natural recharge from rainfall and water from the coke fines settling basin. More important from an environmental impact standpoint, the refinery's underground drainage system was observed to be recovering groundwater and routing it to the wastewater treatment plant. Consequently, there appeared to be no movement of contaminated groundwater off site.

4. Solid Waste Sources

Figure 9 summarizes solid waste sources identified from the sampling program. Sludges accumulated in the drainage and water treatment system accounted for a majority of the solids. Contaminated soils, tank sediments, and spent catalysts accounted for the remaining solids. Spent caustic, although an aqueous solu-

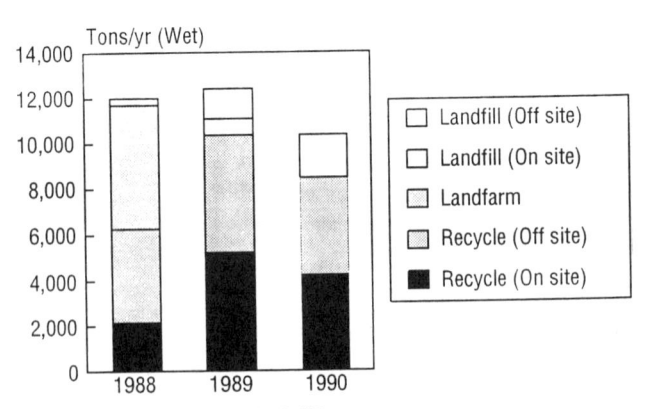

FIG. 10. Solids management.

tion, is usually classified and handled as hazardous waste for waste management purposes. Figures 10 and 11 summarize how these wastes were managed.

Some of the solid wastes, such as spent catalyst and sediment in crude, are by-products of the refining process. Others such as scale in storage tanks and soils swept into the sewer are not directly related to the processing operations. Thus, changing the refining process alone cannot accomplish all the reductions in solid waste generation; changes in other operating practices are required as well. At Yorktown, crude oil naturally contains more than 1000 tons of sediment, which is ultimately deposited in the refinery's oil/water separator or storage tanks. Local soil contributions (about 1000 tons/year) represent a large potential opportunity to reduce solid waste generation.

V. Assessing the Risks Posed by Refinery Releases

To better evaluate pollution prevention options, the project attempted to assess the risks posed to individuals and populations exposed to chemical contaminants released from the refinery. An initial risk assessment analysis was performed to identify chemicals requiring further study, and to establish a baseline by which to judge potential risk reduction opportunities. Since change in exposure to benzene was used as a proxy for evaluating relative risk reductions associated with alternative pollution prevention options, the usual uncertainty associated with risk assessments was not a factor in the option analysis. The uncertainty in absolute risk assessments can arise from multiple sources: the use of animal study results, difficulties with human studies, variation in individual responses to chemical exposures, the impact of differing dose rates, multiple simultaneous exposure to chem-

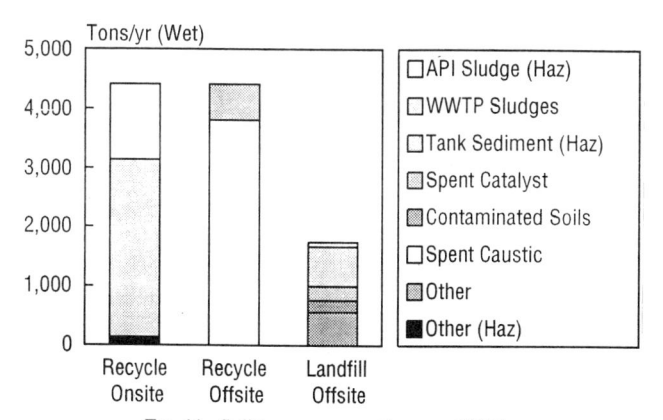

FIG. 11. Solids management by type (1990).

icals, and the use of extrapolation methods to estimate risks from high-exposure populations to low-exposure populations. Therefore, the results of the risk assessment completed for the initial screening should not be interpreted as definitive.

Risk assessment typically begins with a characterization of the risks associated with baseline or current releases. The baseline assessment gives an indication of the potential for human health or ecological risk problems. The predicted changes in emissions and sources are then estimated and the expected risk from the option scenarios is evaluated. The risk evaluation is based on both risk reduction to the most highly exposed individuals and to the exposed population as a whole. Cumulative benefits of risk reduction are estimated by adding the benefits for each risk reduction option.

The project's risk assessment effort followed EPA methods and established agency policy as outlined by the National Academy of Sciences (NAS, 1983) and established in final risk assessment guidelines (EPA, 1986). It involved four steps: (1) hazard identification, (2) determination of dose–response relations, (3) evaluation of human exposure and, finally, (4) characterization of risks.

Boundary conditions for this risk assessment were established so that the risk assessment did not attempt to analyze secondary environmental effects associated with refinery releases such as their contribution to formation of ozone, acid rain, risks associated with occupational exposure, transportation of products or wastes, or the potential for accidental releases.

Screening analysis was conducted for different exposure pathways and chemicals of concern. Using a screening level cutoff of a one-in-a-million excess risk for a 70-year lifetime exposure for the maximally exposed individual (MEI), a set of carcinogenic chemicals was identified for further analysis. For noncarcinogens, MEI exposure levels were compared to health thresholds. If the MEI exposure exceeded the established health threshold, further analysis was done.

A. AIR

Based on the screening analysis and methodology described, nickel, vanadium, methanol, carbon tetrachloride, xylenes, toluene, ethylene, propylene, naphthalene, ethylbenzene, polynuclear aromatic hydrocarbons (PAHs), 1,2,4-trimethylbenzene, and cyclohexane were not evaluated further, since their exposure either posed less than a one-in-a-million excess risk or was below the applicable reference dose.

For MTBE, EPA has not formally established health effects and reference dose. A preliminary estimate of the threshold reference concentration was developed for this risk assessment. Estimated concentrations for the MEI location were 10–15% of the reference concentration. No further evaluation was done for MTBE.

Two other chemicals and one mixture failed this initial screen: benzene, 1,3-butadiene, and VOCs. Benzene is a known carcinogen. It poses a one-in-a-million excess cancer risk at a concentration of 0.12 $\mu g/m^3$ for a lifetime exposure. Air modeling results show the concentration of benzene at the refinery boundary is 2.0 $\mu g/m^3$. At the nearest residence, this concentration is 1.5 $\mu g/m^3$. 1,3-Butadiene is also a known carcinogen. It poses a one-in-a-million excess risk at 0.0036 $\mu g/m^3$ for a 70-year lifetime exposure. Air modeling for this chemical shows a concentration of 0.0057 $\mu g/m^3$ outside the boundary and 0.0050 $\mu g/m^3$ at a residence.

VOCs were present at about 0.2 ppm outside the refinery boundary. VOCs are a complex mixture of hydrocarbons with an unspecified (and variable) composition. In the absence of any reference data that specifies an acceptable concentration of VOCs outside the fence line, it is helpful to compare the concentration data to a common standard used to monitor health in the workplace, the threshold limit value (TLV). The American Conference of Government and Industrial Hygienists (ACGIH) has set a TLV of 300 ppm for workplace exposure to gasoline vapors, another hydrocarbon mixture of unspecified composition (ACGIH, 1990). Since the exposure for workers is different than that of residents outside the refinery, ACGIH TLVs cannot be used to determine exposure limits for the general population; instead, they simply provide a benchmark for this discussion.

There is considerable debate, much uncertainty, and little data on human health effects at the low VOC concentrations expected around the refinery. Recent epidemiology studies of refinery workers, who would be expected to receive a higher exposure to VOCs than the general population, show a lower incidence of total cancer cases than the general population (Wong and Raabe, 1989). However, these studies are not able to differentiate between other potentially confounding factors such as the "healthy worker effect," cigarette smoking, diet, etc. VOC impacts from the refinery were not evaluated further.

B. Surface Water

The refinery has a state water discharge permit covering two outfalls: a combined treated process water mixed with once-through, noncontact cooling water outfall (001), and a stormwater settling basin outfall (002). The surface water analysis used the results from samples of these streams tested for 22 contaminants, total organic carbon, and such physical properties as pH and temperature. Surface water discharge concentrations were generally below the analytic detection limits (Amoco/EPA, 1991a). A screening level analysis identified the potential for risks to either human health or aquatic life based on established federal water quality criteria. This comparison was based on the EPA recommended approach for determining reasonable excursions above water quality criteria (EPA, 1991).

The screening analysis using two data points for these streams shows that copper exceeds EPA criteria for aquatic toxicity. Specifically, the highest copper concentration measured in outfall 002 as 250 µg/liter—about 90 times the marine acute copper criteria of 2.9 µg/liter. The highest copper concentration in outfall 001 exceeded the marine acute copper criteria by 76 times. During 1990 and 1991, the copper concentration in both outfalls have averaged between 66 and 76 µg/liter. Criteria are not the same as standards. However, if federal criteria are adopted as water quality standards in Virginia, these concentrations would constitute an exceedence. It is interesting to note that water from the York River has also exceeded these criteria on occasion. Subsequent analysis on four different occasions in 1992 showed the total recoverable copper concentration in outfalls 001 and 002 averaged 4 and 8 µg/liter, respectively, while the York River averaged 4 µg/liter. Upon evaluation of the data, the Commonwealth of Virginia concluded that no water quality problem existed and recommended that further monitoring was not necessary.

C. Drinking Water

Because the nearest drinking water source is 7 miles from the refinery and the York River is too saline for consumption, drinking water was not considered a potential source of exposure.

D. Groundwater

Groundwater contamination was found to be minimal with no off-site migration (Amoco/EPA, 1991b). Therefore, groundwater appeared to pose little or no risk and was not analyzed further. The sampling program did not find evidence of groundwater contamination from on-site waste disposal.

VI. Measuring Public Perception of the Refinery

To develop another tool that could be valuable in evaluating prevention options, the study team gathered data on the public's perceptions of the refinery and its impact on the local environment. The team believed this information might be useful in characterizing the viability of pollution prevention options. Three activities were undertaken to collect this information: (1) in-depth interviews with 25 thought leaders from state and local governments, community groups, local businesses, educational institutions, and environmental organizations, (2) two focus group meetings, and (3) a telephone survey of 200 households (Amoco/EPA, 1992a).

Probably the strongest conclusion produced by these information gathering activities was that people were generally ambivalent about the refinery, voicing neither major criticisms nor major support. In general, those contacted believed that the refinery complied with environmental laws and that this compliance probably protected the community. Most people felt that there were more pressing problems in the Yorktown area than the refinery. For example, land development, traffic, and sewer and water problems were cited as major quality-of-life concerns. When specifically asked about air pollution, water pollution, and disposal of solid waste, residents indicated that they did have a concern with respect to the oil refinery. These concerns, however, were not strongly felt and not specific.

One important insight gained from the public opinion study was that people obtain information about environmental problems in a random or unstructured way. There was no authoritative source of information about environmental problems accepted as reliable by a majority—or even a significant minority—of people. For example, some people surveyed had concerns about links between the refinery and reduced fishery yields in the Chesapeake Bay despite studies conducted by the Virginia Institute of Marine Science and others showing that refinery effluent was causing no known adverse impact on fish and the aquatic environment (Amoco/EPA, 1992b). This lack of an effective, reliable channel of communication between the refinery and the public could complicate the implementation of new pollution prevention strategies, especially if the public is misinformed about why any obvious change is being made. At the same time, the lack of a reliable communications channel could preclude mustering needed public support for policy or other changes needed to promote pollution prevention.

VII. Developing and Evaluating Options for Reducing Releases

In March 1991, more than 120 representatives from EPA, Amoco, the Commonwealth of Virginia, academic, environmental, and consulting organizations met for a three-day brainstorming workshop in Williamsburg, Virginia. Workshop participants developed options for reducing releases and considered ranking criteria, permitting issues, and obstacles and incentives for implementation. Workshop sessions included a structured review of process synthesis techniques and a more free-wheeling idea generation and discussion session. Participants proposed more than 50 pollution prevention projects for further consideration. Table IV lists all the projects identified.

To meet project schedule and budget constraints, the study team later selected 12 projects for more detailed analysis. Those chosen were felt to (1) be feasible

TABLE IV
POLLUTION PREVENTION OPTIONS IDENTIFIED AT THE WILLIAMSBURG WORKSHOP

Flare
Give away excess gas now flared
Redirect sour water stripper vent gas from flare to sulfur plant
Use gas from crude vacuum unit as fuel rather than as flare
Adjust process conditions to reduce flare gas generation
FCU
Use low attrition catalyst
Use SO_x reduction additive for FCU catalyst
Use more selective catalyst
Capture catalyst fines
Hydrotreat FCU feed
Use oxygen enrichment in regeneration unit
Integrate regeneration energy with air blower
Eliminate fluid-bed reactors
Fugitives
Implement a leak detection and repair (LDAR) program
Reduce barge loading and other transfer/handling losses
Cover sources of volatiles, including double seals on tanks
Track methane as a greenhouse gas
Refinery Water System
Reduce water content of sludge sent to coker
Enclose inside battery limits (ISBL) drains
Contain cleaning fluids to reduce evaporation
Redesign desalter system
Reroute desalter effluent
Keep soils out of sewer
Segregate process water effluent and pretreatment before discharging
Minimize oil/water contact
Use natural gas for stripping in place of steam
Find substitutes for filter aids
Flash difficult emulsions
Optimize sour water system (SWS)
Use stripped sour water as FCU water wash
Energy Integration
Achieve better integration with Virginia Power (VEPCO), including giving away flare gas
Optimize plant-wide energy use
Use oxygen enriched air in furnaces
Other
Perform on-line sampling to improve process control
Remove oxygen from feed streams to eliminate heat exchanger fouling
Recover sulfur in solid instead of a liquid phase
Filter crude to reduce tank sediments
Corporate Ideas
Produce a single grade of gasoline
Eliminate nonbiodegradable products
Desalt at the wellhead and reinject brine and sediment
Use more paraffinic feedstocks

TABLE IV *(Continued)*

Use renewable feedstocks
Institute overall product stewardship
Colocate facilities to maximize recycling
Reprocess used lube oil
Ideas Currently Being Evaluated
Eliminate coker blowdown pond
Move sewer and sewer gas adsorption system above-grade
Segregate process water from rainwater
Enclose or redesign API separator
Redesign product for lower emissions (lower vapor pressure)
Contain waste streams from cleaning operations
General Comments
Track progress in waste minimization to evaluate improvements
Beware of multi-media transfers

---

with current technology, (2) offer significant potential for release reductions, (3) have manageable or no adverse impacts on worker safety, (4) be amenable to more quantitative analysis in the time available, and (5) address concerns in different environmental media. Table V provides a brief description of each project. Figure 12 shows where each option fits into the overall refinery flow.

A. OPTION CHARACTERISTICS

Preliminary material balances and engineering designs were used to analyze each potential option. For each option the following items were determined:

1. *Capital costs.* Estimates with a $\pm 25\%$ accuracy were made for these scoping studies. Additional engineering effort would be required to prepare an estimate with a $\pm 10\%$ accuracy typically needed for management approval.
2. *Operating and maintenance costs.* Costs were estimated as a percentage of total capital cost, with consideration of project complexity. These costs varied between 3 and 6% of total capital.
3. *Liability cost rating.* Each project was evaluated qualitatively for its potential to affect future remediation and catastrophic and product quality liability concerns.
4. *Timeliness.* The number of years needed to complete each project was estimated, subject to current equipment maintenance schedules and operating limitations.
5. *Transferability.* A qualitative assessment was made of the ability to use the project technology at other refineries or in other industries.

TABLE V
Selected Pollution Prevention Engineering Projects

The following projects were identified for further study as a result of the Pollution Prevention Workshop in Williamsburg and subsequent workgroup meetings.

1. **Reroute Desalter Effluent:** Hot desalter effluent water currently flows into the process water drainage system at Combination unit. This project would install a new line and route this stream directly to the API Separator. This reduces volatile losses from the sewer system by reducing process sewer temperature and oil content. Volatile losses at the API Separator increase slightly.
2. **Improve Desalter System:** Evaluate installation of adjunct technology (e.g., centrifuge, air flotation, or other technology) on desalter water stream prior to discharge into the underground process drainage system. This reduces oil and solids waste loads in the sewer system, affecting the wastewater treatment plant and volatile losses from the drainage system.
3. **Reduce FCU Catalyst Fines:** Evaluate possible performance of more attrition-resistant FCU catalyst to reduce fines production. (Subsequent review with catalyst vendors indicated the Refinery was already using the most attrition-resistant catalyst available.) Two other fines reduction options were considered.
3a. **Replace FCU Cyclones:** Assess potential for reducing emissions of catalyst fines (PM10) by adding new cyclones in the regenerator.
3b. **Install Electrostatic Precipitator at FCU:** Assess potential of electrostatic precipitator in reducing catalyst fines (PM10) emissions.
4. **Eliminate Coker Blowdown Pond:** Change operating procedures for coke drum quench and cooldown so that an open pond is no longer needed. This reduces volatile losses from the hot blowdown water.
5. **Install Seals on Storage Tanks:** Double seals or secondary seals will reduce fugitive vapor losses.
    a. Secondary Seals on Gasoline Tanks: Install secondary rim mounted seals on tanks containing gasoline.
    b. Secondary Seals on Gasoline and Distillate Tanks: Install secondary rim mounted seals on tanks containing gasoline and distillate material.
    c. Secondary Seals on All Floating Roof Tanks: Install secondary rim mounted seals on all floating roof tanks.
    d. Option 5c + Internal Floaters on Fixed Roof Tanks: Install secondary rim mounted seals on floating roof tanks and install a floating roof with a primary seal on all fixed roof tanks.
5e. **Option 5d + Secondary Seals on Fixed Roof Tanks:** Install secondary rim mounted seals on all floating roof tanks and then install a floating roof with a primary and secondary seal on all fixed roof tanks.
6. **Keep Soils Out of Sewers:** Use road sweeper to remove dirt from roadways and concrete areas which would otherwise blow or be washed into the drainage system. Develop and install new sewer boxes designed to reduce soil movement into sewer system, particularly from Tankfarm area. Estimate cost for installation on a refinery-wide basis. Both items reduce soil infiltration, in turn reducing hazardous solid waste generation.
7A. **Convert Blowdown Stacks:** Replace existing atmospheric blowdown stacks with flares. This reduces untreated hydrocarbon losses to the atmosphere, but creates criteria pollutants.
7B. **Drainage System Upgrade:** Install above-grade, pressurized sewers, segregating stormwater and process water systems.
7C. **Upgrade Process Water Treatment Plant:** Replace the API Separator with a covered gravity separator and air floatation system. Capture hydrocarbon vapors from both units.

TABLE V *(Continued)*

8. **Change Sampling Systems:** Install flow-through sampling stations (speed loops) where required on a refinery-wide basis. These replace existing sampling stations and would reduce oil load in the sewer or drained to the deck.
9. **Reduce Barge Loading Emissions:** Estimate cost to install a marine vapor loss control system. Consider both vapor recovery and destruction in a flare.
10. **Sour-Water System Improvements:** Sour water is the most likely source of Refinery odor problems. Follow-up on projects previously identified by Linnhoff-March engineering to reduce sour water production, and improve sour water stripping.
11. **Institute LDAR Program:** Institute a leak detection and repair program for fugitive emissions from process equipment (valves, flanges, pump seals, etc.) and consider costs and benefits.
    a. Annual LDAR Program with a 10,000 PPM hydrocarbon leak level.
    b. Quarterly LDAR Program with a 10,000 PPM hydrocarbon leak level.
    c. Quarterly LDAR Program with a 500 PPM hydrocarbon leak level.

6. *Revenues.* Revenues were estimated for those projects where salable materials were recovered. The quantity of recovered material was equivalent to the emissions reduction. All recovered hydrocarbons were valued as gasoline. This tends to overestimate the actual values, since most VOCs, for example, are the lighter portions of gasoline, rather than whole product.
7. *Equivalent annual costs.* These costs were estimated as the sum of annualized capital costs and all fixed and variable expenses (maintenance, operating, taxes, insurance). Future costs were discounted at 10% (EPA rate) or 15% (Amoco rate) to determine their present value.
8. *Pollution prevention mode.* One or more of the pollution prevention modes in the pollution prevention hierarchy (source reduction, recycle, reuse, treatment, and disposal) was assigned based on review, discussion, and consensus among study team members. These classifications were not obvious in several cases and required extended debate.
9. *Net release reduction.* Estimates of emissions reduction (tons/year) vary in accuracy. Additional emissions sampling and more detailed engineering analysis would be needed to improve these estimates. Where possible, generation and transfer of releases in other media were included in estimating the "net" change in release.
10. *Recovery cost.* For liquid hydrocarbons or VOC emissions the equivalent annual cost was divided by the net release reduction volume to determine an average dollar/gallon for each option. This number is equivalent to the price that would have to be charged per gallon of recovered material to recover capital, operating, maintenance, and distribution costs.
11. *Cost effectiveness.* The equivalent annual cost was divided by the tonnage net release reduction to determine a dollar/ton cost effectiveness for each option.

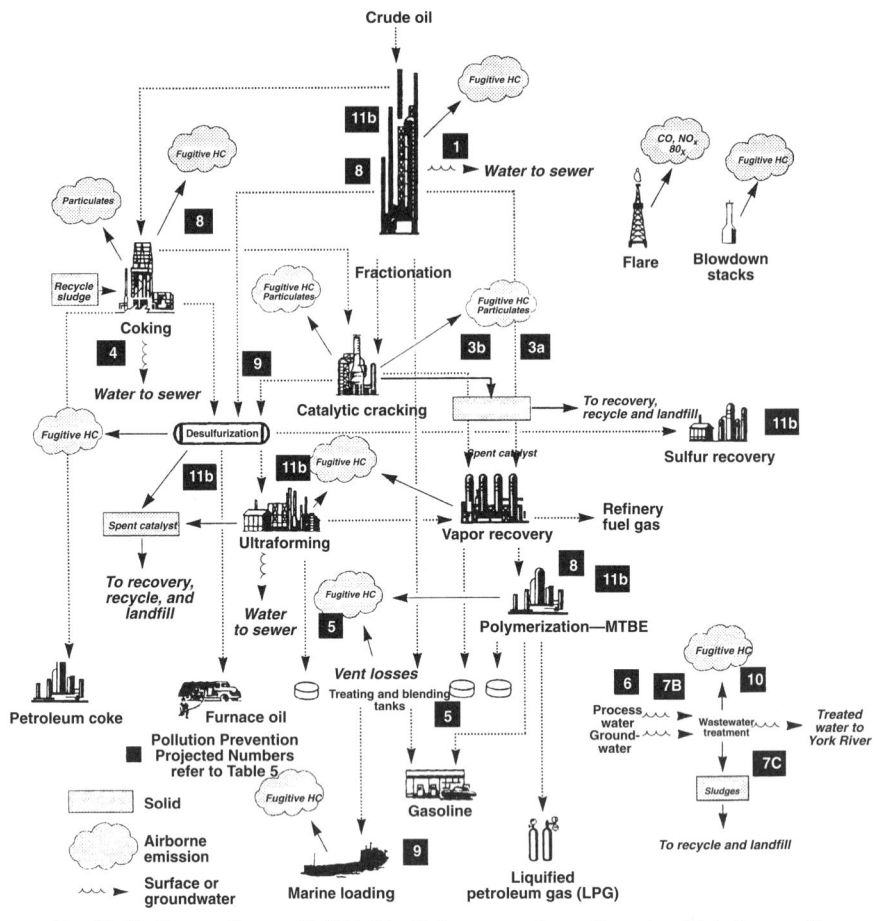

FIG. 12. Yorktown refinery with Table V pollution prevention options placed into the overall refinery flow.

12. *Resource utilization.* Qualitative estimates were developed for each option's effect on raw materials and utilities requirements.
13. *Effects on secondary emissions.* The impact of each project on other emissions was judged qualitatively. For example, increased power requirements would normally increase emissions from utility systems.

B. Ranking the Options

Ranking and prioritizing these options required specific quantitative and sometimes qualitative data about each choice. Table VI provides estimated reduction in re-

TABLE VI
AMOCO/EPA POLLUTION PREVENTION PROJECT

	Option	Net Release Reduction (tons/yr)	Control Efficiency (%)	Timing Years	Pollution Prevention Mode
1	Reroute Desalter Water	52.4	90.0	1–3	Recycle
2	Improve Desalter System	U/D[a]	U/D	U/D	U/D
3a	Replace FCU Cyclones	245.0	48.0	4–7	Recycle/disposal
3b	Install ESP at FCU	442.0	87.0	4–7	Disposal
4	Eliminate Coker B/D Pond	130.0	50.0	1–3	Source reduction
5a	Install Sec. Seals on Gasoline Tanks	474.7	75.0	>7	Source reduction
5b	on Gasoline and Distillate Tanks	482.1	76.0	>7	Source reduction
5c	on all Floating Roof Tanks	541.0	85.0	>7	Source reduction
5d	Option 5c & Floaters on Fxd Tanks	591.7	93.0	>7	Source reduction
5e	Option 5d & Sec. Seals on Fxd Tanks	592.2	94.0	>7	Source reduction
6	Decrease Soils in Drainage Systems	530.0	50.0	4–7	Source reduction
7A	Blowdown System Upgrade	5096.0	98.0	4–7	Treatment
7B	Drainage System Upgrade	112.5	95.0	1–3	Treatment
7C	Water Treatment Plant Upgrade	58.0	95.0	1–3	Treatment
8	Modify Sampling Systems	63.0	100.0	4–7	Source reduction
9	Reduce Barge Loading Emissions	768.0	98.0	1–3	Recycle
10	Sour Water System Improvements	18.0	100.0	1–3	Recycle/treatment
11a	Annual LDAR Program @ 10,000 ppm	319.5	40.0	<1	Source reduction
11b	Quarterly LDAR Program @ 10,000 ppm	510.5	64.0	<1	Source reduction
11c	Quarterly LDAR Program @ 500 ppm	705.5	89.0	<1	Source reduction

[a] U/D = undefined.

leases (tons/year), type of material released, control technology efficiency, expected time required to complete installation of the particular control technology, and type of release management option used for control (source reduction, recycling, etc.).

1. Quantitative Financial Analysis

Table VII summarizes the quantitative financial analysis. For each option, the table shows the net present value (NPV) of the capital cost, the NPV of

TABLE VII
AMOCO/EPA Pollution Prevention Project Financial Summary

	Option	PV of Capital (M$)	PV of O&M (M$)	Annualized Cost (M$/yr)	Cost Effectiveness ($/ton)
1	Reroute Desalter Water	1,000	1,502	329	6,279
2	Improve Desalter System	Undefined	Undefined	Undefined	Undefined
3a	Replace FCU Cyclones	8,300	14,738	3,029	12,363
3b	Install ESP at FCU	9,100	18,153	3,583	8,106
4	Eliminate Coker B/D Pond	2,000	2,807	632	4,862
5a	Install Sec. Seals on Gasoline Tanks	259	426	90	190
5b	on Gasoline and Distillate Tanks	321	531	112	232
5c	on all Floating Roof Tanks	445	734	155	287
5d	Options 5c & Floaters on Fxd Tanks	1,827	3,018	637	1,077
5e	Options 5d & Sec. Seals on Fxd Tanks	2,003	3,306	698	1,179
6	Decrease Soils in Drainage Systems	337	1,207	203	383
7A	Blowdown System Upgrade	5,095	7,303	1,630	320
7B	Drainage System Upgrade	18,800	26,388	5,941	52,809
7C	Water Treatment Plant Upgrade	22,500	33,808	7,403	127,638
8	Modify Sampling Systems	76	129	27	429
9	Reduce Barge Loading Emissions	4,700	7,531	1,608	2,094
10	Sour Water System Improvements	605	909	199	11,056
11a	Annual LDAR Program @ 10,000 ppm	5	695	92	288
11b	Quarterly LDAR Program @ 10,000 ppm	5	1,045	138	270
11c	Quarterly LDAR Program @ 500 ppm	5	1,478	195	276

Notes: All cash flows are discounted at 10%, 15-yr project life. Capital spending for all projects is assumed to begin in 1991. O&M = operation and maintenance costs, depreciation, indirect costs, taxes, and insurance.

operating/maintenance cost minus revenue, the equivalent total annual cost, and the annual recovery cost in dollars/ton. NPV is a useful tool for comparing different options that have different cash flows and different time periods, since all cash flows, present and future, are converted to a current time. The annual cost and recovery cost are in 1991 dollars. Other financial information is on an NPV basis that discounts future cash flows using a 10% discount rate, typical of that used by the EPA for project evaluation purposes. Amoco uses a higher discount rate, 15%, to evaluate cost effectiveness.

2. Risk Exposure

Where possible, the study team used computer modeling tools to evaluate how different options would affect risk exposure. Adequate tools were not available for those options that involved changes in releases of solid waste or surface water discharges. However, an independent risk screening showed that neither surface water nor solid waste presented a significant human exposure pathway. No additional modeling was done for options that might affect groundwater, since groundwater near the site is not a source of drinking water.

As noted before, the most significant refinery emissions were airborne. For those options that involved a change in emissions affecting air quality, impacts were modeled using standard air dispersion techniques. Exposure estimates were developed for three classes of chemicals: (1) benzene, toluene, ethylbenzene and xylene (BTEX), (2) other chemicals reported in the refinery's TRI submissions, and (3) criteria pollutants (SO_2, NO_2, PM_{10}, and CO). Similar modeling techniques were used for all three classes. The project focused on the impact of benzene emissions, since benzene turned out to be the chemical species of greatest concern relative to other releases.

For benzene (and other BTEX compounds) the emissions inventory was used as input to the industrial source complex short term (ISCST) air dispersion model. The model used 1 year of hourly meteorological data collected at the National Weather Service station at Norfolk, Virginia. Norfolk is located about 50 miles from Yorktown, and experiences similar land–sea breeze conditions. A mathematical receptor grid was established around the refinery containing 859 points. Both a fine grid (250-m resolution) and a coarse grid (1000-m resolution) were used, to give better resolution near the refinery where concentrations changed more quickly. The model calculated ground level concentrations at each receptor point for each hour of the year as well as annual average concentrations. About 7.5 million computer calculations were completed to model impacts of each chemical studied for each different control scenario considered. The modeling proved extremely valuable for calculating the refinery's impact on the environment, identifying the sources of greatest impact, and comparing the effectiveness of various options to reduce exposure. Due to the site-specific nature of the emissions data, meteorology, and the release reduction options, the results are relevant for this facility only.

Modeling results are summarized in several ways. First, Fig. 13 presents isopleths of annual average benzene concentrations overlaid on a United States Geological Survey (USGS) topographic map of the refinery area. Each curved line

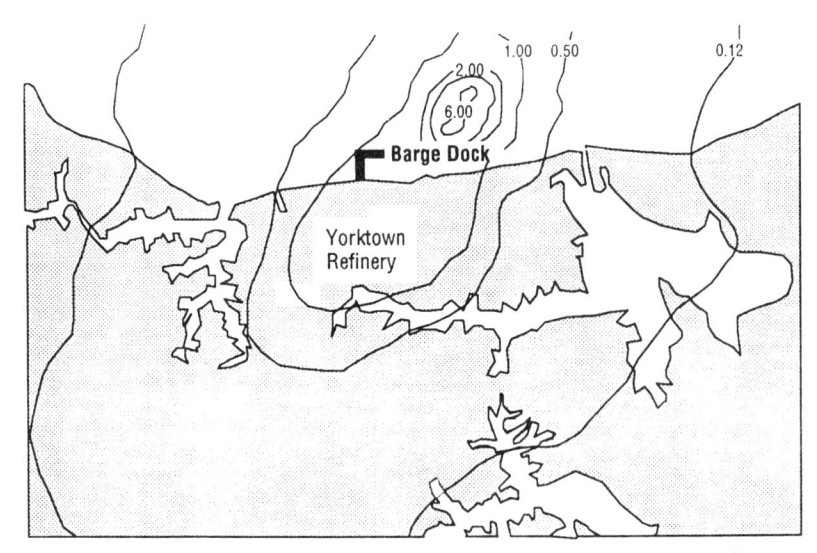

FIG. 13. Baseline emissions for benzene, annual average concentrations ($\mu g/m^3$).

represents a line of constant benzene concentration, much as elevation lines are used on topographic maps.

Second, Table VIII provides maximum predicted ground-level concentrations in micrograms per cubic meter ($\mu g/m^3$) for various receptor locations: within the refinery (9.2), beyond the fence line on the York River (6.2), beyond the fence line on land (1.3), and at a nearby residence (0.6).

Third, additional dispersion modeling was performed to identify source–receptor relationships and define culpable sources (those sources that have the largest contributions to the total impact) for the receptor points identified. These results are illustrated in Fig. 14. For example, at the point of highest concentration

TABLE VIII
MAXIMUM ANNUAL AVERAGE PREDICTED CONCENTRATIONS FOR THE YORKTOWN REFINERY FOR BTEX CHEMICALS

Chemical	Inside Plant Fence Line Conc. ($\mu g/m^3$)	Outside Plant in York River Conc. ($\mu g/m^3$)	Outside Plant Fence Line on Land Conc. ($\mu g/m^3$)	Outside Plant Fence Line at a Nearby Residence Conc. ($\mu g/m^3$)
Benzene	9.20	6.20	1.30	0.64
Toluene	25.50	21.90	5.40	2.37
Ethylbenzene	7.20	5.90	1.60	0.72
Xylene	36.90	5.90	7.20	3.06

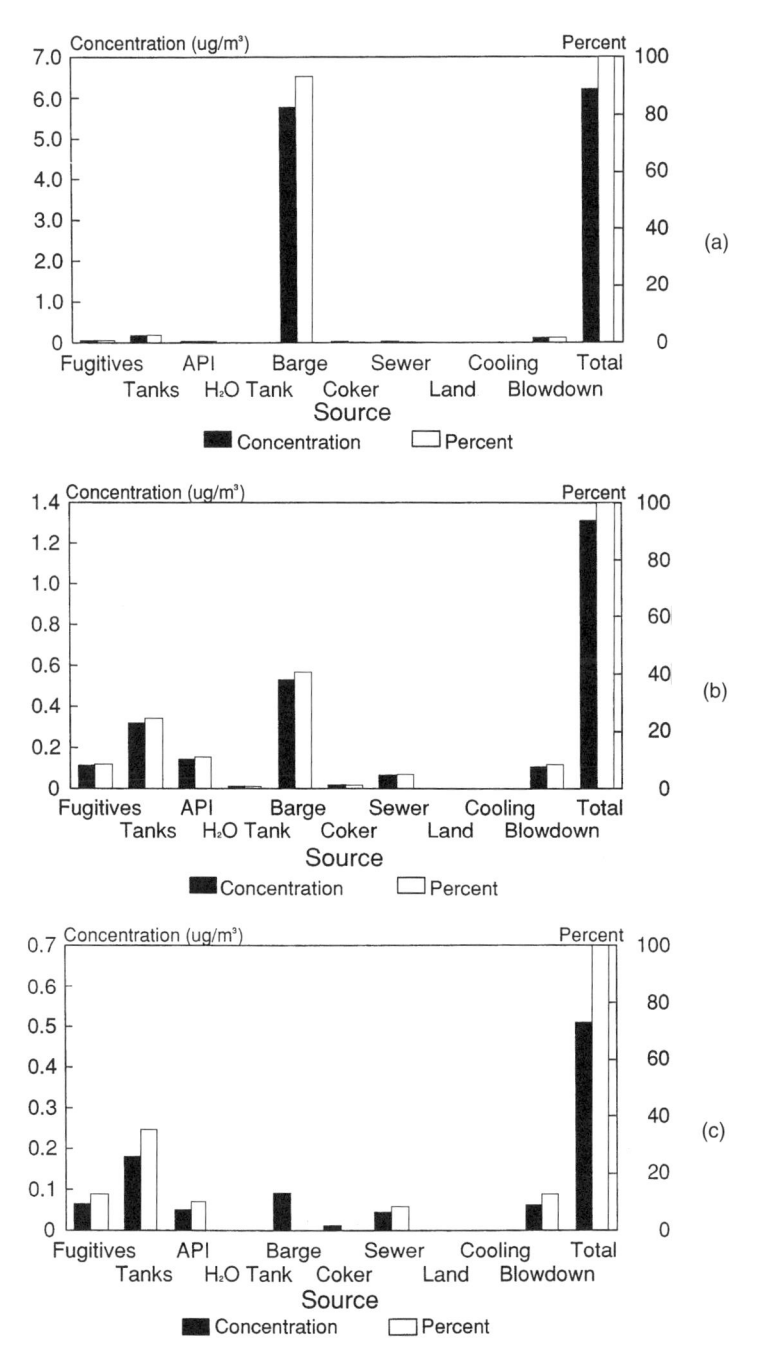

FIG. 14. (a) Source culpability in York River, (b) at fence line on land, and (c) at closest residence.

in the York River, 93% of the impacts resulted from barge loading emissions. In contrast, at a nearby residence, barge loading emissions were responsible for 53% of the total calculated concentration. Storage tank emissions and blowdown stacks accounted for most of the remaining concentration. Barge loading emissions accounted for a significant fraction of the total impacts at both receptor sites.

Changes in emissions resulting from simulating the different pollution prevention options in Table V were also modeled, using identical air dispersion techniques. Again, benzene emissions are discussed here because of their potential health impacts. For each option considered, revised benzene emissions for the affected source(s) were used to calculate a new emissions inventory. For example, recovering barge loading losses could reduce benzene emissions by 11 tons per year. Figure 15 shows a modified histogram reflecting this change. However, the histogram only shows the impact of reduced benzene emissions on the total. It is more helpful to ask how the reduced emissions affect exposure of people outside the fence line. These changes become more apparent when plotted as revised isopleths on the same USGS map used for Fig. 13. Figure 16 shows the new isopleths (shown as "Emissions After Controls") on the original inventory map. The high benzene concentrations around the barge loading area have disappeared. Furthermore, the most outlying isopleth (showing a concentration of 0.12 $\mu g/m^3$) has moved in toward the refinery center. This indicates that the area impacted by refinery emissions has been reduced. Ultimately, the new concentration information can be converted into population exposure and risk estimates. Each pollution prevention option can be viewed and compared in this same way, leading to calculation of changes in relative population risk for each option, compared to current operations at the refinery.

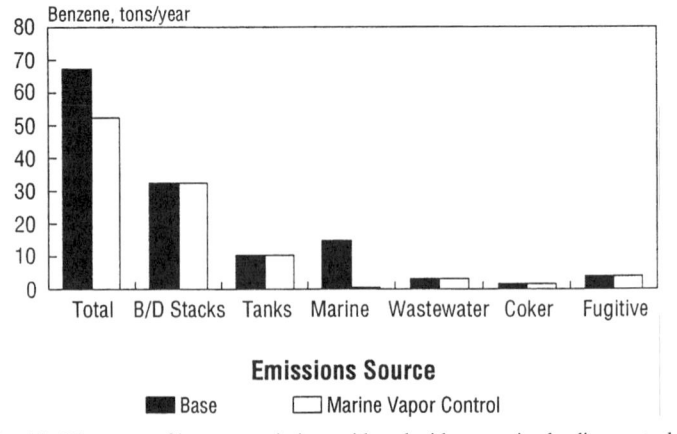

FIG. 15. Histogram of benzene emissions with and without marine loading controls.

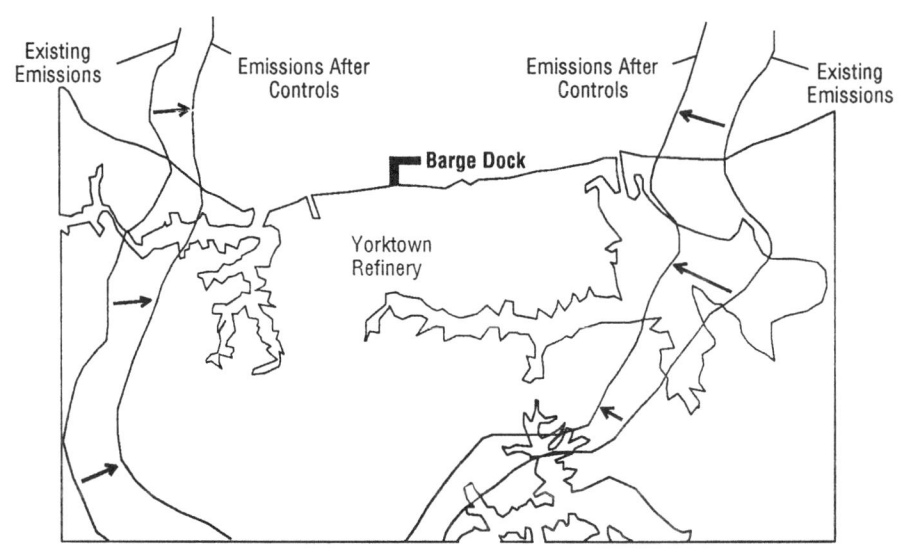

FIG. 16. Effectiveness of barge loading control option for benzene, annual average concentrations.

Changes in benzene exposure were calculated at a nearby residence for each control option. Table IX shows these results. The baseline benzene concentration was 0.26 μg/m^3. The different pollution prevention options reduce this concentration to between 0.61 and 0.28 μg/m^3. Current EPA methodology assumes a linear relationship between the dose of a carcinogen that an individual receives over a lifetime and increased risk of cancer. Based on this conservative (health protective) approach, reductions in an individual's cumulative exposure over a lifetime correlate directly with reductions in carcinogenic risk. By looking at this drop in potential benzene exposure as a surrogate for risk reduction, a more quantitative measure of the effectiveness of each option can be developed. The third column in Table IX shows this exposure reduction. Here the existing facility contributes 100% of the controllable benzene exposure. As benzene concentration at the nearby residence decreases, exposure also drops. Table IX indicates that reducing barge loading emissions has the largest potential to reduce benzene exposure. Many of the other options had small or minimal impact on benzene concentrations at a residence and, therefore, small or minimal potential to reduce relative risk.

Figure 17 shows the changes in the outlying concentration isopleth that result from implementing barge loading controls, adding secondary seals to gasoline storage tanks, and upgrading the blowdown stacks. Qualitatively they show a significant change over a fairly wide area, indicating a potentially effective control strategy. Figure 18 shows a similar plot, but this time the changes reflect the im-

TABLE IX
MAXIMUM ANNUAL AVERAGE BENZENE CONCENTRATIONS AND BENZENE EXPOSURE ASSOCIATED WITH VARIOUS POLLUTION PREVENTION OPTIONS

Option[a]		Benzene Conc. at Nearby Residence ($\mu g/m^3$)	Percent of Benzene Exposure Compared to Base Case	Percent Reduction in Benzene Exposure
0	Base Case	0.62	100	-0-
1	Desalter Control	0.61[b]	99	1
4	Coker Pond Control	0.61	98	2
5a	Secondary Seals on Gasoline Tanks	0.51	82	18
5e	Secondary Seals on all Tanks	0.50	80	20
7A	Blowdown Stacks to Flare	0.55	89	11
7B	Drainage Controls	0.59	95	5
7C	Trtm't Plant Upgrade	0.59	95	5
7C + 7B	Combination	0.56	90	10
9	Capture Barge Loading Losses	0.28	45	55
11a	Annual LDAR	0.61	98	2
11b	Quarterly LDAR	0.60	97	3

[a]See Table V for descriptions.
Notes: 1. Options 5a and 5e cover the range of control which includes option 5c.
2. Options 3a, 3b, 6, 8, and 10 do not affect benzene emissions.

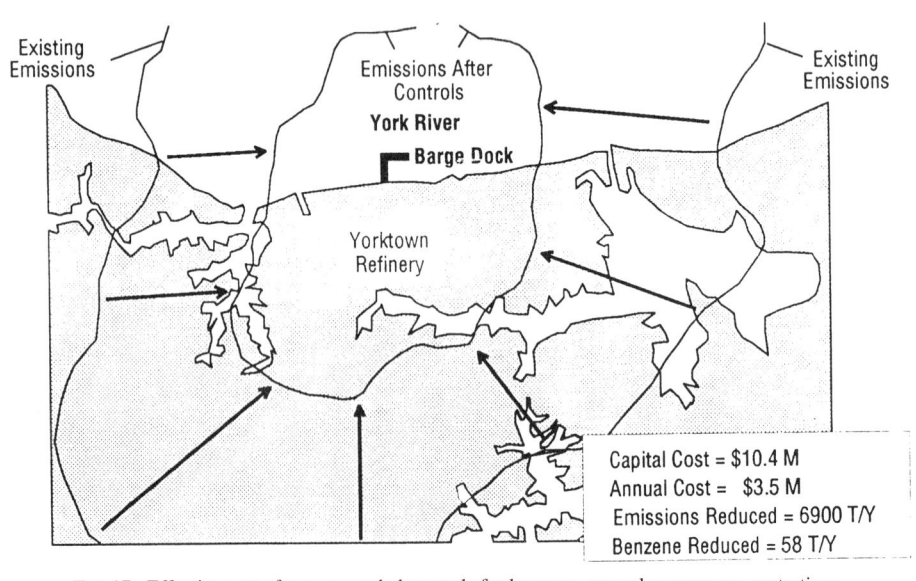

FIG. 17. Effectiveness of recommended controls for benzene, annual average concentrations.

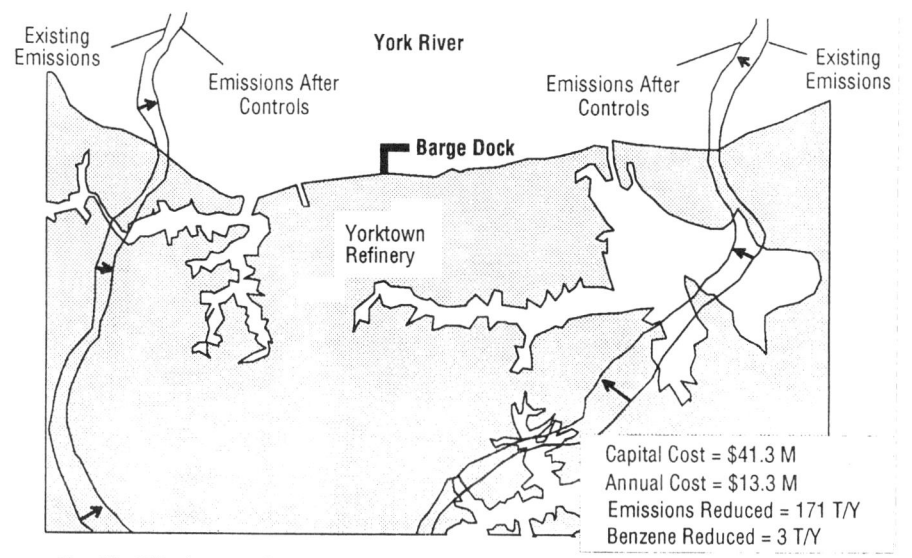

FIG. 18. Effectiveness of mandated sewer controls for benzene, annual average concentrations

pacts of control options for the oil/water separator and underground drainage system. In contrast to Fig. 17, the curves show that almost no impact on air quality would occur in residential areas. This combination of options has a small effect over a narrow area and, therefore, appears to be a relatively ineffective control strategy.

VIII. Ranking Methods and Results

A. Options for Reducing Total Releases

The study team selected 12 options from the 50 identified at the Williamsburg workshop. Important characteristics of the 12 options, and their alternatives, are summarized in Table X. For three options—3, 5, and 11—only one of the several alternatives considered would be implemented. Two options reduce solid wastes (catalyst fines and listed hazardous wastes), while the remaining 10 focus on air emissions (VOC, HC, H_2S, and NH_3). Five of the 12 options employ source reduction to reduce releases. Modifying sampling (option 8) eliminates venting of lines into the environment just prior to sampling. Capital costs range from a low of $10,000 to a high of $22.5 million. Annual costs, based on capital, operating, and maintenance costs at a 10% discount rate, range from $30,000 to $7.4 million.

TABLE X
Amoco/EPA Project Pollution Prevention Option Characteristics

#[a]	Project	Materials	Cost Effective ($/ton)	Benzene Exposure Reduction (%)	Cost Eff BzExRed ($M/%BzE)	Statutory Program	Expected Compliance Year	Impl'n Time (yr)
1	Reroute Desalter	VOC	6,279	1	329			1–3
3a	Replace FCU Cyclones	Cat. fines	12,363	0				4–7
3b	Install FCU ESP	Cat. fines	8,106	0				4–7
4	Elim. Coker Pond	VOC	4,862	2	316	RCRA/CAA	1994	1–3
5a	Sec. Seals-Gasoline Tks	VOC	190	18	5	MACT, Nonatmt Oz	1994	>7
5b	Sec. Seals-Gaso/Dist Tks	VOC	232	18	6	MACT, Nonatmt Oz	1994	>7
5c	Sec. Seals-All FltRfTk	VOC	287	18	9	MACT, Nonatmt Oz	1994	>7
5d	Opt 5c + Flit on FixTk	VOC	1,077	18	35	MACT, Nonatmt Oz	1994	>7
5e	Opt 5d & S. Seal FixTk	VOC	1,179	20	35	MACT, Nonatmt Oz	1994	>7
6	Soils Control	Listed HW	383	0				4–7
7A	B/D Upgrade	VOC	320	11	148	BzNESHAP/Nonatmt Oz	1993	4–7
7B	Drainage Upgrade	VOC	52,809	5	1,188	BzNESHAP/Strmwtr Oz	1994	1–3
7C	Treatment Plit Upgrade	VOC	127,638	5	1,480	BzNESHAP/Nonatmt Oz	1993	1–3
8	Modify Sampling	VOC/HC	429	0		MACT or HON	1995	4–7
9	Barge Loading	VOC	2,094	55	29	MACT, Nonatmt Oz	1994	1–3
10	Sour Water Improvement	H_2S, NH_3	11,056	0				1–3
11a	Ann. LDAR (10,000 ppm)	VOC	288	2	46	Ozone nonattainment	1994	<1
11b	Quart LDAR (10,000 ppm)	VOC	270	3	46	Ozone nonattainment	1994	<1
11c	Quart LDAR (500 ppm)	VOC	276	3	46	Ozone nonattainment	1994	<1

[a] # refers to projects listed in Table V.
Sour water improvement, and materials = H_2S, NH_3.

The cost per ton reduction in environmental releases (cost effectiveness of release reduction) is given for each option in Table X. The most cost-effective options, on the basis of release reductions, are three leak detection and repair (LDAR) alternatives (11a, 11b, and 11c), requiring an average of $278 per ton, and three secondary seal alternatives (5a, 5b, and 5c), which require an average of $236 per ton. Three other options are also low cost, $429 per ton or less. However, the cost for each of the other seven options is $2000 per ton or more. The wastewater treatment plant upgrade represents the least cost-effective option, requiring an annual cost of $127,000 for each ton of VOC recovered. Another way of viewing an option's cost effectiveness is to determine the price at which the recovered material would have to be sold in order to offset the cost of recovery. The recovered material would have to be sold at $0.90 per gallon for the LDAR and secondary seal options to break even, and at $415 per gallon for the most expensive option, treatment plant upgrade. For comparison, the refinery price was about $0.75 per gallon during the study period.

1. Benzene Exposure Reduction Options Ranking

The ranking analysis discussed in the remainder of this section used benzene exposure at a nearby residence as a proxy for the risk associated with population exposure to refinery releases. In Table X, the share each option represents of the total benzene exposure reduction achieved by implementing all options is given in the column labeled "Benzene exposure reduction." The barge loading option accounts for 55% of the benzene exposure reduction attributable to all options. In cost-effectiveness terms, the cost for a 1% benzene exposure reduction ranges from $9000 for secondary seals to $1.48 million for upgrading the wastewater treatment plant.

2. Options and Regulatory Requirements

The statutory requirements addressed by each option are identified in the seventh column of Table X titled "Statutory program." The specified compliance dates for each statutory program are also shown in the table, as are the implementation times for each option. Since the maximum achievable control technology (MACT) requirements have not yet been specified by EPA, projects directed toward those requirements have not yet been undertaken. That being the case, the implementation times identified indicate that options 5, 8, and 9 may have difficulty achieving timely compliance with these regulatory requirements. To proceed with the analysis, requirements that might be imposed under MACT were hypothesized to specify the performance characteristics of the associated pollution prevention options. It should be noted that only those options related with compliance with the Benzene Waste NESHAP (National Emissions Standard for

Hazardous Air Pollutants) requirements deal specifically with benzene release options; other rules address emissions of hydrocarbon and VOCs, not only benzene.

3. Single Criterion Rankings

The project's Peer Review Committee suggested the options be ranked according to a single criterion: risk reduction. In addition to risk reduction, two other single criterion rankings are of interest: total tonnage reduction and cost.

a. Risk Reduction. A risk proxy of benzene exposure at a nearby residence was used to complete ranking of options. Calculated benzene concentrations at a nearby residence were assumed to provide a reasonable indicator for measurement of population exposure, and the exposure reductions achievable by implementing a particular option. Using this measure, and the option characteristics developed by Amoco engineers, several rankings were produced.

The results of ranking the options using benzene exposure reduction as the sole criterion are shown in Table XI. Reducing barge loading emissions is the outstanding option using this metric. No other option comes close to the exposure reduction achieved by reducing barge loading emissions. The other ranking values provide insight into which options generally provide greater exposure reduction. For example, all secondary seal alternatives achieve significant exposure reduction, and the blowdown system upgrade also performs effectively in this regard. Four options achieve no benzene exposure reduction because those options deal with release sources that do not emit benzene.

In reviewing the exposure reduction rankings and other rankings, it is important to bear in mind that the rankings are intended to provide an approximate guide to which options rank near the top with regard to certain criteria and which rank near the bottom. On this basis, options that consistently rank near the top across all criteria felt by the decision maker to be important are generally preferred. Options that receive comparable scores during the ranking process should be considered equivalent independent of their rank. For example, from an exposure reduction perspective Table XI indicates that (1) controlling barge loading emissions is the best single action; (2) installing secondary seals and implementing an upgrade of the blowdown also will achieve beneficial exposure reductions; and (3) that the remaining options achieve minimal or no reduction in benzene exposure.

b. Release Reduction. The results obtained when pollution prevention options are ranked by extent of release reduction are shown in the first set of columns of Table XI. The blowdown system upgrade is the clear winner, reducing releases by more than six times that of the nearest competitor. The remaining options dimin-

TABLE XI
SINGLE CRITERION RANKINGS BASED ON RELEASE REDUCTION AND EXPOSURE REDUCTION

	Release Reduction Option				Exposure Reduction Option		
Rank	Description	#[a]	(tons/yr)	Rank	Description	#[a]	Benzene Expos. Red'n (%)
1	Blowdown System Upgrade	7A	5,096	1	Barge Loading	9	55
2	Barge Loading	9	768	2	Opt 5d & Sec. Seal on Fixed Tanks	5e	20
3	Quarterly LDAR (500 ppm)	11c	706	3	Sec. Seals—Gasoline Tanks	5a	18
4	Opt 5d & Sec. Seal on Fixed Tanks	5e	592	3	Sec. Seals—Gas/Distillate Tanks	5b	18
5	Opt 5c & Floaters on Fixed Tanks	5d	592	3	Sec. Seals—Floating Roof Tanks	5c	18
6	Sec. Seals—Floating Roof Tanks	5c	541	3	Opt 5c & Floaters on Fixed Tanks	5d	18
7	Soils Control	6	530	7	Blowdown System Upgrade	7A	11
8	Quarterly LDAR (10,000 ppm)	11b	511	8	Drainage System Upgrade	7B	5
9	Sec. Seals—Gas/Distillate Tanks	5b	482	8	Treatment Plant Upgrade	7C	5
10	Sec. Seals—Gasoline Tanks	5a	475	10	Quarterly LDAR (10,000 ppm)	11b	3
11	Install FCU ESP	3b	442	10	Quarterly LDAR (500 ppm)	11c	3
12	Annual LDAR (10,000 ppm)	11a	320	12	Annual LDAR (10,000 ppm)	11a	2
13	Replace FCU Cyclones	3a	245	12	Eliminate Coker Blowdown Pond	4	2
14	Eliminate Coker Blowdown Pond	4	130	14	Reroute Desalter	1	1
15	Drainage System Upgrade	7b	113	15	Sour Water System Improvements	10	0
15	Modify Sampling System	8	63	15	Replace FCU Cyclones	3a	0
17	Treatment Plant Upgrade	7c	58	15	Install FCU ESP	3b	0
18	Reroute Desalter	1	52	15	Soils Control	6	0
19	Sour Water System Improvements	10	18	15	Modify Sampling System	8	0

[a] # refers to project numbers from Table V.

ish gradually in terms of release reduction. When compared to the exposure reduction results, all of the highest ranked release reduction options—blowdown system upgrade, barge loading controls, quarterly LDAR program (500 ppm), and double seals on tanks—also rank at the top in terms of exposure reduction.

c. Cost. It is interesting to compare the exposure reduction and release reduction results with the ranking based on cost, shown in Table XII. In terms of annualized costs, modifying the sampling hardware and procedure is the best option, costing three times less than its closest competitor. When evaluating the options, using a single cost criterion alone is not as useful as comparing the cost ranking results with the results based on exposure reduction and release reduction. Modifying sampling ranked high with respect to cost but near the bottom with respect to these other criteria. On the other hand, two options that rank high with regard to exposure reduction and release reduction (secondary seals and quarterly LDAR 500 ppm) also rank well with respect to costs. Barge loading and blowdown system upgrade, which rank near the top from the exposure reduction and release reduction perspective, rank near the bottom from the cost perspective. Based on these three single criterion rankings, the secondary seals and quarterly LDAR options look promising, and, if sufficient funding is available, barge loading and the blowdown system upgrade may be promising as well.

4. Multiple-Criteria Ranking

In some cases, a more integrated multiple-criteria process is desired to help with option selection. For example, the importance attributed to each criterion may be in dispute, and a systematic process may be needed to enable the decision makers to resolve these differences. The study team considered a number of multiple-criteria decision-making techniques for ranking options. The three approaches given greatest attention were (1) the analytical hierarchy process or AHP (Saaty, 1988, 1990); (2) the Kepner–Tregoe approach (Kepner and Tregoe, 1979, 1981), which Amoco has used in reviewing selected corporate decisions; and (3) computation of alternative equivalents (Stokey and Zeckhauser, 1978), which a member of the Peer Review Committee suggested. Ultimately, the study team selected AHP as the ranking methodology since it has proven useful in making decisions involving a large number of diverse criteria and options. As its name implies, AHP devotes a great deal of attention to the process by which the decision is made. Since the Amoco/EPA project involved a diversity of viewpoints at the federal, state, and industrial levels, a systematic process was needed for reaching a consensus or for identifying where and to what extent viewpoints differed. AHP provides such a framework. AHP proceeds by using group discussion to identify criteria, organize them into a hierarchy that embodies relationships

USING RELATIVE RISK ANALYSIS TO SET PRIORITIES 375

TABLE XII
SINGLE CRITERION RANKINGS BASED ON ANNUALIZED COST AND NET ANNUAL CASH FLOW

	Release Reduction Option				Net Cost Option		
Rank	Description	#[a]	Annualized[b] Cost $M	Rank	Description	#[a]	Net Annual Cash Flow[c] $M
1	Modify Sampling System	8	27	1	Quarterly LDAR (10,000 ppm)	11b	(4)[d]
2	Sec. Seals—Gasoline Tanks	5a	90	2	Annual LDAR (10,000 ppm)	11a	(2)
3	Annual LDAR (10,000 ppm)	5b	112	3	Quarterly LDAR (500 ppm)	11c	(1)
4	Sec. Seals—Gas/Distillate Tanks	11a	92	4	Sec. Seals—Gasoline tanks	5a	<(1)
5	Quarterly LDAR (10,000 ppm)	11b	138	5	Modify Sampling system	8	5
6	Sec. Seals—Floating Roof Tanks	11c	195	6	Sec. Seals—Gas/Distillate Tanks	5b	10
7	Quarterly LDAR (500 ppm)	5c	155	7	Soils Control	6	17
8	Sour Water System Improvements	10	199	8	Sec. Seals—Floating Roof Tanks	5c	30
9	Soils Control	6	203	9	Sour Water System Improvements	10	110
10	Reroute Desalter	1	329	10	Reroute Desalter	1	131
11	Eliminate Coker Blowdown Pond	4	632	11	Opt 5c Floaters on Fixed Roof Tanks	5d	242
12	Opt 5c & Floaters on Fixed Roof Tanks	5d	637	12	Eliminate Coker Blowdown Pond	4	246
13	Opt 5d & Sec. Seal on Fixed Roof Tanks	5e	698	13	Opt 5d & Sec. Seal on Roof Tanks	5e	281
14	Barge Loading	9	1,608	14	Barge Loading	9	568
15	Blowdown System Upgrade	7A	1,630	15	Blowdown System Upgrade	7A	734
16	Replace FCU Cyclones	3a	3,029	16	Replace FCU Cyclones	3a	1,158
17	Install FCU ESP	3b	3,583	17	Install ESP FCU	3b	1,548
18	Drainage System Upgrade	7B	5,941	18	Drainage System Upgrade	7B	2,467
19	Treatment Plant Upgrade	7C	7,403	19	Treatment Plant Upgrade	7C	3,120

[a] # refers to project numbers in Table V.
[b] Includes only direct and indirect costs.
[c] Includes both direct and indirect costs, and revenues from recovered materials.
[d] Numbers in parenthesis represent positive cash flow.

among the criteria, and establish priorities (i.e., criteria weights) with respect to an overall goal. AHP has been used in a wide variety of complex decisions. Examples include use by DOE to prioritize hazardous waste remedial efforts at federal energy facilities; use by the Regional Advisory Committee of the National Health Care Management Center to identify problem areas for research affecting health care in the United States; and use for setting priorities in development of a transportation system for the Sudan.

B. Analytical Hierarchy Process

AHP analysis involves five steps. First, identify the overall goal and the important decision criteria. For this project, the goal was to select the most effective pollution prevention options for the refinery.

Second, organize the criteria into a hierarchical structure based on the relationships among criteria and the project objective.

Third, establish the relative significance (weight) of each criterion. This usually is accomplished via a set of pairwise comparisons among the different criteria. In each pairwise comparison, two criteria on the same hierarchical level are directly compared. The decision maker (in this case, the study team) establishes the importance of one criteria relative to the other. All unique pairs of criteria at each level of the hierarchy are compared via such pairwise comparisons until all possible combinations have been compared. AHP then translates the pairwise comparison results into a relative weight for each criterion.

Fourth, evaluate each option within the context of the proposed hierarchy. Overall scores are determined for each option based on its performance on the criteria in the hierarchy. A comparative ranking of options among themselves is thereby established.

Fifth, adjust and/or revise the hierarchy based on information acquired during the preceding steps in the decision-making process. Using sensitivity analyses, decision makers can review the overall contributions of specific criteria and judgments to the final decision; how changes in criteria weights affect outcomes; or how changes in the hierarchical structure influence the decision. This review may lead to altered judgments and/or a revised hierarchy.

1. Identification of Ranking Criteria

An initial list of criteria was generated from the project workplan (Amoco/EPA, 1990) and two brainstorming sessions at the Williamsburg workshop (Amoco/EPA, 1991d). The project workplan provided overall perspective for criteria selection. Criteria identified at the Williamsburg workshop provided a "base" list that was refined at subsequent study team meetings. Initial criteria lists, which were broad in scope, were made more specific as the study team gained knowledge about the characteristics of the options and the availability of data.

Through a process of elimination and refinement, the following criteria were ultimately selected for ranking options based on quantitative (and sometimes qualitative) assessment of the following characteristics:

- *Risk.* Relative benzene exposure reduction.
- *Technical characteristics.* Release reduction (mass); status in pollution prevention management hierarchy (e.g., source reduction versus treatment); transferability of option to other refineries or industries; timeliness of option implementation; secondary emissions.
- *Cost factors.* Resources utilization (raw materials and utilities); capital, operating, and maintenance costs; effects of option implementation on potential remedial, product, and catastrophic liabilities.

2. Development of Hierarchy and Criteria Weights

Hierarchy structure was developed in parallel with refining the criteria list. The study team identified relationships among criteria and constructed a hierarchy to represent these relationships. Within the hierarchical structure, each level is influenced only by the next higher level and can influence only the next lower level. To rank options, each criterion on the hierarchy must be assigned a relative weight. Developing weights involved two steps. First, study team members completed a survey of pairwise comparisons for each set of criteria on the hierarchy. Second, the study team convened an all-day session to review survey results and to revise criteria weights and the hierarchy structure.

3. AHP Ranking Results

Table XIII presents the results of the AHP ranking using the hierarchy and criteria weights developed by the study team. There appear to be three distinct groupings of options: most preferred, least preferred, and a middle ground where no strong preference exists for one choice over another. Two major factors influenced the overall ranking of options: exposure reduction and cost. Technical characteristics determine the rankings within the mid- and low-performance groups.

Reductions in barge loading emissions, which achieves a 55% benzene exposure reduction, receives a ranking score more than two times greater than the next best option. The second group of options, which related to installing secondary seals or upgrading the blowdown system and annual LDAR, achieve significant exposure reductions. The 8 lowest ranked options all have minor or no impact on the benzene exposure to the surrounding human population.

TABLE XIII
AHP Ranking Using Workgroup Weights

Rank	Option Description	#[a]	Score
1	Barge Loading	9	100
2	Sec. Seals—Floating Roof Tanks	5c	43
2	Opt 5d & Sec. Seal on Fixed Tanks	5e	43
2	Sec. Seals—Gas/Distillate Tanks	5b	43
2	Sec. Seals—Gasoline Tanks	5a	43
6	Opt 5c & Floaters on Fixed Tanks	5d	40
7	Blowdown System Upgrade	7a	29
8	Quarterly LDAR (500 ppm)	11c	19
9	Quarterly LDAR (10,000 ppm)	11b	18
10	Annual LDAR (10,000 ppm)	11a	16
11	Drainage System Upgrade	7b	13
12	Treatment Plant Upgrade	7c	12
12	Eliminate Coker Blowdown Pond	4	12
14	Reroute Desalter	1	11
14	Soils Control	6	11
14	Modify Sampling System	8	11
17	Sour Water System Improvements	10	10
18	Replace FCU Cyclones	3a	5
18	Install FCU ESP	3b	5

[a]# refers to project numbers in Table V.

AHP analyses were conducted to compare the results obtained using the criteria weights proposed by Amoco with the results using weights proposed by EPA/Commonwealth of Virginia team members. This analysis suggests how the options might be ranked from an industry outlook as compared with the ranking from a regulator's viewpoint. Despite differences in perspective, the results show that reducing barge loading emissions is the preferred choice for both groups. In addition, while other options change order, the readjustments are minor. Workgroup members from Amoco assigned nearly equal weights to all three criteria, while EPA/Virginia members assigned the highest weight to risk reduction, next highest weight to technical factors, and the lowest weight to cost.

C. Regulatory Requirements and Project Options

As indicated in Table X, 8 of the 12 project options would, if implemented, contribute to meeting current or anticipated regulatory and statutory program requirements. Legal requirements dictate that these options or an equivalent be undertaken at the refinery. The characteristics of these 8 options are summarized in Table XIV,

TABLE XIV
REGULATORY REQUIREMENTS OPTIONS

#[a]	Project	Materials	Annual Cost ($MM)	Release Reductn (tons/yr)	Benzene Exposure Red/n (%)	Statutory Program	Expect. Compl. Year
7C	Treatment Plt Upgrade	VOC	7.40	58	5	Benzene NESHAP/nonattainment	1993
7A	B/D Upgrade	VOC	1.63	5,096	11	Benzene NESHAP/nonattainment	1993
11b	Drainage Upgrade	VOC	5.94	113	5	Benzene NESHAP/stormwater	1994
9	Elim. Coker Pond	VOC	0.63	130	2	RCRA/CAA	1994
7B	Sec. Seals Flt Roof	VOC	0.16	541	18	SIP, nonattainment	1994
4	Modify Sampling	VOC/HC	0.03	63	0	MACT or HON	1995
5c	Quarterly LDAR	VOC	0.14	511	3	Ozone nonattainment	1996
8	Barge Loading	VOC	1.61	768	55	MACT, nonattainment	1996
	Total		17.53	7,279	99		

[a] # refers to project numbers in Table V.

listed in order of compliance year. The eight options at an annual cost of $17.5 million achieve a release reduction of 7300 tons per year and a benzene exposure reduction equaling 99% of that associated with all twelve options. The four options *not* required by current or anticipated regulations include soils control, rerouting the desalter effluent, installing an electrostatic precipitator for fines control at the FCU, and the improving the sour water stripper operation.

For purposes of comparison, an analysis was conducted to assess what options might be selected to achieve comparable release reduction and exposure reduction objectives in the absence of the existing regulatory constraints. To avoid double counting in this analysis, a specific alternative was selected for those options involving multiple alternatives. The alternative options selected were 3b for FCU fines recovery, 5c for secondary seals, and 11b for LDAR. The goal in this analysis was to attain the desired environmental targets—release reduction or exposure reduction—at a lesser cost.

The 12 options are ranked in Table XV with respect to cost effectiveness of release reduction, expressed in dollars per ton. The results indicate that five options—11b, Quarterly LDAR; 5c, Secondary Seals on Storage Tanks; 7a, Blowdown System Upgrade; 6, Soils Control; and 8, Modify Sampling—are the most cost effective with regard to release reduction. Taken together, these five options attain a release reduction of 6700 tons of hydrocarbons and hazardous solid waste per year at an annual cost of $2.2 million. Note that while soils control is a good pollution prevention option since it prevents the generation of hazardous waste, it

TABLE XV
COST-EFFECTIVE RELEASE REDUCTION RANKING

#[a]	Project	Annual Cost ($MM)	Cumul. Annual ($MM)	Release Reductn (tons/yr)	Cumul. Rel. Redn (tons/yr)	Cost Effect ($/ton)	Cumul. Cost Effect ($/ton)
11b	Quarterly LDAR	0.14	0.14	511	511	270	270
5c	Sec. Seals Flt Roof	0.16	0.29	541	1,052	287	276
7A	B/D Upgrade	1.63	1.92	5,096	6,148	320	313
6	Soils Control	0.20	2.13	530	6,678	383	319
8	Modify Sampling	0.03	2.15	63	6,741	429	319
9	Barge Loading	1.61	3.76	768	7,509	2,094	500
4	Elim. Coker Pond	0.63	4.39	130	7,639	4,862	575
1	Reroute Desalter	0.33	4.72	52	7,691	6,279	614
3b	Install FCU ESP	3.58	8.31	442	8,133	8,106	1022
10	Sour Water Improvement	0.20	8.50	18	8,151	11,056	1043
7B	Drainage Upgrade	5.94	14.44	113	8,263	52,810	1748
7C	Treatment Plt Upgrade	7.40	21.84	58	8,321	127,586	2625
	Total	21.84		8,321		2,625	

[a]# refers to project numbers in Table V.

USING RELATIVE RISK ANALYSIS TO SET PRIORITIES 381

TABLE XVI
COST-EFFECTIVE BENZENE EXPOSURE REDUCTION RANKING

#[a]	Project	Annual Cost ($MM)	Cumul. Cost ($MM)	Benzene Exposure Red'n (%)	Cost-Effective Exposure Red. ($M/% exp. red.)
5c	Sec. Seals Flt Roof	0.16	0.16	18	9
9	Barge Loading	1.61	1.76	55	29
11b	Quarterly LDAR	0.14	1.90	3	46
7A	B/D Upgrade	1.63	3.53	11	148
4	Elim. Coker Pond	0.63	4.16	2	316
1	Reroute Desalter	0.33	4.49	1	329
7B	Drainage Upgrade	5.94	10.43	5	1,188
7C	Treatment Plt Upgrade	7.40	17.83	5	1,480
10	Sour Water Improvement	0.20	18.03	0	
6	Soils Control	0.20	18.23	0	
3b	Install FCU ESP	3.58	21.82	0	
8	Modify Sampling	0.03	21.84	0	
	Total	21.84		100	

[a] # refers to project numbers in Table V.

does not reduce air emissions. When compared to the full set of regulatory requirement options, the cost-effective options attain more than 90% of the release reduction at less than 15% of the annual cost. Adding barge loading emissions to the five most cost-effective options achieves 103% of the required tonnage reduction for a little over 20% of the annual cost of the set of options required by regulations. Of this group of six, all options except soils control are required for compliance with current or anticipated regulations.

A similar analysis is shown for exposure reduction in Table XVI. In this case, six options—5c, Secondary Seals on Storage Tanks; 9, Barge Loading; 11b, Quarterly LDAR; 7a, Blowdown System Upgrade; 4, Eliminate Coker Pond; and 1, Reroute Desalter—are much more cost effective than the next two options in terms of benzene exposure reduction. The six options collectively attain 90% benzene exposure reduction of the full set of eight regulatory requirements at about 20% of the annual cost. Four options achieve no reduction in benzene exposure.

The regulatory requirements shown in Table XIV have been or will be developed using administrative procedures. The regulatory development process includes review and comment opportunities for the public and for industry organizations. It is not the intent of the analysis presented here to assess critically all of those regulatory requirements, since the level of evaluative detail here is considerably less. The results presented here merely indicate the possibility that when the collective requirements of the regulations imposed on a given facility are taken into account, granting the industrial organization greater flexibility in how to achieve the designated standards may enable a facility to attain standards at a significantly reduced cost. The cost effectiveness of various pollution prevention or regulatory options varies widely from location to location. In this case, most of the benefits required can be achieved at a fraction of the cost of all the options.

D. SUMMARY OF RANKING RESULTS

The scores achieved by each pollution prevention option under each of the ranking methods are summarized in Table XVII. disregarding minor differences between option scores, the scores achieved under each method are grouped into high, medium, or low categories. The absence of a ranking score under a particular ranking method indicates that option received a low score for that method.

Those options (or alternatives) that received at least a high or medium score under all but one of the rankings are marked with an asterisk (*). These include all five double seal alternatives, blowdown system upgrade, barge loading controls, and the two quarterly LDAR alternatives. By virtue of their favorable ranking under a variety of perspectives, the study team concluded that these four options show the most promise among the twelve different options considered. Note that all four options in this group are required by current or anticipated regulations, and that the blowdown system upgrade is ranked high on release reduction but low on cost ranking.

Three options that fare next best across the ranking protocols are annual LDAR, modification of sampling systems, and soils control. Several options ranked consistently low and were thus least preferred. These included (3a) replacing the FCU cyclones, (7b) the drainage system upgrade, and (7c) the treatment plant upgrade. None of these received a medium or high score. Just above this group, a third tier included options 1, 2, 3b, 4, and 10. The matrix below separates the options into four preference categories.

1. Most Preferred
   5   Install Secondary Seals
   7A   Upgrade Blowdown System
   9   Reduce Barge Loading Losses
   11b,c Quarterly LDAR Program
2. Next Most Preferred
   11a   Annual LDAR Program
   8   Sampling Systems Mod.
   6   Soils Control
3. Next Least Preferred
   1   Reroute Desalter
   3b   Install FCU ESP

TABLE XVII
Option Scores by Ranking Technique

#[a]	Option	Release Reduction	Exposure Reduction	Cost	Cost-Effective Rel Red'n	Cost-Effective Exp Red'n	AHP
1	Reroute Desalter						
3a	Replace FCU Cyclones				M	M	
3b	Install FCU ESP	M			M		
4	Elim. Coker Pond				M	M	
5a	Sec. Seals Gasoline Tks*[b]	M	M	M	H	H	M
5b	S. Seals—Gas/Dist. Tks*	M	M	M	H	H	M
5c	Sec. Seals—All FltRf Tks*	M	M	M	H	H	M
5d	Opt 5c & Flt on Fix Tk*	M	M		H	H	M
5e	Opt 5d & S. Seal Fix Tk*	M	M		H	H	M
6	Soils Control	M		M	H		
7A	B/D Upgrade*	H	M		H	M	M
7B	Drainage Upgrade						
7C	Treatment Pit Upgrade						
8	Modify Sampling						
9	Barge Loading*	M	H	H	H	H	M
10	Sour Water Improvement			M	M		H
11a	Ann. LDAR (10,000 ppm)			M	M		
11b	Quart LDAR (10,000 ppm)*	M		M	H	H	M
11c	Quart LDAR (500 ppm)*	M		M	H	H	M

Note: All options were ranked high (H), medium (M), or low (L). Blank space denotes low (L).

[a] # refers to project numbers in Table V.
[b] Asterisk (*) denotes projects ranked high or medium with only one low ranking.

4 Eliminate Coker Pond
  10 Sour Water Improvements
4. Least Preferred
   3a Replace FCU Cyclone
   7B Drainage System Upgrade
   7C Treatment Plant Upgrade

IX. Implementation of Selected Options: Obstacles and Incentives

The preceding material has identified and ranked a number of technology options that can reduce, eliminate, or change releases from the Yorktown refinery. This section first generally examines obstacles and incentives to implementing pollution reduction initiatives, and then considers specific issues that might apply to implementing five highly ranked options at the Yorktown refinery. The end of this section identifies several general trends gleaned from the specific examination of Yorktown.

A. Background

The Yorktown refinery has operated under state and federal regulations for some 35 years. During this time the standards to be met have changed dramatically. Since the early 1970s, when media-specific environmental regulation began, Amoco's environmental investments have focused on meeting regulatory requirements as they were developed. Thus, investments were made in advanced wastewater treatment in 1974–1976 to comply with Clean Water Act requirements by July 1, 1977. The refinery also improved its sulfur plant to ensure compliance with Clean Air Act requirements. An MTBE unit was built in 1985 to help the refinery meet lead phasedown requirements. A sludge processing unit to handle RCRA listed wastes was constructed in 1987 to allow closure of previously used landfarms. Other facilities were built to meet lower gasoline vapor pressure requirements.

Today, new tanks are replacing surface impoundments previously used to store wastewater. The coke yard is being renovated to reduce windblown dust particles and to minimize potential groundwater and stormwater contamination. Underground sewer systems in the tankfields are being redesigned to meet new benzene waste NESHAP requirements and stormwater regulations. Amoco has funded research at the University of Waterloo in Canada to help identify the mechanism of dioxin formation in catalytic reforming systems. Simultaneously, a multimedia sampling program for dioxins was completed at Yorktown. At the refinery, equipment believed to be responsible for producing dioxins was taken out of service as soon as it could be identified, and an alternative catalyst regeneration process was implemented.

Until quite recently, there was little integrated management of these multiple regulatory requirements because (1) the regulations themselves did not recognize the potential value of such integration, (2) sufficient information about multimedia releases was not available, and (3) relatively short compliance time frames did not allow time for more integrated analysis.

B. Obstacles

There are at least three major obstacles to doing anything different than regulations require (such as earlier implementation, better control, control of more sources, and so on) require: (1) limited resources, (2) poor economic return, and (3) regulatory disincentives.

1. Limited Resources

Doing something over and above actual regulatory requirements often means expending resources either earlier or in greater amounts than otherwise planned. Corporate resources are constantly being prioritized to meet competing demands. Environmental projects that go beyond meeting legal requirements compete with other investments for limited technical manpower and capital. In the refining part of Amoco's business, these competing requirements include the following: (1) sustaining investments to replace worn-out equipment and keep facilities operable; (2) energy conservation programs to reduce manufacturing costs, which coincidentally reduce emissions from combustion sources; (3) safety programs to improve worker safety and meet new OSHA requirements; (4) product reformulations to improve quality and to meet new federal requirements; (5) development of new manufacturing catalysts and processes; and (6) modifications and improvements to existing processes to increase productivity. Therefore, all projects that are not absolutely required for compliance with laws and regulations are evaluated on a profitability basis, and compared with all other investments.

Once investments have been made, they tend to have long operating lives. Most refineries in this country are more than 40 years old. If the investment solves the problem of concern adequately, there is no incentive to address the same problem again later. In this project, the existing end-of-pipe water treatment plant was found to do an excellent job cleaning process water. Discharge quantities are normally well below permitted values. Because waterborne pollutants are handled effectively and the treatment plant has ample capacity, there was little incentive to spend limited project resources developing other ways to reduce or manage the same pollutants.

Compliance with existing and pending regulations consumes substantial manpower. Doing more requires technically trained manpower to invent, evaluate, design, engineer, construct, and start up new equipment or programs. Although Amoco's environmental staff at the refinery has increased by more than 100% in the last two years, staff time is fully committed to understanding and complying with current regulatory requirements. These have typically short-term deadlines and severe penalties for failure; they, therefore, receive top priority in allocating staff time.

2. Return on Investment

Corporate investments are made to earn an adequate return on invested capital for shareholders, ensure economic stability and growth of the corporation, and provide employment for workers. Many of the options identified for the pollution prevention project were found to be uneconomic. That is, the value of the recovered hydrocarbons was simply insufficient to pay for the capital and operating cost of the recovery equipment. Two exceptions were adding secondary seals to gasoline storage tanks and instituting an LDAR program to reduce fugitive emissions. The fact that most of these options are not profitable is not surprising for two reasons: First, product or emission recovery projects that provide positive economic return are probably already implemented for economic reasons; and second, projects that do not have economic returns have already been dropped from consideration.

Amoco's project evaluation approach has usually viewed environmental projects in the limited context of meeting specific regulatory requirements within a fixed time frame. The Yorktown project suggests that a broader, longer term view might reveal some incentives for release reductions beyond meeting current legal requirements. For example, cogeneration of electricity is typically evaluated on the basis of the cost to generate versus the cost to buy steam and power. The lower cost option is usually the option selected.

But cogeneration is also a potential means to capture energy value from VOC emissions that might otherwise require separate investments for recovery, treatment, or destruction facilities. Viewed on a broader geographic basis, cogeneration frequently results in a net reduction in criteria pollutant emissions by replacing low-efficiency power generation with higher efficiency cogeneration. Thus, cogeneration potentially combines emissions control with improved efficiency while avoiding additional treatment and disposal costs.

3. Regulatory Disincentives

Exceeding regulatory requirements could use scarce resources to reduce current emissions without providing any "credit" for these reductions (Levin, 1990).

Many regulatory programs define a baseline period, and measure progress or changes from that base. The need for a Prevention of Significant Deterioration (PSD) Review for an air permit starts with emissions present in August 1977. The 33/50 voluntary reduction program sets 1988 as a base year from which to measure progress. The early reductions provisions of the Clean Air Act Amendments of 1990 use 1987 as a base year. Reducing emissions before required reduces a company's baseline emissions and makes achieving future regulatory targets more difficult and costly, because the most cost-effective reduction would have already been made. Those companies that have made minimal emission reductions are indirectly rewarded because when required to reduce, they can do so more economically than a "more progressive" company that has already made substantial reductions.

a. Clean Air Act Limits. Since 1977, amendments to the Clean Air Act have discouraged industry from making voluntary improvements to a facility because doing so may compromise the facility's future ability to expand or modify processes. Federal and state air regulations generally do not allow "banking" of emission reductions that could be credited toward future facility modifications. In cases where banking programs do exist, significant time restrictions have been imposed on them.

There has been considerable debate on the value, benefits, costs, and administrative procedures for emissions banking (Liroff, 1986). This project has not attempted to resolve these complex issues. It simply points out that from an industrial perspective, the inadequacies of existing banking systems present a disincentive to voluntary emissions reductions. In this context, since most of the options being considered are required by current or anticipated regulations, the concept of banking would involve early implementation of projects.

Most facility modifications require an air quality permit for construction. The time required to obtain this permit could be reduced from a typical 12–18 months to 2–6 months by using eligible emission reduction credits to offset new emissions from the proposed modifications.

Under the 1990 Amendments to the Clean Air Act, this disincentive for voluntary reductions persists. The act requires the refining industry to construct major new facilities to produce reformulated fuels. Since these new facilities must also obtain construction permits, they will, in effect, consume many of the offsets available for other possible changes. It will be increasingly difficult to obtain offsets to modify or expand facilities. One bright spot, although not directly applicable to this situation: On March 29, 1993, the Chicago Board of Trade offered sulfur dioxide emissions as a new tradable commodity. Under a sealed bid system $21.4 million was paid for the right to emit 150,000 tons of sulfur dioxide from electrical utility smokestacks. Amoco has begun discussions with the Environmental Defense Fund to explore a similar concept for VOCs and NO_x as a way to provide incentives to refineries to reduce their emissions.

b. Clean Water Act Limits. A facility that makes voluntary reductions in its waterborne pollutants may find its permit limits permanently changed to these lower values. Indeed, several EPA representatives have commented informally that data from this project showed the refinery is doing such a good job meeting water permit limits most of the time, that it may be appropriate to lower these limits. This approach fails to recognize that equipment failures and operating upsets can occur, causing significant excursions above normal performance. Since the facility is required to meet permit limits under all conditions (or face potential penalties), an operating margin is essential for continuous compliance.

c. RCRA Limits. Regulations governing hazardous wastes under the Resource Conservation and Recovery Act require the toxicity characteristic (TC) test to determine if a substance is hazardous. Once a substance has failed this test and is deemed hazardous, the generator has little economic incentive to spend resources reducing toxicity or mobility of the waste. The waste must still meet the same treatment, storage, and disposal requirements as if the waste still posed the same degree of hazard. There is no credit in terms of reduced regulatory burden for improving waste characteristics. A number of other RCRA-related obstacles have been identified in a recent presentation by Byers (Byers, 1991).

C. GENERAL OBSERVATIONS ON FIVE HIGHLY RANKED OPTIONS

Five highly ranked options were identified by the project:
1. Reduce barge loading emissions.
2. Install secondary seals on storage tanks.
3. Upgrade blowdown systems.
4. Remove soils from sewers.
5. Institute LDAR program.

Specific barriers and incentives for each option are reviewed in the remainder of this section. Table XVIII summarizes the obstacles and incentives for each option discussed. Three themes reappear: First, there is no workable banking system for emissions reductions. Thus, early reductions will frequently "disappear," because they are no longer available to meet pending or anticipated regulatory programs. Also, a facility making such reductions is put at a disadvantage compared to its competition that elects not to make such reductions. Subsequent reductions become increasingly expensive. In effect, the current system "rewards" those who

TABLE XVIII
SUMMARY OF OBSTACLES AND INCENTIVES FOR FIVE POLLUTION PREVENTION OPTIONS

Option	Obstacles	Incentives
1. Reduce Barge Loading Emissions	Safety Regulatory authority Regulatory requirements Permitting Equipment availability	Product recovery Benzene exposure reduction
2. Secondary Seals on Tanks	Tank availability Regulatory requirements No emission banking	Known technology Modest cost Product recovery Benzene exposure reduction Low administrative burden Ability to identify proper sources
3. Upgrade Blowdown Stacks	Engineering complexity Regulatory requirements No emission banking	Benzene exposure reduction
4. Reduce Soil Intrusion	None	Reduced disposal cost and liability Improved treatment plant operation Low administrative burden Modest cost
5. Institute LDAR Program	Manpower Limits No emission banking	Effectiveness Modest cost Low administrative burden Regulatory requirements Timeliness

do little, by leaving them with emissions that are more easily reduced. Similarly, it "punishes" those who do more, since subsequent reductions are usually more difficult and expensive.

Second, few of the projects identified and analyzed in this project offered significant economic incentives. This type of facility is characterized by high mass recoveries (>99%) as an integral part of normal process operations and business practices. Most product or emission recovery projects that have positive economics have already been implemented. For many of the options identified in the project, the cost of further source reduction, recovery, or treating of emissions far exceeded the potential revenue to be gained from material recovered.

Finally, for many of the projects considered, the implementation time required does not match mandated compliance deadlines. Typical refining industry projects take 2 or 3 years to complete under the best of circumstances.

Because the study team initially selected projects that appeared achievable with current technology, we cannot assess the incentives or obstacles for innovative technical solutions. There is no question that inventing, testing, designing, constructing, and starting up new technology takes significantly longer than using standard, well-proven, commercially demonstrated approaches. Many of these issues were addressed in a recent report by the Technology Innovation and Economics Committee of the National Advisory Council for Environmental Policy and Technology (TIE/NACEPT, 1991).

1. Reduce Barge Loading Emissions

Barge loading losses total 800 tons/year of hydrocarbons, including 15 tons/year of benzene. These occur in the York River, around the marine tanker dock and loading facility. The dock extends more than one-half mile into the river from shore. The proposed control scheme would move hydrocarbon vapors emitted during the loading process to an on-shore vapor recovery system or flare. If materials are recovered, they can be recycled to the refinery for redistillation and use in products.

a. Incentives

(i) Potential Revenue. A vapor recovery system for Yorktown's marine loading losses could provide potential revenues of $160,000 per year. No such revenue could be realized if product is flared rather than recovered. Although the potential for some income is an incentive, it is not sufficient in this case to justify this $4.7 million investment. If all the recovered vapors were salable as gasoline, the cost for this incremental gasoline would be more than $6/gallon.

(ii) Benzene Exposure Reduction. Dispersion modeling clearly showed that reduced barge loading emissions had the largest single influence on reducing potential benzene exposure at a nearby residence.

b. Obstacles

(i) Potential Safety Problems. Marine vapor recovery systems collect and move hydrocarbon vapors in a closed system. The safety considerations of this option are slightly different than those of routine hydrocarbon processing, due to the introduction of air into the recovery system. There is always a concern and potential for buildup of air in a recovery system, producing an explosive air/hydrocarbon mixture. For the system considered for Yorktown, safety control hardware represents nearly $1.5 million of the total $4.7 million estimated capital cost.

(ii) Engineering Complexity. Significant engineering details remain to be resolved for an installation at Yorktown. A major concern at this time is how to move hydrocarbons the long distance (approximately 3000 feet) from the dock to an on-shore blower or compressor. Resolving these details while ensuring safe design and operation takes more time than typical compliance deadlines normally

provide. A more ideal solution to marine loading emissions would consider unconventional approaches such as working with barge companies to jointly develop workable, economic ship-board systems. However, solutions involving second- or third-party participation take even more time than single-party answers.

(iii) Regulatory Authority. Process systems dealing with marine vapor recovery must receive U.S. Coast Guard approval before operation. The EPA is required to establish standards for such systems under the Clean Air Act (Title I and III). It is not clear to the regulated community who will have final authority for approval. What may be considered a state-of-the-art system today may not meet tomorrow's regulatory requirements.

(iv) Regulatory Requirements. Title I of the Clean Air Act requires the EPA to establish requirements for marine vapor loading losses to ozone nonattainment areas. Title III of the same act requires control of this emission source to regulate hazardous air pollutants. Standards will therefore be established in the relatively near future for this type of equipment. Until regulatory standards are developed and approved, it will be difficult to construct "acceptable" vapor recovery facilities.

(v) Permitting. A flare for vapor treatment will require an additional air emission permit. If criteria pollutant emissions from burning the recovered vapor exceed PSD levels, an extensive air quality study, and possible control of offsetting emissions, would be required. The permit process could take significant time and technical resources.

(vi) Equipment Availability. A very limited number of suppliers have equipment and control systems with Coast Guard approval. These designs may not have been tested and approved under all conditions. Once marine vapor control becomes mandatory, equipment demand may exceed supply, leading to extended delivery times and higher costs.

2. Install Secondary Seals on Storage Tanks

Vapor losses from tanks account for an estimated 600 tons/year of VOC emissions, including 10 tons/year of benzene. Systems that reduce these losses could recover 500–600 tons/year of VOCs.

a. Incentives

(i) Known Technical Solutions. Secondary seal technology is well developed. Although there are several alternative systems, most are well tested in commercial service.

(ii) Modest Cost. The "average" tank can probably be equipped with a secondary seal system for $20,000. Crude oil tanks would cost significantly more because they are significantly larger. The total cost for sealing a specific number of tanks is a multiple of the single-tank cost. There are no significant economies of scale.

(iii) Reduced Releases, Product Recovery. Vapors that remain in the tank can be recovered as product, provided their recovery does not violate product vapor pressure limits. Since the more volatile compounds typically concentrate in emissions, their recovery may require removing some butane, a light hydrocarbon normally added to gasoline for vapor pressure control. Where butane can be sold or consumed, recovery should not present a serious problem. But in some refineries, excess butane is a low-valued product and cannot be sold economically.

(iv) Reduced Benzene Exposure. Dispersion modeling work showed that storage tank emissions have a measurable influence on exposure, although less than barge loading losses. Thus, reducing these emissions has a modest, direct, and positive impact by reducing potential exposure to surrounding areas.

(v) Minimal Administrative Burden. In the Commonwealth of Virginia, installing a secondary seal on a tank reduces emissions and would not constitute a significant modification to an existing emission source. Thus, no permit action would be required, other than notification of local Air Pollution Control District authorities.

(vi) Opportunity to Identify Proper Sources. While there are more than 80 tanks in the refinery, gasoline storage tanks are responsible for nearly 80% of tank fugitive emissions. At this time, industry is able to select for control those sources that offer the most benefits. This provides the ability to target maintenance and capital spending on a specific subset of tanks.

b. Obstacles

(i) Tank Availability for Maintenance. Each year, 5–10 tanks from the 80 tanks on-site are removed from service, emptied, inspected, and repaired. For safety reasons, secondary seals can only be installed when the tank is empty and hydrocarbon free. Tank availability constrains the implementation of secondary seals, since piping connections among processing units and tankage is fixed, and spare tankage is available for a limited number of variations to the normal product market demands, refinery operations, and crude availability schedule. It would take approximately 10 years for all tanks in the refinery to be available for maintenance and modification work.

(ii) Regulatory Requirements. The Clean Air Act Amendments of 1990 require the application of MACT on certain refinery equipment items. Precise definition of MACT standards for storage tanks has not been completed. Absent a firm definition of requirements, there is a reluctance to commit resources to construct what may be ruled an unacceptable control system. Further, depending on the outcome of current EPA discussions regarding averaging emissions from different sources, it may be more cost effective to overcontrol some emission sources, while undercontrolling others. Thus, it may be advantageous to overcontrol tank emissions and have less control on other fugitive sources.

(iii) Emissions Reduction Banking. As discussed earlier, the lack of a workable hydrocarbon emissions banking program is a disincentive for more rapid implementation.

3. Upgrade Blowdown Stacks

Two blowdown stacks handle process releases during emergency and upset conditions. A third stack handles vent gases released during the cooling cycle at the coker. Emission estimates show these are the largest single-point sources for hydrocarbon releases in the refinery (5200 tons/year), including an estimated 32 tons/year of benzene. The engineering solution considered to control these emissions involves collecting blowdown vent gases in a common system and treating them in one or more new flares(s).

A number of other possible solutions that involved recovery of heat content, cogeneration, or sale to an adjacent Virginia Electric Power Company generating plant were considered in the early stages of this study. Many have been evaluated during the last 10 years by Amoco as part of its energy conservation programs. All had been previously rejected on the basis of difficult, time-consuming contract negotiations, and uncertain, unpredictable energy prices.

a. Incentives
(i) Reduced Off-Site Exposure. Dispersion showed that blowdown stacks contribute to potential off-site exposure to about the same extent that storage tank emissions do. Reducing these releases will also provide a modest reduction in exposure potential.

b. Obstacles
(i) Engineering Complexity. Flare systems are notoriously difficult to size and design because they are required to handle a wide variety of upset conditions. Flare siting and sizing requires that the load be well defined. But definition depends heavily on which scenarios should be included in the system design basis. For ex-

ample, a new flare must typically handle the emergency releases from connected process units which may occur as a result of a storm-induced power failure. Should this same system also be capable of handling simultaneous releases from several different units? Should it be designed to handle releases that might result from a simultaneous power failure in one area of the plant and a fire in a different area?

Each scenario requires detailed analysis of system component size, cost, radiation released, impact on the surrounding ground area, etc. Most analyses require detailed information on existing systems. For older facilities, this is sometimes difficult to locate. Plot space limitations add to the complexity. These issues require substantial engineering time before final design, engineering, and construction can begin. Typical "compliance" deadlines do not normally recognize this time requirement.

(ii) Regulatory Requirements. Clean Air Act amendments may require modification of existing equipment such as atmospheric relief valves. These modifications could affect flare system size and design. Because of the difficult engineering work required to design this type of system, there is a reluctance to conduct detailed studies until regulatory requirements are defined well enough to minimize rework.

(iii) Emissions Reduction Banking. As previously discussed, there is no credit for emissions removed from blowdown stacks against future potential emissions that may occur from facility modifications.

4. Reduce Soil Intrusion into Drainage System

Sampling analysis during this project identified more than 1000 tons/year of soil entering the refinery's underground drainage system. Once there, the solid tends to become oil coated, deposit as sludge in the oil/water separator, and must be handled as listed hazardous waste. At Yorktown, most hazardous waste sludges are recycled to the refinery's coker where hydrocarbons are converted into usable liquid and solid products. A small amount of soil components remains in the coke product.

Engineering solutions to reduce soil intrusion to the drainage system include (1) using a road sweeper to collect soil and catalyst fines from roadways and process areas before they are blown or washed into the drainage system, and (2) modifying sewer box designs, particularly in earthen areas such as the tank field, to keep soil from entering with water runoff.

a. Incentives
(i) Coker Capacity. Reducing solid wastes generated and sent to the coker would initially improve the coke quality and yield of other higher valued prod-

ucts. However, the refinery currently sends some wastes off-site for proper disposal including some tank bottoms and some oil/water separator sludge. It is likely that coker sludge processing capacity made available by reducing soil load would be used to handle other solid wastes. Thus there would be little or no long-term impact on coke quality or other product yields.

(ii) Reduced Disposal Costs and Liability. Handling more waste with on-site treatment reduces the likelihood of transportation and remediation liabilities associated with off-site disposal or treatment of hazardous wastes. Offsite landfill disposal charges are currently about $250/ton and rising. Incineration costs are about $1500/ton for listed hazardous wastes. Reduced off-site disposal would also decrease transportation-related air emissions.

(iii) Improved Wastewater Treatment Plant Operation. Solids in the drainage system tend to collect and foul some wastewater treatment plant equipment. Reducing solids by removing soil would reduce fouling and maintenance needs.

(iv) Minimal Administrative Burden. Since only internal refinery soils are involved, and no additional emission or release sources are contemplated, it seems likely that no permit action would be required, except notification of the Virginia Department of Waste Management.

(v) Modest Cost. Although costs for this project are nearly equal to or exceed current off-site waste disposal costs, disposal costs are escalating rapidly. Potential future disposal cost savings appear attractive.

b. *Obstacles*
No obstacles could be identified for this project.

5. *Institute a LDAR Program*

Process fugitive emissions are about 800 tons/year (including 4 tons/yr of benzene), the second largest source of hydrocarbon emissions at the refinery. A leak detection and repair program involves scheduled inspection of all valves, flanges, pump seals, and other potential leak sources that handle light hydrocarbons, using an organic vapor analyzer to detect leaks. When leaks are found, the inspection team makes an immediate repair attempt and rechecks the component.

If the repair is successful, no further work is necessary. If unsuccessful, a maintenance work order is written for craftsmen to make a second repair attempt within a prescheduled time, usually 15 days. If the second repair is successful, no further work is necessary. If the component cannot be repaired while the equipment is operating, the leak is noted in the maintenance log for repair during the next scheduled shutdown. The EPA has published emission factors for components based on the type of LDAR program used. More frequent inspections have lower emission factors, and therefore lower estimated emissions (EPA, 1985b).

a. Incentives

(i) Effectiveness. Limited testing at Yorktown, which currently has no LDAR program, showed that most of the leaks found could be repaired relatively easily by the inspectors using simple tools (Amoco/EPA, 1991e). Using EPA emission factors, an annual inspection and repair program could reduce estimated emissions by 300 tons/year; a quarterly program could reduce estimated emissions by 500 tons/year.

(ii) Modest Cost. The major cost for an LDAR program is additional manpower. The cost effectiveness is about $275/ton, the second lowest cost per ton of all the options considered. Hydrocarbons recovered have a potential cost of about $0.90 per gallon.

(iii) Low Administrative Burden. The only record keeping is that necessary for adequate internal administration.

(iv) Timeliness. No construction is required, so benefits could be obtained more quickly than with any other option considered.

(v) Regulatory Requirements. LDAR programs are not required in this part of Virginia at the time of the study, but are anticipated to be a part of the MACT requirements of the Clean Air Act Amendments of 1990 and/or the state implementation plan. Similar programs are used in a number of ozone nonattainment areas in other parts of the country. Amoco uses this technique at its Whiting and Texas City refineries. If the Commonwealth of Virginia requirements follow these other programs, no unusual regulatory requirements are expected.

b. Obstacles

(i) Manpower Limitations. As noted earlier, additional manpower would be required for a LDAR program. Initial manpower requirements to start up a new program would probably be higher than final costs.

(ii) Emissions Reduction Banking. Reducing emissions before mandated to do so frequently decreases a facility's baseline emissions. No credit is available for this early decrease, and future emissions reductions become more expensive.

D. FOLLOW-UP TO THE ORIGINAL PROJECT

1. Implementation Status

At the time of this writing, the pollution prevention options identified in the study, as well as other projects not related to the study, are in various stages of implementation. Some of the projects have been engineered and installed, some at little cost and others at high costs. The upgrades to the sewer system and wastewater treatment plant to reduce benzene emissions have been completed. This project involved the construction of a new sewer for process wastewater, built completely aboveground, and the replacement of the existing oil/water separator and flotation system with an aboveground closed unit. The project was installed at a capital cost of $29 million compared to the originally estimated $41 million. The underground sewer remains in service, collecting primarily stormwater, so the passive action of the underground sewer system to prevent groundwater contamination is not inhibited by the new construction. Secondary seals are being added to the gasoline tanks on a multiyear schedule. A survey of the refinery has been partially completed to identify and inventory the thousands of small valves and connectors that are the source of "fugitive" air emissions. The formal LDAR program is currently being evaluated.

Some projects were easier to implement than expected, while others are still stymied by the "system." As a result of some innovative refinery process engineering, the refinery was able to completely eliminate the use of the coker cooling pond (rather than just controlling its emissions), achieving twice the emissions reduction at a fraction of the original cost estimate. This is truly one of the noteworthy outcomes of the project. This is a textbook case where, faced with a costly option, the refinery experts teamed up to identify a cost-effective solution that provided the greatest possible environmental benefit—elimination of the emission source entirely. Costly emission controls were avoided. Furthermore, the reduction was accomplished with far fewer resources than originally estimated.

On the other hand, the project that offered the most risk reduction potential—controlling emissions during barge loading—is still on hold. The capital funds commitment remains in the current investment plan, but the engineering awaits the issuance of the regulations that address this emission source. The refinery cannot afford to risk implementing a system to control the emissions that may not comply with a future technology or performance standard.

2. Subsequent Data Gathering

One of the project findings was the need for good emissions data for development of both environmental regulations and facility-specific pollution prevention plans. Since testing of blowdown stack emissions proved very difficult during the project, emissions were estimated based on AP-42 emission factors. Subsequent to project completion, Amoco continued to investigate other available sampling and analytical techniques to refine blowdown stack emission estimates. As a result of subsequent testing, VOCs from the blowdown stacks are estimated at 100 tons per year (2% of the 5200 tons/year estimated during the project). Total VOC emissions from the Yorktown refinery area are now estimated to be 2800 tons/year.

While these new data do not change the overall findings of the Yorktown project, they do change the relative effectiveness of the blowdown system upgrade option. These new data are critical to the refinery's ability to set priorities for pollution prevention projects.

References

ACGIH. "Threshold limit values," American Conference of Government and Industrial Hygienists, Cincinnati, OH (1990).

AIChE. "Guidelines for Hazard Evaluation Procedures," American Institute of Chemical Engineers, New York, 1985.

Allen, D. T., Cohen, Y., and Kaplan, I. R. "Intermedia Pollutant Transport," Plenum Press, New York, 1989.

Amoco/EPA. "Public Perception Data," prepared by D. Sawaya (Researchable) and S. Swanson (Sanders International) for the Amoco/USEPA Pollution Prevention Project, Chicago, 1992a.

Amoco/EPA. "Ecological Impact Data," prepared by the Virginia Institute of Marine Science, College of William and Mary for the Amoco/USEPA Pollution Prevention Project, Chicago, 1992b.

Amoco/EPA. "Surface Water Data," prepared by S. Baloo, for the Amoco/USEPA Pollution Prevention Project, Naperville, IL, 1991a.

Amoco/EPA. "Groundwater and Soil Data," prepared by C. Cozens-Roberts, V. J. Kremesec, and E. L. Hockman, for the Amoco/USEPA Pollution Prevention Project, Naperville, IL, 1991b.

Amoco/EPA. "Air Quality Data, Volume I," prepared by Radian Corporation for the Amoco/USEPA Pollution Prevention Project, Chicago, 1991c.

Amoco/EPA. "Pollution Prevention Workshop," prepared by H. Klee, Jr., and SAIC for the Amoco/USEPA Pollution Prevention Project, Chicago, 1991d.

Amoco/EPA. "Refinery Release Inventory (DRAFT)," prepared by H. Klee, Jr., for the Amoco/USEPA Pollution Prevention Project, Chicago, 1991e.

Amoco/EPA. "Project Workplan," prepared by H. Klee, Jr. (Amoco) and M. Podar (USEPA) for the Amoco/USEPA Pollution Prevention Project, Chicago, 1990.

API. "Comments on Proposed Emission Standards for Hazardous Waste TSDF's," submitted to United States Environmental Protection Agency, Washington, DC, in response to 56FR 33490, by American Petroleum Institute, Washington, DC, 1991.

API. "Measurement of BTEX Emission Fluxes from Refinery Wastewater Impoundments Using Atmospheric Tracer Techniques," American Petroleum Institute, Washington, DC, 1990.

Byers, R. L. Regulatory barriers to pollution prevention. *J. Air Waste Management Assoc.* **41**, 418–422 (1991).

Cohen, Y., Allen, D. T., Blewitt, D. N., and Klee, H. "A Multimedia Emissions Assessment of the Amoco Yorktown Refinery," prevented at the 84th Annual Air and Waste Management Association meeting, Vancouver, Canada, June 17, 1991.

EPA. "Technical Support Document for Water Quality Based Toxics Control," EPA/505/2-90-001, U.S. Environmental Protection Agency, Washington, DC, 1991.

EPA. "A Compilation of Air Pollutant Emission Factors," Vol. I, Supplement B, 4th ed., September 1988, U.S. Environmental Protection Agency, Washington, DC, 1988.

EPA. "Guidelines for Carcinogen Assessment; Guidelines for Estimating Exposure; Guidelines for Mutagenicity Risk Assessment; Guidelines for Health Assessment of Suspect Developmental Toxicants; Guidelines for Health Risk Assessment of Chemical Mixtures," *Federal Register,* **51,** 185 (1986).

EPA. "VOC Emissions from Petroleum Refinery Wastewater Systems—Background Information for Proposed Standards," EPA-450-3-85-001a, U.S. Environmental Protection Agency, Research Triangle Park, NC, 1985a.

EPA. "A Model for Evaluation of Refinery and Synfuels VOC Emission Data," Vols. I–II and Appendices A–C, prepared by Radian Corporation for United States Environmental Protection Agency, Washington, DC, 1985b.

Kepner, C. H., and Tregoe, B. B. "The New Rational Manager," Kepner-Tregoe, Inc., Princeton, NJ, 1981.

Kepner, C. H., and Tregoe, B. B. "Problem Analysis and Decision Making," Kepner-Tregoe, Inc., Princeton, NJ, 1979.

Levin, M. H. Implementing pollution prevention: Incentives and irrationalities. *J. Air Waste Management Assoc.* **40,** 1227–1231 (1990).

Liroff, R. A. "Reforming Air Pollution Regulation: The Toil and Trouble of EPA's Bubble," The Conservation Foundation, Washington, DC, 1986.

Los Angeles Times, 4/14/88. See also *Chicago Tribune,* 1/25/91, *Houston Post,* 9/3/88, *Philadelphia Inquirer,* 4/29/88, and *Seattle Times,* 2/25/91.

NAS. "Risk Assessment and Management: Framework for Decision Making," National Academy of Science, Washington, DC, 1983.

NRC. "Tracking Toxic Substances at Industrial Facilities: Engineering Mass Balance versus Materials Accounting," National Research Council, National Academy Press, Washington, DC, 1990.

Oge, M. T. Correspondence from M. T. Oge, Deputy Director Economics and Technology Division, U.S. Environmental Protection Agency, Washington, DC, dated June 10, 1988, to J. J. Doyle, Jr., Exxon Company USA regarding reporting requirements for marine transport operations under SARA Title III.

Placet, M., and Streets, D. G. "NAPAP Interim Assessment, Vol. II: Emissions and Controls," U.S. Environmental Protection Agency, Washington, DC, 1989, pp. 1–66.

Saaty, T. "Decision Making for Leaders," Pittsburgh, PA, 1990.

Saaty, T. Decision Making for Leaders: The Analytical Hierarchy Process for Decisions in a Complex World," Pittsburgh, PA, 1988.

Shah, J. J., and Heyerdahl, E. K. "National Ambient Volatile Organic Compounds (VOC) Data Base Update," U.S. Environmental Protection Agency, Washington, DC, 1988.

Stokey, E., and Zeckhauser, R. "A Primer for Policy Analysis," W. W. Norton and Co., New York, 1978.

TIE/NACEPT. "Permitting and Compliance Policy: Barriers to U.S. Environmental Technology Innovations," EPA-101/N-91/001, Report and Recommendations of the Technology Innovation and Economics Committee, U.S. Environmental Protection Agency, National Advisory Council for Environmental Policy and Technology, Office of the Administrator, Washington, DC, January 1, 1991.

Wong, O., and Raabe, G. K. Critical review of cancer epidemiology in petroleum industry employees, with a quantitative meta-analysis by cancer site. *Am. J. Indust. Med.* **15,** 283–310 (1989).

INDEX

A

Acid–base chemistry, very fast chemical reactions, 264
Activity coefficient, 19
ADI methods, *see* Alternating direction implicit methods
AFI methods, *see* Approximate factorization implicit methods
AHP, *see* Analytical hierarchy process
Airborne emissions, petroleum refinery, 341–342
Air pollution, petroleum refinery sources, 332, 348–349, 352–353
Algebraic stress model (ASM), 239, 258
Alternating direction implicit (ADI) methods, 247
Aluminides, combustion synthesis, 92, 101–102, 183, 189, 190–191, 195
Aluminum nitride
  applications, 110, 111
  powders, properties, 109–110
  production costs, 118
Amoco Corp.
  pollution prevention, 330–398
    follow-up, 397–398
    identifying options, 332–335, 351–360
    implementing solutions, 382–396
    inventory of refinery releases, 331–332, 337–351
    modeling options, 363–369
    ranking options, 333–334, 360–384
    regulatory requirements, 378–382
    Yorktown refinery, 335–337, 354–355, 384–385, 388–390
      barge loading emission reduction, 389, 390–391
      blowdown stacks upgraded, 389, 393–394
      LDAR program, 389, 395–396
      schematic diagram, 336
      soil intrusion into drainage system, 389, 394–395
      storage tank secondary seals, 389, 391–393
Analytical hierarchy process (AHP), 333, 376–378
Approximate factorization implicit (AFI) methods, 247
ASM, *see* Algebraic stress model
ASTEC (software), 253
Axial centrifuge, 95
Azeotropic composition, linear algebra, 20

B

Bimolecular reactions, kinetics, 42–44
Bismuth compounds, ferroelectric laminated, 114–115
Blowdown stacks
  airborne emissions, 332, 348
  pollution prevention by upgrading, 393–394
Borides
  combustion synthesis, 89, 92, 99–100, 197
  thermodynamics, 153
  powders, properties, 97
Boron nitride, 110, 111
Bubble columns
  computational fluid dynamics (CFD), 267–270, 271, 298, 300
  discrete bubble model, 300–301
  volume of fluid (VOF) model, 301–311
Bubble point pressure, 19

C

Carbides
  combustion synthesis, 89, 96–99, 115, 186–187
  thermodynamics, 153
  powders, properties, 97

CARPT, *see* Computer-aided radioactive particle tracking
Casting, self-propagating high-temperature synthesis with, 93–95
Cellular models, combustion wave, 130–134
Cemented carbides, gasless combustion synthesis, 97, 98
Centrifugal SHS casting, 94–95, 115
Ceramics
 combustion synthesis, 91, 95, 98, 99, 100
 nitride based, 89, 107–112
 Ti–C–B system, 100
Cermet alloys
 combustion synthesis, 91, 95, 115
 properties, 98
CFDLIB code, 268
Chalcogenides, gasless combustion synthesis, 89, 106
Chemical equilibria
 behavior in neighborhood of, 26–28
 heterogeneous, 28–30
 homogeneous, 22–28
Chemical furnace, 92
Chemically reactive flows, computational fluid dynamics (CFD), 260–265
Chemical reactions; *see also* Reaction networks
 extent of reaction, 4
 heterogeneous chemical equilibria, 28–30
 homogeneous chemical equilibria, 22–28
 independence of, 4
 rate of reaction, 23
 reaction label, 10–11
 residence time distribution (RTD) experiments, 230–231
 very fast, 264
 very slow, 263–264
Chromium carbide, combustion synthesis, 115
Circulating fluidized bed (CVB) reactor
 KTGF study, 296–298
 riser, 276–277
Clean Air Act, petroleum refinery, 387–388
Cocurrent infiltration, 146
Combustion synthesis (CS), 80, 81, 83, 84
 commercial aspects, 117–120
 green mixture preparation, 87–88, 105, 162–169
 infiltration combustion synthesis, 82
 materials synthesized, 96
 aluminides, 92, 101–102, 183, 189, 190–191, 195
 borides, 89, 92, 97, 99–100, 153, 197
 carbides, 89, 96–99, 115, 153, 186–187
 chalcogenides, 89, 106
 diamond composites, 103
 ferroelectric materials, 114–115
 functionally graded materials, 104–106
 high-temperature semiconductors, 91, 113–114
 hydrides, 112
 intermetallic compounds, 89, 91, 102, 103, 106, 107, 153, 170, 171–172, 174
 nitrides, 89, 107–112, 153
 oxides, 112–115
 semiconductors, 91, 113–114
 silicides, 89, 92, 97, 101, 153
 Ti–C–B system, 100
 mechanism for structure formation, 180–182, 203–206
 by time-resolved X-ray diffraction, 195–197
 microstructure of combustion wave, 197–203
 quenching, 183–189
 using model systems, 190–194
 process parameters, 151, 181–183
 combustion wave properties, 86–87
 dilution, 158–162
 gravity and, 177–180
 green mixture density, 163–169
 ignition conditions, 173–174
 initial temperature, 174–176
 particle size, 169–173
 sample dimensions, 176
 thermodynamics, 152–158
 reactor, 88–89
 spin properties, 87
 techniques
 gasless combustion synthesis, *see* Gasless combustion synthesis
 gas–solid combustion synthesis, 82, 107–115
 laboratory techniques, 84–87
 oscillating combustion synthesis, 86–87
 powder production and sintering, 88–89
 reduction combustion synthesis, 82
 self-propagating high-temperature synthesis (SHS), *see* Self-propagating high-temperature synthesis

shock-induced synthesis, 92
solid–gas systems, 107–115
steady SHS process, 86
theory, 120
   cellular models, 130–134
   combustion wave propagation theory, 86–87, 120–127
   filtration combustion, 138–151
   gasless combustion, 120–127, 135–138, 151
   microstructural models, 127–130
   thermal explosion mode, 82
volume combustion synthesis, *see* Volume combustion synthesis
Combustion wave
   propagation
      cellular models, 130–134
      microstructural models, 127–130
      stability, 135–138
   properties, 86–87
   microstructure of, 197–203
   quenching, 183–189
   steady propagation, 86
   temperature profile for preheating zone, 126
   velocity, 120–126
Component label, linear algebra of stoichiometry, 9–10
Computational fluid dynamics (CFD), 227–228, 234
   chemical engineering and, 229, 231–234, 253–254, 314–315
   chemically reactive flows, 260–265
   complex rheology, 259–260
   fluid–solid systems, 273–279
   gas–liquid–solid systems, 279–281
   gas–liquid systems, 267–274
   laminar flows, 254–257
   liquid–liquid systems, 273
   multiphase systems, 265–281
   single-phase systems, 254–265
   turbulent flows, 257–259
   viscoelastic flow, 260
   circulating fluidized beds, 296–298
   bubble columns, 298–311
   laminar entrained flow reactor, modeling, 311–313
   defined, 229, 236
   experimental validation, 282–287
   history, 234–236
   liquid–liquid systems, 273

numerical techniques, 244–247
   multiphase systems, 249–251
   single-phase systems, 247–249
   software, 251–253
theory, 236
   multiphase systems, 242–244
   single-phase systems, 237–241
at Twente University (Netherlands), 287
   discrete particle simulation of gas fluidized beds, 291–296
   two-fluid simulation of gas fluidized beds, 287–291
Computer-aided radioactive particle tracking (CARPT), computational fluid dynamics (CFD), 285, 288
Conservation equations, laminar flow, 242
Conservation of mass problem, 21
Continuous description, linear algebra of stoichiometry, 7–13
Convexity, linear algebra of, 15–17
Cooperative uniform kinetics, CSTR, 51–52
Copper block, for combustion synthesis quenching, 184–185
Countercurrent infiltration, 144–147
Cracking
   kinetics, 44–46
   modeling, 263
Cross-media transport, petroleum refinery pollution, 344
CS, *see* Combustion synthesis
CSTR, kinetics, 50–51, 55–57
CVB riser, *see* Circulating fluidized bed riser

D

Damköhler number, 262
Densification, self-propagating high-temperature synthesis with, 89–92
Diamond composites, gasless combustion synthesis, 103
Diffusion, interference with, 57–59
Digital particle image velocimetry (DPIV), 258, 283–284
Dilution, combustion synthesis and, 158–162
Direct numerical simulation (DNS), 241, 258, 261
Discrete bubble model, 300–301
Discrete description, linear algebra of stoichiometry, 3–7

404 INDEX

Discrete particle simulation, gas fluidized beds, 291–296
DNS, *see* Direct numerical simulation
DPIV, *see* Digital particle image velocimetry

E

Eddy–viscosity model, 239, 240
Effectiveness factor, porous catalysts, 58–60
Effluent discharge, inventory of refining releases, 332
Electrothermography, combustion synthesis studies, 190
ESTET–ASTRID (software), 253
Eulerian–Lagrangian methods, 243
   bubble columns, 267, 268, 269
   multiphase flows, 250–251, 272
Eulerian methods
   bubble columns, 267, 268
   multiphase flows, 249–250
Exact lumping, 32–34
Extent of reaction, linear algebra, 4
Extrusion, self-propagating high-temperature synthesis, 92

F

FC, *see* Filtration combustion
FCC, *see* Fluidized catalytic cracking
FDM, *see* Finite difference methods
Feinberg approach, network topology, 64–66
FEM, *see* Finite element method
Ferroelectric materials, combustion synthesis, 114
Ferrovanadium, 107
FGMs, *see* Functionally graded materials
Fick's law, 238
FIDAP (software), 253
Filtration combustion (FC), 82, 138–149
Finite difference methods (FDM), 246
Finite element method (FEM), 249, 260
Finite volume methods (FVM), 247, 260
First-order intrinsic kinetics, CSTR, 51
First-order kinetics, irreversible, 35–36
FLOW3D (software), 253
Flow patterns, 229–230
Flow reaction, with axial diffusion, 60
FLUENT (software), 253, 264

FLUFIX (software), 253
Fluid dynamics
   computational, *see* Computational fluid dynamics
   history, 234–235
Fluid flow, multiphase systems
   computational fluid dynamics, 242–244
   fluid–solid systems, 274–279
   gas–liquid–solid systems, 279–281
   gas–liquid system, 267–273
   liquid–liquid systems, 273
   numerical techniques, 249–251
   patterns, 229–230
   single-phase systems
      chemically reactive flows, 260–265
      complex rheology, 259–260
      computational fluid dynamics (CFD), 237–241
      laminar flow, 254–257
      numerical techniques, 247–249
      turbulent flow, 257–259
Fluidized bed reactors, computational fluid dynamics (CFD), 275–279
Fluidized catalytic cracking (FCC), riser-type reactors, 277
Fluid–solid systems, 273
   computational fluid dynamics (CFD)
      fluidized bed reactors, 275–279
      packed-bed reactors, 273, 275
Forced gas permeation, 149–150
Frank–Kamenetskii approximation, 124
Functionally graded materials (FGMs), 104–106
FVM, *see* Finite volume methods

G

Gamma distribution, 10, 54
Gamma-ray densitometry, computational fluid dynamics (CFD), 286
Gas fluidized beds
   discrete particle simulation, 291–296
   two-fluid simulation, 287–291
Gasless combustion synthesis, 82, 96–107
   combustion wave propagation, 120–126
   green mixture for, 162–165
   materials synthesized by, 96
      aluminides, 101–102
      borides, 89, 99–100
      carbides, 89, 96–99

INDEX

chalcogenides, 89, 106
diamond composites, 103
functionally graded materials, 104–106
intermetallic compounds, 89, 102, 103, 107
silicides, 89, 101
Ti–C–B system, 100
thermodynamics, 154
Gas–liquid solid systems, 279
slurry rectors, 280
trickle-bed reactors, 279
Gas–liquid stirred tank reactors, computational fluid dynamics (CFD), 270, 272
Gas–liquid systems, 267
computational fluid dynamics (CFD), 267–273
bubble columns, 267–270, 271
gas–liquid stirred tank reactors, 270, 272
Gas pressure combustion sintering, 92
Gas–solid combustion, 82
green mixture for, 165–169
hydrides by, 112
initial temperature, 174
nitrides by, 107–112, 189
oxides by, 112–115
quenching, 185
sample diameter, 176
thermodynamics, 156–158
General micromixing model (GMM), 261
Gibbs–Duhem equation, 62
Gibbs free energy, 23–24, 28–29
Global convexity, 16
Global system models, 232–233
GMM, see General micromixing model
Gravity, combustion synthesis and, 177–180
Green mixture
for gasless systems, 162–165
for gas–solid systems, 165–169
preparation, 87–88, 105
Grid generation, 245
Groundwater contamination, petroleum refinery, 332, 344, 354
Group IV metal carbides, gasless combustion synthesis, 89, 96–99
Group Va carbides, combustion synthesis, 99

H

Hafnium carbide, combustion synthesis, 91, 96
Hafnium nitride, properties, 108

"Hard sphere" model, 277
Hazardous waste, petroleum refinery, 339, 341
Heterogeneous chemical equilibria, 28–30
High-temperature superconductors, combustion synthesis, 91, 113
Homogeneous chemical equilibria, 22–28
Hot isostatic pressing (HIP), self-propagating high-temperature synthesis, 92
Hydrides, combustion synthesis, 112
Hydrocarbides, synthesis, 112
Hydrocyclones, 272
Hydronitrides, synthesis, 112

I

Infiltration combustion synthesis, 82, 138
Interaction by exchange with the mean model (IEM), 261
Intermetallic compounds
combustion synthesis, 91, 102, 106, 171–172, 174
thermodynamics, 153
hydrides, 112
"solid" flame systems, 170
Inventory, of refinery releases, 331–332, 337–351

K

k–ϵ model, 258, 259
Kinetics
of combustion synthesis
combustion wave propagation models, 126–135
combustion wave velocity, 120–126
filtration combustion, 138–149
forced gas permeation, 149–150
stability of gasless combustion, 135–138
of multicomponent mixtures, 2
bimolecular reactions, 42–44
cooperative uniform kinetics, 51–52
cracking, 44–46
first-order intrinsic kinetics, 51
irreversible first-order kinetics, 35–36
irreversible nonlinear kinetics, 37–40
lumping, 2, 13–14, 30–34
number density function, 48–49
parallel reactions, 41–42

Kinetics *(Continued)*
  polymerization, 46–47
  reaction networks, 40–47
  reversibility, 40
  sequential reactions, 41–42
Kinetic theory of granular flow (KTGF), 276, 296
KIVA I and II (software), 253

L

Laminar entrained flow reactor (LEFR), modeling, 311–313
Laminar flow
  computational fluid dynamics (CFD), 254–257
    around bluff bodies, 256
    in ducts, 254–256
    in stirred vessels, 256
  conservation equations, 242
Laminated metallic foils, combustion synthesis studies, 190–191
Large eddy simulation (LES), 241, 258, 262
Laser Doppler anemometry (LDA), computational fluid dynamics (CFD), 283, 288
Laser-induced fluorescence, computational fluid dynamics (CFD), 284, 288
Lattice–gas automata (LGA), 278
Layer-by-layer combustion, 165, 166
LEFR, *see* Laminar entrained flow reactor
Liquid–liquid systems, computational fluid dynamics (CFD), 273
Local convexity, 16
Loop reactors, 272
Lumping
  defined, 10, 13–14, 30–32
  exact lumping, 32–34
  notation, 14
  reaction network, 47

M

Marker and cell (MAC) method, 248, 251
Maximum mixedness reactors, kinetics, 52–54
Maximum segregation reactors, kinetics, 52
MELODIF (software), 253
Merization system, 6
Metallic nitrides, combustion synthesis, 107–108

Metastable set, 17
Method of moments, 20–22
Mg–Ni intermetallic compound, 106–107
Microbalance models, 232–233
Microgravity, combustion synthesis and, 179–179
Microheterogeneous cell model, combustion wave, 133
Micromixing models, 262–263
Microstructural models, combustion wave, 127–130
Mixed Eulerian–Lagrangian method, 273
Modeling, 232–233
  combustion synthesis
    cellular models, 130–134
    microcellular models, 127–130
    microheterogeneous cell models, 133
    reactant interaction, 190–194
  cracking, 263
  experimental work and, 233–234
  global systems models, 232–233
  laminar entrained flow reactor (LEFR), 311–313
  microbalance modeling, 232–233
  pollution prevention options, 363–369
  Reynolds stresses, 238, 239
  turbulent flow, 239–241
Molecular dynamics, 278
Molybdenum carbide, combustion synthesis, 115
Molybdenum silicide
  combustion synthesis, 92, 101, 119
  properties, 97
Molybdenum–silicon system, "solid" flame systems, 172
Moment methods, 262
Monomolecular system, 30
Multicomponent mixtures, 1–3
  Feinberg approach, 64–66
  kinetics, 2
    bimolecular reactions, 42–44
    cooperative uniform kinetics, 51–52
    cracking, 44–46
    first-order intrinsic kinetics, 51
    irreversible first-order kinetics, 35–36
    irreversible nonlinear kinetics, 37–40
    lumping, 2, 13–14, 30–34
    number density function, 48–49
    parallel reactions, 41–42
    polymerization, 46–47
    reaction networks, 40–47

reversibility, 40
    sequential reactions, 41–42
  linear algebra of stoichiometry, 2
    continuous description, 7–13
    discrete description, 3–7
  mathematical concepts, 66–69
  network topology, 64–66
  orthogonal complement, 7, 25–26, 61
  reactors, 49–50, 60–61
    CSTR, 50–51, 55–57
    interference with diffusion, 57–60
    maximum mixedness reactors, 52–54
    maximum segregation reactors, 52
    multiplicity of steady states, 55–57
    nonisothermal reactors, 54–55
    plug flow reactors, 57
  thermodynamics, 2
    heterogeneous chemical equilibria, 28–30
    homogeneous chemical equilibria, 22–28
    phase equilibria, 14–22
Multiphase systems, computational fluid dynamics (CFD), 242–244
  fluid–solid systems, 273–279
  gas–liquid–solid systems, 279–281
  gas–liquid systems, 267–273
  liquid–liquid systems, 273
  numerical techniques, 249–251
Multiple steady states, 55–57

N

Natural permeation combustion, 138, 139
NEKTON (software), 253
Net present value (NPV), pollution prevention options, 361–363
Network topology, Feinberg approach, 64–66
Nickel aluminides, combustion synthesis, 92, 101–102, 183, 189, 190–191, 195
Niobium–boron mixtures, "solid" flame systems, 170
Niobium carbides
  gasless combustion synthesis, 99
  properties, 97
Niobium nitride, 107
  properties, 108
Niobium silicide, combustion synthesis, 92
Nitride-based ceramics, combustion synthesis, 107–112

Nitrides
  combustion synthesis, 89, 107–112
  thermodynamics, 153
  properties, 108
Nonideal liquid solutions, multicomponent mixtures, 19–20
Nonisothermal reactors, kinetics, 54–55
Nonlinear kinetics, irreversible, 37–40
Nonmetal carbides, combustion synthesis, 99
Nonmetallic nitrides, combustion synthesis, 108–112
NPV, *see* Net present value
Nuclear magnetic resonance imaging, computational fluid dynamics (CFD), 284
Nucleation, 16
Number density function, 48–49
Numerical grid generation, 245

O

Orthogonal complement, 7, 25–26, 61
Oscillating combustion synthesis, 86–87
Oxides, combustion synthesis, 112–115

P

Packed-bed reactors, computational fluid dynamics (CFD), 273, 275
Parallel reactions, kinetics, 41–42
Particle–foil experiments, combustion synthesis studies, 191–194
Particle image velocimetry (PIV), 283–284, 288
Particle tracking, computational fluid dynamics (CFD), 285
Partition coefficient, 20
PDF, *see* Probability density function
Petroleum refinery
  Amoco Corp. Yorktown refinery, 335–336, 354–355, 384–385, 388–390
    barge loading emissions reduced, 389, 390–391
    blowdown stacks, upgraded, 389, 393–394
    LDAR program instituted, 389, 395–396
    soil intrusion into drainage syst4em, 389, 394–395
    storage tank secondary seals installed, 389, 391–393

Petroleum refinery *(Continued)*
  follow-up, 397–398
  inventory of refinery releases, 331–332, 337–351
  pollution prevention options
    identifying, 332–335, 351–360
    implementation, 382–396
    modeling, 363–369
    ranking, 333–334, 360–384
    regulatory requirements, 378–382
Phase equilibria, 14–15
  approximate methods, 20–22
  convexity, 15–17
  nonideal liquid solutions, 19–20
  Raoult's law, 17–19
PHOENICS (software), 252
Phosphides, gasless combustion synthesis, 106
PIV, *see* Particle image velocimetry
Plug flow reactor (PFR), 57
Pollution prevention, 330–331
  Amoco Corp. Yorktown refinery, 335–337, 354–357, 384–385, 388–390
    barge loading emissions reduced, 389, 390–391
    blowdown stacks upgraded, 389, 393–394
    LDAR program instituted, 389, 395–396
    soil intrusion into drainage system, 389, 394–395
    storage tank secondary seals installed, 389, 391–393
  follow-up, 397–398
  inventory of refinery releases, 331–332, 337–351
  options
    identifying, 332–335, 351–360
    implementation, 382–396
    modeling, 363–369
    ranking, 333–334, 360–384
    regulatory requirements, 378–382
POLYFLOW (software), 253, 259
Polymerization reactions, kinetics, 46–47
Porous catalysts, effectiveness factor, 58–60
Prandtl–Kolmogorov model, 240
Prandtl's mixing length model, 239, 240
Probability density function (PDF), 263
Pseudomonomolecular systems, 38

Q

Quadrature method, 20–22
Quantitative financial analysis, pollution prevention options, 361–363
Quenching, in combustion synthesis, 183–189

R

Radial centrifuges, 94–95
Radioactive particle tracking (RPT), computational fluid dynamics (CFD), 285, 288
RAMPANT (software), 253
Ranking options, refinery pollution prevention, 333–334, 360–384
RANS equations, *see* Reynolds averaged Navier–Stokes equations
Raoult's law, 17–19
Rate of reaction, 23
Reaction coalescence, 189
Reaction label, 10–11
Reaction networks, 40–41, 74
  bimolecular reactions, 42–44
  cracking reactions, 44–46
  irreversible kinetics, 35–40
  parallel and sequential reactions, 41–42
  polymerization, 46–47
Reduction combustion synthesis, 82
Reference components, 7
Refinery release inventory, 331–332, 337–351
Residence time distribution (RTD) experiments, chemical reactions, 230–231
Resource Conservation and Recovery Act, 388
Reversibility, kinetics, 40
Reynolds averaged Navier–Stokes (RANS) equations, 262
Reynolds number, defined, 241
Reynolds stresses, modeling, 238, 239
Reynolds stress model (RSM), 239, 258
RPT, *see* Radioactive particle tracking

S

Selenides, gasless combustion synthesis, 106
Self-propagating high-temperature synthesis (SHS), 81–84
  with casting, 93–95
  centrifugal SHS casting, 94–95, 115

INDEX 409

commercial applications, 118–119
  with densification, 89–93
  with extrusion, 92
  history, 83
  with hot isostatic pressing (HIP), 892
  initiation of, 173
  materials synthesized by
    cermet alloys by, 98, 99
    diamond composites, 103
    ferroelectrics, 114–115
    magnesium nitride, 106–107
    niobium nitride, 107
    silicides, 101
    superconductors, 113
  powders, properties, 96–97, 108–111
  reactor, 88–89
  steady SHS process, 86
  thermite-type SHS, 115–116
  TRXRD studies, 195
Semiconductors, combustion synthesis, 91, 113–114
SEPRAN (software), 270
Sequential reactions, kinetics, 41–42
SGS models, *see* Subgrid scale models
Shock-induced synthesis, 92
Shock waves, for combustion synthesis quenching, 185
SHS, *see* Self-propagating high-temperature synthesis
Silicides
  combustion synthesis, 89, 92, 101
    thermodynamics, 153
  powders, properties, 97
Silicon carbide
  combustion synthesis, 92, 99, 119
  properties, 97
Silicon nitride
  applications, 110
  combustion synthesis, 108, 109, 118, 119
Silicon nitride ceramics, properties, 11
SIMPLE algorithm, 247
Simplified marker and cell (SMAC) method, 248, 251
Single-phase systems, computational fluid dynamics (CFD), 237–241
  chemically reactive flows, 260–265
  complex rheology, 259–260
  laminar flows, 254–257
  numerical techniques, 247–249

turbulent flows, 257–259
viscoelastic flow, 260
Slurry reactors, computational fluid dynamics (CFD), 280
"Soft sphere" model, 277
"Solid" flame systems, 170
Solid–gas systems, combustion synthesis (CS) in
  hydrides by, 112
  nitrides by, 89, 107–112
  oxides by, 112–115
  green mixture for, 165–169
  initial temperature, 174
  quenching, 185
  sample diameter, 176
  thermodynamics, 156–158
Solid waste, petroleum refinery, 339, 341, 344, 350–351
Steady propagation, combustion synthesis wave, 86
Steady SHS process, 86
Steady states, multiplicity of, 55–57
Stirred tank reactors, gas–liquid, computational fluid dynamics (CFD), 270, 272
Stoichiometry, 2
  linear algebra, multicomponent mixtures, 3–14
Subgrid scale (SGS) models, 262
Subsurface contamination, petroleum refinery discharge, 344, 349–350
Sulfides, gasless combustion synthesis, 106
Superadiabatic condition, combustion synthesis, 149–150
Superconductors, combustion synthesis, 91, 113
Surface combustion, combustion synthesis, 165–166
Surface water, petroleum refinery pollution, 342–344, 349, 353–354

T

Tantalum–boron mixtures, "solid" flame systems, 170
Tantalum–carbon mixtures, "solid" flame systems, 170
Tantalum nitride
  combustion synthesis, 107
  properties, 108

Tantalum–nitrogen system, "solid" flame systems, 170
Thermal explosion mode, combustion synthesis (CS), 82
Thermodynamics
  combustion synthesis, 152–158
    gasless systems, 154
    gas–solid systems, 156
    reduction-type systems, 154–156
  multicomponent mixtures
    heterogeneous chemical equilibria, 28–30
    homogeneous chemical equilibria, 22–28
    phase equilibria, 14–22
Three-phase fluidized beds, computational fluid dynamics (CFD), 280
Ti–C–B system, gasless combustion synthesis, 100
Ti–C–Ni system, quenching, 187–188
Ti–CO system, gasless combustion synthesis, 102
Ti–Fe system, gasless combustion synthesis, 102
Time-resolved X-ray diffraction (TRXRD), combustion synthesis studies, 195–197
Ti–Ni powders, combustion synthesis, 102
Titanium aluminides, VCS synthesis, 102
Titanium boride
  combustion synthesis, 91, 92, 99–100, 119–120
  functionally graded materials with, 105, 106
  properties, 97
Titanium carbide
  combustion synthesis, 91, 92, 93, 96–99, 115, 163, 186–187, 197
  commercial production, 119
  powder properties, 97
Titanium nitride
  combustion synthesis, 107
  properties, 108
Titanium–nitrogen system, "solid" flame systems, 172
Titanium silicide
  combustion synthesis, 92, 101, 197
  properties, 97
Tomography, computational fluid dynamics (CFD), 285–286
Toxic Release Inventory (TRI), petroleum refinery, 331–332, 345–347
Transition metal carbides, gasless combustion synthesis, 99
Transition metal nitrides, properties, 108
Transmission electron microscope (TEM), combustion synthesis studies, 191–192
Transport phenomena, 229
Trickle-bed reactions, computational fluid dynamics (CFD), 279
TRXRD, *see* Time-resolved X-ray diffraction
Tungsten carbide, combustion synthesis, 115
Turbulent flow
  computational fluid dynamics (CFD), 257–259
  modeling, 239–241
Two-fluid simulation, gas fluidized beds, 287–291
Two-phase flow, flow regime classification, 266

U

Unconditional convexity, 16
UPWIND (software), 267

V

Vanadium nitride, combustion synthesis, 107
Viscoelastic flows, computational fluid dynamics (CFD), 260
Volume combustion synthesis (VCS), 81, 84
  materials synthesized by, 88, 101
Volume of fluid (VOF) model, 251, 268, 270, 282, 301–311

W

Wastewater, petroleum refinery, 339
Water pollution, inventory of refining releases, 332
Weissenberg number, 260

Y

Yttrium powders, properties, 113–114

Z

Zirconium carbide, gasless combustion synthesis, 97
Zirconium nitride
  combustion synthesis, 107, 183
  properties, 108

CONTENTS OF VOLUMES IN THIS SERIAL

Volume 1

J. W. Westwater, *Boiling of Liquids*
A. B. Metzner, *Non-Newtonian Technology: Fluid Mechanics, Mixing, and Heat Transfer*
R. Byron Bird, *Theory of Diffusion*
J. B. Opfell and B. H. Sage, *Turbulence in Thermal and Material Transport*
Robert E. Treybal, *Mechanically Aided Liquid Extraction*
Robert W. Schrage, *The Automatic Computer in the Control and Planning of Manufacturing Operations*
Ernest J. Henley and Nathaniel F. Barr, *Ionizing Radiation Applied to Chemical Processes and to Food and Drug Processing*

Volume 2

J. W. Westwater, *Boiling of Liquids*
Ernest F. Johnson, *Automatic Process Control*
Bernard Manowitz, *Treatment and Disposal of Wastes in Nuclear Chemical Technology*
George A. Sofer and Harold C. Weingartner, *High Vacuum Technology*
Theodore Vermeulen, *Separation by Adsorption Methods*
Sherman S. Weidenbaum, *Mixing of Solids*

Volume 3

C. S. Grove, Jr., Robert V. Jelinek, and Herbert M. Schoen, *Crystallization from Solution*
F. Alan Ferguson and Russell C. Phillips, *High Temperature Technology*
Daniel Hyman, *Mixing and Agitation*
John Beek, *Design of Packed Catalytic Reactors*
Douglass J. Wilde, *Optimization Methods*

Volume 4

J. T. Davies, *Mass-Transfer and Interfacial Phenomena*
R. C. Kintner, *Drop Phenomena Affecting Liquid Extraction*
Octave Levenspiel and Kenneth B. Bischoff, *Patterns of Flow in Chemical Process Vessels*
Donald S. Scott, *Properties of Concurrent Gas–Liquid Flow*
D. N. Hanson and G. F. Somerville, *A General Program for Computing Multistage Vapor–Liquid Processes*

Volume 5

J. F. Wehner, *Flame Processes–Theoretical and Experimental*
J. H. Sinfelt, *Bifunctional Catalysts*
S. G. Bankoff, *Heat Conduction or Diffusion with Change of Phase*

George D. Fulford, *The Flow of Liquids in Thin Films*
K. Rietema, *Segregation in Liquid–Liquid Dispersions and Its Effect on Chemical Reactions*

Volume 6

S. G. Bankoff, *Diffusion-Controlled Bubble Growth*
John C. Berg, Andreas Acrivos, and Michel Boudart, *Evaporation Convection*
H. M. Tsuchiya, A. G. Fredrickson, and R. Aris, *Dynamics of Microbial Cell Populations*
Samuel Sideman, *Direct Contact Heat Transfer between Immiscible Liquids*
Howard Brenner, *Hydrodynamic Resistance of Particles at Small Reynolds Numbers*

Volume 7

Robert S. Brown, Ralph Anderson, and Larry J. Shannon, *Ignition and Combustion of Solid Rocket Propellants*
Knud Østergaard, *Gas–Liquid–Particle Operations in Chemical Reaction Engineering*
J. M. Prausnitz, *Thermodynamics of Fluid–Phase Equilibria at High Pressures*
Robert V. Macbeth, *The Burn-Out Phenomenon in Forced-Convection Boiling*
William Resnick and Benjamin Gal-Or, *Gas–Liquid Dispersions*

Volume 8

C. E. Lapple, *Electrostatic Phenomena with Particulates*
J. R. Kittrell, *Mathematical Modeling of Chemical Reactions*
W. P. Ledet and D. M. Himmelblau, *Decomposition Procedures for the Solving of Large Scale Systems*
R. Kumar and N. R. Kuloor, *The Formation of Bubbles and Drops*

Volume 9

Renato G. Bautista, *Hydrometallurgy*
Kishan B. Mathur and Norman Epstein, *Dynamics of Spouted Beds*
W. C. Reynolds, *Recent Advances in the Computation of Turbulent Flows*
R. E. Peck and D. T. Wasan, *Drying of Solid Particles and Sheets*

Volume 10

G. E. O'Connor and T. W. F. Russell, *Heat Transfer in Tubular Fluid–Fluid Systems*
P. C. Kapur, *Balling and Granulation*
Richard S. H. Mah and Mordechai Shacham, *Pipeline Network Design and Synthesis*
J. Robert Selman and Charles W. Tobias, *Mass-Transfer Measurements by the Limiting-Current Technique*

Volume 11

Jean-Claude Charpentier, *Mass-Transfer Rates in Gas–Liquid Absorbers and Reactors*
Dee H. Barker and C. R. Mitra, *The Indian Chemical Industry–Its Development and Needs*
Lawrence L. Tavlarides and Michael Stamatoudis, *The Analysis of Interphase Reactions and Mass Transfer in Liquid–Liquid Dispersions*
Terukatsu Miyauchi, Shintaro Furusaki, Shigeharu Morooka, and Yoneichi Ikeda, *Transport Phenomena and Reaction in Fluidized Catalyst Beds*

CONTENTS OF VOLUMES IN THIS SERIAL   413

Volume 12

C. D. Prater, J. Wei, V. W. Weekman, Jr., and B. Gross, *A Reaction Engineering Case History: Coke Burning in Thermofor Catalytic Cracking Regenerators*
Costel D. Denson, *Stripping Operations in Polymer Processing*
Robert C. Reid, *Rapid Phase Transitions from Liquid to Vapor*
John H. Seinfeld, *Atmospheric Diffusion Theory*

Volume 13

Edward G. Jefferson, *Future Opportunities in Chemical Engineering*
Eli Ruckenstein, *Analysis of Transport Phenomena Using Scaling and Physical Models*
Rohit Khanna and John H. Seinfeld, *Mathematical Modeling of Packed Bed Reactors: Numerical Solutions and Control Model Development*
Michael P. Ramage, Kenneth R. Graziano, Paul H. Schipper, Frederick J. Krambeck, and Byung C. Choi, *KINPTR (Mobil's Kinetic Reforming Model): A Review of Mobil's Industrial Process Modeling Philosophy*

Volume 14

Richard D. Colberg and Manfred Morari, *Analysis and Synthesis of Resilient Heat Exchanger Networks*
Richard J. Quann, Robert A. Ware, Chi-Wen Hung, and James Wei, *Catalytic Hydrometallation of Petroleum*
Kent David, *The Safety Matrix: People Applying Technology to Yield Safe Chemical Plants and Products*

Volume 15

Pierre M. Adler, Ali Nadim, and Howard Brenner, *Rheological Models of Suspensions*
Stanley M. Englund, *Opportunities in the Design of Inherently Safer Chemical Plants*
H. J. Ploehn and W. B. Russel, *Interactions between Colloidal Particles and Soluble Polymers*

Volume 16

Perspectives in Chemical Engineering: Research and Education

Clark K. Colton, *Editor*

Historical Perspective and Overview

L. E. Scriven, *On the Emergence and Evolution of Chemical Engineering*
Ralph Landau, *Academic–Industrial Interaction in the Early Development of Chemical Engineering*
James Wei, *Future Directions of Chemical Engineering*

Fluid Mechanics and Transport

L. G. Leal, *Challenges and Opportunities in Fluid Mechanics and Transport Phenomena*
William B. Russel, *Fluid Mechanics and Transport Research in Chemical Engineering*
J. R. A. Pearson, *Fluid Mechanics and Transport Phenomena*

Thermodynamics

Keith E. Gubbins, *Thermodynamics*
J. M. Prausnitz, *Chemical Engineering Thermodynamics: Continuity and Expanding Frontiers*
H. Ted Davis, *Future Opportunities in Thermodynamics*

Kinetics, Catalysis, and Reactor Engineering

Alexis T. Bell, *Reflections on the Current Status and Future Directions of Chemical Reaction Engineering*
James R. Katzer and S. S. Wong, *Frontiers in Chemical Reaction Engineering*
L. Louis Hegedus, *Catalyst Design*

Environmental Protection and Energy

John H. Seinfeld, *Environmental Chemical Engineering*
T. W. F. Russell, *Energy and Environmental Concerns*
Janos M. Beer, Jack B. Howard, John P. Longwell, and Adel F. Sarofim, *The Role of Chemical Engineering in Fuel Manufacture and Use of Fuels*

Polymers

Matthew Tirrell, *Polymer Science in Chemical Engineering*
Richard A. Register and Stuart L. Cooper, *Chemical Engineers in Polymer Science: The Need for an Interdisciplinary Approach*

Microelectronic and Optical Materials

Larry F. Thompson, *Chemical Engineering Research Opportunities in Electronic and Optical Materials Research*
Klavs F. Jensen, *Chemical Engineering in the Processing of Electronic and Optical Materials: A Discussion*

Bioengineering

James E. Bailey, *Bioprocess Engineering*
Arthur E. Humphrey, *Some Unsolved Problems of Biotechnology*
Channing Robertson, *Chemical Engineering: Its Role in the Medical and Health Sciences*

Process Engineering

Arthur W. Westerberg, *Process Engineering*
Manfred Morari, *Process Control Theory: Reflections on the Past Decade and Goals for the Next*
James M. Douglas, *The Paradigm After Next*
George Stephanopoulos, *Symbolic Computing and Artificial Intelligence in Chemical Engineering: A New Challenge*

The Identity of Our Profession

Morton M. Denn, *The Identity of Our Profession*

Volume 17

Y. T. Shah, *Design Parameters for Mechanically Agitated Reactors*
Mooson Kwauk, *Particulate Fluidization: An Overview*

Volume 18

E. James Davis, *Microchemical Engineering: The Physics and Chemistry of the Microparticle*
Selim M. Senkan, *Detailed Chemical Kinetic Modeling: Chemical Reaction Engineering of the Future*
Lorenz T. Biegler, *Optimization Strategies for Complex Process Models*

Volume 19

Robert Langer, *Polymer Systems for Controlled Release of Macromolecules, Immobilized Enzyme Medical Bioreactors, and Tissue Engineering*

J. J. Linderman, P. A. Mahama, K. E. Forsten, and D. A. Lauffenburger, *Diffusion and Probability in Receptor Binding and Signaling*

Rakesh K. Jain, *Transport Phenomena in Tumors*

R. Krishna, *A Systems Approach to Multiphase Reactor Selection*

David T. Allen, *Pollution Prevention: Engineering Design at Macro-, Meso-, and Microscales*

John H. Seinfeld, Jean M. Andino, Frank M. Bowman, Hali J. L. Forstner, and Spyros Pandis, *Tropospheric Chemistry*

Volume 20

Arthur M. Squires, *Origins of the Fast Fluid Bed*
Yu Zhiqing, *Application Collocation*
Youchu Li, *Hydrodynamics*
Li Jinghai, *Modeling*
Yu Zhiqing and Jin Yong, *Heat and Mass Transfer*
Mooson Kwauk, *Powder Assessment*
Li Hongzhong, *Hardware Development*
Youchu Li and Xuyi Zhang, *Circulating Fluidized Bed Combustion*
Chen Junwu, Cao Hanchang, and Liu Taiji, *Catalyst Regeneration in Fluid Catalytic Cracking*

Volume 21

Christopher J. Nagel, Chonghun Han, and George Stephanopoulos, *Modeling Languages: Declarative and Imperative Descriptions of Chemical Reactions and Processing Systems*

Chonghun Han, George Stephanopoulos, and James M. Douglas, *Automation in Design: The Conceptual Synthesis of Chemical Processing Schemes*

Michael L. Mavrovouniotis, *Symbolic and Quantitative Reasoning: Design of Reaction Pathways through Recursive Satisfaction of Constraints*

Christopher Nagel and George Stephanopoulos, *Inductive and Deductive Reasoning: The Case of Identifying Potential Hazards in Chemical Processes*

Keven G. Joback and George Stephanopoulos, *Searching Spaces of Discrete Solutions: The Design of Molecules Possessing Desired Physical Properties*

Volume 22

Chonghun Han, Ramachandran Lakshmanan, Bhavik Bakshi, and George Stephanopoulos, *Nonmonotonic Reasoning: The Synthesis of Operating Procedures in Chemical Plants*

Pedro M. Saraiva, *Inductive and Analogical Learning: Data-Driven Improvement of Process Operations*

Alexandros Koulouris, Bhavik R. Bakshi, and George Stephanopoulos, *Empirical Learning through Neural Networks: The Wave-Net Solution*

Bhavik R. Bakshi and George Stephanopoulos, *Reasoning in Time: Modeling, Analysis, and Pattern Recognition of Temporal Process Trends*

Matthew J. Realff, *Intelligence in Numerical Computing: Improving Batch Scheduling Algorithms through Explanation-Based Learning*

Volume 23

Jeffrey J. Siirola, *Industrial Applications of Chemical Process Synthesis*
Arthur W. Westerberg and Oliver Wahnschafft, *The Synthesis of Distillation-Based Separation Systems*
Ignacio E. Grossmann, *Mixed-Integer Optimization Techniques for Algorithmic Process Synthesis*
Subash Balakrishna and Lorenz T. Biegler, *Chemical Reactor Network Targeting and Integration: An Optimization Approach*
Steve Walsh and John Perkins, *Operability and Control in Process Synthesis and Design*

Volume 24

Raffaella Ocone and Gianni Astarita, *Kinetics and Thermodynamics in Multicomponent Mixtures*
Arvind Varma, Alexander S. Rogachev, Alexander S. Mukasyan, and Stephen Hwang, *Combustion Synthesis of Advanced Materials: Principles and Applications*
J. A. M. Kuipers and W. P. M. van Swaaij, *Computational Fluid Dynamics Applied to Chemical Reaction Engineering*
Ronald E. Schmitt, Howard Klee, Debora M. Sparks, and Mahesh K. Podar, *Using Relative Risk Analysis to Set Priorities for Pollution Prevention at a Petroleum Refinery*

ISBN 0-12-008524-0